Oceanic Basalts

Oceanic Basalts

Edited by

P.A. FLOYD
Department of Geology
University of Keele
Staffordshire

Springer Science+Business Media, LLC

16 15 14 13 12 11 10 9 8 7 6 5 4 3 2 1

Originally published by Blackie and Son Ltd in 1991

First published 1991

British Library Cataloguing in Publication Data

Oceanic basalts.
I. Floyd, P.A.
552.09162

ISBN 978-0-216-92697-4 ISBN 978-94-011-3042-4 (eBook)
DOI 10.1007/978-94-011-3042-4

In the USA and Canada

ISBN 978-0-216-92697-4

Library of Congress CIP data available

Typesetting by Thomson Press (India) Ltd., New Delhi

Preface

Basalt is the most voluminous of all the igneous rocks. Extensive field, experimental, petrographic and geochemical studies of basalt have provided us with a considerable understanding of igneous petrogenesis, plate tectonics, and crust-mantle interaction and exchange. One important aspect of geology that has developed over the last few decades is the study of oceanic basalts. The ocean basins cover about two thirds of the earth's surface and are floored by a basement of oceanic basalt that is continuously undergoing generation at spreading centres and destruction at subduction zones, a process which throughout geological time is recognized as the principal means of generating new crust.

The study of oceanic basalts enables us to understand better the generation and recycling of crustal materials (including the continental crust), and the exchange between oceanic crust and seawater via hydrothermal activity. Compositional variations displayed by oceanic basalts provide windows into the mantle, and the identification of isotopically-distinct mantle reservoirs demonstrates that the source of oceanic basalts is heterogeneous and is controlled by convection and reservoir interactions within the mantle.

The Deep Sea Drilling Project (DSDP) and Ocean Drilling Program (ODP) have been instrumental in providing *in situ* basaltic materials from which it has been possible to formulate and test hypotheses of crustal generation, growth and alteration. However, most ocean drilling has only penetrated the top few hundred metres of the oceanic crust and, apart from Hole 504B in the Costa Rica Rift (about 1250 m penetration), we have yet to sample a complete crustal section. It is important to obtain data on the composition and structure of the whole oceanic crust and not just the basaltic upper layers, in order to ratify the seismically-defined crustal stratigraphy and to test the ophiolite model of the crust.

The importance of oceanic basalts in the development of the earth is quite clear, although most of the rapidly accumulating data on oceanic basalts is available only in specialist journals or within DSDP and ODP publications. In contrast, continental basalts have received wide coverage in textbooks. This new book will provide earth science practitioners and postgraduate students with a summary of oceanic basalts and the oceanic crust, and a view of current ideas and interpretations of compositional variations and key processes.

The book is divided into four sections: Part I (Structure) outlines the geophysical structure of the oceanic crust and the methods used to survey and sample the ocean floor. Recent ideas on ophiolites as analogues for the oceanic crust are reviewed and compared. Part II (Processes) is concerned with the primary and secondary processes involved in the generation of oceanic basalts, and draws on field, experimental, petrographic and chemical data. Part III (Environments) considers basalts in different oceanic settings—major oceans, back-arc basins, intraplate oceanic islands, and seamounts. Part IV (Sources) briefly reviews the stable isotopic composition of basalts and takes us into the source region via mantle-derived peridotites.

P.A.F.

Acknowledgements

From a personal position as editor, I would like to acknowledge not only the expertise of the contributors to this book, but the freely-given time devoted by workers actively involved in basalt research in their specific fields. I would not have made much progress, however, without the help of the following colleagues, who provided reviews of chapters and sections of text, as well as general comment and advice: J.R. Cann, C.S. Exley, M.F.J. Flower, G.J. Lees, P. Nixon, J.A. Pearce, G. Rowbotham, A.D. Saunders, J. Tarney, R.N. Thompson, J.A. Winchester and T.L. Wright. Finally, without the continuous encouragement and helpful guidance provided by the publishers and domestic support given by my wife Margaret, this book would not have reached the press at all.

D. Elthon acknowledges support by grants from the Texas Advanced Research Program and the National Science Foundation.

M.F.J. Flower thanks Thomas Wright of the Hawaii Volcano Observatory for his careful review and comments.

J. Natland gratefully acknowledges support from US National Science Foundation grants NSF OCE-83-08696 and NSF OCE-85-10526.

A.D. Saunders and J. Tarney acknowledge discussions with Mark Allen and Mike Norry, and thank Peter Baker for use of unpublished South Sandwich Island basalt data.

C.L. Walker acknowledges discussions with Bob Thompson, Roger Searle and Andy Saunders.

The contributors also acknowledge the following individuals and publishers who have given permission for the reproduction of copyright material in the following figures and tables: Fig. 2.6 Director, USGS; Fig. 2.7 S. Cande; Fig. 2.9 Director, Ocean Drilling Program; Fig. 3.2 J.G. Slater and B. Parsons (and with permission from *Journal of Geophysical Research*); Figs. 3.3 and 3.8 R.S. White (and with permission from the Geological Society of London); Figs. 3.4 and 3.9b R.S. White and D. McKenzie (and with permission of *Scientific American*); Figs. 3.6 and 3.7a R.S. White (and with permission of *Geology*); Fig. 4.3 S.J. Lippard (and with permission of the Geological Society of London); Fig. 5.4 M.R. Perfit and D.J. Fornari (and with permission of the *Journal of Geophysical Research*); Fig. 7.1 D.C. Presnall and J.D. Hoover (and with permission of the Geochemical Society); Fig. 7.3 E.R. Oxburgh (and with permission of Princeton University Press); Fig. 7.7 J.A. Whitehead (and with permission from *Nature*, MacMillan Magazines Ltd); Fig. 7.8 M.P. Ryan (and with permission from the Geochemical Society); Table 8.1 J.K. Bohlke (and with permission from the *Journal of Geophysical Research*); Table 8.2 J.C. Alt and J. Honnorez (and with permission from *Contributions to Mineralogy and Petrology*, Springer-Verlag); Table 8.4 and Fig. 8.5 D. Elthon (and with permission from J. Wiley Ltd); Table 8.9 and Fig. 8.3 G. Thompson (and with permission from Academic Press); Fig. 8.2 R.A. Hart (and with permission from Deep Sea Drilling Project); Fig. 8.4 T.W. Donnelly (and with permission from Deep Sea Drilling Project); Fig. 8.6 J.R. Cann (and with permission from

American Geophysical Union); Figs. 8.7 and 8.8 M.J. Mottl (and with permission from the Geological Society of America); Figs. 8.9 and 8.10 J.C. Alt (and with permission from the *Journal of Geophysical Research*); Fig. 9.1 K. Burke and J.T. Wilson, I.G. Gass (and with permission from *Scientific American* and *Journal of Geophysical Research*); Figs. 9.2 and 9.3 D.J. Fornari (and with permission from *Nature*, MacMillan Magazines Ltd and *Journal of Geophysical Research*); Fig. 9.4 P. Vogt (and with permission from *Journal of Geophysical Research*); Figs. 9.5 and 9.12 J.G. Moore, D.A. Clague (and with permission from Geological Society of America and Geological Society of London); Figs. 9.6 and 9.10 H. Staudigel (and with permission from *Journal of Geophysical Research* and Elsevier); Fig. 9.8 BVSP (and with permission from Lunar and Planetary Institute); Fig. 9.9 M. Loubert (and with permission from Elsevier); Fig. 9.11 G.R. Davies and A. Zindler (and with permission from Geological Society of London and Elsevier); Fig. 9.13 F. Frey, H. Staudigel and BVSP (and with permission from Elsevier and Lunar and Planetary Institute); Fig. 9.14 S.H. Richardson and S. Humphris (and with the permission of Elsevier); Fig. 9.15 F. Frey, J. Mahoney and M. Storey (and with the permission of DSDP, the Geological Society of London and *Nature*, MacMillan Magazines Ltd); Fig. 9.16 J.-G. Schilling, M.F.J. Flower and W.M. White (and with the permission of Elsevier and Springer International); Fig. 9.17 A. Zindler, J.F. Allan and D. Fornari (and with the permission of the American Geophysical Union and *Nature*, MacMillan Magazines Ltd); Fig. 10.4 P. Fryer and D.M. Hussong (and with permission of DSDP); Fig. 11.9 J. Sinton, D. Hey and F. Duennebier (and with the permission of the Geological Society of America); Fig. 12.2 J. Mahoney (and with permission from *Journal of Geophysical Research*); Fig. 13.1b J. Francheteau and R.D. Ballard (and with permission from *Earth and Planetary Science Letters*, Elsevier); Figs. 13.3b, 13.4a and 13.8 S.P. Jakobsson (and with permission from *Acta Naturalia Islandica*); Fig. 13.12a R.K. O'Nions and R.J. Pankhurst (and with permission from *Earth and Planetary Science Letters*, Elsevier).

Contributors

Dr R. Batiza Hawaii Institute of Geophysics, University of Hawaii at Manoa, Honolulu, Hawaii, USA.

Professor J.R. Cann Department of Earth Sciences, University of Leeds, Leeds, UK.

Professor D. Elthon Department of Geosciences, University of Houston, Texas, USA.

Dr R.A. Exley VG Isotech Ltd., Aston Way, Middlewich, Cheshire, UK.

Dr M.F.J. Flower Department of Geological Sciences, University of Illinois, Chicago, Illinois, USA.

Dr P.A. Floyd Department of Geology, University of Keele, Staffordshire, UK.

Dr S. Lewis Branch of Pacific Marine Geology, US Geological Survey, Menlo Park, California, USA.

Dr M. Menzies Department of Geology, Royal Holloway and Bedford New College, Egham, Surrey, UK.

Dr J. Natland Scripps Institute of Oceanography, La Jolla, California, USA.

Dr A.D. Saunders Department of Geology, University of Leicester, University Road, Leicester, UK.

Dr J.G. Spray Department of Geology, University of New Brunswick, Fredericton, Canada.

Professor J. Tarney Department of Geology, University of Leicester, University Road, Leicester, UK.

Professor G. Thompson Woods Hole Oceanographic Institute, Woods Hole, Massachusetts, USA.

Dr C.L. Walker Department of Geological Sciences, University of Durham, South Road, Durham, UK.

Professor R.S. White Bullard Laboratories, Department of Earth Science, University of Cambridge, Madingley Road, Cambridge, UK.

Contents

PART I STRUCTURE

1 Introduction and the ophiolite model

JOE CANN

1.1 Historical perspectives

The islands of the ocean basins were familiar to early geologists. They were well acquainted with the volcanoes of Iceland, the Azores, the Canaries and the Mediterranean islands. When Darwin landed on St Paul's Rocks in the equatorial Atlantic from the *Beagle* in 1831, he could recognize the anomaly of the peridotite mylonites that he found there. These mylonites are the only subaerial outcrop of an active oceanic transform fault zone, although that phrase could only have been used after 1965. However, early nineteenth century geologists were much less clear about what lay below the surface of the sea. This is not surprising as the first reliable deep-sea sounding was not made by Sir James Clark Ross until 1842. Even now it is difficult to convey to land-bound lay people the great depth of the oceans, the fundamental differences between continents and oceans, and the constant renewal of the ocean floor by seafloor spreading and subduction, when their concepts are bounded by a wrinkled sea surface viewed from the air. In the early nineteenth century such ideas lay beyond everyone's grasp.

The systematic investigation of the ocean floor began with the laying of the first trans-oceanic cables from Ireland to Newfoundland in the late 1850s and 1860s. In the centre of the Atlantic appeared a broad rise, Telegraph Plateau, which further soundings before the end of the century showed to be part of a chain of submarine mountains running down the centre of the Atlantic Ocean. From this Mid-Atlantic Ridge, one of the cable ships, grappling for a broken end of cable, brought up a piece of basalt which was the subject of the first paper on oceanic basalts, and was also the first paper in volume 1 of the new *Mineralogical Magazine* (Hall, 1876).

By this time, HMS *Challenger* had set out on her 4 year circumnavigation of the globe (1872–1876), which set the foundations for the new sciences of oceanography and marine geology. She brought up the first manganese nodules and the first samples of red clay, and also pieces of glassy basalt, most of which had been thoroughly transformed by seafloor weathering

(Murray and Renard, 1891). The discovery of basalts on the deep ocean floor seems to have convinced geologists that they could conveniently regard oceanic islands, with their basalt lavas, as representative outcrops of the ocean floor, and for many decades there were very few samples of basement obtained from the deep oceans. The important exception was a series of dredges made during the John Murray expedition to the Indian Ocean, which recovered the first hydrothermally altered basalts from the ocean floor (Wiseman, 1937).

After the Second World War there was a new upsurge of interest in the ocean floor. Seismic experiments showed that oceanic crust is very different in character from continental crust; precision echo sounding delineated the worldwide system of mid-ocean ridges, the crests of which coincide with a chain of shallow earthquakes; and magnetic surveys discovered large magnetic anomalies striped parallel to the mid-ocean ridges. What could the composition of the thin oceanic crust be? How could it generate the magnetic stripes? What might the deep fracture zones that cut through the ridges be? On this scale, the oceanic islands began to seem more like anomalies of the oceanic floor than basement outcrops. Geophysicists began to be interested in the materials of the ocean floor and started new dredging campaigns.

Maurice Ewing pioneered this new phase with a series of dredge hauls from near 30°N on the Mid-Atlantic Ridge in the late 1940s (Shand, 1949; Quon and Ehlers, 1963), and others followed suit with dredging in the Atlantic, Pacific and Indian Oceans. It rapidly became clear that ocean floor basalts are a distinctive class. Engel and Engel (1963) showed that they are extremely low in potassium (and in other incompatible elements) and are very different geochemically from the basalts of oceanic islands. Gast (1965) demonstrated from analyses of strontium isotopes that oceanic floor basalts must be derived from a part of the mantle depleted in rubidium (and hence presumably in the other incompatible elements) over time spans comparable to the age of the continental crust. It became clear that mid-ocean ridge basalts are as geochemically distinctive as the oceanic crust is geophysically distinctive, and that they are products of a very particular environment.

1.2 Oceanic lithospheric processes

It became clear in the 1960s that this distinctive environment is that of seafloor spreading. New oceanic lithosphere, crust on top of mantle, is created continuously at mid-ocean ridges at rates of 2–20 $\mathrm{cm\,y^{-1}}$. It moves away across the ocean basins and is then destroyed in the subduction zones marked by the deep ocean trenches. About 3 $\mathrm{km^2}$ of new crust is created each year, enough to renew the whole ocean floor in 100 million years, so that the oldest ocean floor is probably less than 200 million years old, an order of magnitude younger than the continental crust.

The ocean crust is created by igneous activity. As the lithospheric plates move apart, mantle rises from below to fill the space between, and as it does so it undergoes partial melting. The partial melt (of basaltic magma) rises through the residue from melting (solid peridotite) and from this melt the crust is formed. The mechanism by which this happens is still not agreed, but the currently most popular model gives an important role to a crustal magma chamber within which cumulates form and from which lava flows are fed through dykes. This model is consistent with the structure of ophiolite complexes, that is, tectonic slices of basaltic and peridotitic rock that have been thrust onto continents during ancient mountain-building episodes. A sheeted dyke complex, made up entirely of dykes intruding dykes, is characteristic of many ophiolites and demonstrates graphically an origin by some kind of seafloor spreading. Also characteristic is a structure of extrusive rocks overlying sheeted dykes, which in turn overlie gabbros overlying peridotite. This structure corresponds in general with the observed seismic structure of the oceanic crust. In the magma chamber model the gabbros are cumulates from the magma chamber, the peridotites are residual mantle after partial melting and the lavas and dykes are melts derived from the magma chamber. Geophysical evidence supports the presence of a magma chamber at fast-spreading mid-ocean ridges, but the picture may be more complicated at slow-spreading mid-ocean ridges such as the Mid-Atlantic Ridge, where any magma chambers present may be small and short-lived.

Although the oceanic crust is constructed from igneous material, it is subsequently modified by tectonic and hydrothermal processes. The tectonic processes are most apparent in rifted ridges such as the Mid-Atlantic Ridge, where the spreading axis is marked by a rift valley with a relief of up to 2 km. Newly created crust on the floor of the median valley is elevated by a staircase of faults, which produce major tectonic rotations and disrupt the simple crustal structure as well as providing channels for the rise of serpentinite diapirs through the crust. On unrifted ridges such as the East Pacific Rise, tectonism also plays an important role. Swarms of fissures mark the spreading axis in some places, and faults play an important role in the construction of the abyssal hills of the seafloor.

Tectonism is also important in modifying the permeability structure of the crust, which in turn is one of the controls on hydrothermal circulation. The most spectacular manifestation of hydrothermal circulation is the black smoker hot spring activity at spreading centres, in which tens to hundreds of megawatts of power are emitted from each field of hot springs by water up to 350°C. The water can be shown to be normal deep ocean seawater which has penetrated the crust to a depth of 1–2 km, where it has been heated and has reacted with the rock. The resulting hot solutions are highly acid, enriched in dissolved hydrogen sulphide, iron, copper, zinc and manganese, and depleted in sulphate and magnesium. The solutions precipitate iron, copper and zinc sulphides as they emerge at the seafloor, partly as solid deposits and partly as finely divided particles that make up the black smoke.

Such a profound transformation of the water is matched by an equally profound transformation of the rocks. The chemical reactions produce metamorphism and metasomatism of the basaltic crust in the greenschist facies, with the bulk addition of magnesium and loss of other metals, although the transformation is not a simple process.

The source of heat for this hydrothermal circulation and the accompanying metamorphism is directly or indirectly, the axial magma chamber. Water may circulate close to the magma and extract heat through a thin uncracked lid, or it may penetrate into hot, newly crystalline gabbro through a network of fine cracks. In either instance, the high temperature water–rock interaction is related to the magmatic activity as a source of heat and to tectonism as a source of permeability. Within the zone of intense geological activity that marks the spreading centres, all three processes proceed together, profoundly interlinked, over the few hundreds of thousands of years that it takes to create and modify a piece of oceanic crust.

Outside the belt of intense activity, no more than 20 km wide and probably much narrower, the now mature oceanic crust spreads slowly away from the axis. Tectonism is now a minor process, and hydrothermal activity is slow, affecting mainly the upper part of the crust and gradually producing low temperature ocean floor weathering. Most of the igneous activity (by volume) is confined to the axial region, but off-axis and mid-plate volcanism can be important. Near the spreading axis this takes the form of small seamounts, geochemically closely related to the axial activity. On older crust larger seamounts or oceanic islands can be built, and these are usually geochemically very different from the axial volcanism. They have hot-spot characteristics, with enriched contents of incompatible elements relative to mid-ocean ridge basalts, and are derived from a very different mantle reservoir. These major volcanic structures can be rejuvenated tens of millions of years after their initial activity, and may be the sites of several phases of magmatism. Broad areas of the ocean floor may also be the site of renewed igneous activity, producing lava flows and sills intercalated with later sediments, covering areas of tens of thousands of square kilometres. The causes of this magmatic rejuvenation are still obscure.

All of these complex processes of creation, modification and evolution can be traced in the marginal basins that are formed during back-arc spreading behind the volcanic arcs that mark subduction zones. Magma compositions are more or less modified by interaction with products of the subduction zone, but the physical processes appear to be the same. This has been reinforced by the discovery of highly active hydrothermal systems at the spreading axes of several marginal basins of the west Pacific.

1.3 Concluding statements

This introduction has attempted to convey some of the history of research on oceanic basalts, and also the rich and complex relationships between oceanic basalts and the processes which shape the oceanic crust. If, for convenience, oceanic basalts are treated as phenomena in themselves, it is important at the same time to remember the links with other components of the ocean floor system and the complex dynamics of this important geological environment.

2 Surveying and sampling the ocean floor

STEPHEN LEWIS

2.1 Introduction

Surveying and sampling the oceanic crust requires specialized tools and techniques because, in general, oceanic crust is found only at great depths beneath the oceans. Basaltic lavas erupted from centres such as Iceland or Hawaii reach above the surface of the sea, but often display chemical and petrological characteristics which make them significantly different from the basalts which comprise 'normal' oceanic crust. For these reasons, techniques and equipment have been developed to make the rocks which form the foundations of the oceans accessible.

Surveying the ocean floor involves many specialized and sophisticated techniques, but most of these have their roots in the ancient technology of sounding the depths of the oceans. Modern methods use sound energy rather than lead lines to measure the depth of the oceans, and very large areas of the ocean floor can now be continuously and rapidly mapped; however, the basic principles and goals of such measurements remain unchanged.

The first goal of most ocean surveying programmes is to continuously measure the depth of the seafloor along the track of the survey ship as it steams in a grid pattern and to construct a bathymetric contour map from the resulting series of point depth measurements. The bathymetric map, a precise analogue of the topographic maps that show elevations on land, defines the topography of the seafloor, delineating the submarine canyons, abyssal plains, seamounts and other features of the ocean floor. The topography of the seafloor is the product of a wide range of active marine processes, such as sedimentation, ocean current activity and submarine volcanism, and a good bathymetric map of a region can therefore provide important indications of the kinds of geological processes that have helped to shape the ocean floor. Such maps can serve as a tool for making preliminary predictions about the kinds of rocks that may form the ocean floor. The better the quality of the bathymetric map, defined by criteria such as its accuracy and the resolution of individual features, the more information can be provided to the marine researcher. Recent technological advances in the hardware used for mapping the ocean floor give a greater accuracy of measurement, more closely spaced measurements to improve resolution and simultaneous decreases in the amount of time (and hence cost) required to conduct surveys.

The actual sampling process consists of several phases. The process begins with the definition of the specific scientific objectives of a research programme based on existing data such as bathymetric maps or other acoustic imagery data, followed by the identification of a general target region of the seafloor. The next planning phase involves the tentative selection of specific sampling locations, based on the interpretation of existing bathymetric maps, seismic reflection profiles, or other information from the target region. Researchers then select the sampling equipment and techniques that they will use, based on factors including the expected seafloor age, morphology and sediment thickness of the target region.

Shipboard work begins by briefly surveying the tentative sample sites by acoustic techniques to confirm or improve the accuracy and resolution of existing maps, followed by final site selection immediately before commencing sampling operations. The sampling hardware is deployed and oceanic sediments, perhaps with igneous basement rocks, are recovered. If the sampling effort fails, the shipboard scientists must decide whether or not to make another attempt at that site. The samples are usually described and archived soon after recovery, and specific samples are selected for shipboard analysis. The remainder of the recovered material is packaged for transport to shore based laboratories when the research vessel reaches port at the end of the expedition.

Each phase of the process of surveying and sampling the oceanic crust will be discussed in the following sections.

2.2 Surveying the ocean floor

Surveying the ocean floor has been an important activity since seafarers began shipping goods from port to port. Charting the approaches to harbours and the channels between islands is vital for the safe passage of commercial vessels. The original method of seafloor surveying probably consisted of the use of a simple weighted rope along which distance intervals were marked. The 'leadsman' perched in the bows of a vessel as it slowly approached land, throwing the weighted line a few yards ahead of the ship. He could determine the depth of the water from the amount of rope that had run out through his hands when the weight hit the seafloor. The leadsman periodically called out the depth of the water to the pilot and captain nervously pacing the fantail. Such bathymetric measurements were restricted to shallow waters because the length of rope required to reach the seafloor in the deep ocean could not be handled manually.

Measuring the depth of the oceans for scientific purposes was first undertaken in the nineteenth century, the best example of which is the famous

Challenger expedition of 1872–1876. A system was devised for use aboard the *Challenger* which was used repeatedly for successful soundings in the deep sea. The hardware consisted of long Italian hemp sounding lines of 1 in (2.54 cm) circumference, with a breaking strength of 1400 lb (635 kg). Lengths of hemp 120 fathoms long (1 fathom = 1.83 m) were spliced together to form a continuous sounding line 3000 fathoms long, which was stored on a reel winch. Coloured marks were woven into the sounding line every 25 fathoms, with 'the 25 and 75 fathom marks being white, the 50 fathom marks red, and the 100 fathom marks blue' (Thomson and Murray, 1891). Cast iron weights of 300–400 lb (136–181 kg) were attached to the end of the sounding line (Figure 2.1) and the apparatus was rigged out over the side of the *Challenger*. Between 400 and 500 fathoms of line were carefully paid out, and then the sounding weight was allowed to fall freely to the seafloor while crew

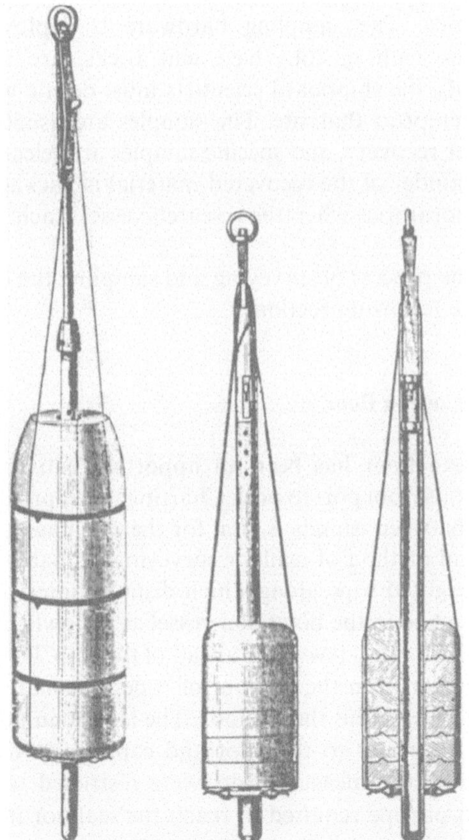

Figure 2.1 Sounding machines used during the voyage of *HMS Challenger*. Variable amounts of weight could be added to the machines, and when the weights reached the seafloor they were automatically released from the end of the rope.

members counted the coloured marks on the line as it passed down into the water. The time interval required for each 100 fathoms of rope to pay out was recorded, and when that interval suddenly increased, the sounding weight was judged to have landed on the seafloor. The sounding weight took between 40 and 50 min to fall 3000 fathoms to the seafloor. After the release of the weights from the end of the sounding line, a steam-powered winch was used to haul in the line.

Soon after the *Challenger* left port on its 3 year expedition, sounding systems were perfected that used wire rather than rope for bathymetric soundings. The greatest advantage that wire sounding lines had over rope lines was that the sink rate was much faster with wire, as a result of its smaller diameter, smoother surface, and hence lower resistance through the water. The need to survey underwater telegraph cable routes across the Atlantic in the last part of the nineteenth century provided the motivation for developing faster deep-sea sounding capabilities, so that more depth measurements could be made in a given period of time. In spite of this, the time required to make a single depth measurement by mechanical means, as long as several hours in the deep ocean, remained an important limitation of traditional sounding methods.

2.3 Acoustic systems

Research involving the transmission and detection of sound energy through water, stimulated by antisubmarine warfare during the Second World War, led to the development of echo sounding techniques for bathymetric mapping (Graham, 1987). As its name implies, the principle of the echo sounding technique is to measure the time required for a sound impulse emitted by a surface ship to propagate down through the water, reflect from the seafloor, and return to the surface ship (Figure 2.2). Independent knowledge of the velocity at which sound travels in water (approximately $1500\,m\,s^{-1}$) allows the calculation of water depth from the observed travel time of the sound impulse. In practice, a clock device triggers an electromechanical transducer mounted on the hull of the ship to emit a sound impulse. The echo returning from the seafloor is detected, amplified, and either recorded by a paper precision depth recorder (Figure 2.3), or digitally sampled by a computer and stored on magnetic tape or other electronic media for later analysis. This process is repeated every few seconds as the ship steams along its track, producing a bathymetric profile.

Acoustic bathymetric sounding has numerous important advantages over the older mechanical sounding technique. One of the most important of these is that the depth measurement process is nearly continuous; each individual measurement requires only the time needed for a sound wave to make the round-trip from the ship to the seafloor and back. The two-way

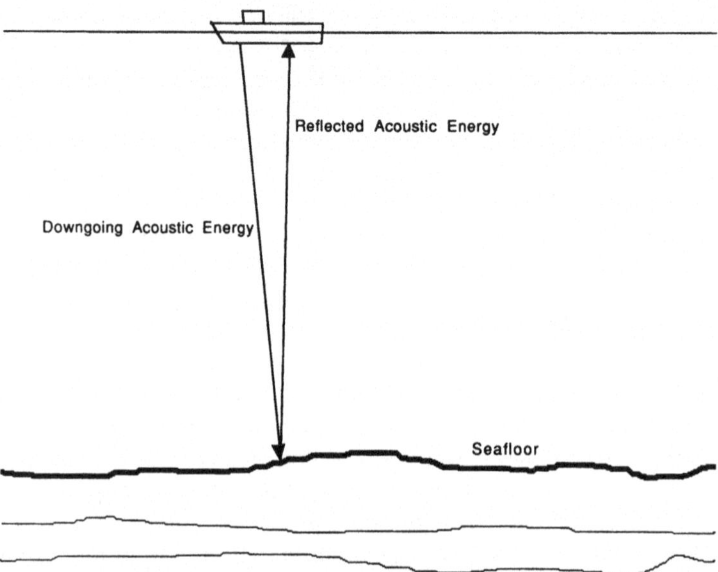

Figure 2.2 Schematic diagram of the ray-path geometry for sound waves used for seismic reflection profiling.

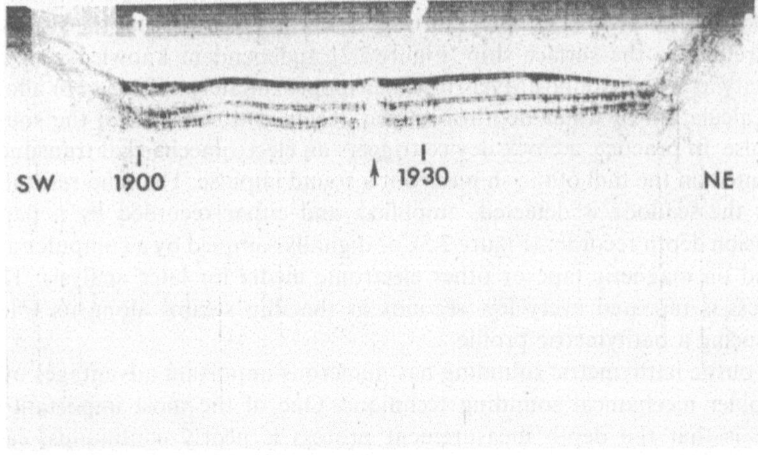

Figure 2.3 Precision depth recorder profile using acoustic energy at 3.5 kHz. The strong echo from the seafloor is followed by weaker reflections from sedimentary horizons below the seafloor.

travel time of a sound impulse in a water depth of 5000 m is only 6.7 s. Thus, in the example, a new measurement can be made every 7 s as the ship steams along. The resulting seismic reflection profile represents a two-dimensional view of the seafloor beneath the ship. Another important advantage is that the accuracy of acoustic bathymetric measurements is much greater than with wireline methods.

The accuracy of depth measurement depends on the ability of the instruments to measure the time required for the sound wave to travel to the seafloor and return, and on independent knowledge of the velocity of sound through water. Time measurements, which routinely can be made with uncertainties of less than millisecond, introduce uncertainties of < 1 m in acoustic depth measurements. The velocity of sound in water depends on the density of a particular water mass, which in turn depends most strongly on the temperature and salinity of the water. As these parameters can vary in both time and space, corrections to water depths determined using an average velocity of sound in water (1500 m s^{-1}) must be made to achieve the greatest possible accuracy. Compilations of water sound velocity profiles in numerous regions of the world's oceans (Matthew's Tables; Carter Tables) are used to make the corrections. Bathymetric maps often specify whether the depths are uncorrected (determined assuming 1500 m s^{-1} velocity of sound in water) or corrected (determined by applying a correction for deviations from 1500 m s^{-1} in the velocity of sound).

Seismic reflection profiling systems can do much more than simply measure the depth of the ocean. If the system emits an acoustic signal of high enough energy, a fraction of the sound that reaches the seafloor penetrates into the sediments rather than being completely reflected back to the sea surface. The acoustic energy that travels into the strata beneath the seafloor can then be reflected back to the receivers at the surface. The travel path length of sound that penetrates the seafloor is longer than that reflected from the bottom, and the deeper reflections therefore arrive back at the sea surface later than the reflections from the seafloor. Continuous profiling of this type produces a sub-bottom seismic reflection profile that can be used to measure the thickness and stratigraphic character of sediments overlying the oceanic crust, to identify regions of outcropping crustal rocks on the seafloor, and to determine the nature of the layering of the oceanic crust itself.

Implicit in the discussion of bathymetric measurements is the idea that these measurements are made at specific locations on the surface of the earth. This is particularly true for the widely spaced soundings made by *Challenger* in the nineteenth century, but it is equally true for the continuous echo sounding conducted aboard research vessels today. Acoustic soundings are actually a series of closely spaced individual depth measurements. It is clearly essential to know the locations of depth measurements, and the accuracy of navigation is as important as the accuracy of the depth measurement itself. As research focuses increasingly on small seafloor features, such as the hydrothermal vents associated with the mid-ocean ridge system, navigational accuracy becomes critical to successful marine operations. Additionally, as our seafloor mapping tools become increasingly sophisticated and capable of higher resolution of the seafloor, the correct location of small seafloor features with respect to neighbouring features in addition to their position in a global geographic coordinate system becomes increasingly important. The advent of extremely accurate satellite based navigation systems has improved navigational accuracy from several hundreds of metres of uncertainty in the 1950s to as little as several tens of metres today. This improvement in navigational accuracy provides the foundation for the modern mapping systems discussed in the following sections.

2.3.1 *Multibeam swath bathymetric systems*

The acoustic echo sounders developed in the 1940s emitted a single sound beam into the water and detected the reflected arrival back at the sea surface. The operation of this type of device during marine surveying results in a map showing the track of the ship, along which are positioned numerous depth measurements. Depth measurements are only made directly beneath the ship, and the depth of the seafloor not traversed by the ship is not measured. A great deal of survey time is required to place ship tracks close enough to each other to ensure that important bathymetric features are not overlooked.

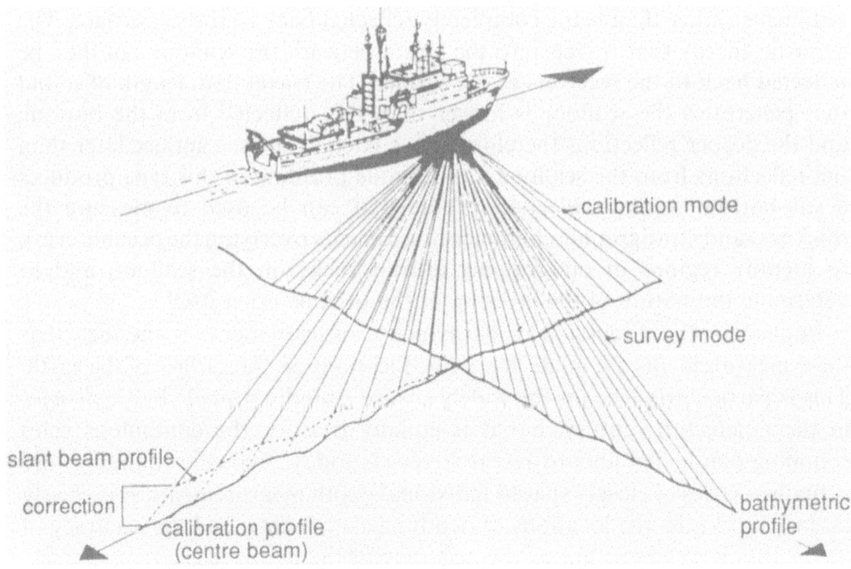

Figure 2.4 Beam geometry for swath bathymetric mapping.

The development of multibeam swath bathymetric systems during the 1960s and 1970s addressed the problem of only making bathymetric measurements directly beneath the ship. Swath bathymetric systems transmit up to 59 individual narrow sound beams in a fan-shaped geometry that extends to either side of the ship (Figure 2.4). A complex array of transducers mounted on the hull of the ship detects the returning sound energy and calculates the water depths and cross-track distances to the reflection points for each sound beam.

These computer-controlled systems produce a swath of bathymetric measurements, centred on the ship's track, during surveying (Tyce, 1987). The width of a single bathymetric swath can be as large as twice the water depth, or up to 10 km in deep ocean waters. The very small distances between measurement points within a swath produces very high resolution bathymetric maps. The ship's survey track can be designed so that neighbouring bathymetric swaths abut, producing 100% bathymetric coverage of the survey region. The swath bathymetric data can be combined with navigational data from the ship's sensors by onboard computers to produce a bathymetric contour map in real time. Bathymetric relief images can also be produced by computer manipulation of the bathymetric data (Figure 2.5). This capability to map the seafloor and analyse the resulting bathymetric charts at sea soon after acquisition is important for subsequent sampling operations.

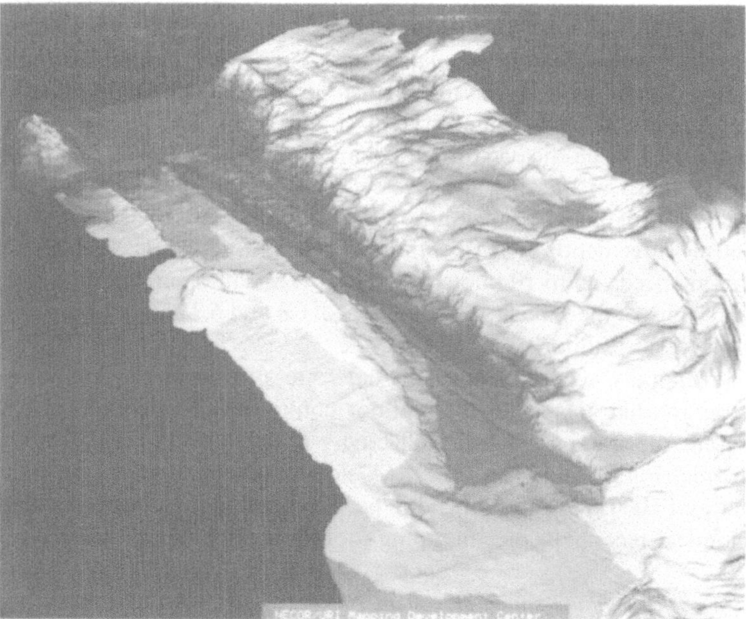

Figure 2.5 Computer-produced bathymetric image based on SeaBeam swath bathymetric mapping system from the Chile Margin Triple Junction region. Image produced by S. Cande, S. Lewis, J. Miller and S. Ferguson.

2.3.2 *Signal amplitude measurements: side-scan sonar*

The seafloor mapping techniques discussed so far are all based on the principle of measuring the travel time of an emitted sound impulse to determine the water depth. Another property of the reflected sound pulse, that of the strength or amplitude of the returning echo, can also be measured. Numerous side-scan sonar seafloor mapping systems detect and measure the strength of the acoustic signal backscattered from the seafloor to the sides of the sonar vehicle producing a sonar image based on the acoustic properties of the seafloor material. Side-scan sonar images can be reminiscent of aerial photographs of land, but they are acoustic images, not light images.

Side-scan sonar systems usually consist of a torpedo-shaped vehicle that is towed behind the research vessel, and electronic recording and imaging hardware in the ship's laboratory. Side-scan sonar systems can be divided into two basic types: shallow-towed systems intended to map large regions of the seafloor in relatively short times with high tow speeds and very wide swath widths; and deep-towed systems which map narrower swaths at slower speeds than the shallow-towed systems, but generally produce much more detailed images of the seafloor. The former systems are ideally suited for reconnaissance style surveys of large regions, whereas the latter are designed for use in more narrowly focused topical studies (Kappel and Normark 1987).

2.3.2.1 *Long-range side-scan sonar systems* The best example of long-range side-scan sonar systems is the British developed GLORIA (*G*eological *LO*ng-*R*ange *I*nclined *A*sdic) system (Figure 2.6). The GLORIA system consists of a large, heavy torpedo-shaped tow vehicle which is deployed from the stern of the operating vessel using a special gantry system. The vehicle, approximately 8 m long and weighing about 1820 kg, is towed at 50–60 m water depth 300 m behind the research vessel. The vehicle can be towed at speeds up to 10 knots. The GLORIA vehicle emits sound impulses to both sides of the vehicle at frequencies of 6.3 and 6.7 kHz, and can ensonify swaths of the seafloor up to 60 km wide.

The digitally recorded acoustic data are processed by computer into photograph-like images of the seafloor composed of individual pixels that represent the average acoustic reflectivity of areas on the seafloor 50 m on each side (Figure 2.7). The GLORIA side-scan sonar system can image as much as 17 000 km^2 of ocean floor per day, making it a very effective and economical reconnaissance mapping tool (Somers *et al.*, 1978).

2.3.2.2 *Intermediate-range side-scan sonar systems* An extremely popular and capable intermediate-range side-scan sonar system is the SeaMarc II system, operated by the Hawaii Institute of Geophysics (Blackinton *et al.*, 1983). The SeaMarc II system is a shallow-towed vehicle, similar in

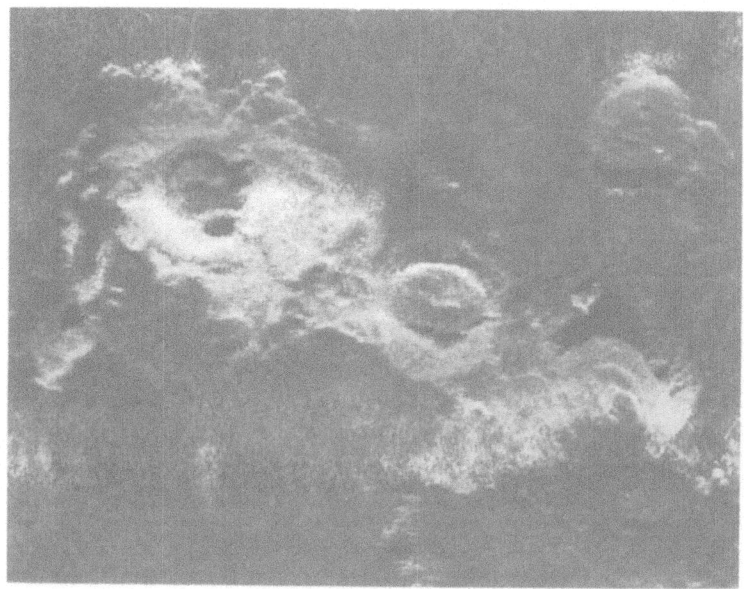

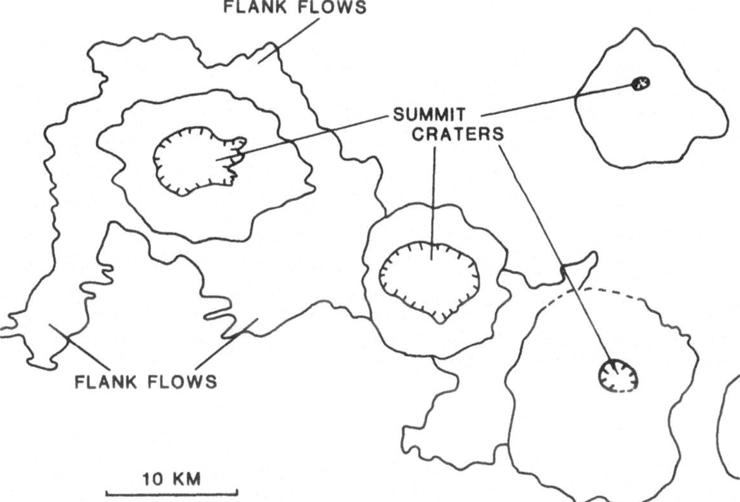

Figure 2.6 GLORIA sonar mosaic from offshore California. This image was produced by assembling many overlapping swaths of GLORIA imagery as shown in Figure 2.4 to produce a sonar image of a wide region of the seafloor. Interpretation of the image shows a series of small seamounts with summit craters. Published in: *Atlas of the Exclusive Economic Zone, Western Conterminous United States*, United States Geological Survey Miscellaneous Investigation Series I-1792 (1984).

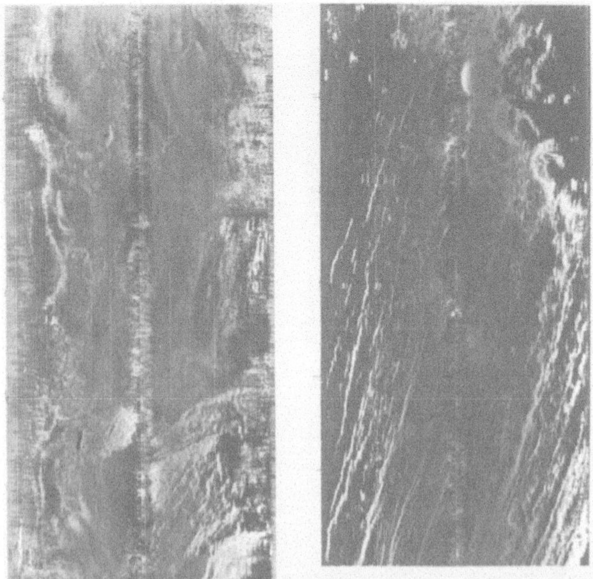

Figure 2.7 Side-scan sonar record from the GLORIA system. The centre of the image represents the ship's track over the seafloor, with acoustic images both to the right and left of the vehicle. Light areas represent regions of the seafloor with strong acoustic returns, whereas dark regions represent regions of weaker returns.

configuration to the GLORIA side-scan system. The SeaMarc system uses higher acoustic frequencies than GLORIA, roughly 12 kHz, and can therefore produce higher resolution seafloor images. Typical processed SeaMarc II images are composed of pixels nominally 10 × 10 m. The SeaMarc II swath width is a maximum of 10 km, narrower than that of GLORIA. The high survey speeds, however, still allow large seafloor areas to be quickly surveyed.

The SeaMarc II side-scan sonar system can make bathymetric measurements in addition to producing sonar imagery. This dual capability is presently unique to the SeaMarc II system.

2.3.2.3 *Short-range side-scan sonar systems* Many short-range high resolution side-scan sonar systems are presently in use, but one of the most advanced is the new SeaMarc IV system, developed at the Lamont-Doherty Geological Observatory of Columbia University. High resolution sonars are towed near the seafloor, typically a few tens to a few hundreds of metres above the seafloor. Slow tow speeds, in the range 1–2 knots, are required to tow the vehicle near the seafloor. These slow speeds, coupled with swaths of up to 6 km width, make these tools most suitable for carefully surveying specific targets that may have been identified using the side-scan devices discussed earlier.

The high frequency acoustic signals used in the SeaMarc IV vehicle (30 and 72 kHz) produce very high resolution sonar images of the seafloor, with resolutions of the order of 0.5 m. The deep-towed side-scan sonar systems therefore provide the closest image of the seafloor using acoustic methods. Very small sampling targets such as hydrothermal vent fields can be identified and located using these deep-towed side-scan systems.

2.4 Deep-sea photography

Underwater camera systems are widely used for exploring the seafloor, both from deep-towed vehicles tethered to surface ships, and from remotely piloted vehicles controlled from manned submersibles or surface ships. Small-scale features, often previously located using acoustic methods, can be photographed in great detail. The main difficulty with optical devices in the deep ocean is the need to provide illumination for photography. The high power consumption of lights limits the range that cameras can see underwater to a few tens of metres at most, which in turn limits the use of underwater photography to investigating small, specific targets.

2.5 Geological information from bathymetric mapping: the Chile Triple Junction region

A SeaBeam swath bathymetric image of the Chile Triple Junction region is shown in Figure 2.5. Here, an active spreading ridge between the Nazca and Antarctic plates is being subducted beneath the South American plate. This bathymetric image was produced by acquiring overlapping swaths of bathymetric measurements, with the position of the research vessel determined primarily by the Global Positioning System satellite data. High quality navigation, together with accurate swath bathymetric measurements, provides a detailed bathymetric map of the region from which important geological interpretations can be made.

The Chile trench reaches a maximum depth of 3440 m in the triple junction region, in a location where the Darwin Fracture Zone intersects the spreading ridge. This relatively shallow depth compared to other trenches around the world results from the young age of the subducting oceanic lithosphere (zero age at the triple junction), and the thick sediments that overlie much of the oceanic crust near the triple junction. The oceanic crust seaward of the trench averages about 2700 m deep, but reaches depths as shallow as 1940 m. The bathymetric fabric of the oceanic crust is strongly lineated parallel to the spreading ridge, defined by linear normal fault scarps with relief of up to 400 m. One prominent normal fault scarp appears to truncate a volcanic seamount on the seaward side of the rift valley, on the Antarctic plate, resulting in a split seamount. The portion of the original seamount inferred to have been formed on the landward side of the rift valley, on the Nazca plate, has been subducted beneath the South American margin.

The ridge axis shallows southward towards the triple junction from its maximum depth of 3440 m to about 2800 m at the triple junction itself. The rift valley contains numerous small circular seamounts 60–200 m in height, with basal diameters between 0.5 and 2 km. These seamounts are common along the rift valley floor near the Darwin Fracture Zone, and they probably represent small volcanic centres along the ridge axis, which are progressively buried by clastic sediments transported north along the rift valley–trench axis from the triple junction.

Normal fault scarps in the oceanic crust landward of the rift axis appear to be subdued relative to those seaward of the rift. Faults landward of the spreading centre are less continuous than those seaward of the rift, resulting in a much less well defined bathymetric fabric in the oceanic crust at the base of the inner trench slope.

Ten small upper plate seamounts were mapped in the survey; their basal diameters range between 2 and 4 km, and they stand between about 100 and 300 m above the surrounding seafloor. No samples have been recovered from these features, but their well developed conical morphology is suggestive of a volcanic origin, or perhaps diapirism involving either serpentinite or mobile mud. One upper plate seamount exhibits a subsidiary bathymetric ridge which extends southward and downslope from the main body of the seamount; this ridge may represent a lava, serpentinite or mud flow originating from the summit of the seamount.

The landward trench slope between water depths of about 500 and 2000 m is dominated by large, broad submarine canyons generally trending perpendicular to the trench axis. These canyons are about 2–8 km wide, and are typically 500–1000 m deep. They often exhibit smaller tributary canyons or gullies at the heads of the main canyons, suggesting headward erosion of these features. One canyon shows two unusual features: it contains an isolated circular depression about 1 km in diameter and over 100 m deep and the north wall is a strongly linear and steep scarp trending perpendicular to the trench and the regional bathymetry. The canyon wall is as much as 400 m high, with local slopes as steep as 38°. Both the linearity and steepness of the canyon wall suggest that it represents a fault scarp trending perpendicular to the trench.

These geological interpretations of the Chile Triple Junction region based on high quality bathymetric data can be confirmed with other kinds of data, such as seismic reflection profiles. They show that the morphology of the seafloor, when mapped at a high level of accuracy, can provide as much insight into seafloor geological processes as geomorphologic studies on land.

2.6 Selecting sampling targets

The selection of sampling sites is governed by two important considerations: the goals of the scientific research and the choice of sampling equipment to be used. These factors are dependent on each other. For example, a research project to determine the spatial distribution of the uppermost strata of oceanic crust would probably use rock dredging as the most effective sampling technique. If a research project was targeted at determining the composition of the deep levels of oceanic crust, ocean drilling techniques might be necessary because the deeper crustal layers may not outcrop on the seafloor where they can be sampled with surface methods. In addition, each type of sampling equipment usually provides the best results when used in specific seafloor environments. It is therefore important to have as much information about the region around the sampling targets as possible.

Bathymetric maps are very important for selecting sampling locations, in addition to being valuable for guiding the actual sampling operation. The advent of swath bathymetric systems has greatly facilitated the production of high resolution bathymetric maps. With the correct computer hardware on board, maps can be plotted and used for the selection of sample sites just hours (or even minutes) prior to commencing sampling operations. Similarly, the acoustic images produced by side-scan sonar systems can directly identify rock outcrops on the seafloor because rocks are very efficient reflectors of sound energy compared to sediment-covered areas. Side-scan sonar imagery can therefore provide a very good map of the distribution of basement outcrop, information important in planning an expedition to identify sampling targets at sea. Vertical-incidence seismic reflection profiles also reveal the depth below the seafloor of particular lithological or stratigraphical horizons, and can be used to identify areas where a specific target horizon is close enough to the seafloor to be within reach of a specific sampling technique.

2.7 Sampling methods

Four principal sampling methods are commonly used in the deep sea to recover basement rocks. These are dredging, coring, drilling and direct sampling by submersibles. Each method has advantages and disadvantages, and the choice of which to use is dependent on the scientific objectives, the expected local seafloor geology and the resources that can be devoted to the project.

2.7.1 *Dredging*

Dredging is the method most widely used for recovering rocks from the deep seafloor, and has certainly been responsible for recovering the largest volume of samples of any sampling technique. Dredging has been used since the earliest days of ocean exploration, including the *Challenger* expedition of 1872–1876 (Figure 2.8). Since then, research vessels have occupied thousands of dredge stations, and have recovered many tons of seafloor samples. Perhaps more than any other oceanographic research technique, the hardware and techniques of dredging have changed very little since the first scientific dredge sample was recovered.

The modern version of the dredge consists of a strong welded steel frame, typically 50 cm × 1.5 m, that holds open a sturdy bag, itself often constructed

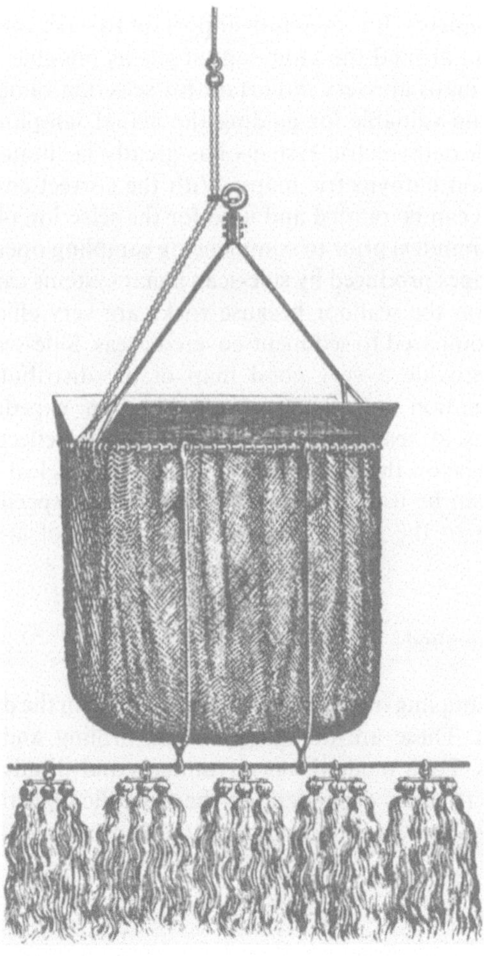

Figure 2.8 Dredge bag used aboard *HMS Challenger* in the 1870s.

of steel chain. Hinged attachment arms are fastened to the frame, which in turn are connected to the heavy steel wire that lowers the dredge to the seafloor. A length of wire at least several hundred metres in excess of the water depth is spooled off the winch as the ship drifts or is held accurately in position over the sampling site. When the dredge has reached the seafloor along with the extra wire, or 'scope', the winch brake is set and the ship steams slowly or is allowed to drift, slowly dragging the dredge along the seafloor, picking up material in its bag. In regions of steep topography, where basement rocks are most likely to outcrop, the dredge is generally pulled uphill along the seafloor; otherwise the dredge may 'kite' up off the bottom and fail in its recovery attempt.

The dredge is very successful in picking up talus or loose rubble from the seafloor. Debris piles at the base of a steep scarp are very good dredge targets. When strong outcrops are encountered the dredge can often break off a piece of rock from the outcrop, the best possible outcome, or the outcrop can prove to be stronger than the dredge or the dredge wire, and too strong a pull by the research vessel can result in the parting of the wire and the loss of the dredge. Spare dredges are carried aboard because some dredges can be expected to be lost on the seafloor during a dredging cruise. The heavy dredge wire is more difficult and expensive to replace. In order to safeguard the wire, a weak link, consisting of a swivel or a short piece of wire weaker than the main dredge wire, is inserted between the dredge and the wire. If the dredge encounters an outcrop that it cannot break, and manoeuvering the ship at the surface fails to free the dredge, then the weak link parts, and all that is lost is the dredge itself.

One of the few significant improvements in dredging technology since the nineteenth century is the ability to accurately determine the location of the dredge independent of the vessel at the surface using acoustic methods. An acoustic pinger, a battery-operated sound source, is fastened to the dredge wire a few meters above the dredge itself. The acoustic range from the dredge pinger to a network of separate transponders placed by the surface vessel can be calculated by measuring the travel times of the sound impulses from the dredge to the transponder network. Simple triangulation determines the exact location of the pinger, and hence the dredge itself, on the seafloor. Using these acoustic methods, it is possible to determine the exact location of a dredge sample that is recovered. Recovering samples from several short dredge deployments on the face of an escarpment, for example, can reveal the stratigraphy of the rock units exposed on the seafloor, whereas a single long dredge haul up the entire escarpment would produce a mixed bag of samples, the relative stratigraphic positions of which may not be easily reconstructed. However, the best method for recovering samples of the oceanic crust and simultaneously preserving their stratigraphic relationships is deep-sea coring and scientific drilling.

2.7.2 *Coring*

Coring techniques are most commonly used to sample deep-sea sediments, where piston coring techniques can successfully recover cores of unlithified sediment as long as 15 m. Specialized coring hardware, often called dart cores, can be used to sample igneous basement rocks. The dart core method differs from piston coring in that the dart core barrel, the pipe in which the sample will be recovered, is very short, often only 15–20 cm long, and the weight used to drive the core barrel is as heavy as that used for long piston cores, as much as 900 kg. In practice, the dart core is lowered very rapidly by a winch to within 20–30 m of the seafloor, and is then allowed to freefall. The impact of the core barrel with the seafloor is energetic enough to break off small pieces of igneous rock, if any are reached. The core barrel is often destroyed, whether a sample is recovered or not.

Dart coring can be a very fast operation if a high speed winch is used. Many locations along a transect can be sampled in a relatively short time. This method has been very successfully used in regions where dipping strata outcrop along the seafloor, and the rocks recovered along a sampling transect can be used to determine the stratigraphic succession of the seafloor exposure. Shortcomings of the dart coring method include the small sample size that can be recovered, and the frequent failure to recover bedrock samples. However, dart coring is a simple and cost-effective system for recovering submarine basement rocks.

2.7.3 *Drilling*

The techniques developed on land for drilling oil wells have been successfully applied to drilling in the deep ocean for scientific purposes. Whereas many of the details of on-land drilling techniques are different for scientific drilling, many of the basic principles remain the same. A long drill string, made up of a number of individual sections of drill pipe that screw together, is lowered from a tall derrick positioned amidships through a hole in the bottom of the ship to the seafloor. The lower end of the drill string is equipped with a drill bit that will both cut through the rock and sediment encountered and allow the passage of a core sample through the middle of the bit into the centre of the hollow drill string. The drill string is rotated by a motor on the ship while it is lowered into the seafloor. Drilling fluid, usually seawater, is pumped from the drilling ship down through the drill string to the bit to carry away the cuttings from the bottom of the hole. Coring pipes are lowered down by wireline through the drill pipe between periods of rotary drilling to acquire core samples of the strata penetrated by the hole.

Drilling wells in deep water is an extremely difficult undertaking, and might be referred to as an art rather than a science. Many of the proven methods developed in the petroleum industry cannot be used because of the inability to circulate drilling fluids, the cost and time constraints, or because of the scientific emphasis on taking core samples rather than purely drilling wells. Scientific drilling in the Ocean Drilling Program (ODP) is therefore a specialized part of the larger drilling community. The ODP has borrowed technology and techniques from the commercial drillers, but much of the hardware and many of the methods used have been developed by the programme to meet the specialized needs of scientific ocean drilling. The discussion that follows is focused on the Ocean Drilling Program and the DV *JOIDES Resolution*, the drill ship most important to modern marine geological research.

2.7.3.1 *General procedures used on the DV* JOIDES Resolution The *JOIDES Resolution* is equipped with a large drilling derrick positioned amidships, reaching a height of 62 m above the water-line. Beneath the derrick is a 7 m diameter moonpool through which the drill string is lowered. The travelling block assembly on the derrick includes a heave compensator that decouples the vertical movement of the ship on the waves from the drill string.

When the vicinity of a drill site is reached, an acoustic beacon is dropped and freefalls to the seafloor. The seismic reflection equipment used for the final location of the site is retrieved from the sea and the ship returns to the site and the positioning thrusters and hydrophones are lowered to the dynamic positioning configuration. The dynamic positioning computer system is then locked onto the beacon signal, and the ship commences drilling operations. Satellite navigation data are received while the ship is on station so that the actual drill site is located accurately.

2.7.3.2 *Drilling operations* As the ship stabilizes in the dynamic positioning mode, the bottom hole assembly and the drill string are prepared for lowering to the seafloor. The bottom hole assembly commonly consists of a 6.2 × 25.0 cm roller cone bit, one 21 cm outer core barrel, eight 9 m drill collars and three joints (sections) of heavy drill pipe. The drill string is maintained in tension during drilling; the purpose of the drill collars is to provide weight at the bottom of the string. The driller supports most of the weight of the drill string during drilling, and the remainder represents the downward pressure on the bit. The bottom hole assembly is then attached to 27 m three-joint stands of drill pipe. The exact configuration of the bottom hole assembly is varied to suit the expected hole conditions, with heavier assemblies rigged for coring hard rock and lighter configurations for coring soft sediment.

The drill string is then run down to within about 20 m of the seafloor, based on the 12 kHz PDR, corrected for the expected sound velocity in water at the site. Sophisticated pipe-handling equipment is used to lift the pipe out of its storage rack, connect the new section to the drill string, and lower the drill string towards the seafloor. It is important to run the pipe as quickly as possible, especially in deep water where several thousand metres of drill string must be run before the seafloor is reached. Great care is taken to ensure that the screw joints between pipe stands are tightened to the correct torque (45 000 ft lbs). A special hydraulic machine called the mechanical roughneck tightens each pipe joint to the correct torque.

The top drive, the large electric motor that actually rotates the pipe during

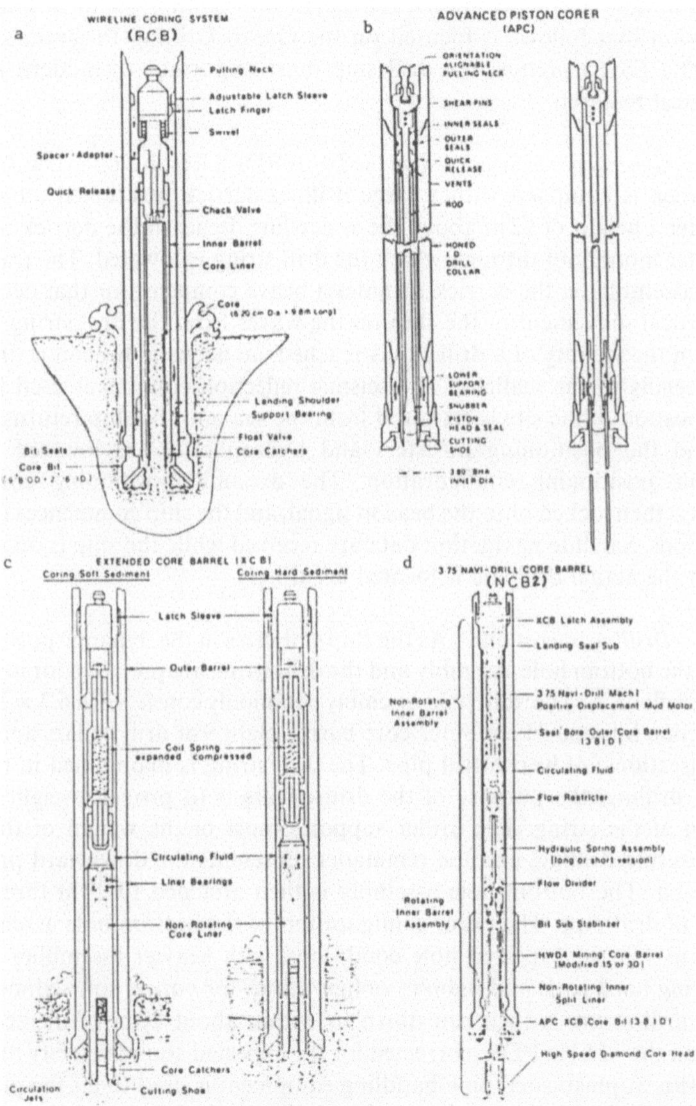

Figure 2.9 Schematic cross-section of coring systems used by the Ocean Drilling Program. (a) Wireline coring system (RCB); (b) advanced piston corer; (c) extended core barrel; and (d) Navidrill core barrel.

drilling, is attached to the travelling block and the entire assembly is mated to the top of the drill string. Contact with the seafloor, known as spudding in, is then carefully executed. If the shipboard scientists wish to recover the uppermost surface sediments, then a mudline core is recovered by lowering the drill string to within about 3–7 m of the seafloor and taking an advanced piston core (APC) (Figure 2.9b). The core barrel will be roughly half-full of sediment in a successful mudline core recovery. Drilling is accomplished by rotating the entire drill string with the top drive motor, which is capable of exerting 41 000 ft lbs of torque. Additional pipe joints are added to the top of the drill string as drilling progresses.

One major difference between drilling at sea and on land is that the ocean is not calm. The drilling platform heaves up and down with the ocean swell even when the ship's dynamic positioning keeps it above the hole on the seafloor. In rough weather, the ship can heave as much as 10–12 m. This motion is directly transmitted to the drill string and drill bit if the drill string is not stabilized.

A drill string heave compensator is routinely used during rotary drilling operations to isolate the drill string from the heave of the ship. The heave compensator consists of large pneumatic cylinders that are integral to the travelling block in the derrick. The amount of drill string weight supported by the compensator can be varied simply by adjusting the air pressure in the cylinders. As the ship heaves due to wave motion, the pistons in the air cylinders also heave up and down, cushioned by the pressurized air in the cylinders. Good heave compensation contributes to the recovery of undisturbed material, and it reduces the rate of wear on the bearings in the drill bit, helping to prolong the life of the bit. The heave compensator also allows drilling operations to proceed in bad weather conditions, thus reducing the amount of weather-related down time.

Seawater is used for circulating around the drill bit. Water is pumped down through the pipe, exits at the bit at the bottom of the hole, and then flows upward outside the drill string to the seafloor, carrying cuttings away from the bottom of the hole. The water flow must be monitored very carefully. The upward flow-rate must be fast enough to carry away the cuttings, but too much circulation can wash out weakly consolidated strata, causing caving of the hole, or can wash away material from the bottom of the hole and prevent recovery in cores. Because water is much less dense than traditional drilling fluids (mud), the sink rate of rock cuttings is faster than in mud. Therefore the upward flow-rate of the water must be faster than if mud were used as a drilling fluid. Occasionally, 20–50 bbl. slugs of mud (freshwater bentonite) are used to assist in hole cleaning when shale, chert, or basalt chips are sloughing off the walls of the hole. Coring operations are carried out on a continuous basis or as spot cores during drilling.

After the total depth is reached in a hole, the bit is released from the bottom of the drill string and left in the bottom of the hole. This opens the bottom of the drill string enough to allow the passage of the logging tools. The drill string is pulled up until the bottom of the string is positioned at the base of the uppermost competent formation, typically about 100 m below seafloor, and the logging tools are run. Following logging, the drill string is retrieved, the ship is configured for underway operations, and the transit to the next site begins. Meanwhile, the shipboard scientists complete their shipboard analyses of the samples collected.

The amount of core disturbance varies between the different coring techniques. Cores obtained using the APC technique are relatively undisturbed, and are the most desirable, but these can only be recovered to 200–300 m sub-bottom in soft sediment. Extended core barrel (XCB) (Figure 2.9c) cores typically show more disturbance than APC cores, but can be recovered in considerably more lithified sediment than APC cores. When APC refusal is reached, then XCB cores are used. Both types of core can be taken through the same kind of drill bit.

When very lithified sediments or igneous basement rock are encountered, rotary coring and drilling techniques must be used. The rotary bit is a very different device from the bits typically used for APC or XCB coring. Swapping bits requires a round trip of the drill string, bringing all the pipe back on board, changing the bottom hole assembly, running the pipe back down to the seafloor again, and re-entering the old hole or spudding in for a new hole. When this procedure is anticipated, the first hole drilled at the site (the A hole) is cored using the APC or XCB bottom hole assembly as deep as the sediments will allow. The drill string is then retrieved, the rotary bit assembly is deployed, and a second hole (the B hole) is quickly drilled without coring to the depth of the A hole. Rotary drilling and coring commences again until the depth required is reached. There are typically two reasons for starting a second hole in such circumstances: hole stability decreases with time, and a re-entry cone must be set in order for a hole to be re-entered after the drill string is completely removed from the hole.

Re-entry cones (large cones and 'mini-cones') are set on the seafloor when it is anticipated or planned that re-entry will be necessary. For example, very deep penetration into oceanic crust requires the replacement of worn bits with new bits as drilling progresses. Bit changes can be carried out only by 'tripping' the drill string. In such circumstances, a re-entry cone is set, and the new bit and drill string can be manoeuvered back into the hole by carefully moving the ship while viewing the seafloor and the bottom of the drill string with a television camera lowered down along the pipe. When the bit is over the re-entry cone, the drill string is quickly but carefully lowered back into the hole, the camera is hauled back up to the surface, and drilling re-commences.

2.7.3.3. *Coring and drilling equipment for igneous rocks*: (a) *Rotary coring system* The rotary coring system (RCB) is used for drilling into the igneous basement rocks of the oceanic crust. This configuration places the core barrel above the central annulus between the roller cones of a rotary bit (Figure 2.9a). The bit cuts a circular zone out of the formation, leaving a column of rock that passes between the rollers and into the core barrel. The formation must be indurated enough to be self-supporting for the few centimetres between the cutting surface and the bottom of the core barrel. Often the amount of circulation required for the adequate removal of cuttings can be too vigorous for the preservation of the unsupported core, and recovery is poor. The RCB system allows fluid circulation at the cutting surfaces of the drill bit, and hence is at its best in well indurated rocks.

(b) *Navidrill core barrel* The Navidrill coring system (Figure 2.9d) is presently under development by the ODP. It is designed to improve the recovery of undisturbed material in lithified rocks, such as basaltic basement and other hard formations. The Navidrill coring system consists of a slimline core barrel, similar to that used in the mining industry, rotated at high speed (up to 850 rev min^{-1}) by a downhole mud motor. The entire assembly is lowered down the normal drill string and recovered by wireline, identical to the APC and XCB coring systems. The downhole mud motor uses seawater pumped down the drill string to rotate the diamond bit drill rod, the drill string itself does not rotate during Navidrill drilling.

The Navidrill system is presently configured to recover a core of 61 mm diameter up to 4.5 m long. The Navidrill penetrates ahead of the main core bit so that the slimline core is cut from undisturbed material. The core is recovered in a plastic or split metal core liner. After core recovery, the drill string with its rotary bit is rotated to drill down over the Navidrill pilot hole, and the entire process is repeated. Development work is continuing on the Navidrill system, with recent tests conducted on the ODP Engineering Leg 124E in the western Pacific. The Navidrill should greatly increase the rate of core recovery in igneous basement rocks, in fractured formations, and in sequences characterized by alternating hard and soft layers.

2.7.4 *Submersible sampling*

The most detailed sampling and the best direct observation of the ocean floor can be achieved by submersible diving. The great depths of most of the world's oceans preclude scuba diving for direct observation of the seafloor, and only a few submersibles have the required depth range. The special capabilities of a submersible make their role in the exploration and sampling of the ocean floor important. One of the most widely used and successful research submersibles is the DSV (deep submergence vehicle) ALVIN, owned by the United States Navy and operated by the Woods Hole Oceanographic Institution.

ALVIN is a small submersible, just 7.6 m long. The crew of the submersible,

Figure 2.10 Section through ALVIN, showing the pressure sphere for personnel.

one pilot and two scientific observers, are located in a spherical pressure sphere at the front of the vehicle (Figure 2.10). From the pressure sphere they can observe their surroundings through three small viewing ports, and they can operate remote sampling arms, cameras and other devices. ALVIN has a maximum depth capacity of 4000 m, which puts much of the mid-ocean ridge system within reach.

Batteries supply electrical power for the operation of the propulsion motors and scientific equipment. Owing to power limitations, ALVIN, has a cruising speed of just 1 knot, and a cruising range underwater of 8 km. These limitations, coupled with the short range of visual observation from the submersible, makes ALVIN a tool for detailed exploration and sampling of specific, well identified targets, rather than a tool for the exploration of unknown regions.

The front of ALVIN is covered by an array of sampling arms, cameras, lights and a large sample basket, into which rocks are placed for transport back to the surface. ALVIN is launched and recovered from the RV *Atlantis II*, a 64 m research vessel operated by the Woods Hole Oceanographic Institution. ALVIN is launched and recovered by a large gantry mounted on the fantail of the *Atlantis II*. Divers are required to assist in the launch and recovery procedure.

Using the remote manipulator arms of the submersible, scientists can conduct very precise and detailed sampling programmes, in principle as detailed and well documented as at an outcrop on land. The difference between submersible sampling and outcrop geology on land is that different methods must be used to extrapolate the detailed observations made from the submersible to a larger region. The mapping methods discussed previously in this chapter provide the information that establishes a geological context for the small-scale and detailed exploration that can be conducted from a research submersible.

2.8 Concluding statements

1. The first mapping techniques to be used in an unexplored region of the ocean should provide a broad overview of a large area, so that specific and scientifically important targets can be identified for further high resolution surveys. Intermediate-scale mapping and imaging techniques can then be used which provide a more detailed picture of the seafloor, but over a smaller area. The sampling techniques most appropriate for the lithologies present and the scientific objectives of the research programme can then be used.
2. The techniques of making bathymetric measurements have improved immeasurably since the days of lowering lead lines over the sides of vessels. Acoustic echo sounders can make very accurate depth measurements from moving surface ships, with recent developments including computer-controlled multibeam swath bathymetric mapping systems that can make depth measurements over a broad swath of the ocean floor. The most important benefit derived from the use of swath bathymetric systems is the near-complete coverage of the ocean floor and the accurate and reliable mapping of very small features on the seafloor.
3. Side-scan sonar seafloor mapping techniques using GLORIA and Sea-Marc are able to produce images based on the acoustic reflectivity of the seafloor at particular frequencies. Lower sound frequencies are most effective for mapping very wide swaths to the sides of the towed vehicle, whereas higher frequencies provide a much more detailed image of the seafloor.
4. The primary methods used for recovering igneous basement rocks from the deep sea are coring, drilling, dredging and submersible sampling. Coring can be a very rapid operation, requiring little ship time, but usually recovers only small basement samples. Dredging can recover very large volumes of loose materials, but it can often be difficult to actually break samples from outcropping seafloor exposures. As a result of this, reconstructing the stratigraphic relationships of seafloor sequences can be difficult. Deep-sea drilling is the best way of determining the stratigraphy of rock units, and it provides a means of sampling materials that do not outcrop on the seafloor. Drilling in deep water is a very specialized technique that requires a dedicated vessel, a great deal of technological and engineering development, and is very expensive to undertake. A similar level of technological development and support is required to support the operation of manned submersibles, but sampling from a small submarine is the closest approach currently available to the traditional methods of field geology in the marine realm.

3 Structure of the oceanic crust from geophysical measurements

ROBERT WHITE

3.1 Introduction

Most of the oceanic crust has never been sampled, either areally or in depth. Only a handful of drill holes penetrate more than a few hundred metres into the igneous crust, and even the deepest drill hole, DSDP 504B (Becker *et al.*, 1989), only reaches a depth of a little over 1 km into the basement, barely one fifth of the way to the base of the crust. In general, our knowledge of the *in situ* structure and petrology of the oceanic crust is restricted to dredges and drill samples that barely scratch the surface.

Against this background geophysical measurements at sea have provided an extremely powerful method of investigating the deep structure of the oceanic crust and the processes, both tectonic and magmatic, that operate in the ocean basins. Whereas the covering of deep water acts as a curtain separating the geologist with his hammer from the object of his studies, it is in many ways a boon to the geophysicist, enabling him to traverse large areas with little obstruction by topography or by barriers of politics or land ownership. Geophysical studies in the oceans were thus able to provide the basis for plate tectonic theories by the large-scale mapping of seafloor spreading magnetic anomalies. Simple mapping of the seafloor depth has revealed how the lithosphere thickens and develops with age as it moves away from the spreading centres. As there is little sediment cover in the deep oceans, bathymetric mapping and side-scan sonar are able to reveal the tectonics and structure of the spreading axis in a way impossible on land where the underlying structure is generally obscured by sedimentation or removed by erosion.

In the sections that follow, the contribution of geophysics to understanding the large-scale structure of the ocean basins is discussed. Attention is then focused on studies which give information on the tectonic and magmatic processes operating on the oceanic ridges and in the areas of off-axis volcanism that form oceanic islands.

3.2 Ocean basins

A major contribution of geophysics to understanding the ocean basins was the documentation of the way in which lithospheric plates move on the surface of the earth (Cox and Hart, 1986). As the plates move apart in an ocean basin, the ductile asthenospheric mantle which lies beneath them wells up to fill the space created along the zone of separation (Figure 3.1, top). Partial melting of the upwelling mantle occurs as it decompresses and crosses the solidus (Figure 3.1, bottom). The melt bleeds upwards rapidly, eventually, solidifying above the mantle to form the oceanic crust with a typical thickness

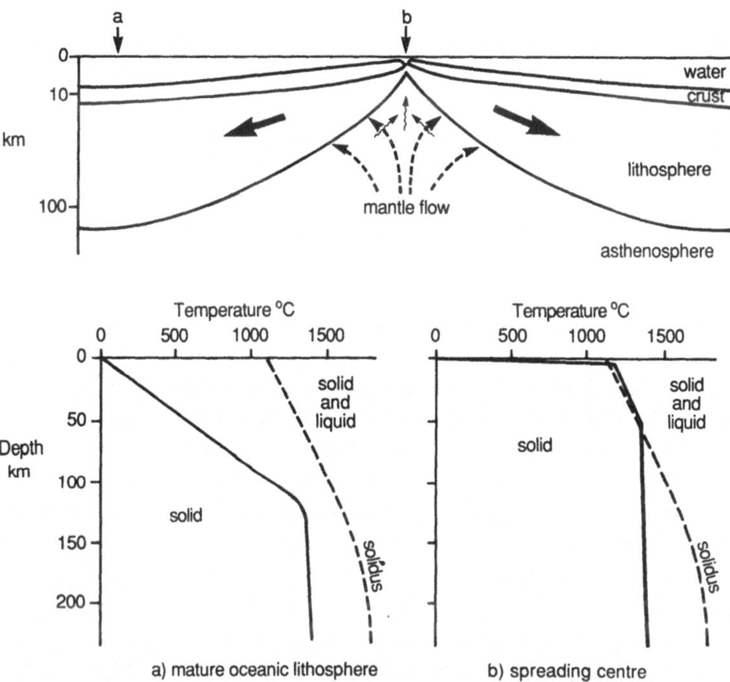

Figure 3.1 Top diagram shows schematic cross-section of an ocean basin. As the plates move away from the spreading axis, the lithosphere thickens by conductive cooling and subsides to maintain isostatic equilibrium. (a) Oceanic lithosphere older than about 80 Ma has a thickness of about 125 km, with small-scale convection at the base preventing a further increase in lithosphere thickness. The mantle is well below the solidus, so no melt is produced. (b) Beneath the oceanic spreading centre the asthenospheric mantle wells up passively as the plates move sideways [broken arrows in top diagram]. Upwelling mantle with a normal potential temperature of 1280°C crosses the dry solidus at about 50 km depth and begins to melt. As it continues to rise it melts further until an average of 25% of the mantle is melted. The partial melt from all depths [wavy arrows in top diagram] is extracted rapidly upwards and focused to the spreading centre by corner flow and buoyancy forces, there cooling to form the oceanic crust with typical MORB composition. If the asthenospheric mantle is at higher temperatures than normal, it crosses the solidus at a greater depth as it rises upwards and considerably higher volumes of melt are produced (see Figure 3.4).

of 6–7 km. Separation of the lithospheric plates and the concomitant generation of new oceanic crust occurs along a narrow zone in the centre of the ocean basins called the spreading centre.

As the newly created crust moves away from the spreading centre and cools below the Curie point, it becomes permanently magnetized in the earth's magnetic field (Cox and Hart, 1986). The uppermost 500 m of fine-grained basalts are the most highly magnetic, with the underlying coarser-grained basalts being less magnetized and the lower crustal coarse-grained gabbros only very poorly magnetized. As the polarity of the earth's magnetic field reverses irregularly with a frequency of up to several times every million years, the acquired thermo-remanent magnetization also reverses. This creates a series of stripes of alternating polarity parallel to the spreading axis, similar to a bar code. The magnetic stripes can be easily detected with total field magnetometers, towed either from ships or, with some loss of resolution, from aeroplanes. The irregularity of the reversals makes it possible to distinguish the stripes one from another, and to number the reversals uniquely. The same sequence of reversals can be identified in ocean basins throughout the world, making it possible to determine the relative ages and spreading rates of the ocean basins. The skewness of the magnetic anomalies has been used to estimate the geomagnetic latitude of the crust at the time of its formation.

The magnetic anomaly identifications do not themselves provide an absolute age for the seafloor. By drilling into the basaltic basement as part of the Deep Sea Drilling Project, the age of the magnetic anomalies can be fixed either by radiometric dating of the basement or through the biostratigraphy of the immediately overlying sediments. From the resultant magnetostratigraphic timescale, which extends back to 165 Ma, the age, spreading direction and rate of separation of vast tracts of the ocean floor can be determined simply by towing a magnetometer from a ship. This was a major achievement which underpinned the formulation of the theory of global plate tectonics.

Another cornerstone of plate tectonics also came from geophysical studies of the oceanic crust, through earthquake seismology. The location of seismicity delineates the active boundaries of the lithospheric plates: the spreading centres where plates are diverging; collision zones where plates are converging; and the strike-slip boundaries where two plates are sliding past each other. Perhaps more important still, the creation of a uniform global network of earthquake seismometers, the Worldwide Standardized Seismography Network, made it possible to determine the fault-plane solutions of remote earthquakes which give information on the type of fault, its strike and dip, and the amount of energy released by the earthquake. This allows a detailed study of the tectonics of crustal deformation in remote regions of the ocean basins without ever having to observe or map the outcrop.

Before leaving the large-scale features of whole ocean basins and moving on to a more detailed discussion of the oceanic crustal structure, it is necessary to review the structure of the oceanic lithosphere and underlying asthenosphere because these control both the large-scale tectonics and the magmatism which generates the oceanic crust. The lithosphere is the rigid outer layer of the earth. Beneath mature oceanic crust it attains a thickness of about 125 km. The upper 6–7 km is the oceanic crust, and the remainder consists of mantle. Beneath the lithosphere lies the asthenosphere, which is formed of vigorously convecting mantle (White, 1988a).

At the oceanic spreading centres the lithosphere is stretched to almost zero thickness. As it moves sideways away from the spreading centre, the underlying asthenospheric mantle loses heat vertically by conduction. Cooling of the asthenospheric mantle causes it to become denser and more rigid, thus converting to lithosphere. The lithosphere therefore thickens away from the spreading centre, in a manner that can be modelled very simply using thermal conduction equations. There are two results that can easily be observed geophysically. The first is that in order to maintain isostatic equilibrium the seafloor depth increases with age as the mantle cools and converts to denser lithosphere. The increase in depth is proportional to the square root of age. The second is that the heat flow through the seafloor decreases as the lithosphere ages.

Global measurements of seafloor depth and heat flow confirm the simple model of lithospheric thickening with age, but with two significant departures

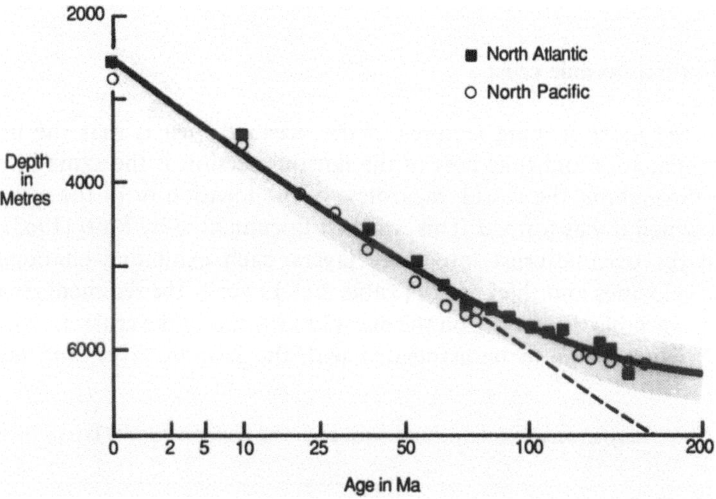

Figure 3.2 Relationship between mean depth and the square root of age for the North Atlantic and North Pacific, after Sclater and Parsons (1981). The shaded area represents an estimate of the error in the original points used to determine the mean data. The broken line is the expected subsidence for the simple one-dimensional lithosphere cooling model, with the departure from it beyond 80 Ma shown by the solid line. This departure is a result of the onset of small-scale convection in the thermal boundary layer at the base of the lithosphere.

from the simple cooling model. The main difference from the simple model is that beyond about 80 Ma age the depth does not continue to increase (Figure 3.2) and the conductive heat flow does not continue to decrease. The rate of change of both the seafloor depth and the heat flow values flattens out. The explanation for this is that there is small-scale convection in the thermal boundary layer between the overlying rigid plate, or mechanical boundary layer, and the underlying vigorously convecting mantle (Sclater and Parsons, 1981). The thermal boundary layer of small-scale convection is unstable and prevents the plate thickness increasing indefinitely. As will be seen in the next section, the thermal structure of the lithosphere and the asthenosphere is crucial in controlling the generation of the oceanic basaltic crust.

The other main departure from the simple cooling model is near the spreading axis, where measurements of the heat loss by conduction through the surface of the crust fall short by a factor of two or more from those expected from the simple lithospheric cooling model (Sclater and Parsons, 1981). The reason is that there is vigorous hydrothermal circulation through approximately the top 2 km of the igneous crust which removes huge amounts of heat by advection (Lister, 1972) and quenches the upper crust. Such vigorous hydrothermal circulation is relatively short lived, but has a significant effect in modifying the petrology of the basaltic crust through which it passes and in the generation of large sulphide and ore bodies on the seafloor.

3.3 Normal oceanic crust

One of the most striking features of the oceanic crust is that the normal seismic structure and thickness of the igneous section is the same in ocean basins throughout the world, regardless of the location or of the spreading rate at which it was formed. This was first documented by Raitt (1963), who divided the oceanic crust into three layers, each exhibiting characteristic seismic velocities and thicknesses (Table 3.1). Layer 1, the sedimentary layer, is highly variable depending on the age and location of the crust. In contrast, layer 2, which came to be associated with the basaltic layer, and layer 3,

Table 3.1 Velocity structure of the oceanic crust assuming uniform velocity layers, from Raitt (1963)

Velocity layer	Velocity ($km\,s^{-1}$)	Thickness (km)
Layer 1 (sedimentary)	About 2	Variable
Layer 2	5.07 ± 0.63	1.71 ± 0.75
Layer 3	6.69 ± 0.26	4.86 ± 1.42
Layer 4 (mantle)	8.13 ± 0.24	

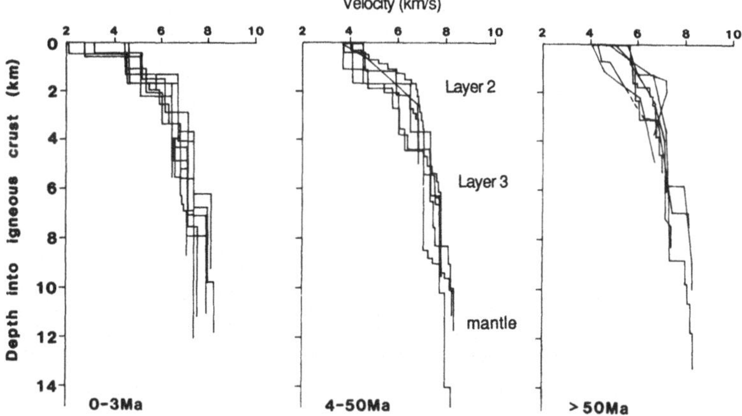

Figure 3.3 Velocity–depth profiles from the Atlantic Ocean through normal oceanic crust, away from the influence of hot-spots and fracture zones. Only those profiles constrained by synthetic seismogram modelling are shown. Diagram from White (1984).

thought to consist of intrusive gabbros, exhibit a remarkably similar structure throughout the world.

More detailed investigations through the subsequent two decades confirmed the global consistency of oceanic crustal structure. Better interpretation methods, made possible by synthetic seismogram modelling using computers, showed that the seismic structure could be better described by velocity gradients than by uniform velocity layers (Figure 3.3; from White, 1984). In layer 2, the upper basaltic layer, the seismic velocity increases from as low as $3\,km\,s^{-1}$ at the top of basement to about $6\,km\,s^{-1}$ in the middle of the crust. There is a simple explanation for this. The velocity gradient in layer 2 is caused not by major petrological changes but by the decrease in porosity with depth. As the crust ages, the seismic velocity at the top of the basement increases due to infilling of the numerous cracks, fissures and pore spaces by secondary minerals such as calcite. The velocity gradient in the lower crust (layer 3) is much smaller, reflecting homogeneity of composition and much decreased porosity compared to the upper crust.

Detailed studies of the crust–mantle transition using wide-angle seismics shows that it varies in character from a sharp transition between crustal and mantle velocities, to a gradient zone or a series of alternating high and low velocity layers up to 2 km in total thickness. By comparison with ophiolite sections preserved on land the transition zone is thought to represent alternating layers of mafic and ultramafic material formed by injection of melt as sills at the base of the crust. The underlying mantle often exhibits horizontal seismic anisotropy, which may be caused by alignment of the olivine crystals in the mantle flow under the spreading centre.

Unfortunately, the compressional wave seismic velocity itself cannot be used to determine uniquely the composition of the rock because widely varying rock types (such as salt and basalt) may exhibit similar seismic velocities. If the shear wave velocity can also be determined, the constraints on possible rock types may be tighter (Spudich and Orcutt, 1980), but there are at present few reliable shear wave measurements of *in situ* oceanic crust.

The consistency of normal crustal structure indicated by seismic studies demands a consistent mechanism for generating the igneous rock which is independent of spreading rate variations of more than an order of magnitude. The key to this is that at the spreading centre the lithosphere is stretched to zero thickness, and the melt must be derived ultimately from the asthenospheric mantle as it passively wells up and decompresses. If the spreading rate is doubled, twice as much mantle wells up, twice as much melt is generated, and as it has to fill twice the space, it ends up solidifying to form the same thickness of crust. The oceanic crust is formed by an average of about 25% melting of the mantle source.

The volume and composition of the melt generated by decompression can be calculated using parameterizations developed by McKenzie and Bickle (1988) of the pressure and temperature conditions under which small samples melt in the laboratory. By extrapolating to the pressure and temperature conditions in the earth (Figure 3.1), the incremental partial melting of the upwelling mantle can be modelled. Almost all the melt bleeds rapidly to the surface, leaving only tiny amounts in the matrix. The total amount of melt generated by decompression is extremely sensitive to the initial temperature of the asthenospheric mantle. An increase of as little as 100°C above the normal potential temperature of 1280°C more than doubles the volume of melt (Figure 3.4).

The consistency of oceanic crustal thickness of 6–7 km therefore points to a global uniformity of normal asthenospheric potential temperature of 1280 ± 20°C (the potential temperature is the temperature the mantle would have if brought to the surface adiabatically without melting). This global consistency is not surprising; the upper mantle convects vigorously and the plates move across it at relatively high speeds, so it would be hard to maintain areas of different temperatures beneath any particular region. As is discussed in section 3.6, in areas where there are thermal plumes in the mantle such as beneath Hawaii or Iceland, considerably more melt is generated by decompression, as expected from Figure 3.4. As the centre of the Iceland hot-spot is approached, the mantle temperature increases and consequently the oceanic crustal thickness produced at the spreading centre increases (Klein and Langmuir, 1987; White and McKenzie, 1989b). Beneath Iceland itself the crustal thickness reaches 25–30 km, indicative of mantle potential temperatures of more than 1500°C at the centre of the plume, some 250°C above the normal asthenospheric temperature.

Geophysical studies of the normal structure of oceanic crust are thus crucial in constraining the volumes of melt produced at spreading centres and the

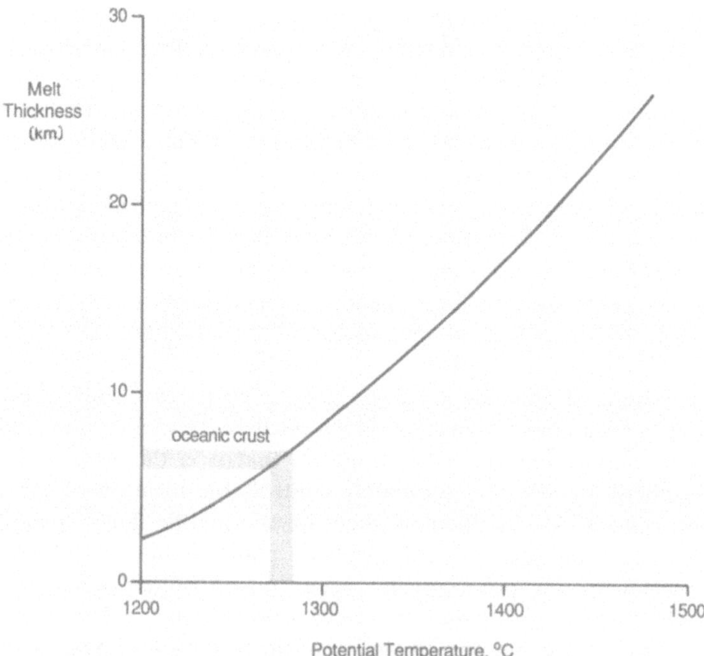

Figure 3.4 Thickness of melt produced by adiabatic decompression of upwelling mantle beneath an oceanic rift as a function of asthenospheric potential temperature. The normal temperature of 1280°C generates an average normal crustal thickness of 6–7 km. The temperature is of the order of 250°C hotter in thermal plumes in the mantle. Curves from White and McKenzie (1989b) based on parameterizations of McKenzie and Bickle (1988).

primary controls on the mechanism by which partial melting of the mantle occurs.

3.4 Spreading centres

As the igneous crust is newly formed at the spreading centres, there is very little sediment cover and simple mapping of the bathymetry of the seafloor reveals much about the tectonics and extrusive igneous processes. The main geophysical tools for this are multibeam echo sounding and side-scan sonar (see Chapter 2). Two major results have been found in normal segments of the spreading centres away from fracture zones. The first is that the extrusive volcanism occurs along a very narrow zone, typically only 1–2 km wide. The active zone is often marked by shallow seafloor volcanoes. The second major observation is that the main tectonic activity is normal faulting, which occurs on the flanks of the median valley on slow spreading ridges. Most of the normal faulting occurs within about 10 km of the spreading axis.

The narrowness of the zone of extrusion on the seafloor is at first sight a puzzle because the asthenospheric mantle upwelling which generates the melt must occur over a fairly broad region, at least several tens of kilometres wide. There is no doubt that such a narrow extrusion zone is typical of the past history of the ocean ridges and is not just a peculiarity of the present day structure mapped along active spreading centres. This conclusion comes from modelling the seafloor spreading magnetic anomalies; if the extrusive basalt had been produced over a broader region, then the magnetic stripes, which may be as little as 1 km wide, would have been so severely degraded by admixture of normal and reversed polarity basalts that they would not have been recognizable. By modelling magnetic reversal transitions, Schouten *et al.* (1982) estimate the extrusion zone to have been less than a few kilometres wide.

Explanations of why the extrusion zone is so narrow while the mantle upwelling is broad have been sought by modelling the mantle flow under the spreading centres. The modelling suggests that it is the deep flow in the asthenospheric mantle that ultimately controls the location of the seabed extrusive centres. As the ductile asthenospheric mantle flows upwards and sideways under the separating lithospheric plates at the axis of the ocean basin, partial melting occurs over a broad region (Figure 3.1). The corner flow of the asthenospheric mantle creates a pressure field which focuses the melt strongly towards the spreading axis as the melt moves upwards through the matrix (Spiegelman and McKenzie, 1987), provided the viscosity of the mantle is of the order of 10^{20}–10^{21} Pa s. Such viscosities are probably too high. Another explanation suggested by Scott and Stevenson (1989) is that the buoyancy forces generated by partial melting can drive secondary circulation beneath the ridge axis, which concentrates the melt production beneath the axis. This effect is most significant for mantle viscosities of less than 10^{19} Pa s and for low spreading rates, and can only cause moderate focusing which cannot account entirely for the extreme narrowness of surface volcanism.

The actual fluid mechanics of the mantle motion under the spreading centre is much more complicated than is incorporated in the simple models discussed above, although it is probable that refinement of the models will lead to a greater understanding of the processes at work. In particular, the study of two-phase flow as the melt separates from the matrix suggests that the melt travels through porosity waves, which may account for both the episodicity of extrusions and their spatial distribution.

The location of the melt within the crust beneath the spreading axis has long been a matter for debate. Some of the most popular early models (Cann, 1974; Pallister and Hopson, 1981) considered that a large magma chamber exists beneath the spreading axis. This was thought to be continually replenished by fresh magma from below to balance the loss of fractionated basalts extruded upwards and the freezing of melt at the edges of the magma chamber as the crust moved sideways away from the axis. Because of its geometric shape this became known as the 'infinite onion' model. Seismic experiments across ridge axes have failed to detect the broad crustal zone of zero shear-wave velocities that would be created by such large crustal level magma chambers. Based on geophysical evidence the most probable scenario is that in which magma is intruded into the crust in smaller sill-like bodies, rather than into a large hole filled with molten rock.

Direct geophysical evidence of molten rock under the spreading axes has been derived from seismic reflection profiling, particularly across the East Pacific Rise (Detrick *et al.*, 1987) and the Valu-Fa ridge in the Lau Basin (Morton and Sleep, 1985; Collier and Sinha, 1990). High amplitude reflectors imaged at depths of between 1.5–4 km beneath the seafloor have been interpreted as caused by melt in the crust (Figure 3.5). The lateral extent of the high amplitude reflectors reaches a maximum of about 5 km, but is usually less than this. A continuous reflector has been traced more than 40 km along the Valu-Fa ridge across the entire length of a detailed survey and it is clear that such along-axis continuity is not uncommon. Wide angle seismic experiments and careful analysis of the reflection polarity confirm that the high amplitude reflection is caused by the top of a low velocity zone which is typical of that produced by molten rock (Collier and Sinha, 1990).

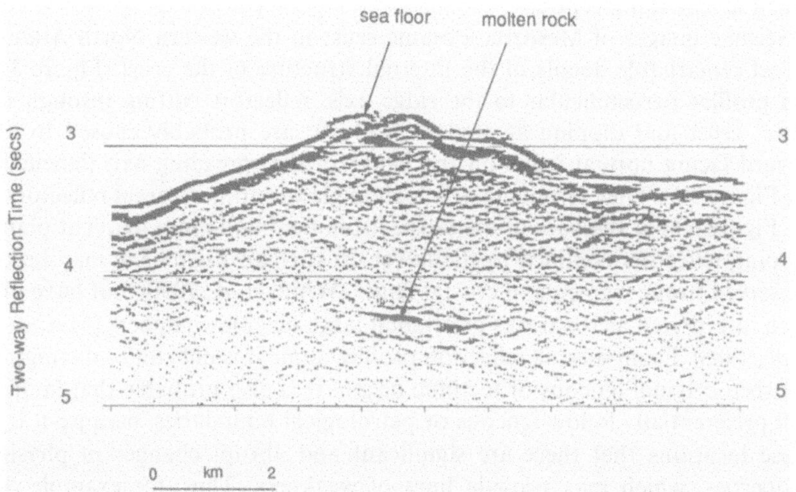

Figure 3.5 Migrated seismic reflection profile across the central Valu-Fa spreading centre in the Lau Basin showing a high amplitude reflector at 4.2–4.3 s two-way travel time approximately 3 km beneath the seafloor, caused by a sill-like intrusion of melt in the crust. Vertical exaggeration is 2.7 times at the sea bed. Profile is a true amplitude section from Collier and Sinha (1990)

Large changes in the seismic properties of rock are caused by the presence of melt, generating strong seismic reflectors. Even larger changes are caused in the resistivity. The presence of even a few percent of partial melt can reduce the bulk resistivity of the rock by two orders of magnitude, provided it resides in interconnected pore spaces. Newly developed active source electromagnetic sounding experiments focused on the spreading ridge axes promise to give further details of the amount of melt present in the crust and of its distribution (Sinha *et al.*, 1990).

A broad region of depressed seismic velocities is found in the lower crust beneath the partial melt reflector (Harding *et al.*, 1989). This is diagnostic of a region of hot crust, as is to be expected in this area of new, intruded crust. Beneath the spreading axis itself, the Moho, or base of the crust, is not developed, but normal crustal structure is attained within a short distance off-axis as the crust moves laterally and cools (Fowler, 1976).

The internal structure of the crust can be imaged using seismic reflection profiles, although in the vicinity of the rugged terrain of the ridge axis the scattering of seismic energy by the seafloor prevents or degrades coherent reflections from within the crust. Over older crust where sedimentation has buried the relief and therefore reduced the irregular impedance contrast at the seafloor, considerably better images can be recorded on seismic profiles. As the majority of the tectonic and magmatic processes which generate the crust occur close to the spreading axis, with little modification thereafter, profiles across old oceanic crust can be used to study the structure imposed near the spreading centre.

Seismic images of Mesozoic oceanic crust in the western North Atlantic reveal remarkable details of the internal structure of the crust (Figure 3.6). On profiles perpendicular to the ridge axis, reflectors cutting through the entire crust and dipping at angles of 30–40° are probably caused by the inward facing normal faults on the flanks of the spreading axis (labelled 2 on Figure 3.6). They terminate at a high angle to sub-horizontal reflectors (4 on Figure 3.6) at the base of the crust. Prominent and more frequent planar dipping reflectors confined to the lower crust (1 on Figure 3.6) may be the preserved lower segments of crustal faults (White *et al.*, 1990), but have also been interpreted as the traces of original large magma chambers (McCarthy *et al.*, 1988). From seismic reflection profiles alone it is difficult to distinguish between original igneous or tectonic origin. It is also probable that faulting will preferentially follow igneous or petrological boundaries, because it is in these locations that there are significant and abrupt changes in physical properties, which may provide lines of weakness. Thus, for example, the sub-horizontal Moho reflectors (4 on Figure 3.6) certainly occur near a major petrological change from crustal mafic rocks to underlying ultramafics, but may also mark the depth at which the extensional faulting becomes horizontal.

Layered sub-horizontal reflectors imaged in the lower crust and the upper

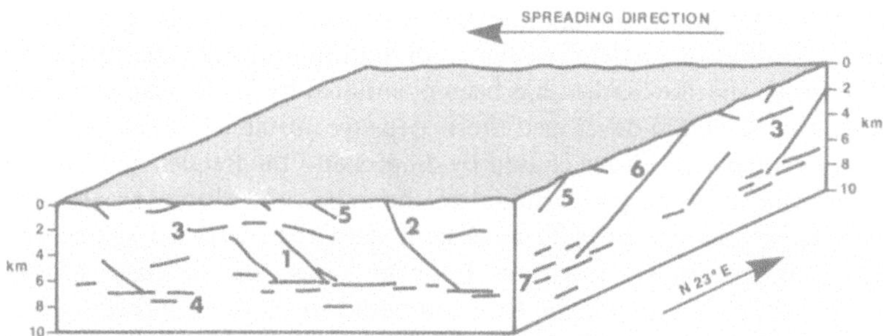

Figure 3.6 Schematic diagram of the internal structure of the oceanic crust imaged by multichannel seismic profiles in the western North Atlantic away from the influence of fracture zones (from White *et al.*, 1990). The axes of the diagram are orientated parallel and perpendicular to the spreading direction. See text for discussion of numbered features.

mantle (7 on Figure 3.6) are probably caused by igneous intrusions as sills. Shallower, discontinuous reflectors 1–3 km below the basement (3 on Figure 3.6) may mark the greatest depth of penetration of the vigorous hydrothermal circulation at spreading axes which generates black smokers and plumes through the seafloor (White *et al.*, 1990). Finally, whole crustal failure causes planar reflectors dipping at 15–30° in a direction approximately parallel to the ridge axis (6 on Figure 3.6). This failure is probably caused by compressional thermoelastic stresses generated on the flanks of the ridge axis as the lithosphere cools and thickens (White *et al.*, 1990).

As studies of oceanic spreading centres became more detailed in the 1980s, it became clear that although the ridge spreading is continuous and smooth, when viewed on a timescale of a million years, there is considerable variability both temporally and spatially on smaller scales. On slow spreading ridges such as the Atlantic (less than 30 mm y^{-1} spreading rate), the extrusive episodes are episodic on a time-scale of typically 10^4–10^5 years. On fast spreading ridges such as those in the Pacific Ocean, the time-scales are a factor of two or three times shorter. Vigorous hydrothermal vents are very short-lived; about 100 years may be typical.

The spreading centres are also segmented spatially. Fracture zones with large offsets provide a natural segmentation length of 40–80 km on slow spreading ridges. On fast spreading ridges, the fracture zones may be hundreds of kilometres apart, but the ridges are nevertheless still segmented on a scale of around 100 km by independent magmatic cells. The boundaries between the segments may show only minor offsets, with overlapping spreading centres at the ends.

There are sometimes, although not always, distinct petrological differences between basalts from adjacent segments (Langmuir *et al.*, 1986). This segmentation at the crustal level is probably caused by separation at deeper levels feeding magma in separate batches to crustal levels. One possible mechanism, that of Rayleigh–Taylor gravitational instabilities in the rising melt to form separate diapirs, has been postulated by Whitehead *et al.* (1984). Schouten *et al.* (1985) developed these ideas quantitatively to show that the spacing of intrusion centres caused by diapirs and the frequency of intrusion events should both be proportional to the cube root of the spreading rate. These predictions are broadly consistent with the increased segmentation length and the higher intrusion frequency seen on the fast spreading (100–200 mm y^{-1}) East Pacific Rise compared to the slow spreading (about 25 mm y^{-1}) Mid-Atlantic Ridge.

3.5 Fracture zone structure

Fracture zones mark places where the spreading centres are offset laterally. The active portions of the faults, between the offset spreading centres, are called transform faults. Their offsets vary from less than 10 km to more than 100 km, with several prominent large offset examples in the equatorial Atlantic. Physiographically, they are often marked by deep, narrow troughs in which the top of the basement is up to 2 km or more deeper than on the adjacent crust. Earthquake fault-plane solutions show that the strike-slip faulting extends through the lithosphere in the transform faults.

When they were first discovered, it was hoped that the steep scarps on either side of the fracture zone valley would present an ideal section in which to sample the vertical stratigraphy of the top few kilometres of oceanic crust. Unfortunately, these hopes were dashed when submersible studies showed that the walls of the fracture zone valleys were not clean cuts through the crust, but were produced by multiple small throw normal faults. Not only did these shed voluminous talus which masked the underlying rock, but they caused the top few hundred metres of rock to be repeatedly downthrown, thus making it impossible to sample deeper sections of the crust even where the country rock was exposed.

Subsequent geophysical mapping of fracture zone crust, particularly by wide-angle seismic reflection and refraction studies, and by gravity measurements, have shown that the crust under fracture zones is highly anomalous. Its seismic structure is nothing like that of normal oceanic crust and it is generally very thin, sometimes as little as only 1–2 km (White *et al.*, 1984). Fracture zones therefore do not provide a window into normal oceanic crustal structure. The thinning from normal crustal thicknesses of 6–7 km usually occurs gradually over distances of 10–30 km towards the fracture zone (Figure 3.7). There is often a concomitant gradual increase in the depth of the basement beneath the sea surface. These gradual changes are found regardless of whether the offsets across the fracture zones are small or large. They probably represent the tailing off of magma supply towards the ends

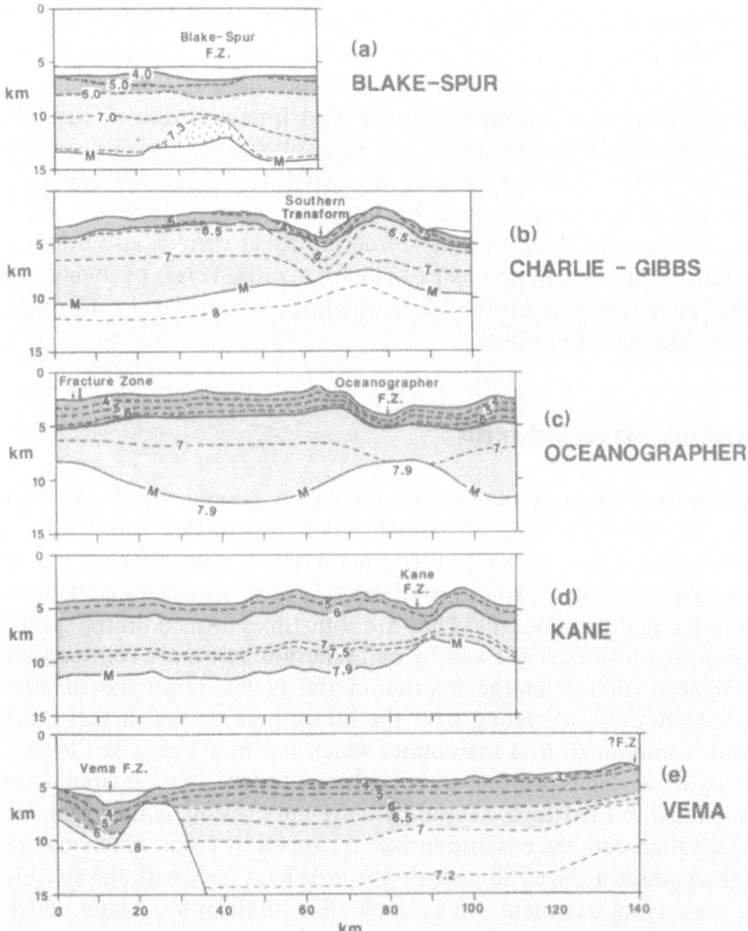

Figure 3.7 Cross-sections redrawn at the same scale showing the velocity–depth structure across fracture zones. The profiles are parallel to, or along, the spreading centre. Stippled areas show extent of igneous crust, with M denoting the Moho where it is developed (note that the Moho has not yet formed on the profile along zero age crust adjacent to the Vema Fracture Zone). The dark stipple shows areas with seismic velocities less than 6 km s^{-1} and the light stipple those areas with crustal velocities above 6 km s^{-1}. Note the long wavelength crustal thinning over distances of 20–30 km from both small offset (e.g. Fracture Zone (FZ) I) and large offset (e.g. Oceanographer, Charlie-Gibbs, Kane) fracture zones. (a) Blake-Spur FZ (White *et al.*, 1990); (b) Charlie-Gibbs FZ from line 10617 (Whitmarsh and Calvert, 1986); (c) Oceanographer FZ from line B (Sinha and Louden, 1983); (d) Kane FZ from line EXP81 (Abrams *et al.*, 1988); and (e) Vema FZ along 0 Ma crust (Louden *et al.*, 1986).

of the individual spreading segments, with the intrusion centre lying approximately mid-way between adjacent fracture zones (Schouten and White, 1980; White, 1984).

Under the fracture zones themselves, the crust may thin to as little as 1 km. The seismic structure is characterized by the absence of a normal oceanic layer 3 (see section 3.3), and generally exhibits a high velocity gradient that can be explained by the highly faulted and fractured upper crust of the tectonized fracture zone region (White *et al.*, 1984). Clearly normal accretionary processes do not act in the fracture zone, and it is likely that much of the crustal section has been derived by intrusion laterally from the adjacent spreading segments. Beneath several fracture zones there is also evidence of serpentinization of the uppermost 2–3 km of mantle. This is probably caused by water penetration down the faults which in the transform zone must cut right through rigid lithosphere.

3.6 Oceanic islands and swells

The interiors of oceanic plates are marked, in general, by the absence of igneous or tectonic activity, in complete contrast to the plate boundaries. There is one major exception to this general observation, which is that huge volcanic edifices, often projecting above sea level to generate islands, are found in the middle of oceans. They are sometimes formed on the spreading axes and then transported away by plate motion, but more commonly they are emplaced off-axis in the interior of the plates. Once the islands are emplaced, they subside along with the lithosphere on which they sit. It is therefore common to find seamounts which are now below sea level, that originally were islands projecting above the sea surface. In these circumstances they often exhibit flat tops, caused by wave-cut erosion as they sank below sea level. Often, too, the basaltic surface is capped by coral, which may reach great thicknesses if the coral growth upwards kept pace with the subsidence of the underlying basement. An example of a coral-capped island underlain at depth by basalts is Bermuda in the western North Atlantic.

As with the crust formed at oceanic spreading centres, the only major source of the huge amounts of igneous rock emplaced in mid-plate islands is the underlying asthenospheric mantle. It was postulated that there were 'hot-spots' in the mantle beneath intra-oceanic islands which were responsible for the excess volcanism in those locations (Wilson, 1963a; Morgan, 1971, 1981). Hot-spots around the world move only slowly with respect to one another because they are caused by thermal anomalies in the deep mantle, whereas the rapid plate motions across the top of them may create chains of islands and seamounts as the plates move across the underlying hot-spots. The age of the islands and seamounts increases uniformly along the chain if the plate motion continues uniformly above the hot-spot. In detail, hot-spots move slowly with respect to one another, at rates of up to 10–30 mm y^{-1} (Molnar and Stock, 1987; Sager and Bleil, 1987).

Geophysical measurements across hot-spots confirm that they are caused by thermal plumes in the asthenospheric mantle (Courtney and White, 1986; Detrick *et al.*, 1989). The heat flow above the plume reaches a maximum of 25% above the normal heat flow through the oceanic lithosphere in the case, for example, of the Cape Verde swell (Figure 3.8). In addition to causing increased heat flow, the mantle plume also dynamically uplifts the seafloor

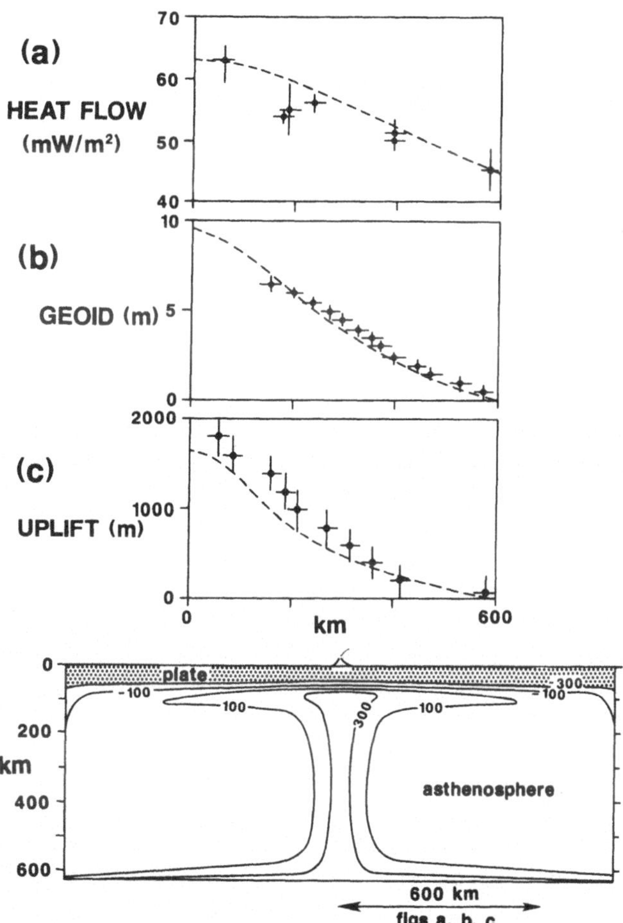

Figure 3.8 Model of the temperature anomalies in the mantle (bottom panel), and the resultant predicted heat flow (a), geoid (b) and bathymetry (c) anomalies along a radius across the Cape Verde swell in the eastern Atlantic. From White (1989), based on work by Courtney and White (1986). Crosses show observational constraints. Temperature anomalies in the bottom panel are labelled in °C with respect to the mean asthenosphere temperature.

by as much as 2 km. As the hot material rising in the 200–250 km diameter mantle plume is deflected by the overlying plate, it creates a region up to 2000 km in diameter of abnormally hot asthenospheric mantle which in turn creates a broad swell of elevated seafloor and a concomitant geoid anomaly of similar dimensions (Figure 3.8).

The oceanic islands sit in a narrow region at the centre of the broad swell. They are formed by decompression melting of mantle rising in the central plume, and therefore are found only in the region directly above the typically 200–250 km diameter central plume. Their petrology and geochemistry reflect their asthenospheric mantle source, with some additions from the lithosphere picked up as the melt moves through it on its way upwards. In addition, the lowermost part of the lithospheric mantle may be partially melted by heat from the mantle plume beneath it. The mantle plume only rises to the base of the overlying plate before it is deflected sideways and so decompression melting only occurs over a depth interval of a few tens of kilometres immediately beneath the plate. This generates only relatively small percentages (typically between 3 and 7%) of partial melt compared with the 25% melting that occurs beneath oceanic spreading centres where the mantle wells up almost to the surface. As the flow-rates in the mantle plume are high, the total rate of production of melt in the central plume is also high as new mantle continues to flow upwards and generate melt.

Mantle plumes exist beneath continents as well as oceans. If a continent breaks up above the broad thermal anomaly in the mantle caused by a hot-spot, then huge volumes of melt can be created extremely rapidly by decompression melting of the hot mantle welling up passively to shallow depths beneath the rift. Instead of generating 6–7 km of melt as is found in rifts above normal temperature mantle, a temperature anomaly of 150°C will cause the generation of more than 15 km of melt (Figure 3.4). This process was responsible for generating the Deccan flood basalts (Figure 3.9a) as massive volumes of melt generated beneath the rift poured out onto the adjacent continent. As the Seychelles continental block rifted away from mainland India at 66 Ma above the mantle thermal anomaly generated by the Réunion hot-spot, some 2×10^6 km^3 of basalt were generated in less than 0.5 million years (White and McKenzie, 1989b). This was caused by upwelling of the broad region of hot mantle into the lithospheric rift. Subsequently the Indian plate drifted northwards, and melt generated continuously in the rising central plume of the Réunion hot-spot bled upwards to create the thick igneous crust of the Lagos–Chaccadive Ridge and its surmounting islands (Figure 3.9b). The spreading centre of the central Indian Ridge subsequently split the trace of the Réunion hot-spot, leaving the youngest portion of the hot-spot trace, the Mascarene Plateau, separated from the older portion, the Lagos–Chaccadive Ridge, by the continued seafloor spreading at the central Indian Ridge. At present the hot-spot plume lies beneath the Réunion volcanic island.

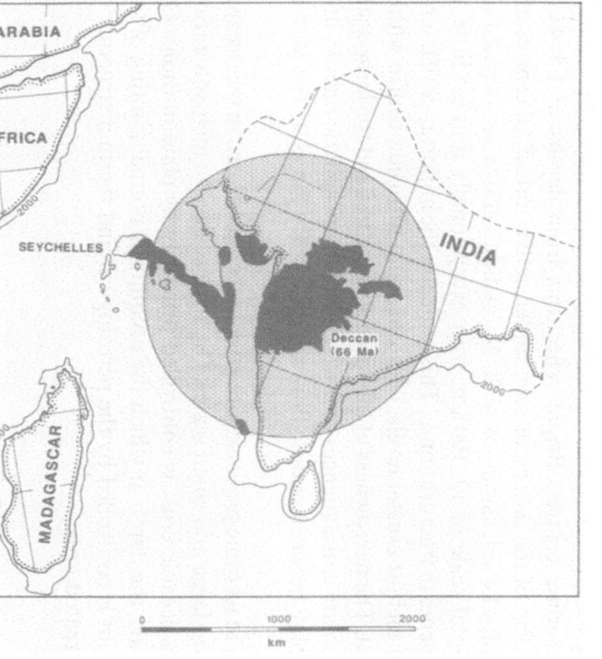

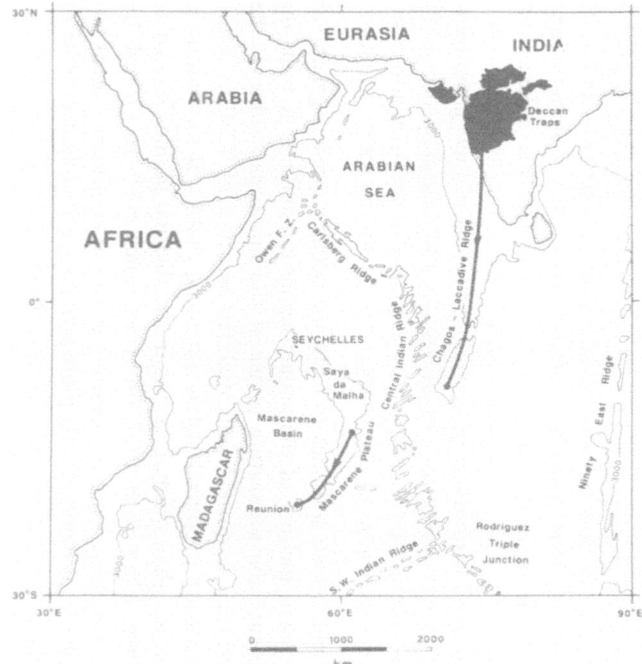

Figure 3.9 Reconstruction of the north-west Indian Ocean at approximately 65 Ma shortly after the onset of rifting. Solid areas show the extent of the Deccan plateau basalts and of contemporaneous offshore basaltic volcanism. The circle shows the extent of anomalously hot asthenospheric mantle around Réunion plume at the time of rifting. Equal area projection is centred on the plume location. (b) Present geography of the west Indian Ocean showing the trail of volcanic ridges and islands left by the Réunion plume as the Indian plate migrated northwards above it. Diagrams from White and McKenzie (1989b).

3.7 Concluding statements

Petrological samples can only be recovered from the uppermost sections of the *in situ* oceanic crust. Geophysical measurements provide a powerful means of determining the framework from which the samples are taken, and particularly of measuring the crustal structure and its lateral variability. They also provide strong constraints in constructing models of melt generation and migration. The over-riding control on igneous and tectonic processes in the ocean basins comes from the temperature and flow of the asthenospheric mantle. The main conclusions of this chapter can be summarized as follows:

1. The partial melt which solidifies to form the oceanic crust is generated by decompression melting of the asthenospheric mantle as it wells up passively beneath the separating plates at the mid-ocean ridge. If the asthenosphere has a normal potential temperature of 1280°C, some 6–7 km of crust are produced at the spreading centres. If the mantle is 200–250°C hotter than normal, as it is in the thermal plumes known as hot-spots, then more than four times as much melt is formed by decompression melting when a spreading centre or continental break-up crosses the plume. When a mantle plume underlies the intact interior of an oceanic plate, then the mantle rises only to the base of the plate before being deflected laterally and thus generating by dynamic uplift a swell up to 2000 km in diameter. Melt is generated by decompression only in the 200–250 km diameter central plume, from whence it bleeds rapidly upwards to form intra-oceanic islands.
2. Seismic methods reveal details of the internal structure of the oceanic crust at spreading centres. Bright reflections from depths of 1.5–4 km beneath the top of the basement seen under active spreading centres are caused by sill-like intrusions of melt. Spreading centres are segmented on a horizontal scale of 40–100 km, with exceptionally thin (as little as 1 km) crust beneath fracture zones. This indicates the importance of lateral transport of melt at crustal levels, with one main intrusion centre within each segment. The narrowness of the spreading centres and the large-scale segmentation are consequences of flow in the asthenospheric mantle, and of the interplay between mantle flow and buoyancy forces as the mantle melts.
3. Magmatic and tectonic processes are intimately interrelated in the oceanic crust, both on large and small scales. On large scales the melt generation which forms oceanic crust is controlled primarily by plate motions. On a smaller scale, the depth at which the ubiquitous normal faulting of the crust soles out is controlled by the petrological and thermal structure of the newly created crust.

4 Structure of the oceanic crust as deduced from ophiolites

JOHN SPRAY

4.1 Introduction

The determination of the structure of the oceanic crust has been achieved mainly by indirect methods because the ocean floor is relatively inaccessible and there are limitations to the depths penetrable by drilling (currently about 1 km below the seafloor). Our current knowledge of crustal structure is based on three sources of information: (1) rock samples obtained by submersibles and from oceanic islands, drill core and dredge hauls; (2) geophysical data; and (3) subaerially exposed fragments of ocean crust and upper mantle, that is, ophiolites (Chapter 1).

Rock sampled from *in situ* oceanic crust is predominantly basaltic, although minor volumes of plutonic lithologies and their metamorphic equivalents have been retrieved from fracture zones and as xenoliths in alkali basalts of oceanic islands (Chapter 15). However, with the exception of *in situ* basalts, none of these lithologies can be used to indicate the nature of an undisturbed oceanic stratigraphy. In contrast, geophysical studies have been of critical importance in establishing the origin and structure of the oceanic crust. For example, the discovery of magnetic stripes in the uppermost layer of basalt led to the verification of the seafloor spreading hypothesis (Vine and Matthews, 1963), and variations in seismic velocities have revealed the layered nature of the crust (Raitt, 1963).

Notwithstanding the contributions made by geophysicists, it is ophiolite studies that have put the meat on the seismic skeleton. Field, petrological and geochemical studies of these on-land oceanic fragments have provided a wealth of information on the workings of the oceanic lithosphere.

Our present knowledge of oceanic crustal structure is based primarily on a combination of both ophiolite and geophysical data. Using this combined approach it has been possible to ascribe lithologies to seismic velocities, propose mechanisms for the creation of new crust at spreading centres and formulate petrogenetic models for the generation of oceanic basalts. This chapter specifically concentrates on the contributions that ophiolite studies have made to our knowledge of the igneous and metamorphic state of the oceanic crust.

4.2 Alpine-type peridotites: variants and nomenclature

The term ophiolite has had a lengthy history, having first been used in Europe by Brongniart (1813) for certain green, serpentine-rich mélanges in allusion to their snake-like appearance ('ophi' is Greek for serpent). During the nineteenth century the term evolved to embrace radiolarian chert, volcanics, dolerite and gabbro as well as peridotite, such that it defined a distinctive association of predominantly igneous lithologies. The concept of a kindred relationship was subsequently consolidated by the work of Steinmann (1927), after whom is named the 'Steinmann Trinity' (the radiolarite–pillow basalt–serpentinite association).

At about the same time in the United States, Benson (1926) proposed the term 'alpine-type' to describe peridotite bodies occurring within orogenic belts that he considered were intrusive into geosynclinal sediments. In fact, both the European and American geologists were attempting to describe the same phenomenon, although US workers tended to treat the basic and ultrabasic components as being unrelated. The two schools converged in the 1960s and the term 'alpine-type peridotite' was coined to describe consanguineous ultrabasic–basic associations of which ophiolites were a member (Wyllie, 1967).

Until the advent of plate tectonics, the origin of ophiolites remained problematic. Were they emplaced into the continental crust as intrusions or as tectonic slices? Brunn (1959) and Gass (1968) were among the first geologists to directly equate ophiolites with ocean crust and mantle. Subsequently, evidence for thrusting came from field studies of, amongst others, Greek (Moores, 1969), Papuan (Davies, 1971) and various Appalachian (Dewey and Bird, 1971) ophiolites and it was realized that they represented tectonically transported fragments of oceanic lithosphere (Coleman, 1971; Church, 1972).

At about the same time that ophiolites were playing an increasingly significant role in the development of plate tectonics theory, an attempt to formally define the term was made at a Penrose conference held in 1972 (Anonymous, 1972). This was an important meeting and it is worth quoting in full the recommended definition made by the conference participants:

'Ophiolite refers to a distinctive assemblage of mafic to ultramafic rocks. It should not be used as a rock name or as a lithologic unit in mapping. In a completely developed ophiolite the rock types occur in the following sequence, starting from the bottom and working up: ultramafic complex, consisting of variable proportions of harzburgite, lherzolite and dunite, usually with a metamorphic tectonic fabric (more or less serpentinized); gabbroic complex, ordinarily with cumulus textures commonly containing cumulus peridotites and pyroxenites and usually less deformed than the ultramafic complex; mafic sheeted dyke complex; mafic volcanic complex, commonly pillowed.

Associated rock types include (1) an overlying sedimentary section typically including ribbon cherts, thin shale interbeds, and minor limestones; (2) podiform bodies of chromite generally associated with dunite; (3) sodic felsic intrusive and extrusive rocks.

Faulted contacts between mappable units are common. Whole sections may be missing. An ophiolite may be incomplete, dismembered, or metamorphosed, in which case it should be called a partial, dismembered, or metamorphosed ophiolite. Although ophiolite generally is interpreted to be oceanic crust and upper mantle the use of the term should be independent of its supposed origin.'

Twenty years later, geologists know considerably more about ophiolites, but the Penrose definition remains useful. One drawback is that it does not readily accommodate the spectrum of ultrabasic–basic complexes that occur within the alpine-type peridotite family, of which ophiolites are merely an end-member. In order to assess the current status of ophiolites as guides to the structure of oceanic crust it is prudent to consider them in a broader context.

Figure 4.1 shows a simplified depiction of the alpine-type peridotites in terms of their gross internal structure and petrology. The ends of the block represent the two main subtypes: lherzolitic (also known as orogenic root zone and high temperature peridotite) and ophiolitic (also known as

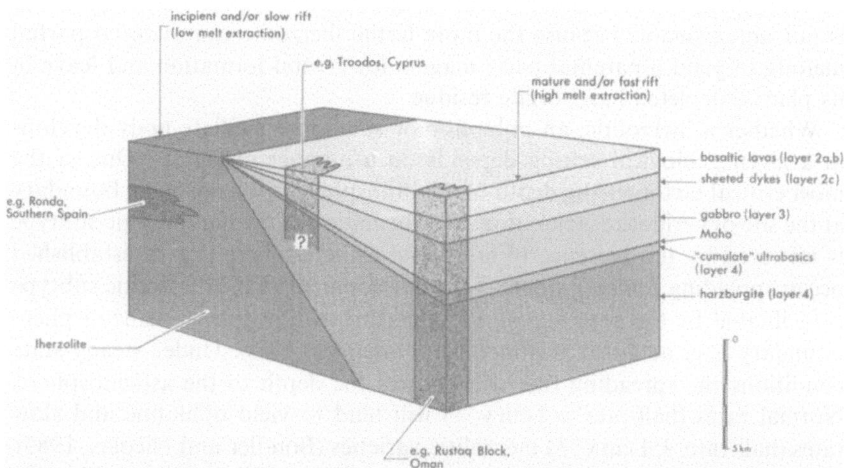

Figure 4.1 Idealization of alpine-type peridotite variants based on degree of melt extraction and plate velocity. The constituent components and their corresponding geophysical layers for the ophiolite end-member are shown on the right. The Rustaq Block of the Semail ophiolite of Oman is a representative end-member (Lippard *et al.*, 1986), whereas the Troodos complex is intermediate (the effects of additional pillow lava units are omitted from this diagram) and the Ronda peridotite is wholly lherzolitic. Plate velocity and/or rift maturity increases from left to right along the length of the block diagram. Scale is approximate.

harzburgitic or low temperature peridotite). The lherzolite subtype is exemplified by the peridotite bodies of the western Mediterranean and western Alps (e.g. Lherz, southern France; Ronda, southern Spain; Beni Bouchera, northern Morocco; Lanzo, Italy). These form large (up to about 1000 km^3) slab- or pip-shaped allochthonous or parautochthonous massifs usually related to major intra-continental fault zones or aborted continental rifts (incipient oceans). They are dominated by garnet-, spinel- or plagioclase-bearing olivine–orthopyroxene–clinopyroxene assemblages and so constitute four-phase lherzolites. The harzburgitic subtype is typified by such ultrabasic–basic bodies as the Troodos ophiolite of Cyprus and the Semail ophiolite of Oman (Figure 4.1). These comprise olivine + orthopyroxene mantle units with overlying crusts of bulk basaltic composition (i.e. basaltic lavas, sheeted dykes, gabbro and, immediately beneath the geophysical Moho, cumulate ultrabasics). These tend to form discontinuously exposed, allochthonous linear belts (up to several hundreds of kilometres long, $\leqslant$50–100 km wide and $\leqslant$15–20 km thick) associated with past or recent plate boundaries.

These two subtypes are merely end-members of a spectrum of peridotite types. The majority of alpine-type peridotites are combinations of lherzolite and harzburgite overlain by variable thicknesses of basic crust (Nicolas, 1989). Figure 4.1 shows that the pure lherzolite end-member possesses no crust (e.g. the Ronda massif of Southern Spain), whereas the pure ophiolite end-member carries a 5–7 km thick crust (e.g. the Rustaq Block of the Semail ophiolite of Oman). The diagram also implies that the amount of lherzolite present is inversely proportional to the thickness of crust developed. This implication is not unreasonable because the more fertile lherzolite can undergo partial melting to yield a parental basic magma for crustal formation and leave in its place a depleted harzburgite residue.

Whether a lherzolitic, an ophiolitic or some intermediate body develops in a given geological setting depends on a number of factors. One of the most critical factors is the depth of the lithosphere–asthenosphere boundary at the site of peridotite generation. The formation of the harzburgitic subtype is favoured by the presence of high level asthenosphere (e.g. at established ocean spreading centres), whereas the development of the lherzolitic subtype is facilitated by the presence of a less perturbed lithosphere–asthenosphere boundary (e.g. at failed continental or incipient rifts). Under steady-state conditions the spreading rate will control the depth to the asthenosphere. Normal rates (half-rate $> 1\ cm\ y^{-1}$) will tend to yield ophiolitic and slow rates (half-rate $< 1\ cm\ y^{-1}$) lherzolitic varieties (Boudier and Nicolas, 1985).

As different types of ultrabasic–basic complex can be generated within the alpine-type peridotite family, it is possible to have sequences that do not possess the requisite components of a Penrose ophiolite. For example, they may lack a fully developed crustal sequence or exhibit a sheeted sill rather than a sheeted dyke complex, e.g. Port Sal, southern California (Hopson and Franno, 1977), Canyon Mountain, Oregon (Ave Lallemant, 1976) and Xigaze,

Tibet (Girardeau *et al.*, 1985). These differences may indicate corresponding variations in the structure of oceanic crust related to diverse sites of generation. However, before the implications of this aspect of ophiolite studies are discussed, the ophiolite–oceanic crust seismic connection must be considered.

4.3 Seismic comparisons between oceanic crust and mantle

One of the fundamental reasons for equating certain ophiolites with oceanic lithosphere rests with the results of seismic refraction studies. By the early 1960s marine geophysicists had acquired sufficient data to construct a velocity profile through the oceanic crust and upper mantle (Raitt, 1963). The conclusion was that the crust is layered and that individual layers tend to be of a consistent thickness. Numbers and names were given to each of these layers as follows: layer 1, sediments of varying thickness; layer 2, oceanic basement; layer 3, oceanic layer; and, via a sharp velocity increase known as the Mohorovicic discontinuity (Moho), layer 4, upper mantle. Subsequently, new and more sophisticated methods allowed these layers to be further divided such that, by the late 1970s, layers 2a, b and c and layers 3a and b could be distinquished, although there appear to be no sharp boundaries between these sublayers (e.g. Ewing and Houtz, 1979; Spudich and Orcutt, 1980).

In a critical study, Salisbury and Christensen (1978) determined the velocity structure of the Bay of Islands ophiolite and concluded that it was indistinguishable from that of normal oceanic crust. The results were derived by reconstructing the velocity structure of selected samples in the laboratory and so compensation was made for the lower seismic velocities shown by ophiolites (due to various emplacement and weathering effects). Along with the earlier work of Poster (1973) and Pedersen *et al.* (1974), the results of Salisbury and Christensen (1978) allowed a direct comparison to be made between oceanic crustal structure based on seismic velocities and ophiolite structure based on petrology. It also enabled lithologies to be equated with seismic velocites and confirmed earlier held views that, beneath the layer 1 sediments, the oceanic crust probably consists of the following: fractured basalt (layer 2a), massive basalt with minor dykes (layer 2b), dykes (or possibly sills) with minor basalt (layer 2c), isotropic gabbros (layer 3a) and cumulate and/or sheared gabbros with minor ultrabasics (layer 3b). There is also evidence that the seismic layers coincide with varying states of metamorphic overprinting (discussed later in section 4.5) which approximately correspond to the igneous structure (Salisbury and Christensen, 1978).

Figure 4.2 shows layer thickness comparisons between oceanic crust (6.0–7.5 km thick) and seven stratigraphic Penrose-type ophiolites. It is noteworthy that none of these ophiolites apparently possess supra-Moho sequences thicker than that of the ocean crust. However, the actual crust–mantle

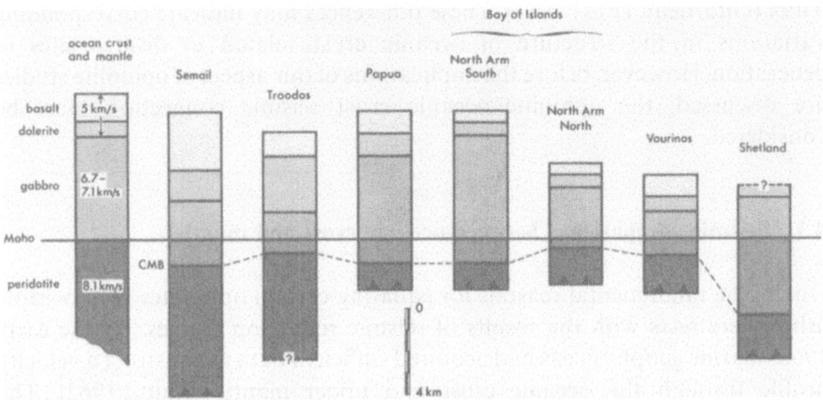

Figure 4.2 Simplified section through the oceanic crust and mantle compared with sections of seven ophiolites using the Moho as a reference horizon. Approximate seismic P-wave velocities are shown for the different oceanic layers (Raitt, 1963; Harrison and Bonatti 1981). The basalt horizon corresponds to seismic layers 2a and b, the dolerite sheeted intrusives to layer 2c, gabbro to layers 3a and b and peridotite to layer 4. Ophiolite thicknesses obtained from the following: Semail (Lippard *et al.*, 1986); Troodos (Wilson, 1959); Papua (Davies, 1971); Bay of Islands North Arm South (Casey, 1980) and North (Rosencrantz, 1980); Vourinos (Moores, 1969); Shetland, UK (Prichard, 1985; Spray 1988). CMB represents the crust–mantle boundary. Scale is approximate.

boundary in ophiolites can lie beneath the layer 3 to 4 transition because ultrabasic cumulates may form the base of the crust. Consequently, these rocks will exhibit virtually identical velocities to those of the underlying mantle peridotites. As the geophysically defined Moho need not coinicide with the crust–mantle boundary (CMB), some workers refer to the true CMB as the petrological Moho (Greenbaum, 1972). Placing the ultrabasic cumulates in their rightful position above the CMB increases the crustal thicknesses of the seven ophiolites, especially Shetland (Figure 4.2). No real comparison can be made between oceanic and ophiolitic crustal thicknesses because we do not know the thickness of the ultrabasic cumulates in the oceanic lithosphere. For this reason the geophysical Moho remains the only viable reference horizon. In spite of the different methods of determining oceanic versus ophiolitic structure and the possible confusion this may cause, Figure 4.2 suggests that the overall constitution of oceanic and ophiolitic crust is similar. This similarity is the main reason for using ophiolites as windows for elucidating the structure of oceanic crust.

Similarities aside, ophiolite studies also show that there is considerable variation in crustal thickness and internal structure from ophiolite to ophiolite. Whereas some geophysicists have used this to argue against the value of using ophiolites for comparisons with oceanic crust, the variation in crustal structure shown by ophiolites is probably an indication of a real variation in oceanic crustal structure. This inference is supported by the results of marine geological, geophysical and topographic studies of the oceanic crust and floor which reveal a much more complicated picture than previously assumed. Despite an apparent global consistency in so-called 'normal' crustal structure and thickness (see Chapter 3), there is considerable diversity possible depending on whether the crust is developed at a normal spreading centre, in a back-arc or marginal basin, or is modified by intra-plate volcanism to yield oceanic islands, seamounts or plateaux, or whether it was generated near a fracture zone. With this in mind, it should not be surprising to find that ophiolites do not conform to the simple models proposed for the structure of the oceanic crust.

4.4 Implications for magmatic processes occurring at oceanic ridges

Figure 4.3 shows an idealized section through an ophiolite. The diagram is based on the Semail ophiolite of Northern Oman (Lippard *et al.*, 1986), which is now considered by many geologists to be the archetypal ophiolite. It conforms to the Penrose definition (Anonymous, 1972) and serves as a reference model for other ophiolites. Although there are other examples that have been cited as being representative of a complete ophiolite sequence, such as Troodos and the Bay of Islands, the Semail ophiolite is the largest, one of the more intact, best exposed and most studied examples. With the exception of the metamorphic sole occurring along the base of the ophiolite, which will not be considered further here because it is related to the obduction process (Spray, 1984), Figure 4.3 shows a simplified petrology and structure of the ophiolite. Metamorphic effects that overprint some of these features will be considered in the next section. The diagram will be discussed from the base up.

The lowest unit of the ophiolite is a basal banded unit consisting of sheared peridotites. The zone comprises mylonitized and serpentinized dunite–harzburgite–lherzolite whose fabric parallels that of the underlying thrust. The development of this unit is related to the ophiolite obduction process such that thrust-related structures are superimposed on the mantle fabrics displayed by the bulk of the overlying mantle unit.

The lower half to two thirds of the Semail ophiolite consists of mantle peridotites, with lherzolite forming the lower kilometre or so and harzburgite the remainder. These rocks constitute the geophysicist's layer 4. Both peridotite types display pronounced tectonite fabrics defined by a compositional banding or gneissosity. This fabric is generally not coincident with the mylonite fabric of the underlying banded unit. Alternating layers of olivine and orthopyroxene form the bulk of the foliation, but it is orthopyroxene and spinel that form visibly prominent layers in outcrop. The mantle peridotites show a complex tectonic history that typically involves more than one phase of deformation that is attributed to solid-state asthenospheric flow

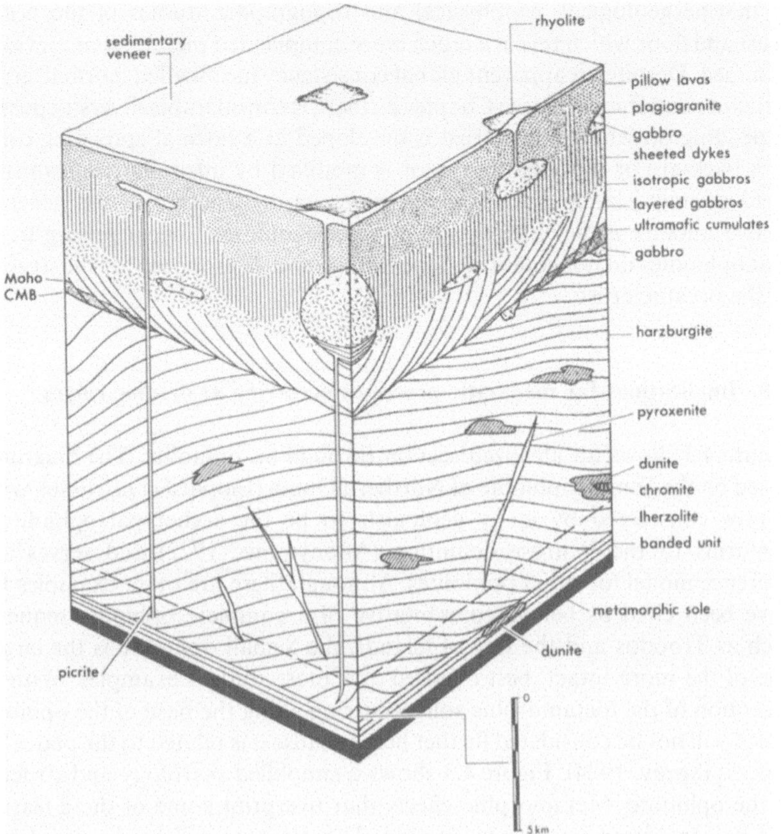

Figure 4.3 Idealized section through an ophiolite showing internal lithologies and structure. Pyroxenite dykes, other feeders and picrite dykes and dunite bodies are shown enlarged for clarity. Note that the Moho and CMB (crust–mantle boundary) need not be coincident. Scale is approximate. Based on the Semail ophiolite and modified after Lippard *et al.* (1986).

(Bartholomew, 1983). The mantle peridotites thus provide a unique view of the tectonics of the upper mantle.

In addition to lherzolite and harzburgite, a number of other lithologies occur in the mantle unit. These include dunite pods, some of which may enclose chromitite bodies of economic value, and various dykes and sills of ultrabasic or basic composition. Some of these may have been related to melt channels that fed the higher level magma chambers of the overlying crust, i.e. they may be the residues of parental melts or the trapped parental melts to oceanic basalt (Spray, 1982, 1989). Others were responsible for the development of picrite and so-called 'late intrusive' crustal complexes (Figure 4.3; Lippard *et al.*, 1986).

The boundary between the mantle unit and the overlying oceanic crust proper is marked by the petrological Moho (shown as CMB in Figure 4.3). As discussed above, this contact is not coincident with the geophysical Moho (layer 4 to 3 boundary) if cumulate ultramafic rocks are present at the base of the crustal sequence. The cumulate ultramafics form the base of the 'layered series' which consists of, in addition to the ultramafics, a centimetre-scale rhythmically layered, predominantly gabbroic sequence of the order of 2 km thick (although its thickness in the Semail and many ophiolites is highly variable). The attitude of the layering is only rarely parallel to the plane of the petrological Moho. The layering of these rocks is attributed by most workers to magmatic cumulate processes occurring in response to mineral settling within a magma chamber. The development of various small-scale sedimentary structures (e.g. cross-bedding, load casts, flame structures and slump folds) accompanying this layering testifies to the operation of current activity that would be compatible with the existence of a magma chamber. The sequence is thought to correspond to the geophysicist's layer 3b.

In most complete ophiolite sequences a series of isotropic or massive gabbros and related intermediate to acid intrusives occur between the top of the layered series and the overlying sheeted dykes. These are believed to constitute the geophysicist's layer 3a. They have been referred to as 'high level intrusives' and can reach several hundreds of metres in thickness, although their thickness can be highly variable and they may even be absent. These rocks lack the characteristic layering of the previous series, and hence they are seismically distinguishable, even though they may be of the same mineralogy. The associated intermediate to acid rocks, collectively referred to as 'plagiogranites' by Coleman and Peterman (1975), are believed to represent late-stage residual liquids formed by the low pressure crystal fractionation of subalkaline low potassium tholeiitic magma under hydrous conditions. In other words, they may be considered as late-stage silicic differentiates that accumulated towards the top of a magma chamber.

One of the most characteristic features of the complete ophiolite sequence is the sheeted dyke complex which typically forms a 0.5–2 km thick unit (Figure 4.3). This lies between the isotropic gabbros and the basal pillow lavas and is believed to constitute the geophysicist's layer 2c. It consists of virtually 100% vertical dykes of dolerite that, by their very nature, must have been intruded in a tensional environment. As such the presence of a sheeted dyke complex is considered to be strong evidence for ophiolites having formed at oceanic spreading centres. Individual dykes are typically 0.5–1 m thick and, towards the top of the sheeted sequence, give way over a distance of several metres to pillow lavas and late-stage plutonic rocks. The dykes are considered to have originated from the underlying isotropic gabbros via a transition unit where somewhat thicker dykes (up to 6 m thick) emanated from the top of magma chamber. At the same time, the dykes acted as feeders for the overlying pillow lavas. Dyke trends and dyke chilling statistics have been used to infer the orientation of the palaeospreading centre and even the sense of spreading relative to a given ophiolite body (e.g. Kidd and Cann, 1974; Kidd, 1977), although some workers would urge caution in using the facing direction of chilled margins for the latter purpose (Gass and Smewing, 1981).

A feature of the Semail and certain other ophiolites is the development of late intrusives. These are cross-cutting, multi-intrusive plutonic complexes that post-date the layered and isotropic gabbros, the sheeted dykes and pillow lavas. In the Semail ophiolite they occur in two forms, either as relatively large (up to 10 km^2 in outcrop) gabbro–norite–plagiogranite bodies (an example of which is shown along the facing corner of Figure 4.3) or as smaller (< 1 km diameter) peridotite–gabbro complexes. The origin of these late intrusions is not clear; the gabbro–norite–plagiogranite type typically shows evidence of crystallization from a hydrous magma probably related to off-axis magmatism, whereas the peridotite–gabbro type may be the result of mobilizing the underlying layered ultramafics and gabbros while these were still partially molten.

The uppermost part of a complete ophiolite sequence is represented by predominantly tholeiitic pillow lavas and lava flows. In the Semail ophiolite these are up to 2 km in thickness and are accompanied by minor amounts of andesitic and rhyolitic extrusives (Figure 4.3). This unit forms part of the geophysicist's layer 2, with the upper 500 m of the pillow lavas typically showing a lower seismic velocity if younger than 50 Ma (designated layer 2a). This effect has been attributed to the presence of fractures and breccias which 'heal' in time. If the rocks are sufficiently young, this velocity difference enables the geophysicist to distinguish layer 2a from the bulk of the underlying lavas referred to as layer 2b.

The above description of the petrology and structure of an ophiolite serves to illustrate how ophiolites can be used to infer the petrology and structure of the oceanic crust. Ophiolite models have been particularly important in helping to constrain the probable magmatic processes occurring at spreading centres. The presence of some form of magma chamber is implied from studies of the ultramafic and gabbroic cumulates, even though the size and life expectancy of these chambers may be variable and dependent on the tectonic setting and maturity of the spreading centre. Detailed studies of the crustal sections of ophiolites, and in particular the layered series, have enabled geologists to reconstruct the likely shape and size of magma chambers and to determine how they work (Browning, 1984).

One important area that uses the ophiolite model as a guide to oceanic crustal structure concerns the quest for establishing the parental magma to tholeiitic basalt. Is oceanic basalt itself a primary magma or is it derived by the low pressure fractionation of some unknown parent? Although this is a controversial subject (Chapter 6), attempts have been made to estimate the bulk composition of the oceanic crust by averaging the compositions of the constituent lithologies according to their thicknesses. In this way, a parental magma composition from which an oceanic basalt could have been derived can be estimated. For example, Elthon (1979) used such data from a number of well studied ophiolites to suggest that the primary melt segregating beneath an oceanic ridge contains about 18% MgO (i.e. is picritic) and not the 9–11% MgO typical of the tholeiitic basalts that form the upper part of the oceanic crust. His conclusions were supported by the recognition of MgO-rich dykes intruding cumulate rocks in the Tortuga ophiolite of Chile. A similar mass-balance calculation was made for the Semail ophiolite (summarized in Lippard *et al.*, 1986) which yielded a primary magma composition with 15.5% MgO. These studies have lent credence to the then controversial deductions made by O'Hara (1965) over 25 years ago.

One of the complications to be aware of when using ophiolites for understanding magmatic processes at ocean ridges concerns the effects of static and, in particular, dynamic metamorphism within the crust. This will be evaluated in the next section.

4.5 Implications for metamorphic processes occurring in the oceanic crust

As we are primarily concerned with oceanic basalts, consideration will be limited to those metamorphic processes seen within the crust. Figure 4.4 provides a simple summary of both static and dynamothermal metamorphic effects exhibited within rocks above the crust–mantle boundary in ophiolites. These effects can be divided into two: (1) seafloor metamorphism of the upper 2–4 km of the crust due to hot rock–water interaction (Chapter 8) and (2) sub-solidus recrystallization and deformation of igneous phases within layer 3 due to asthenosphere-induced shear. A third type of metamorphism, related to regional tectonometamorphic overprinting and general retrogresssion of the crustal sequence, also affects certain ophiolites. This is related to obduction and post-emplacement processes and can usually be distinguished from the effects of oceanic metamorphism by virtue of its pervasive regional nature; the effects of metamorphism extend beyond the actual ophiolite.

The upper 2–4 km of the crust are overprinted by the essentially static retrogression of igneous phases ranging from weathering effects (brownstone facies) to amphibolite facies at the top of layer 3 (Figure 4.4). This reflects the presence of an unusually steep geothermal gradient (around 200°C km^{-1}) which is interpreted to be the result of convectionally-driven hot rock–water interaction in the vicinity of a magma source (Spooner and Fyfe, 1973; Chapter 8). This explanation provides further evidence in support of ophiolites having been generated at some form of submarine spreading centre where a relatively high level chamber could exist. The overlying pelagic sediments are unaffected by this overprinting, also indicating that the metamorphism must occur while the ophiolite is *in situ* and close to a magma source.

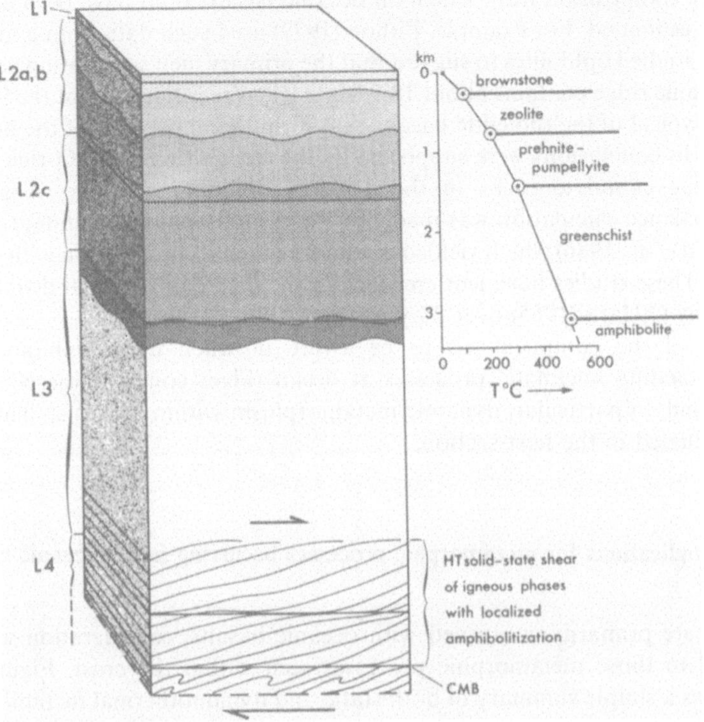

Figure 4.4 Simplified section through the crustal component of an ophiolite showing the effects of static and dynamic metamorphism as discussed in text. Layer 1, sediments; layer 2a, fractured basalt overlying layer 2b, massive basalt with dykes and layer 2c, dykes with massive basalt; layer 3, gabbros; layer 4, peridotite. Note that the CMB (crust–mantle boundary) is shown within layer 4. Arrows indicate the sense of asthenosphere-induced shear imposed on the lower crust and imply that the palaeospreading centre was located to the right of the section.

The metamorphism is such that original igneous mineral shapes and textures are preserved. The metamorphic grade ranges from the so-called brownstone facies (Cann, 1979), a form of submarine weathering at $T < 100°C$ within the upper 50–100 m of the basalts, through zeolite (about 100–200°C) and prehnite–pumpellyite (about 200–300°C) facies for the remainder of the pillow lavas. The lower part of the pillow lavas and the underlying sheeted dykes typically exhibit greenschist facies metamorphism (about 300–500°C) and the top of the gabbros low amphibolite facies effects ($T > 500°C$). Beyond the upper few hundred metres of the gabbros it appears that hydrothermal circulation is prevented due to the more impermeable nature of these massive lithologies (i.e. the gabbros do not possess the requisite fractures and fissures to facilitate the flow of hot water). Some of the diagnostic metamorphic minerals developed in each of these facies include the following: low temperature clays especially certain illites and smectites (brownstone facies);

heulandite, stilbite, mesolite, celadonite, laumontite, albite, chlorite and calcite (zeolite facies); prehnite, pumpellyite and epidote (prehnite–pumpellyite facies); actinolite and epidote (greenschist facies); and calcic plagioclase (oligoclase to andesine) and hornblende (amphibolite facies). With the exception of the amphibolite facies, many of the metamorphic minerals occupy cavities, veins and fractures within the lavas and dykes. Some workers use different facies schemes from those used here to describe these metamorphic effects (for example, the brownstone and prehnite–pumpellyite facies may not feature), but all essentially reflect the consequences of imposing a steep thermal gradient on rock in the presence of seawater. One of the more important effects of this metamorphic overprinting concerns the change in bulk chemistry that is caused due to the metasomatic exchange of certain ionic species between rock and seawater (Chapter 8). This process leads to the generation of 'spilites' or metabasalts in pillow lavas through the introduction of Na^{+}, Ca^{2+} and Mg^{2+}, seen in the development of albite, pumpellyite/epidote and chlorite, respectively. An additional effect pertains to the hydrothermal leaching and redeposition of base metals (e.g. Mn, Fe, Co, Ni, Cu, Zn, Ag, Au and Pb) in potentially economic amounts (Coleman, 1977).

Significantly, the laboratory seismic studies carried out by Salisbury and Christensen (1978), discussed in section 4.3, indicate that the layered seismic structure of both ophiolites and the oceanic crust may be due primarily to metamorphically induced density changes which, in general, correspond to the overall igneous structure.

In addition to the predominantly static metamorphic/metasomatic effects seen in the upper 2–4 km of the crust, a second type of metamorphism involving high temperature ductile deformation is recognized within the gabbros and ultramafic cumulates of many ophiolites (Figure 4.4). These effects have been known for some years (Thayer, 1963, 1980; Christensen and Salisbury 1975), yet their significance has not always been appreciated, particularly by igneous petrologists. These dynamothermal effects take the form of (1) sub-solidus deformation and recrystallization of igneous phases such that a mineral layering or gneissosity is generated at high temperatures ($T > 700°C$) and (2) localized hydration of gabbros to amphibolite ($T < 700°C$) and ultramafics to serpentinite ($T < 600°C$) within shear zones. These effects are commonly developed towards the base of the gabbros and within the underlying ultramafic cumulates, although amphibolitized zones may occur throughout the gabbros. The high temperatures required for the ductile deformation and/or recrystallization of igneous phases indicate a geological setting in the proximity of a spreading centre. The ductile behaviour of the igneous phases could then be explained as a syndeformation cooling phenomenon, such as the shearing of a semi-solid mush. Away from the heat or magma source, successively lower temperatures would allow for hydration and the formation of the high grade amphibolites (brown amphibole with andesine to labradorite plagioclase). The development of this high grade tectonometamorphism is attributed to asthenosphere-induced shear extending up into the crust. Low angle tectonite fabrics seen in ophiolites should therefore not be considered as being restricted to the mantle sequence. In addition, there is some evidence for certain plagiogranites being generated by amphibolite anatexis within these low angle shear zones rather than by the fractionation of a basic magma (Flagler and Spray, 1991). Significantly, geophysical evidence in support of the occurrence of strong sub-horizontal reflectors in the lower oceanic crust comes from recent seismic data obtained from the North Atlantic by White *et al.* (1990). One important consequence of the recognition of lower crustal shear zones is that estimates of the bulk chemistry of oceanic crust based on ophiolite sections may underestimate the true thickness of 'cumulate' ultramafics and gabbros due to shear-induced thinning. Studies of these high temperature metamorphic and deformation effects in the lower crust offer an exciting area for research in the future.

4.6 Concluding statements

1. Ophiolites have played a fundamental role in helping geologists to elucidate the structure of the oceanic crust. This is because the available geophysical and petrological evidence strongly supports an origin for ophiolites as displaced fragments of oceanic lithosphere. As a result of this, and in conjunction with geophysical evidence, ophiolites have enabled a realistic igneous and metamorphic petrological cross-section of the upper 10–15 km of the oceanic lithosphere to be deduced; a section that could not otherwise be determined due to its inaccessibility.
2. With regard to understanding the generation of new oceanic crust at spreading centres, ophiolites have provided evidence for the existence of magma chambers and yielded information as to how these magma chambers might work. It has also been possible to more fully evaluate the critical relationship between oceanic basalts sampled from the seafloor and oceanic islands and their remote mantle source, because ophiolites can provide more complete petrological sections between these parent–daughter lithologies.
3. One important conclusion that arises from an appraisal of ophiolites is that it is the alliance of different disciplines that facilitates the greatest progress in our understanding of geological processes. In the case of the oceanic crust and mantle, it has been the combination of seismic studies, marine geology and traditional field and petrological work that has resulted in our present level of understanding of the workings of the oceanic lithosphere and of the origins of oceanic basalts.

PART II PROCESSES

5 Mineralogy and crystallization of oceanic basalts

JAMES NATLAND

5.1 Introduction

Basalt was once termed the 'universal earth magma' (Daly, 1903). Basalt erupted at spreading ridges comes closest to being a universal igneous substance on our planet, covering nearly two-thirds of the earth's surface beneath the oceans. Three decades ago, the lavas of the oceanic ridges were virtually unknown and unsampled. Today, major portions of the ocean ridge system in the Atlantic, Pacific and Indian Oceans have been dredged, drilled and sampled by submersible. An early impression of almost monolithic uniformity of composition (e.g. Engel and Engel, 1963; Muir and Tilley, 1964; Engel *et al.*, 1965; Aumento, 1967; Miyashiro *et al.*, 1969; Kay *et al.*, 1970) gave way successively to understanding that: (1) source regions are geochemically and isotopically distinct (e.g. Corliss, 1970; Schilling, 1973; Hart and Schilling, 1973); (2) the degree of differentiation varies from place to place (e.g. Bass, 1971; Scheidegger, 1973; Clague and Bunch, 1976; Christie and Sinton, 1981); and (3) there is a range in the bulk compositions of parental magmas (e.g. O'Hara, 1968a; Stolper, 1980) which shows a systematic relationship to axial depths, geoidal signature and crustal thickness (Bryan and Dick, 1982; Dick *et al.*, 1984; Klein and Langmuir, 1987; McKenzie and Bickle, 1988).

In the 1920s, Bowen (1928) endorsed mineralogical criteria as the most rational basis for the classification of igneous rocks, owing to the obvious genetic link between the minerals in the rocks and the phase equilibria which control differentiation processes. However, successive advances in our understanding of basalts from spreading ridges have each been based much more on rock compositions than mineralogy. Thus, these rocks are most often considered from a geochemical rather than a classical petrological point of view, and the most widely used classifications (involving terminology such as N-MORB, E-MORB, P-MORB, FeTi basalt, MgCa basalt and ferrobasalt) are based on chemical criteria.

Nonetheless, the most appropriate comparisons to experimental phase equilibria, based on differentiation and partial melting processes, are still the compositions of the liquid and mineral phases in natural volcanic rocks. A balanced perspective on the petrogenesis of basalts from spreading ridges must therefore include an understanding of their mineralogy. The terms tholeiite and alkalic olivine basalt have rigorous definitions and considerable precedent in experimental petrology (Yoder and Tilley, 1962), thus, abyssal tholeiite and abyssal alkalic basalt are used here to denote depleted and enriched basalts, respectively, from the ocean floor. The term spreading-ridge basalt is also employed and includes both abyssal tholeiitic and alkalic basalts rather than the sometimes erroneous geographical connotation of mid-ocean ridge basalt (MORB). For example, the latter term obviously should not be applied to depleted basalts, identical to those from spreading ridges, which are found on seamounts (Batiza and Vanko, 1984) or in back-arc basins (Hawkins and Melchior, 1985).

This chapter and Chapter 12 (dealing with aspects of Indian Ocean basalts) summarize the mineralogy and crystallization histories of spreading-ridge basalts, chiefly the predominant abyssal tholeiites. The study of crystallization histories begins with the petrographic identification of minerals in the rocks, and proceeds to careful assessment of crystal morphologies, crystallization sequences and mineral relationships. Powerful analytical tools such as electron and ion microprobes can then be used to outline the complexity of processes which go into the coalescence of parental basalts, and which influence their subsequent differentiation. The data obtained from these instruments make it clear that, although distinctive types of basalts may be readily identified from spreading ridges using thin sections and mineral compositions, the rocks are amenable neither to simple schemes of classification nor to petrogenetic interpretation.

The difficulty is that very few of the basalts crystallized under conditions even close to equilibrium. This is for two reasons. Firstly, the majority erupted at depths of 2–4 km, encountering bottom water temperatures of only a few degrees Celsius. The margins of the lava flows and pillows are therefore invariably quenched to glass, and even the interiors of thicker eruptive cooling units experienced high rates of cooling uncommon in subaerial circumstances. At such extreme cooling rates, crystallization proceeds as best it can, but kinetic processes predominate (Bryan, 1972; Kirkpatrick, 1979). The minerals commonly observed in the groundmass or mesostasis of pillow lavas therefore do not represent liquidus mineral assemblages.

Secondly, many of the basalts are porphyritic. Phenocrysts, megacrysts and glomerocrysts of plagioclase and olivine, and sometimes also clinopyroxene or chromian spinel, can be fairly abundant in abyssal tholeiites. Such minerals were originally interpreted to reflect crystallization and zoning in thermally stratified crustal magma chambers (Bryan and Moore, 1977), but subsequent detailed studies of porphyritic basalts have revealed patterns of magma coalescence and mixing which cannot all be related to shallow differentiation processes. Many of the minerals are probably the products of polybaric crystallization in diverse magma strains beginning in the mantle, and they provide important clues about the compositions of near-primary magmas close to the melting stage. Few phenocrysts are directly related to the host glass compositions. Petrogenetic interpretation must therefore start with the recognition that all spreading-ridge basalts contain disequilibrium assemblages of minerals with compositions determined largely by the combined effects of crystallization kinetics and magma mixing.

This chapter considers abyssal basalts as they appear under the microscope (quench crystallization) and the low pressure controls within the crust on magmatic differentiation (liquid lines of descent). Chapter 12 can be considered a companion chapter which focuses on the mineralogical aspects of basalts from the Indian Ocean and deals with distinctions between parental basalts and the origin of phenocrysts, megacrysts and glomerocrysts observed in many abyssal tholeiites. These bear on the nature of the melting process and the coalescence of primary magmas beneath spreading ridges.

5.2 Quench textures: the consequences of supercooling

Basalts erupted under water require a different basis for description than slowly cooled subaerial basalts. Discounting phenocrysts and microphenocrysts which are readily identified, a thin section of a pillow margin may contain glass without any minerals at all, or a matrix which is obviously partially crystalline but still too fine grained to allow the identification of minerals using petrographic criteria. Still other parts of pillow lavas may be sufficiently crystalline to identify several minerals in patches, but only plagioclase and a cryptocrystalline matrix between the patches. This entire variation can occur over distances as little as 5–10 cm from a pillow rim.

Kirkpatrick (1979) provided a systematic basis for evaluating pillow margins petrographically, using inferences from programmed cooling experiments conducted primarily on lunar basalts, together with theories of crystallization and nucleation. With extrusion on the seafloor, the rate of cooling varies sharply across a pillow rim, and this together with the properties of the liquid establishes whether crystals will form, which crystals will form, the sizes and morphologies they will assume and modal proportions. Towards the pillow interiors, with lesser undercooling, crystallinity is high and the minerals coarser. Crystal shapes, mineral compositions and crystallization sequences resemble those obtaining near-equilibrium. Closer to pillow rims, extremely small crystals with spherulitic and dendritic morphologies predominate. Commonly described variolites are actually types of spherulites consisting mainly of aggregates of radiating plagioclase fibres and acicular crystals set in glass or cryptocrystalline matrix.

Figure 5.1 shows the effects of undercooling on crystallization sequences in

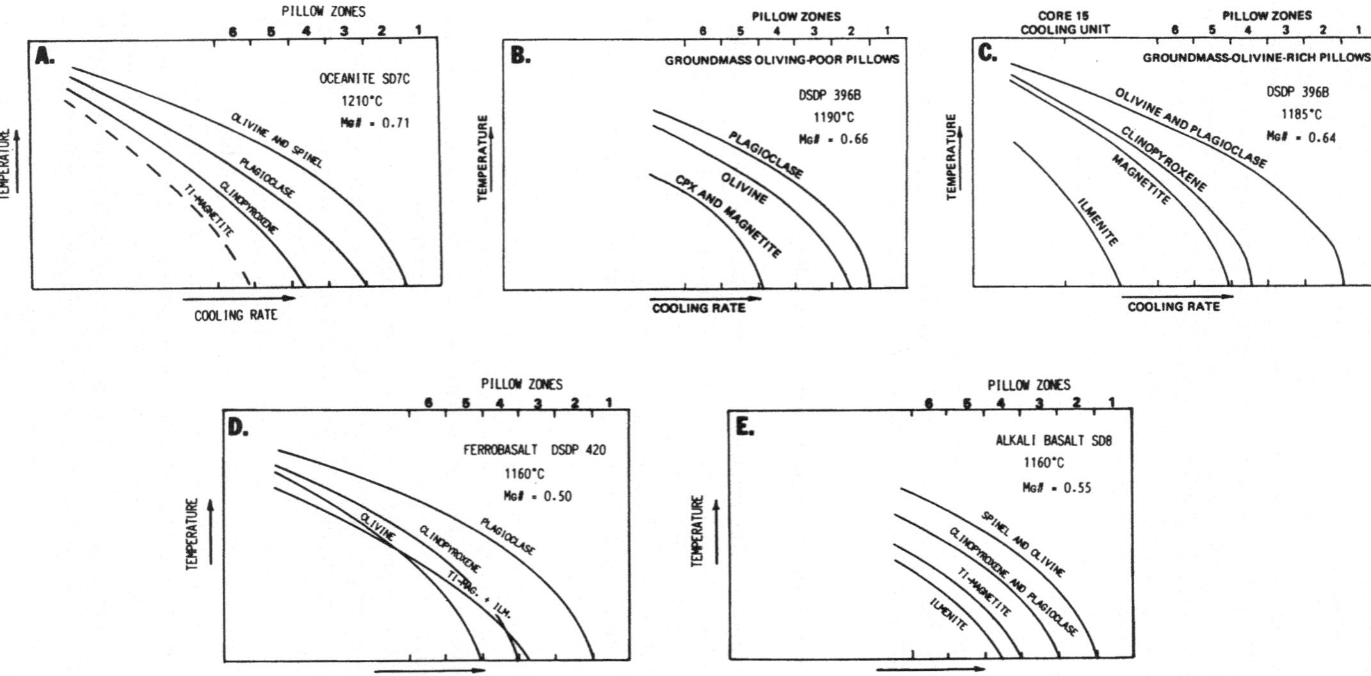

Figure 5.1 Schematic temperature versus cooling rate for five pillow types sampled from spreading ridges. A–D are abyssal tholeiites and E is an alkalic basalt. Zones, from Kirkpatrick (1979), are (1) glass; (2) isolated olivine dendrites and/or plagioclase spherulites; (3) coalesced olivine dendrites and/or plagioclase spherulites with areas of residual glass in between; (4) fully coalesced plagioclase spherulites with well-defined boundaries; (5) bow-tie or sheaf plagioclase spherulites with diffuse boundaries; (6) plagioclase microlites with dendritic clinopyroxene and skeletal titanomagnetite in between. Liquidus temperatures were estimated from FeO and MgO abundances in glasses (Roeder, 1974).

five basalts: (a) a tholeiitic picrite (Natland, 1980a), (b) two moderately evolved olivine- and plagioclase-phyric basalts (Kirkpatrick, 1979), (c) a typical highly fractionated East Pacific Rise ferrobasalt (Natland, 1980a) and (d) an alkalic basalt dredged from the Siqueiros Fracture Zone near the East Pacific Rise (Batiza *et al.*, 1977; Natland, 1989). The rocks represent most of the compositional range of spreading-ridge basalts, and the first four (all tholeiites) are given in order of magnesium number, which is used as an index of differentiation (Figure 5.1). The vertical axes give the relative temperatures of mineral crystallization and the horizontal axes the distances from pillow rims in terms of the six crystallization zones defined by Kirkpatrick (1979) based on textures and crystal morphologies. Good descriptions and illustrations of crystal morphologies and their relationship to undercooling based on experimental studies are given for olivines by Donaldson (1976) and for plagioclases by Lofgren (1971, 1974, 1980). Sequential photomicrographs of pillow margins from glassy rims inwards are illustrated by Kirkpatrick (1979) and Natland (1979, 1980a).

The outermost part (zone 1) of each pillow type in Figure 5.1 is glass which may or may not contain phenocrysts or microphenocrysts. The initial effects of crystallization nearest the pillow rims produce olivine dendrites in the picrite, isolated plagioclase spherulites and dendritic olivines in the olivine tholeiites and plagioclase spherulites alone in the tholeiitic ferrobasalt. These crystals form at temperatures well below the equilibrium crystallization temperatures, which are given schematically along the left-hand vertical axes. Both olivine and chromian spinel occur as quench phases in the alkalic basalt, but only olivine has tiny dendritic extensions to indicate continued crystal growth under quench conditions.

Moving into the pillow interiors, spherulites and dendrites coalesce, and additional minerals are added to the crystallization sequences, as defined by the curves in each diagram. If a curve falls completely to the left of some imaginary line drawn vertically to represent one distance from the pillow rim (corresponding to a specific cooling rate, decreasing from right to left in each diagram), the particular mineral or minerals designated cannot crystallize. If some portion of a curve falls to the right of the imaginary vertical line, the mineral can form, but only at temperatures less than equilibrium temperatures, given by the intersection of the curve with the vertical line. The determined mineral compositions do not match equilibrium compositions. For example, isolated plagioclase spherulite fibres tend to be more sodic than crystals formed near equilibrium, and they are surrounded by narrow zones of evolved liquid driven directly away from the composition of the crystallizing feldspar (Dowty, 1980). The width of the zone is controlled by rates of diffusion of the different cations to and away from the forming crystals and depends on such factors as melt viscosity, which varies with (rapidly diminishing) temperature.

Towards the pillow interiors, the temperatures at the onset of crystallization of each mineral increase, approaching those at equilibrium. The crystal morphologies of the earliest formed minerals tend toward acicular and elongate forms, and further inside the pillows (at lesser cooling rates) to tabular or euhedral morphologies unless the crystals interfere with each other. Later formed margins to minerals in largely crystalline pillow interiors are still usually either dendritic (e.g. clinopyroxene) or skeletal (titanomagnetite), even in the centres of very large pillows (0.5–1.0m).

One important result of the comparisons given in Figure 5.1 is that it is still possible to estimate the approximate extent of differentiation of an abyssal tholeiite using petrographic criteria. An abundance of olivine dendrites near glass, for example, specifies that an abyssal tholeiite is almost certainly magnesian in composition, which may be further supported by the presence or abundance of olivine phenocrysts and accessary chromian spinel. The predominance of plagioclase spherulites coupled with the presence of small plagioclase and clinopyroxene microphenocrysts indicates that the abyssal tholeiite is fairly evolved. Iron-rich abyssal tholeiites (ferrobasalts) tend to have dark plagioclase spherulites caused by the presence of extremely tiny titanomagnetite crystals between individual plagioclase fibers (Natland, 1980a). Such oxide minerals are not present in the same crystallization zones of the olivine tholeiites and picrite. If thin sections are available from both pillow rims and the interiors of the same eruptive cooling units, then distinctive petrographic identities can be established through a range of undercoolings, even for basalts with only subtly different glass compositions. The distinctions may be based only on differences in the sizes and proportions of spherulites near glass, corresponding differences in the lengths of acicular plagioclase needles in flow interiors (Natland, 1979), and spacings between plagioclase dendrites at known distances from glassy rims (Kirkpatrick, 1979). Nevertheless, such criteria can provide an important guide to identifying chemically different basalts sampled in a vertical section in the oceans by drilling, or on land in an ophiolite.

One special consequence of crystallization at high undercooling concerns the occurrence of titanomagnetite in tholeiitic ferrobasalts. As discussed in the following section, titanomagnetite is not a liquidus phase in the abyssal tholeiite differentiation sequence until fairly high abundances of iron and titanium are achieved in residual liquids by segregation of olivine, plagioclase and clinopyroxene. However, all pillow basalts, ferrobasalts especially, are magnetized and contain opaque oxides, and this is a fundamental cause of magnetic anomalies in the ocean crust. Where ferrobasalts are especially abundant, unusually high amplitude magnetic anomalies may occur (Anderson *et al.*, 1975; Vogt, 1979).

In pillow interiors, the occurrence of titanomagnetite may be considered to result from the continued crystallization of silicate minerals until the interstitial liquids become saturated with the oxide mineral, which forms as skeletal crystals in a mesostasis. As a result of extended crystallization, the mesostasis is an evolved residuum with a composition substantially richer in iron and titanium than the bulk rock. Titanomagnetite therefore crystallizes as a late-stage mineral in any sufficiently large pillow interior, even if the quenched margin is picritic glass.

However, skeletal titanomagnetite in pillow interiors is often too coarse grained to provide the stable magnetization of pillow basalts. This is because large crystals contain several magnetic domains with magnetization vectors pointing in different directions and reducing the net magnetization by partially cancelling each other out (Marshall and Cox, 1971). In tholeiitic ferrobasalts, single-domain crystals, all of which align with the earth's magnetic field at the time the lavas freeze, are obviously the very tiny crystals seen between the fibres of isolated plagioclase spherulites (Kirkpatrick's (1979) Zones 2–4 shown in Figure 5.1), occurring within a few centimetres of the glass rim. In these spherulites, oxide crystallization follows directly after the crystallization of plagioclase fibres (rather than the cotectic crystallization of three silicate phases). It occurs because the glass is already enriched in iron and titanium, and is quickly driven to titanomagnetite saturation by the crystallization of any silicate or combination of silicates, even in the narrow zones of melt diffusion around the spherulites which formed under conditions of extreme undercooling. The circumstance of single-silicate (plagioclase) control in spherulites is shown in Figure 5.1d by having the curve for the crystallization of titanomagnetite cross those for olivine and clinopyroxene near the pillow rim, and approach that of plagioclase. This effect is possibly enhanced by, or even results from, an increased oxidation state following the incorporation of seawater into slightly porous pillow interiors during the crystallization of these minerals (Christie *et al.*, 1986). The unmistakeable consequence is that the greater magnetization of tholeiitic ferrobasalts than olivine tholeiites results not just from more abundant iron and titanium, but because of the formation of more abundant crystals of single-domain iron–titanium oxides near the pillow rims resulting from disequilibrium crystallization at extreme undercoolings.

In the alkalic basalt (Figure 5.1e), chromian spinel and euhedral-skeletal olivine are present in glass. Despite this, there are only tiny dendritic extensions on olivine proceeding into the pillow interiors, and none on spinel. Spinel and olivine are followed in the less rapidly cooled spherulitic portions of the rock by the successive crystallization of fibrous clinopyroxene, spherulitic-acicular plagioclase, and tiny grains of titanomagnetite and ilmenite. The oxide minerals give the rock a very dark appearance. The presence of ilmenite is related to the high TiO_2 content of the basalt, and the appearance of clinopyroxene before plagioclase to low CaO (equivalently, low CaO/Al_2O_3).

How are the liquidus minerals in pillow basalts determined petrographically? The sequence of crystallization of spherulitic and dendritic crystals in the more crystalline portions of pillows provides an indication, but such minerals do not have the appropriate compositions. Small, equant or tabular crystals in glass can be inferred, by analogy to the morphologies of minerals produced in programmed cooling experiments, to have formed at fairly small undercoolings. Such minerals probably crystallized prior to extrusion in liquids barely antecedent in composition to host glasses. These, and the exterior zones on phenocrysts (not dendritic extensions) usually have nearly identical compositions and are the best approximations to liquidus minerals that can be determined from natural basalts. Divergences from equilibrium compositions are successively more extreme for minerals forming later in a crystallization sequence at a given cooling rate, and at greater cooling rates (distance from left vertical axes in Figure 5.1). Rare, truly aphyric basalts may provide no phenocrysts approaching liquidus compositions, although the equilibrium sequence can be estimated using pillow interiors.

5.3 Primary mineralogical controls on fractionation pathways of abyssal tholeiites

Consideration of the large-scale magmatic processes acting in the oceanic crust places us almost exclusively within the domain of the abyssal tholeiites. Alkalic basalts occur very rarely at spreading ridges and usually at structural offsets (Batiza *et al.*, 1977; Langmuir *et al.*, 1986; Thompson *et al.*, 1989). More typically, alkalic melts are probably supplied in increments to axial magma chambers where they mix with the far more ubiquitous depleted tholeiitic magmas (Natland, 1989). This and the next two sections consider the mineralogical controls on the liquid line of descent of abyssal tholeiites, whereas a following section deals with alkalic differentiation trends based on samples from seamounts, where evolved sequences have been found.

5.3.1 *General course of fractionation in abyssal tholeiites: influence of magma chamber mixing*

Left alone, abyssal tholeiite cooled slowly in the oceanic crust will experience crystallization differentiation during which minerals will appear in a particular sequence, in equilibrium with successively more fractionated liquids. The successive liquids define fractionation pathways which can be established in an idealized sense by crystallization experiments in the laboratory. Figure 5.2 shows the results of one set of experiments (from Walker *et al.*, 1979) in a pseudo-ternary projection from plagioclase of the normative components olivine (Ol), diopside (Di) and silica (SIL). The hachured field encloses the projection of nearly 2000 basalt glass compositions, most of which correspond closely to the 1 atm cotectic boundary of liquids saturated in plagioclase, clinopyroxene and olivine. As the position of the cotectic boundary is influenced by pressure, Walker *et al.* (1979) considered the above close

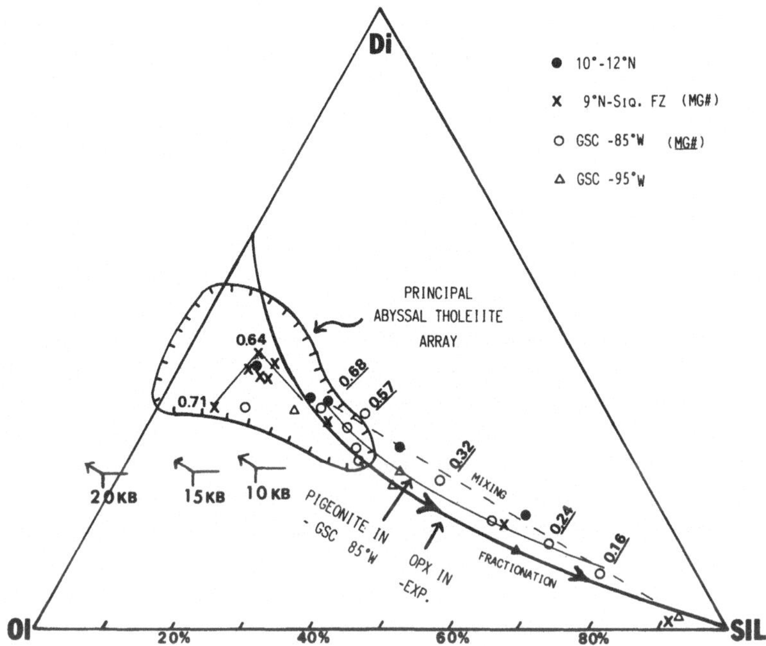

Figure 5.2 Proportions of normative olivine (Ol), diopside (Di), and SiO_2 (SIL) for analyses of primitive and fractionated eastern Pacific basalt glasses listed in Table 5.1, calculated using the algorithm of Walker *et al.* (1979). High-pressure pseudo-invariant points are from Stolper (1980). The experimental fractionation trend of Walker *et al.* (1979) is given by the solid curved line. Natural compositions are displaced toward Di (thin curved line). A possible mixing trend between basaltic and silicic compositions is shown (dashed line), magnesium numbers of selected glass compositions are positioned next to appropriate data points.

relationship to indicate that most abyssal tholeiites experience low pressure fractionation in which liquids are saturated in plagioclase. The physical evidence in the rocks is that most abyssal tholeiites contain plagioclase and either, or both, olivine and clinopyroxene, and are thus multiply saturated in these silicate phases.

At the lower right corner of Figure 5.2, the experiments establish that low-Ca pyroxene supplants olivine late in the crystallization sequence. Very few abyssal tholeiite glasses actually project into this portion of the diagram, although the tendency for the natural liquids to evolve to the point where low-Ca pyroxene is stable is verified by the occurrence of pigeonite associated with glassy mesostasis in some highly crystalline interiors of thicker lava flows (Thompson and Humphris, 1980). Walker *et al.* (1979) believe that the rarity of lavas evolved to this degree may be related to axial magma chambers which are regularly replenished with primitive (unfractionated) magma, thus preventing such extensive fractionation from occurring and ‘perching’ typical basalt liquids at fairly magnesian compositions. At some places this process is extremely efficient and there is little variability among fairly magnesian lava types (Natland *et al.*, 1983; Stakes *et al.*, 1984). The general picture of abyssal tholeiite differentiation is that most of it occurs at low pressure in crystal magma chambers, where liquids are multiply saturated with silicate assemblages including plagioclase. The course of differentiation leads to more silica-saturated residua, but repetitive mixing prevents highly evolved liquids from developing.

Nevertheless, sampling on ridge crests is now extensive enough for localities to be identified where basalts are more fractionated than elsewhere, and where a few are actually sufficiently evolved to contain low Ca pyroxene. Ferrobasalts in which glasses contain more than (say) 13% total iron as FeO* and 2.5–3.5% TiO_2 contents are fairly abundant at intermediate to fast spreading ridges in the eastern Pacific (Bass, 1971; Scheidegger, 1973; Clague and Bunch, 1976; Morel and Hekinian, 1980), and along elevated, slow-spreading ridges close to hot-spots such as the Reykjanes Ridge near Iceland (Schilling, 1973b; Sigurdsson, 1981; Schilling *et al.*, 1983). They are not yet known from rifted segments of slowly spreading ridges in either the Atlantic or Indian Oceans (Natland, 1980b; Bloomer *et al.*, 1989; Natland *et al.*, in press). The general correspondence appears to be that differentiated basalts erupt more commonly where axial magma chambers are substantial in size and long-lived or permanently established, regardless of spreading rate, as inferred from geophysical data and axial topography (Rosendahl, 1976; Orcutt *et al.*, 1976; Natland, 1980b; Detrick *et al.*, 1987).

Detailed studies of locations in the eastern Pacific now suggest that ridge segmentation plays a strong role in the distribution of highly evolved lavas. The East Pacific Rise and Galapagos Spreading Centre are both offset by major transform faults and numerous smaller discontinuities such as propagating rifts and overlapping spreading centres. Ferrobasalts have been dredged from a number of these locations and their proportion in the crust appears to be unusually large, especially when based on the mapping of abundant high amplitude magnetic anomalies which result from the presence of unusually magnetized (iron-rich) basalts (Anderson *et al.*, 1975; Christie and Sinton, 1981; Sempere and Macdonald, 1986). The cause of high magnetization in ferrobasalts was discussed in the previous section.

Why ferrobasalts are so abundant at such locations is a complex matter, although it may be related to the lateral propagation of dykes from centres of magma injection into older crust, which enhances fractionation (Christie and Sinton, 1981; Sinton *et al.*, 1983). Alternatively, or in addition, physical isolation of pockets of magma at shallow levels in rift systems allow differentiation to proceed without mixing with primitive basalt (Natland, 1980b; Perfit *et al.*, 1983). Even so, most isolated magma bodies must still be closely linked to the magma plumbing system as ferrobasalts are obviously purged to the seafloor by more primitive basalts which displace them in the crust. They are usually closely associated spatially with less fractionated olivine tholeiites, even being obtained in the same dredge haul (Natland, 1980b).

5.3.2 *Advanced differentiation*

There are several locations where lavas even more evolved than ferrobasalts have been discovered. These include the following examples: the eastern end of the Galapagos Rift in the Panama Basin, eastern Pacific, where the Rift reaches the Ecuador Fracture Zone at about 85°W (Perfit *et al.*, 1983); a portion of the same spreading centre near the tip of a propagating rift at 95°W (Byerly, 1980; Clague *et al.*, 1981); the southern end of the segment of the East Pacific Rise at 9°N (Langmuir *et al.*, 1986; Natland *et al.*, 1986); and north of the Clipperton Fracture Zone at about 10°N on the East Pacific Rise (Thompson *et al.*, 1989). Iceland, a subaerial segment of the Mid-Atlantic Ridge (Chapter 13) where the oceanic crust is greatly thickened, also has fairly abundant silicic lavas associated with central volcanoes (Carmichael, 1964; Sigurdsson and Sparks, 1981), although here we will only consider submarine occurrences where the crust is of normal thickness and the silicic lavas erupt along normal rift segments.

Table 5.1 compares the compositions of basalt glasses from the four locations in the eastern Pacific mentioned above. For each location, a least-fractionated basalt, a highly iron-enriched ferrobasalt, and one or more siliceous glass compositions are included. These compositions fall approximately along the 1 atm cotectic in the projection used in Figure 5.2 and thus define the most extensive low pressure differentiation sequences known from glassy lava samples in the ocean basins. At two of the four locations (95°W on the Galapagos Rift; 9°N on the East Pacific Rise), lavas reach rhyodacitic compositions, with up to 70% SiO_2 contents, but there are few intermediate rock types. The 85°W Galapagos Rift suite has nearly a full range of compositions reaching about 64% SiO_2 and has been carefully studied mineralogically (Perfit and Fornari, 1983). This suite is used here as a case study to consider the mineralogical controls on extensive abyssal tholeiite differentiation.

The glass compositions in Table 5.1 demonstrate for each location that fractionation to the most iron-enriched ferrobasalts first causes substantial enrichments in FeO*, TiO_2, Na_2O, P_2O_5 and S, with reductions in CaO, Al_2O_3 and MgO, but little change in SiO_2 contents. On the basis of least-squares computer calculations, this is considered to result from the fractionation of plagioclase, olivine, lesser clinopyroxene and no other minerals (Clague and Bunch, 1976; Perfit *et al.*, 1983). About 60–70% crystallization occurs to produce ferrobasalt liquid and this represents a range of cooling from about 1210°C (corresponding to the most primitive compositions thought to be supplied from the mantle to axial magma chambers) to less

Table 5.1 Glass compositions for extended differentiation series. 9°N (Natland, 1989 and new data); 10–12°N (Thompson *et al.*, 1989), East Pacific Rise; 85°W (Perfit *et al.*, 1983); and 95°W (Melson *et al.*, 1976, Byerly, 1980), Galapagos Spreading Centre

	Sample location														
	9°N									10–12°N					
	SD7-C	SD4-1	R9-1Top	PROT41	R14-1	R-9AND	R-9$_{Dac1}$	R-9$_{Dac2}$	FeTi Mix[a]	70-A	48-A		48-25 Mix[b]	48-25	48-2
SiO_2	49.15	50.38	50.73	50.63	50.41	57.01	71.19	69.50	50.0	49.80	51.31	50.45	50.50	53.70	59.39
TiO_2	0.90	1.33	2.03	2.33	2.59	1.72	0.45	0.39	2.5	1.29	2.42	2.55	2.66	2.28	1.37
Al_2O_3	17.58	15.24	13.98	13.23	13.10	12.64	11.97	14.55	11.6	15.97	13.40	13.21	13.03	13.29	13.02
FeO*	8.01	9.32	11.93	13.60	13.82	12.42	6.03	5.50	16.3	9.24	12.99	13.56	13.64	12.27	9.64
MnO	0.149	0.17	0.29	0.234	0.256	0.28	0.12	0.15	0.35	—	—	—	—	—	—
MgO	9.58	8.37	6.86	6.05	5.46	2.75	0.77	0.42	4.1	8.15	5.58	5.63	5.38	4.58	2.87
CaO	12.14	12.47	11.28	10.07	9.68	6.89	3.10	3.24	8.9	12.30	10.05	10.07	10.37	8.90	6.42
Na_2O	2.38	2.59	2.91	3.28	3.47	3.65	4.17	4.54	3.2	2.50	3.13	3.10	3.31	3.52	4.37
K_2O	0.016	0.058	0.16	0.223	0.303	0.55	1.13	1.15	0.2	0.10	0.23	0.20	0.18	0.34	0.74
P_2O_5	0.072	0.169	0.179	0.268	0.29	0.408	0.20	0.03	0.6	0.13	0.21	0.24	0.23	0.20	0.22
Total	100.04	100.10	100.35	99.90	99.38	98.32	99.13	99.47	97.7	99.38	99.33	99.33	99.30	99.08	98.04
S	1070	1160	1270	1400	1780	1020	—	140	1510	—	—	—	—	—	—
Cl	25	100	210	—	1360	2850	—	4100	2150	—	—	—	—	—	—
Mg#	0.712	0.651	0.543	0.430	0.450	0.315	0.209	0.137	0.42	0.646	0.463	0.463	0.461	0.436	0.382

Table 5.1 (Continued)

	Sample location											
	85°W							95°W				
	100-5B	998-4B	996-1B	1002-4B	999-1B	1001-1C	994-1E	C98 (D5)	C32 (D6)	998 (D6)	D64 (D6)	996 (D6)
SiO_2	51.40	51.06	50.34	53.50	56.70	59.22	64.28	50.12	51.94	52.35	57.06	70.77
TiO_2	1.33	2.04	3.47	2.50	1.91	1.86	1.22	1.06	2.49	2.46	1.76	0.61
Al_2O_3	14.17	13.40	11.69	11.82	11.42	11.74	12.26	16.52	12.43	12.52	13.48	12.30
FeO*	11.43	13.67	17.13	15.05	16.30	12.66	11.05	8.71	15.97	15.72	12.12	5.30
MnO	0.19	0.22	0.21	0.26	0.29	0.25	0.23	—	—	—	—	—
MgO	7.37	6.23	4.37	3.48	1.80	1.89	1.04	8.62	4.87	4.37	2.75	0.43
CaO	11.39	10.40	8.93	8.51	6.90	6.32	5.14	12.41	8.95	9.16	6.87	2.92
Na_2O	2.42	2.47	2.83	3.01	3.48	3.66	4.13	2.19	2.60	2.76	3.31	4.14
K_2O	—	0.13	0.24	0.31	0.39	0.46	0.58	0.06	0.33	0.28	0.58	1.30
P_2O_5	0.14	—	0.37	0.58	0.65	0.53	0.21	0.09	0.20	0.25	0.17	0.05
Total	99.84	99.62	100.38	99.02	99.84	98.59	100.00	99.78	99.72	99.90	98.13	97.82
S	1560	2100	2520	1920	1760	1320	800	—	—	—	—	—
Cl	600	900	—	3100	4300	5000	5100	—	—	—	—	—
Mg#	0.572	0.486	0.336	0.324	0.186	0.236	0.163	0.072	0.387	0.366	0.320	0.144

[a] R-9 AND = 0.36 DAC1 + 0.64 FeTi Mix

[b] 48-25 = 0.17 DAC2 + 0.83 48-25 mix (mix SiO_2 values assumed)

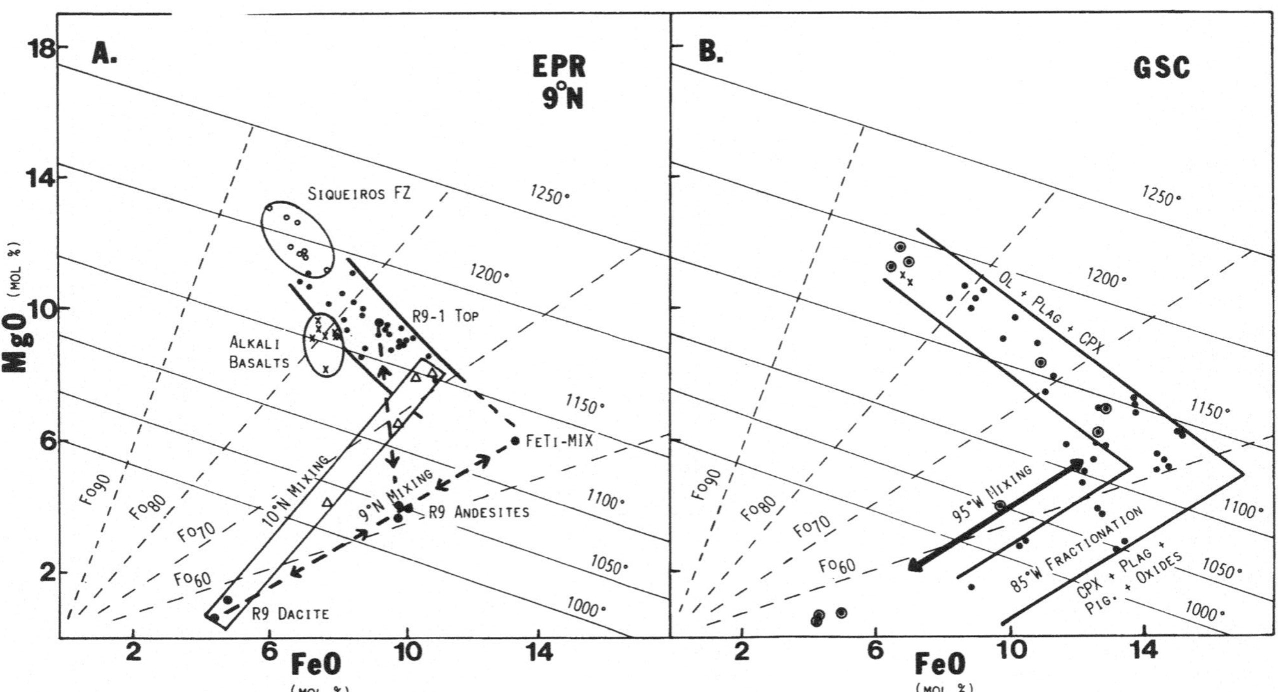

Figure 5.3 FeO* versus MgO (mol. %) for basaltic and silicic glasses from (A) 9°N on the East Pacific Rise, and (B) the Galapagos Spreading Centre. Lines of constant temperature and olivine isopleths are from Roeder (1974). Symbols in (A) are: open circles – Siqueiros Fracture Zone; dots – all other 9°N lavas; X – alkali basalts; open triangles – 10°N compositions from Table 5.1. Individual samples identified are as in Table 5.1. In (B) symbols are: dots – 85°W glasses; filled circles – 95°W glasses; X – Costa Rica Rift glasses.

than 1150°C (Figure 5.3; see also temperature estimates in Perfit and Fornari, 1983; Wilson *et al.*, 1988).

More extended differentiation results in SiO_2 enrichment, continued increases in alkali abundances, continued decreases in CaO, Al_2O_3 and MgO, and accompanied now by decreases in FeO*, TiO_2, P_2O_5 and S. The final residual liquids have estimated liquidus temperatures of 1000–1100°C (Figure 5.3). The total extent of calculated crystallization for the 85°W Galapagos Rift suite is more than 80% (Perfit *et al.*, 1983). The decreases in FeO* and TiO_2, in P_2O_5 and in S require fractionation of the additional phases titanomagnetite plus ilmenite apatite and sulphides, respectively. In addition, petrographic evidence indicates that in these later stages of differentiation, olivine is supplanted by low-Ca pyroxene (in this case pigeonite). Figure 5.4 shows the calculated proportions of fractionating mineral phases determined by least-squares calculations using glass and mineral phase compositions for successive stages of differentiation of the Galapagos Rift suite (Perfit *et al.*, 1983). Close correspondence of calculated to natural glass compositions for both major oxides and trace elements, as well as recent experimental studies (Juster *et al.*, 1989), support this low pressure fractionation model.

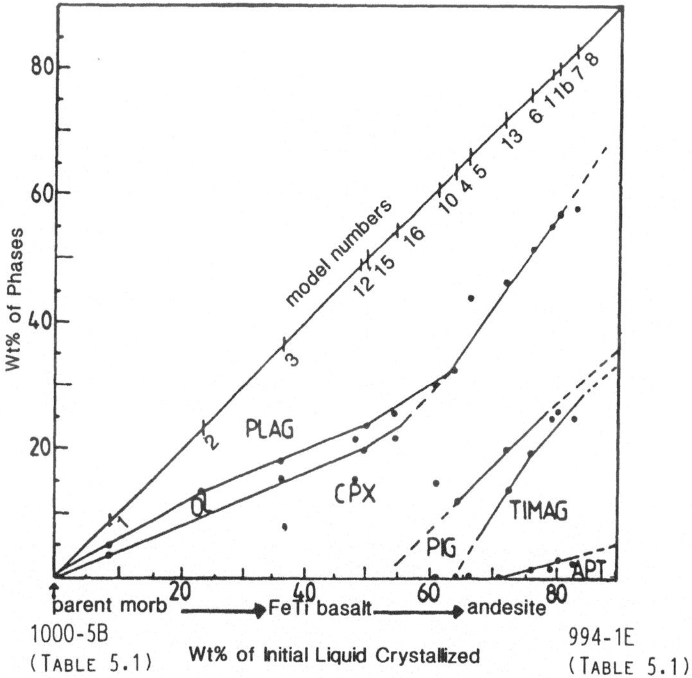

Figure 5.4 Modelled (least-squares) proportions of fractionating phases during differentiation versus weight percent of initial liquid crystallized for a magmatic lineage from parental Galapagos Rift abyssal tholeiite to dacite (64% SiO_2 contents), from Perfit and Fornari (1983).

Least-squares fractionation models are point-to-point estimates and thus do not allow for continuous variations of mineral compositions or liquids as differentiation proceeds. The smaller the incremental differences between glass and mineral compositions used in the calculations, the more precise the results are likely to be. However, Figure 5.3 shows that olivine is expected to vary continuously in composition from about Fo_{90} in the least fractionated basalt along the East Pacific Rise, to about Fo_{60} in the most evolved ferrobasalt at both the East Pacific Rise and the Galapagos Rift. This corresponds fairly well to olivine compositions actually measured by electron microprobe in the basalts. Other mineral compositions are not so simple. Figure 5.5 compares the magnesium numbers of natural glasses with those of liquids calculated to be in equilibrium with olivines (Roeder and Emslie, 1970; Roeder, 1974) and clinopyroxenes in the same samples, using the mineral

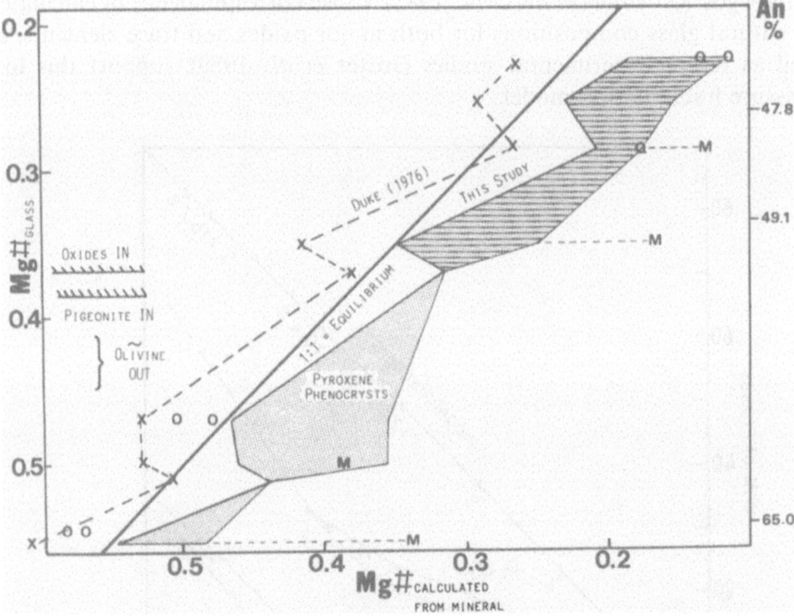

Figure 5.5 Calculated magnesium numbers of liquids using mineral compositions versus actual magnesium numbers of host glasses for Galapagos Rift 85°W samples (Perfit and Fornari, 1983). The relationship of Roeder and Emslie (1970) was used for olivine compositions, and a modification of that of Duke (1976) for clinopyroxenes. A solid line on the right of the diagonal links the most magnesian clinopyroxenes in each sample based on the modified relationship. This can be compared with the dashed line to the left of the diagonal, which is based on the unmodified equation. The modification assumes that clinopyroxene follows olivine in the crystallization sequence, as observed petrographically. Letter identifications are: O – olivines; Q – quench crystals in glass; M – microlites. Shaded areas give ranges for clinopyroxene phenocrysts in individual samples. Petrographically observed locations of olivine out, oxides in, and pigeonite in, for successive liquids, are given to the left. Corresponding plagioclase compositions (An %) in host glasses are shown to the right.

data of Perfit and Fornari (1983) for the 85°W Galapagos Rift suite. In general, the olivines predict liquid magnesium numbers closely, and the most magnesian clinopyroxenes give results matching those of olivines. However, the majority of clinopyroxenes give estimates of lower magnesium numbers (more iron-rich liquids) in many samples. This almost certainly is the consequence of the crystallization of many of these pyroxenes at heightened undercoolings, varying from sample to sample, and in each sample with distance from the glassy margins, as discussed previously. Obviously, for least-squares computational purposes, the best results will follow from the use of mineral compositions which show these effects the least.

Figure 5.5 also shows that clinopyroxene as well as olivine tends to become more iron-rich as crystallization differentiation proceeds. Plagioclases correspondingly become more sodic, as indicated along the right side of the figure. In sequence, pigeonite replaces olivine in the crystallization sequence and iron–titanium oxides join the liquidus assemblage. Such variations are similar to those classically observed among other basaltic differentiation series (Kuno, 1968; Wager, 1968).

5.3.3 *Role of oxygen fugacity and oxide minerals*

Juster *et al.* (1989) provide experimental evidence at controlled oxygen fugacities of the role of iron–titanium oxides in late-stage abyssal tholeiite differentiation at the Galapagos Spreading Centre. Christie *et al.* (1986) showed that the crystallization of most abyssal olivine tholeiites and ferrobasalts occurs at oxygen fugacities 2–3 log units below the nickel–nickel oxide (NNO) buffer. However, because olivine and pyroxene extract primarily divalent iron from melts, the Fe^{3+}/Fe^{2+} ratio increases in successively more fractionated liquids. Juster *et al.* (1989) argue that this process is insufficient to raise oxygen fugacity to the NNO buffer, which is about where oxide minerals join the liquidus in their experiments for melts with about 4.5% MgO contents. They propose that the oxygen fugacities were enhanced by reaction with rocks in the shallow crust for NNO to have been reached.

Modelling of the tholeiitic differentiation of the Thingmuli volcano, Iceland (Ghiorso and Carmichael, 1985), showed that fractionation along a buffer [whether NNO or fayalite–magnetite–quartz (FMQ)] requires that the magmatic system be open, first to lose oxygen and counteract the tendency for silicate fractionation to increase the Fe^{3+}/Fe^{2+} ratio and then (after oxide phases have joined the liquidus) to gain oxygen. Interaction with crustal rocks is suggested as one mechanism to accomplish this (Ghiorso and Carmichael, 1985). If, however, the system is closed, the fractionation of silicate phases will continue, oxygen fugacities will continually increase and oxide minerals will not join the liquidus assemblage until the residual melts have very low MgO contents. This feature is also indicated by the calculations of Juster *et al.* (1989). The result is very extended high iron differentiation, the classic Fenner trend (Fenner, 1929, 1931). An important aspect of the extended high iron differentiation is that silicic differentiates cannot form as early, or at elevated temperatures, as in the more buffered open systems. Moreover, the compositions of late-forming oxide minerals in the closed system, and their liquidus proportions, may not be sufficient to arrest the increase in oxygen fugacities (Ghiorso and Carmichael, 1985) or reverse it, as at 85°W (Juster *et al.*, 1989).

Closed system conditions might prevail in the deeper ocean crust. Juster *et al.* (1989) argue that higher pressure fractionation than at 85°W operated to produce the rhyodacites observed at 95°W. It was concluded that at 95°W, the rare, intermediate lavas are hybrids between rhyodacite and basalt rather than representing a continuum along either open or closed system liquid lines of descent (double arrow in Fig. 5.3). The rock suite is thus fundamentally bimodal, basaltic and rhyodacitic, in contrast to the spectrum of silicic lava types sampled at 85°W. The question remains, why are there no intermediate lavas along a liquid line of descent at 95°W? The explanation may be related to the buoyancy of low density rhyodacite in basaltic magma and its tendency to accumulate near the top of magma chambers, as in Iceland (Yoder, 1973; Sigurdsson and Sparks, 1981). Along a rift system at a spreading ridge, laterally or vertically injected basalt dykes encountering small, silicic magma bodies coalesced near the top of the principal magma chamber complex (upper layer 3) would displace and partially mix with them hoisting silicic melt to the seafloor.

An example of precisely this sequence of events is offered by the ferroandesite listed in Table 5.1 from 9°N along the East Pacific Rise. The sample (R-9AND) is actually from a composite lava flow with ferroandesite in the interior and a 4 cm thick basaltic carapace. The interior of the flow contains scattered crystals of clinopyroxene (magnesium number 0.81) and plagioclase (An_{66-68}) identical in composition to minerals in the basalt carapace, yet the groundmass contains siliceous patches with contents of up to 66% SiO_2. The rhyodacite composition listed in Table 5.1 (R-9DAC) is from a glass bleb in another nearly identical ferroandesite from the same dredge haul. Calculations suggest that the principal basaltic component required to produce the ferroandesite by mixing with the rhyodacite had up to 16% FeO* and 2.5% TiO_2 content, with magnesium number of about 0.35 (FeTi–MIX; Table 5.1). Lavas of such extreme iron enrichment have not been sampled from the East Pacific Rise, but they have been found at the Galapagos Spreading Centre (Figure 5.3). The actual FeO* and TiO_2 contents of the iron-rich mixing component were probably higher than these estimates, because no account has been taken of the proportion of the carapace basalt (R9-1TOP; Table 5.1) in the mixing calculation.

Basaltic magmas with 16–18% FeO* have been produced experimentally under strongly reducing conditions using an abyssal tholeiite starting material (Dixon and Rutherford, 1979). In the experiments, iron enrichment was not arrested, but proceeded until liquids with up to 22% FeO* were produced. This is as predicted by thermodynamic modelling of closed system fractionation (Ghiorso and Carmichael, 1985). Such extreme iron-rich compositions also represent the late-stage liquid line of descent at the Skaergaard intrusion, which ultimately produced a granophyric residuum (Wager, 1960; Wager and Brown, 1967) very similar in composition to the rhyodacites at 9°N and 95°W. At 9°N the ferroandesite is thus a complex hybrid of its entraining basalt, rhyodacite and a very iron-rich liquid (Figure 5.3) which may have formed during closed system, extended high iron differentiation in the deeper oceanic crust. The hybrid has no minerals approaching equilibrium with its own composition.

A similar calculation shows that an andesite with 53.7% SiO_2 at 10°N on the East Pacific Rise is also probably a hybrid between rhyodacite and moderately iron-enriched basalt (48-25 MIX in Table 5.1), very similar to basalts dredged from the immediate vicinity. Both it and the dacite from 10°N plot to the left of the general 85°W liquid line of descent shown in Figure 5.3, as they should if mixing occurred. Thus, at three of the four regions considered, mixing between basalt and very silicic melt is the most probable mechanism for the formation of intermediate rocks, rather than fractionation, which probably followed an extended high iron trend in the deeper oceanic crust.

5.3.4 *Apatite and sulphides*

The influences of apatite and sulphides during the differentiation of abyssal tholeiites have not been assessed experimentally. This is primarily a result of the problems in dealing with phases which are modally insignificant in experimental charges. The natural lavas, however, provide qualitative information.

The very late onset of apatite crystallization can be monitored by using the P_2O_5 content of lavas. Phosphorus nearly behaves as an ideal incompatible element during the fractionation of silicates in basaltic liquids (Anderson and Greenland, 1970). In the 85°W Galapagos Rift suite (Figure 5.6) it increases systematically with degree of fractionation until theoretically estimated apatite saturation is reached in dacitic melts with about 2% MgO content (Harrison and Watson, 1984; Juster *et al.* 1989). With the commencement of apatite fractionation the P_2O_5 content of more siliceous residues drops rapidly. Hanson (1989) has pointed out that this requires a systematic change in the solid–melt partition coefficients for phosphorus during the late stages of differentiation, as the proportion of phosphorus in apatite is fixed by stoichiometry, and at constant K_D phosphorus abundances in apatite-saturated melts should remain constant. Intermediate lavas at 95°W on the Galapagos Spreading Centre and at 9°N and 10°N on the East Pacific Rise have

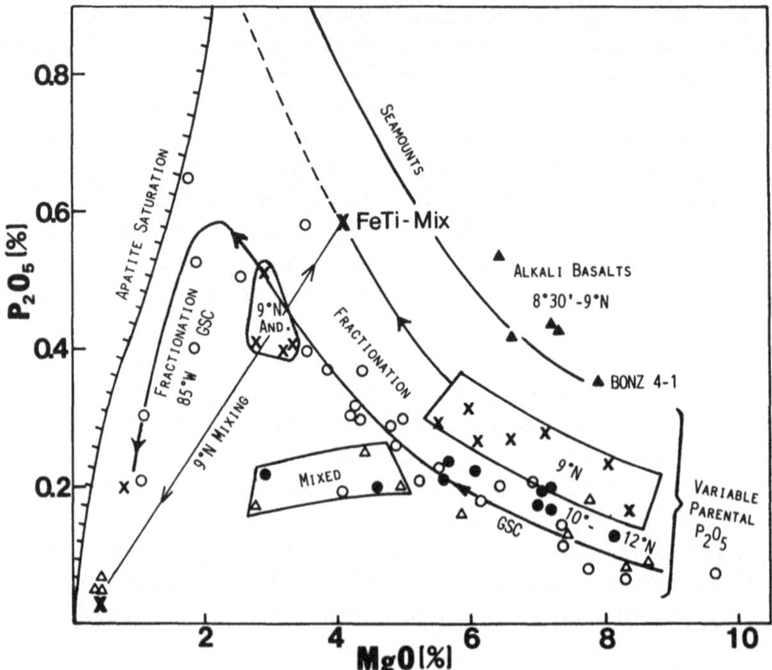

Figure 5.6 P_2O_5 versus MgO contents for lavas of the Galapagos Spreading Centre and the East Pacific Rise at 9°N and 10°N using data sources listed in the text, and following Juster *et al.* (1989). Symbols are as in Figure 5.2. The apatite saturation surface of Harrison and Watson (1984) intersects the diagram at the hachured line. Hybrid lavas at 9°N and 10°N fall below the average trends for increasing P_2O_5 during fractionation, and the apatite saturation surface. Likely mixing end components for 9°N are indicated by arrows. A trend for seamount lavas (Table 5.2) is also shown.

reduced P_2O_5 abundances at MgO contents of $>2\%$, supporting the interpretation that they are hybrid compositions (Figure 5.6).

Sulphide spherules occur in the most evolved (ferroandesitic and dacitic) members of the Galapagos Rift suite, intergrown with oxide microphenocrysts (Perfit and Fornari, 1983). Similar spherules often occur without the associated oxide minerals in less differentiated abyssal tholeiites and ferrobasalts, and as decorations in vesicles (Mathez, 1976, 1980; Perfit and Fornari, 1983). The spherules represent immiscible sulphide droplets segregated from basaltic magmas (Wager *et al.*, 1957). They may be homogeneous quenched sulphide liquid in basaltic glasses, or partially crystallized multiphase aggregates in more slowly cooled pillow interiors. The phases are solid solutions involving pyrrhotite (FeS), pentlandite (NiS), and chalcopyrite or cubanite (Cu–Fe sulphides) end-members (Czmanske and Moore, 1977; Francis, 1980). More Ni-rich globules occur in primitive abyssal tholeiites, whereas sulphides in ferrobasalts and ferroandesites approach pure pyrrhotite in composition (Perfit *et al.*, 1983). However kinetic factors also play a part in the crystallization of minerals from liquid sulphide spherules subjected to the same range of undercooling as associated silicate liquids in a pillow margin.

Although most abyssal tholeiites appear to be saturated in sulphide regardless of their stage of differentiation, the sulphur abundance actually increases in residual glasses from about 1000 to 2000 ppm until ferrobasalt compositions are reached, as the segregation of sulphides is more than matched by the differentiation of silicate minerals (Mathez, 1976, 1980; Czamanske and Moore, 1977). However, sulphide droplet segregation is evidently accelerated in late differentiation at the time of onset of the crystallization of iron–titanium oxides, which radically reduces the iron contents of glasses (Perfit *et al.*, 1983). As iron oxide is considered to buffer sulphide contents according to the reaction.

$$FeO + S^{2-} = FeS + O^{2-} \qquad \text{(Mathez, 1976),}$$

sulphur contents in glass can increase as long as the proportion of iron (Fe^{2+}) increases in residual liquids during differentiation. The reaction also contributes to increasing the oxygen fugacities of residual liquids during differentiation. The onset of crystallization of iron–titanium oxides at 85°W both decreased the iron contents and reversed the trend of increasing oxygen fugacities (Juster *et al.*, 1989). The demand for oxygen by oxides forced the Mathez (1976) reaction to the right, increasing the segregation of sulphides. This resulted in the reduction of sulphur in residual silicic glasses (Table 5.1).

The combined effect of the crystallization of iron–titanium oxides and the enhanced segregation of sulphide droplets is the cause of late-stage silica enrichment in the ferroandesites and dacites in the Galapagos Rift 85°W suite. No combination of silicate minerals precipitating by themselves would produce such an enrichment, nor could it cause the corresponding decreases in FeO*, TiO_2 and S abundances. In closed system, extended high iron differentiation, oxide crystallization is deferred, and sulphur abundances may reach an extreme in very iron rich residual liquids (perhaps 3000 ppm in melts with 20% or more FeO* contents). Consequently the very late formation of oxide minerals in the eventual closed system evolution of rhyodacitic liquids may produce an unusual concentration of sulphides associated with oxides in cumulates at this stage of differentiation (Natland *et al.*, in press).

5.4 Liquid immiscibility and the significance of melt densities

In addition to the eruption of lavas with different buoyancies, another explanation for the basalt–rhyodacite bimodality at 95°W and 9°N may follow from greatly extended high iron differentiation. In the course of the experiments of Dixon and Rutherford (1979), when residual liquids attained about 22% FeO*, two silicate liquids, one siliceous, the other very rich in iron and titanium, separated immiscibly. This process has been documented in essentially anoxic iron-rich lunar basalts (Roedder, 1979) and is suggested for some terrestrial occurrences such as the Skaergaard intrusion (McBirney and Nakamura, 1974) where extremely high FeO* abundances were reached. Immiscible silicic and iron-rich droplets were also observed in the highly fractionated mesostasis of the coarse-grained interior of a single 9 m thick abyssal tholeiite lava flow drilled near the Mid-Atlantic Ridge (Sato, 1979).

In Dixon and Rutherford's (1979) experiments, the iron-rich liquids existing just prior to segregation were intermediate compositions along the liquid line of descent. For example, they have low magnesium numbers, comparable to those of andesites at 85°W, although they retain low SiO_2 contents and in this respect still resemble basalts. In contrast, the immiscible silicic liquids have about 70% SiO_2 and closely match the compositions of the rhyodacites at 9°N on the East Pacific Rice and 95°W on the Galapagos Spreading Centre, as well as granitic dykelets obtained from fracture zones of the Mid-Atlantic Ridge and the Central and South-west Indian Ridges (Engel and Fisher, 1975; Miyashiro & Shido, 1980; Robinson *et al.*, 1989). If there are only liquids having either about 50% SiO_2 or 70% SiO_2 available, any andesitic or dacitic eruptive composition must necessarily be a hybrid.

Whether this type of liquid immiscibility generally occurs in the oceanic crust is very difficult to establish. Obviously, whenever open system processes operate to put oxide minerals on the liquidus at a comparatively early stage (as at 85°W) immiscibility has not operated. The principal evidence for liquid immiscibility, should it follow from closed system fractionation, lies in the deeper gabbroic sections of the oceanic crust which are almost entirely inaccessible except to drilling. However, the same general bimodality of compositions should exist even if silicic liquids form by fractional crystallization at the very end of extended high iron differentiation. This is as interpreted, for example, for the melanogranophyres of the Skaergaard intrusion (Wager and Brown, 1967), based on the experimental work of Bowen and Schairer (1935) and Osborn (1959). Until oxide minerals join the liquidus, residual liquids will not increase in SiO_2 contents. If oxide crystallization is this late, then SiO_2 enrichment will occur very abruptly, during the last 1–3% of crystallization of the magma, making the proportion of intermediate liquids (having 53–64% SiO_2 contents) very small.

At this stage, the contrasting liquid densities become very important. Densities of very iron-rich liquids produced experimentally (Dixon and Rutherford, 1979) are 3.0–3.2 $g\,cm^{-3}$ [estimated using the procedure of Bottinga *et al.*, 1982] and because they will actually sink into cumulate mats composed of silicate minerals will probably not be able to erupt. The coalescence and crystallization of such dense melts within cumulates at the floors of magma chambers may provide one explanation for the origin of Ti-ferrogabbros in absyssal gabbro suites (Natland *et al.*, in press). The estimated densities of rhyodacite liquids, in contrast, are only 2.0–2.2 $g\,cm^{-3}$, and would be buoyant in all basaltic and even andesitic melts. Their potential for mixing with basaltic magmas and conveyance to the seafloor is thus considerable.

5.5 Mantle–crust environments controlling oxygen fugacity

Although interaction with crustal rocks may have accelerated the crystallization of oxide minerals at 85°W, compared with other differentiation series, such as continental, island arc and oceanic island tholeiites, the suite attained fairly substantial iron enrichments in ferrobasalts (to 18% FeO*) prior to the formation of silicic differentiates (Clague *et al.*, 1981). As suggested earlier, even greater iron enrichment probably occurs in the gabbroic portion of the oceanic crust. This is because abyssal tholeiite crystallization initially occurs under the highly reducing conditions documented by Christie *et al.* (1986), at or near the fayalite–magnetite–quartz buffer. Such low oxidation states are a consequence of both the properties of parental magmas supplied from the mantle (an indication of mantle properties beneath spreading ridges), and of the poor, but variable, ability of bottom waters to penetrate through hydrothermal into magmatic systems with their original oxygen abundances intact.

Indeed, hydrothermal interactions may actually strip oxygen from circulating saline fluids as evidenced by the amounts of sulphides that precipitate at hot smoker vents at ridge axes (Chapter 8). Such fluids evidently enter the deepest axial magmatic systems at some stage as indicated by the great abundances of chlorine even in the ferroandesite and rhyodacite glasses at 9°N (Table 5.1) which were previously inferred as forming during closed system fractionation at the deep levels in the ocean crust. Either chlorine-rich fluids interact with the magmas directly, or hydrated, amphibolized oceanic crust is partially assimilated (Michael and Schilling, 1989). However, this need not necessarily increase the oxygen fugacities in magmas. Chlorine abundances thus confirm inferences that magmatic differentiation at spreading ridges is not a simple closed system process (O'Hara and Mathews, 1981), not even in the deeper crust where extended high-iron differentiation is favoured. Rather than presume that the unbuffered evolution of oxygen fugacities during such differentiation requires a closed system (Ghiorso and Carmichael, 1985), it is more accurate to say that reactions with crustal rocks or fluids also occur, but that the environment is non-oxygenating.

The course of late differentiation, which is controlled strongly by the crystallization of oxide minerals, probably depends primarily on where in the crust it takes place with respect to the depth of penetration of hydrothermal systems. The further from the ocean floor (and closer to the mantle), the more nearly magmas will sustain extended high iron differentiation and attain a state of silicate liquid immiscibility. However, in arc systems or ophiolites formed in supra-subduction environments, magmas may leave the mantle with intrinsically higher oxygen fugacities and high-iron differentiation will be suppressed before the effects of interaction with hydrothermally modified crustal rocks are superimposed. In the oceanic crust, differentiation is dominated by the crystallization of plagioclase, olivine and clinopyroxene. Late-stage siliceous differentiates are very rare, representing less than about 1% of eruptive compositions even in the eastern Pacific. The rarity of extreme differentiates, the distinctive control of iron–titanium oxides in producing them, and their compositions, should all be considered in comparing oceanic crust to ophiolites.

Finally, a note of caution should be made concerning the correspondence of mineral compositions in late-stage Galapagos Rift differentiates to minerals found in gabbroic rocks such as those recently described from the South-west Indian Ridge (Bloomer *et al.*, 1989; Natland *et al.*, in press). The gabbros clearly crystallized from abyssal tholeiite parental magmas, and include facies in which low-Ca pyroxenes, ilmenite, magnetite, assorted sulphide minerals and apatite all occur. However at the Ocean Drilling Program Hole 735B (South-west Indian Ridge), both of the oxide minerals joined crystallization sequences much later than in the Galapagos Rift suite, sulphide segregation was correspondingly deferred, and there are no gabbros with SiO_2 abundances higher than 54% (Robinson *et al.*, 1988), indicating that none crystallized from andesitic melts along a liquid line of descent. Rare silicic dykelets probably formed immiscibly, with complementary iron-rich liquids crystallizing to produce Ti-ferrogabbros (Natland *et al.*, in press). The compositions of silicate minerals by themselves thus do not necessarily indicate the course of the liquid line of descent; for Hole 735B, this was established by the petrographic observation of when oxide minerals crystallized with respect to silicates.

The high proportions of fractionated gabbros in the fracture zones stands in stark contrast to the absence of equally fractionated lavas on Indian Ocean Ridges, and poses an important problem in understanding the mechanics of magma distribution in crust produced at slowly spreading ridges. The later stages of differentiation are clearly more important in the evolution of plutonic assemblages in the deep oceanic crust than in its eruptive carapace. Increased understanding of the shallow differentiation of abyssal tholeiites will probably follow from an integrated study of such gabbroic rocks and their associated basalts.

5.6 Alkalic magmatic lineages on seamounts

Both abyssal tholeiites and alkalic basalts have been dredged from seamounts built on young crust in the eastern Pacific (Chapters 9 to 11), together with a variety of 'transitional' basalts thought to be mixes of the two (Batiza *et al.*, 1977; Batiza and Johnson, 1980; Batiza and Vanko, 1984). The overall suite, including the presence of alkalic basalts matches the mineralogical and chemical variability exhibited by the East Pacific Rise. The seamount alkalic basalts, for example, are comparatively enriched in $^{87}Sr/^{86}Sr$ (Zindler *et al.*, 1984; Natland *et al.*, 1986; Natland, 1989) and are similar chemically and mineralogically to that described here from the Siqueiros Fracture Zone (Figure 5.1E).

Engel *et al.* (1965) postulated that seamount alkalic basalts are derived from parental abyssal tholeiites by differentiation processes acting in columns of magma within conduit systems of the volcanoes. Although isotopic systematics now rule out this parental relationship, one aspect of the hypothesis is still worthy of consideration. The 'average' oceanic alkalic basalt of Engel *et al.*

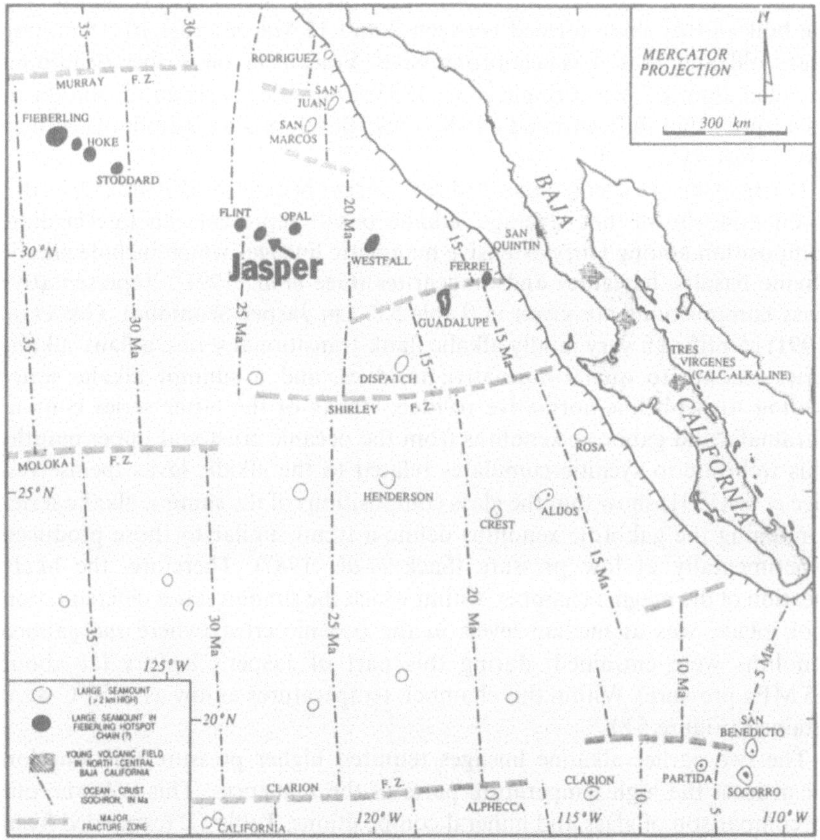

Figure 5.7 Location of Jasper Seamount along the Guadalupe–Fieberling chain in the eastern Pacific, modified from Lonsdale (1989). Fracture zones and crustal isochrons, based on magnetic anomalies, are also shown.

(1965) is, as they properly recognized, of a fairly fractionated composition, with lower MgO and higher total alkali contents than most of the alkalic basalts subsequently sampled from seamounts near the East Pacific Rise or from the rise itself (Chapter 11, Tables 11.4 and 11.7). It is also more oxidized than abyssal tholeiites. What is the correct parent to 'average' alkalic basalt and how is its differentiation accomplished?

The more fractionated composition of the average alkalic basalt of Engel *et al.* (1965) compared to alkalic basalts from seamounts near the East Pacific Rise stems from the different types of volcanoes sampled in the earlier and later studies. The average combines analyses of alkalic rocks obtained primarily from seamounts off the west coast of North America, most of which are larger than the young seamounts near the present East Pacific Rise. Volcanoes along the Guadalupe–Jasper–Fieberling chain (Figure 5.7), for example, reach very shallow depths. Fieberling itself is a guyot with a flat, eroded summit (Carsola and Dietz, 1952). Radiometric ages establish that the bulk of this chain formed between 7 and 17 Ma on crust 10–15 million years old and that it is age-progressive. Volcanism on Jasper Seamount spanned almost 7 Ma (Pringle *et al.*, 1991; Lonsdale, 1991) and it should be no surprise that differentiated alkalic rocks developed on seamounts with so long a history.

Fairly extensive dredging on and near Jasper Seamount (Figure 5.7) in this chain now shows that 'average' alkalic basalt represents an intermediate composition among fairly extensive magmatic lineages which include alkalic olivine basalts, hawaiites and mugearites (Gee *et al.*, 1991). Representative glass compositions are given in Table 5.2. On Jasper Seamount, Gee *et al.* (1991) identified a very mildly alkalic flank transitional series, a flank alkalic series leading to quartz-normative residues, and a summit alkalic series leading to nepheline-normative residue. Rocks of the latter series contain ultramafic and gabbroic xenoliths from the oceanic crust and upper mantle, plus wehrlitic to syenitic cumulates related to the alkalic lavas themselves. Gee *et al.* (1991) show that the glass compositions of the summit alkalic series containing the gabbroic xenoliths define a trend similar to those produced experimentally at low pressure (Sack *et al.*, 1987). Therefore, the likely location of the magma chamber, within which the summit series differentiation took place, was at median levels in the oceanic crust (where the gabbro xenoliths were entrained) during this part of Jasper's history (at about 1.5 MPa pressure). Within this chamber, temperatures as low as 1050°C were attained (Figure 5.8).

The two earlier alkaline lineages required higher pressure fractionation for at least the high temperature parts of the sequences. This is borne out by comparison of glass and mineral compositions; Table 5.2 (expanded from Gee *et al.*, 1991) summarizes the compositions of microphenocrysts found in glass.

The crystallization sequence of principal minerals in all three series are: olivine plus a spinel phase (Ti-bearing chromian spinel in the least fractionated

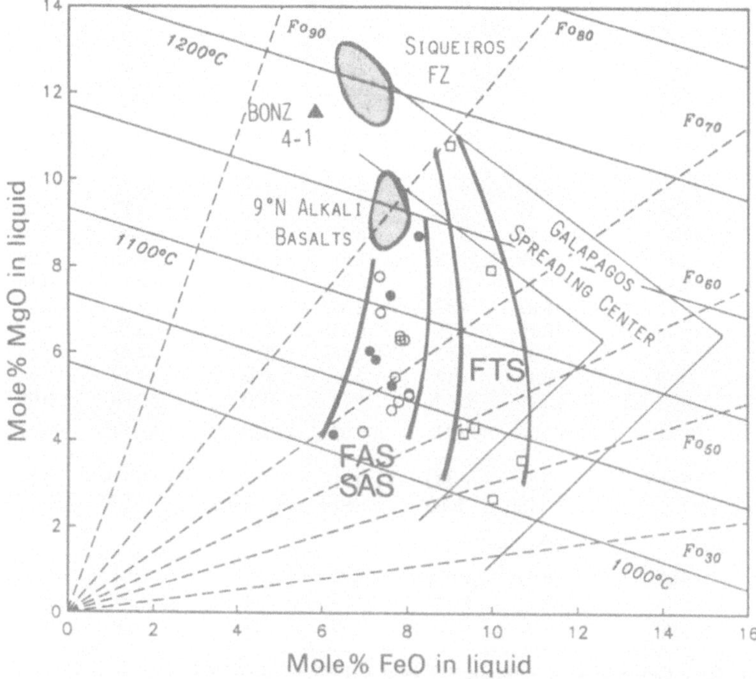

Figure 5.8 FeO versus MgO (mol. %) for basalt glasses from Jasper seamount and vicinity, compared with the trend for the Galapagos Spreading Center from Figure 5.3B, and data fields for parental abyssal tholeiites (Siqueiros FZ) and alkali basalts from the East Pacific Rise. Two general lineages, corresponding to the Flank Transitional Series (FTS) and the Flank and Summit alkalic series (FAS and SAS) from Jasper seamount (Gee *et al.*, 1991), occur, neither of which achieves significant iron enrichment.

basalts, chromian titanomagnetite in the others), sodic plagioclase in hawaiites and mugearites, ilmenite in a few hawaiites, and apatite in one of the mugearites. Sulphide globules are especially abundant in the hawaiites. The sequence of crystallization and mineral compositions are similar in all the series and differ from abyssal tholeiites in three regards: the absence of clinopyroxene; the occurrence of oxide minerals (prominence of chromian titanomagnetite and presence of ilmenite); and the late crystallization of plagioclase. All of these features were apparent from the crystallization at high cooling rates of the alkali basalt from the Siqueiros Fracture Zone (Figure 5.1E), but here the minerals are closer to being in equilibrium with host glasses which represent fairly complete differentiation sequences.

Gee *et al.* (1991), however, show that these minerals are not adequate to explain changes in glass compositions within the lineages by crystallization differentiation. The most important additional requirement is that a substantial amount of clinopyroxene fractionation, combined with olivine and spinel, is necessary to account for the simultaneous changes in CaO/Al_2O_3, MgO and

Table 5.2 Representative seamount alkalic series glasses[1]

	High Mg#	Jasper flank transitional				Jasper flank alkalic	
	BONZ 4-1	JS5-1	JS3-1	JS14-1	JS19-1	JS2-2	JS2-3
SiO_2	49.18	49.94	49.11	54.42	49.09	51.95	55.61
TiO_2	1.91	1.41	2.86	2.29	2.49	2.13	2.00
Al_2O_3	17.63	16.19	15.51	14.91	17.06	17.90	17.09
FeO	8.17	11.79	12.68	14.91	17.06	8.90	7.84
MnO	0.18	0.16	0.19	0.24	0.20	0.16	0.16
MgO	7.95	6.82	4.87	2.55	5.42	3.63	2.50
CaO	10.98	10.19	8.76	6.23	9.57	6.54	5.17
Na_2O	3.20	2.91	3.52	4.44	3.89	4.71	5.18
K_2O	0.79	0.45	1.12	2.01	1.36	2.19	3.04
P_2O_5	0.35	0.19	0.63	1.18	0.55	0.88	0.92
Sum	100.11	100.05	99.27	100.05	100.28	98.99	99.47
Mg#[2]	0.669	0.545	0.443	0.310	0.513	0.458	0.398
S (ppm)		1130	1000	910	580	230	180
		Normative[3]					
Q	0.00	0.00	0.00	1.24	0.00	0.00	0.00
Hy	5.69	13.44	8.54	19.37	0.00	1.21	6.64
Ne	0.00	0.00	0.00	0.00	2.80	0.00	0.00

[1] Data from Gee *et al.* (1991) except BONZ 4-1, 1471-1-2 and 1471-5-1.
[2] Average repeatability (twice standard deviation) of 5 measurements on different spots, 20 seconds counting time each 15 Ma beam current.
[3] Computed assuming $Fe^{2+}/(Fe^{2+} + Fe^{3+}) = 0.86$.

Cr contents, even in the summit alkalic series. Some evidence for clinoproxene fractionation at depth is provided by a few wehrlitic cumulates found as xenoliths in lavas of the summit alkalic series. These also contain oxide minerals and orthopyroxene.

The minerals observed in the glasses evidently have very little to do with controlling alkalic series differentiation. They are primarily an expression of late-stage crystallization, possibly only of post-eruptive processes. This is suggested at one location by the recovery of highly fragmental breccia with a glass composition identical to the margins of vesicular basalts obtained very close by (Table 5.3). The glass in the breccia contains no microphenocrysts whereas the basalts have abundant small crystals of olivine and spinel. The breccia was evidently produced by a type of submarine fire-fountaining during the one eruption which also produced the lavas. This suggests that the process of vesiculation (nucleation of dissolved volatiles) itself, which was more extensive in the lava samples than in the breccias, may have augmented crystallization in the lavas. More significantly, the aphyric glass breccias suggest that alkalic series magmas can arrive at volcanic summits with excess heat, indicating rapid (non-adiabatic) ascent from deep magma hearths where

Table 5.2 (Continued)

Jasper summit alkalic			Engel *et al.*	Jasper 1471	Satellite 1471	
JS2-1	JS5-1	JS12-1	Ave. AOB	1-2 pillow	5-1 breccia	Estimated error
47.63	48.11	53.02	47.61	50.08	49.66	0.40
2.76	3.35	2.28	2.88	3.05	3.04	0.12
17.25	16.98	17.25	18.10	15.96	15.77	0.18
9.27	9.68	8.69	9.55	9.22	8.96	0.22
0.17	0.21	0.19	0.16	0.16	0.14	0.04
4.74	3.76	2.54	4.81	5.62	5.26	0.18
10.17	8.54	5.65	8.69	9.13	9.38	0.18
4.22	2.91	5.52	4.01	3.48	3.61	0.18
1.95	2.60	3.67	1.67	1.31	1.29	0.12
0.64	0.91	1.13	0.92	0.66	0.76	0.12
98.80	98.05	99.95	98.40	98.66	97.97	
0.515	0.446	0.377	0.511	0.558	0.549	
930	—	250		470	—	40
0.00	0.00	0.00	0.00	0.00	0.00	
0.00	0.00	0.00	0.00	10.20	5.90	
8.76	4.12	5.96	3.20	0.00	0.00	

differentiation actually occurred. The surficial crystallization in effect results from a type of undercooling produced by outgassing (Lipman and Banks, 1987).

A number of features all point to fractionation of the alkalic series under fairly oxidative conditions (NNO buffer): the involvement of oxide minerals in most stages of development of the liquid lines of descent, the lack of pronounced iron enrichment during differentiation (Figure 5.8), the presence of oxide minerals in the wehrlitic cumulates, the steady decline of sulphur abundances during differentiation (Table 5.3) and the occurrence of quench ilmenite and magnetite in some glasses. As critical portions of the fractionation took place in the upper mantle and lower crust, this has to reflect a property of the parental magmas rather than interaction with altered rocks. This means in turn that the mantle sources of the parental, enriched alkalic basalts beneath these seamounts were more oxidized than the sources of abyssal tholeiites (Mattioli *et al.*, 1989).

In the final analysis, the existence of extensive alkalic magmatic lineages on these seamounts is a consequence of their formation on fairly old, cool and thick oceanic lithosphere. Liquidus temperatures as low as 1100°C were attained at some level in the lithospheric upper mantle, which once, at the ridge axis, was at about 1250°C at its very top (below the base of the magma chamber). The absence of extensive alkalic magmatic lineages on seamounts

Table 5.3 Near-liquidus mineral compositions of seamount glass groups[1]

Glass group	Mg#[2]	Spinel Mg#[2]	Spinel Cr#[3]	Ol Fo	Plag An	Cpx	Ti-mag mol% Usp.	Other
				Hop Sing Seamount				
Bonz 4-1	66.9	71.5	32.3	88.3	—		—	—
				Jasper Seamount				
Flank transitional series								
JS 5-1	54.5	55.4	40.0	82.6	70.3	—	—	
JS 3-1	44.3	—	—	76.4	60.5	—	58.1	
JS 14-1	31.0	—	—	57.0	47.4	—	65.0	Ilm.
JS 4-2	21.1	—	—	41.8	41.0	—	70.0	Ap.
Flank alkalic series								
JS 19-1	51.3	—	—	82.4	62.9	—	—	
JS 5-3	40.7	—	—	76.8	64.5	—	—	
JS 2-3	39.8	—	—	—	57.7	[4]$En_{41}Wo_{43}Fs_{16}$	51.4	
Summit alkalic series								
JS 2-1	51.5	49.7	36.7	82.4	70.3	—	—	
JS 5-4	44.6	30.5	46.9	80.5	64.1	—	—	
JS 13-1	44.3	30.4	48.1	81.3	65.0	—	—	
JS 9-1	44.3	—	—	81.3	65.0	—	—	
JS 6-1	38.6	—	—	78.7	57.1	—	56.1	
JS 12-1	37.7	—	—	73.2	53.9	—	56.4	Hbl. xenocryst

[1] Compositions are from euhedral or near-euhedral crystals in glassy margins. Fo and An are maximum values in each sample.
[2] Calculated assuming $Fe^{2+}/[Fe^{2+} + Fe^{3+}] = 0.86$ for glasses; for spinels, after calculation of Fe^{3+} from stoichiometry.
[3] Cr# = Cr/[Cr + Al].
[4] Slightly rounded, probably a xenocryst.

near the East Pacific Rise thus can be attributed to the lack of long conduit systems through a portion of upper mantle and oceanic crust, within which ascending magmas could episodically stagnate and differentiate to such low liquidus temperatures. Engel *et al.* (1965) were correct in emphasizing the importance of the magma conduit system in the evolution of 'average' alkalic basalt. However, mineral and glass compositions show that the height of the volcano above the seafloor is only a small part of the total column that must be involved.

5.7 Concluding statements

This chapter has described the crystallization at high cooling rates of tholeiitic and alkalic basalts from the ocean's spreading ridges and seamounts, the mineralogical controls on differentiation in the ocean crust and the evolution of liquid lines of descent.

1. A representative crystallization sequence for abyssal tholeiites was illustrated by a suite of basalts from the Galapagos Spreading Centre at 85°W. This suite initially experienced fractionation of plagioclase, olivine and clinopyroxene under reducing conditions to the point of formation of iron-enriched ferrobasalt. Eventually, the nickel–nickel oxide buffer was reached and oxides joined the liquidus. Subsequent differentiation was controlled along or near NNO by plagioclase, clinopyroxene, low Ca pyroxene, oxides, sulphides and apatite, resulting in silica-enriched ferroandesites and dacites. The onset of oxide crystallization was accelerated by interactions with altered crustal rocks, which increased the oxygen fugacities. Oxide crystallization in turn enhanced immiscible sulphide segregation. Apatite crystallization followed, reducing the P_2O_5 contents in the most siliceous lavas. All minerals except apatite, are represented by phenocrysts and microphenocrysts in the quenched lavas, but not all of them have liquidus compositions.
2. Even more extended high-iron differentiation (the classic Fenner trend) can occur where interactions with crustal rocks are restricted to non-oxygenating environments, perhaps in the central portions of principal magma chambers forming the gabbroic layer of the oceanic crust. This may lead to the segregation of separate immiscible rhyodacitic and iron-rich liquids. The latter are too dense to erupt, but mixing between basalts and buoyant rhyodacites possibly produced in this way occurred at 95°W on the Galapagos Spreading Centre and at 9°N and 10–12°N on the East Pacific Rise, resulting in hybrid ferroandesites.
3. Alkalic magmatic lineages leading to hawaiite, mugearite and trachyte, develop on seamounts built on eastern Pacific crust of 10–20 Ma. Olivine, spinels (magnesio-chromite, Cr-titanomagnetite), sulphide globules and in some cases ilmenite, are quench phases, with sodic plagioclase and apatite joining these in the more evolved rocks. However, liquid lines of descent, which do not attain significant iron enrichment, are controlled at or near the NNO by polybaric crystallization involving these phases and clinopyroxene in long conduit systems extending into the cooled and thickened lithospheric upper mantle.

6 Experimental phase petrology of mid-ocean ridge basalts

DON ELTHON

6.1 Introduction

Experimental studies of the phase petrology of mid-ocean ridge basalts (MORBs) have focused principally on the determination of mineral compositions in equilibrium with basaltic liquids under a variety of magmatic conditions. These experimental studies generally fall into two major groups, those at low pressures (generally at 1 atm (1 atm = 101 325 $N m^{-2}$), but sometimes 2–3 kbar (1 bar = 10^5 $N m^{-2}$)) and those at high pressures (generally about 8–30 kbar).

The overall objective of the low pressure (1 atm) experimental studies has been to determine liquid lines of descent for MORBs in shallow magma chambers in the oceanic crust. This objective has been met by determining the compositions of coexisting minerals and glasses in experimental studies and comparing these results to phenocrysts and glasses that occur in MORBs. Although some important gaps persist in the experimental studies, the general aspects of the low pressure crystallization of MORBs are well known.

The principal objective in high pressure studies of MORBs and related compositions has been to determine the compositions of possible primary MORB magmas, which are those liquids produced by the partial melting of the sub-oceanic mantle. By definition, primary magmas have not changed in composition following their segregation from the mantle. Another important objective is the detailed knowledge of how these magmas crystallize at high pressures once they have separated from the mantle, i.e. what is the nature of high pressure liquid lines of descent? With respect to these objectives, there is no general consensus on the interpretation of the available data and they remain amongst the most contentious subjects in MORB petrology.

The experimental procedures, results and interpretations of these studies are discussed in this chapter. The significance of conclusions obtained from these experimental studies extend beyond the immediate questions of deducing liquid lines of descent and the nature of primary MORB magmas. These experimental data affect other important problems such as the temperature distribution in the upper mantle (Oxburgh, 1980), the formation of the oceanic lithosphere (McKenzie and Bickle, 1988), the interpretation of some aspects of the origin and identification of major element mantle heterogeneities (Langmuir and Hanson, 1980; Dick *et al.*, 1984), and the possibility of using basalts to understand the geochemistry of planetary interiors (BVSP, 1981).

6.2 Experimental studies at 1 atm

6.2.1 *Experimental techniques*

Most recent experimental studies of oceanic basalts at 1 atm have been conducted using the Pt-wire loop technique in a vertical tube furnace similar to that shown in Figure 6.1. The sample to be melted is mounted on a Pt-wire loop (Presnall and Brenner, 1974) which is suspended in a vertical ceramic tube through which flows a mixture of gases (generally CO and CO_2) that buffer the oxygen fugacity (fO_2) at a specified value (Nafziger *et al.*, 1971). The use of Pt-wire loops minimizes the loss of Fe from the sample and enables

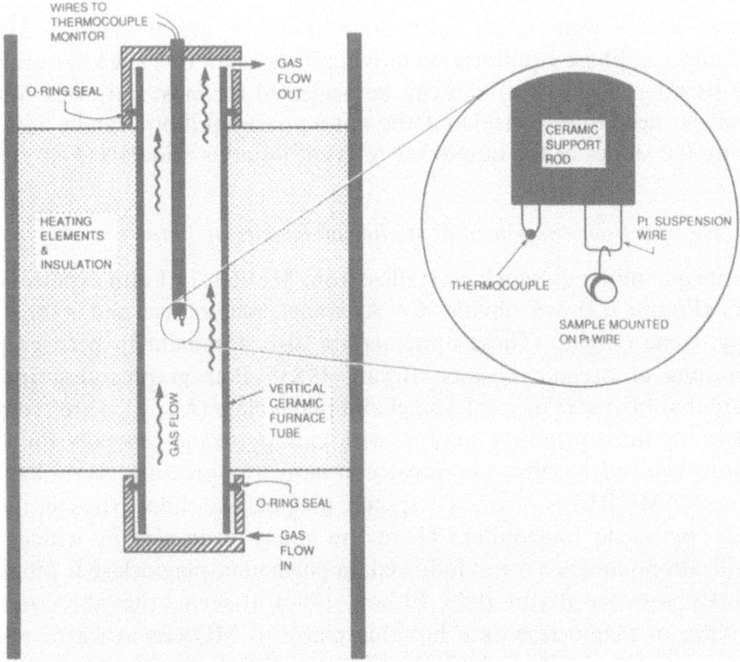

Figure 6.1 Schematic diagram of a 1 atm gas-mixing furnace assembly. The furnace assembly is about 25–40 cm across, and the ceramic support rod shown in the inset is about 0.5 cm in diameter. An excellent description of these furnaces and related gas-mixing equipment is given in Huebner (1971, 1987).

the maximum interaction between the sample and the flowing gas. In some instances, the Pt-wire loop is presaturated with Fe to reduce the loss of Fe to essentially zero (Walker *et al.*, 1979; Grove, 1981). Sodium, however, is generally partially lost (up to 60%) from the sample into the gas using this technique, particularly at high temperatures, high gas flow-rates, and when the ratio of the surface area to the volume of the sample is large (Donaldson, 1979; Corrigan and Gibb, 1979). Tormey *et al.* (1987) have developed a technique to minimize this Na loss by placing a Na_2O–SiO_2 mixture in with the CO–CO_2 gas mixture. This procedure increases the partial pressure of sodium in the flowing CO–CO_2 gas to the point where only a small amount of sodium (about 7%) is lost from the sample.

The 1 atm experiments are usually 24–200 h in duration, after which the samples are quenched. The samples are generally mounted in epoxy for study with an electron microprobe, which is used to analyse the phases present. In the more complete and systematic studies of phase relationships in oceanic basalts (Bender *et al.*, 1978; Walker *et al.*, 1979; Grove and Bryan, 1983; Tormey *et al.*, 1987), the compositions of almost all phases are reported. However in many studies, particularly the earlier ones, only a small number of microprobe analyses of phases are reported.

Although these experiments are conducted at 1 atm pressure, magmatic processes which occur in crustal magma chambers are at about 1–3 kbar. The changes in phase equilibria on moving from 1 atm to a few kbar are very minor (Bender *et al.*, 1978) and can be neglected for most purposes. Where these effects need to be considered, the 1 atm phase equilibria can be adjusted by using the slopes of the liquidi for relevant liquidus minerals (Fisk, 1984).

6.2.2 *Results from experimental studies of basalts at 1 atm*

The principal minerals which crystallize from MORBs in 1 atm experimental studies (Figure 6.2) are olivine, Cr–Al spinel, plagioclase and a high Ca clinopyroxene (augite). These minerals are also dominant in petrographic descriptions of oceanic basalts (Bryan, 1983). Petrographic descriptions indicate that olivine (Fo_{80-90}), spinel and plagioclase (An_{75-90}) are typically found in the most primitive basalts, with clinopyroxene generally found in the more evolved basalts. The most common low pressure crystallization sequence of MORBs is olivine (+ spinel), plagioclase, clinopyroxene, then a low Ca pyroxene (pigeonite). There are many examples in which this crystallization sequence is not followed; in particular, plagioclase is often the liquidus phase (see Bryan 1983; Elthon, 1984). It seems that the common occurrence of plagioclase as a liquidus phase in MORBs at 1 atm results from an excess amount of plagioclase (greater than cotectic proportions) in the populations of MORB phenocrysts and from partial resorption of plagioclase during magma mixing (Elthon, 1984).

Experimental studies indicate that there is an olivine + plagioclase + liquid

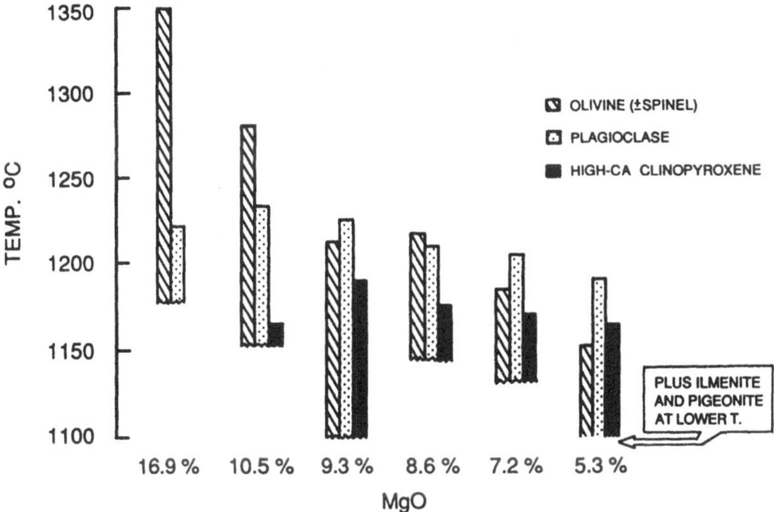

Figure 6.2 Representative 1 atm experimental results for the crystallization of various MORB-type samples. Note that olivine ($\pm$ spinel) crystallizes first in the primitive samples (MgO > 9.5%), but that the other basalts crystallize plagioclase and olivine near their liquidus, with high Ca-clinopyroxene at lower temperatures. In very evolved samples (Fe- and Ti-rich basalts), pigeonite and ilmenite crystallize. These experiments were conducted on Pt-wire loops with the oxygen fugacity set at the QFM buffer (Elthon, unpublished data).

cotectic at 1 atm in which these two minerals co-crystallize over an extended temperature interval of up to about 75°C (Walker *et al.*, 1979; Grove and Bryan, 1983), which had been previously inferred from phenocryst assemblages in MORBs (Miyashiro *et al.*, 1979). Clinopyroxene, olivine and plagioclase co-crystallize over an extended temperature range until a low Ca pyroxene (pigeonite) begins to crystallize and olivine begins to disappear via a reaction relationship (Walker *et al.*, 1979). At lower temperatures, plagioclase, pigeonite, magnetite/ilmenite, amphibole, apatite and eventually silica crystallize from ferrobasalts, andesites and highly evolved silicic magmas (Perfit and Fornari, 1983).

The interpretation of liquid lines of descent is often best evaluated qualitatively using pseudo-liquidus phase diagrams. A wide variety of these diagrams have been developed in the last two decades (O'Hara, 1968; Green *et al.*, 1979; Walker *et al.*, 1979; Elthon, 1983, Grove *et al.*, 1982), based on the early work of Yoder and Tilley (1962) and Coombs (1963). These pseudo-liquidus phase diagrams involve recalculating the major element analysis of a sample into four end-members (typically olivine, plagioclase, clinopyroxene and silica) that constitute the apices of a tetrahedron. The compositions of glasses can be projected from one apex of the tetrahedron

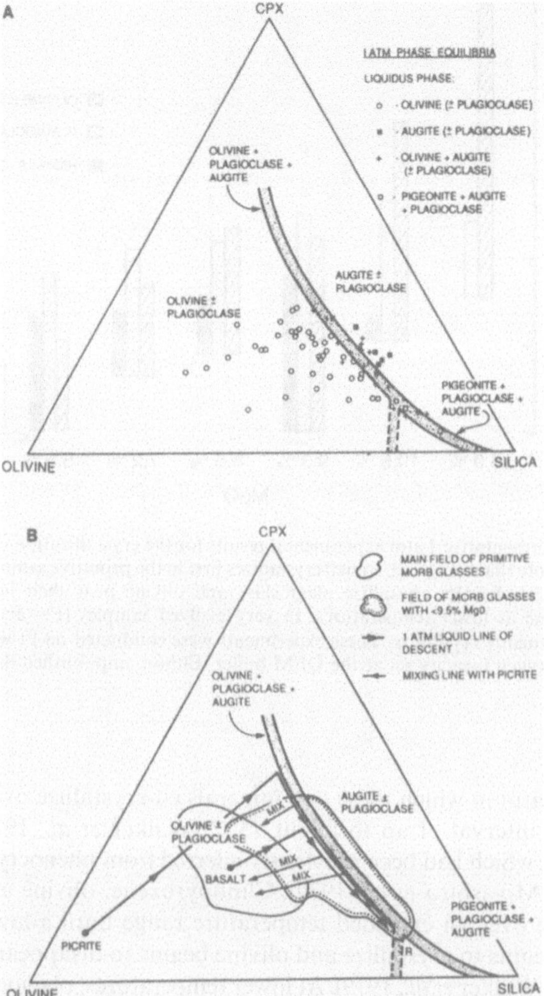

Figure 6.3 Phase equilibria results on MORBs projected (Elthon, 1983) onto a portion of the cpx–olivine–silica plane. Data are from the literature, but only slightly modified from Walker *et al.* (1979). In (A) the liquid compositions from experiments are projected onto the plane, where the different symbols indicate the phases in (apparent) equilibrium with the liquid. In (B) the fields for both primitive and more evolved MORBs are shown. The 1 atm liquid lines of descent for a picritic and a basaltic magma are shown along with some mixing paths. The apices of the triangle shown are (top) $cpx_{60}ol_{10}sil_{30}$, (left) $cpx_{10}ol_{60}sil_{30}$ and (right) $cpx_{10}ol_{10}sil_{80}$.

onto the opposing face and the phase equilibria can be used to determine the locations of multiply-saturated phase boundaries (Figure 6.3A).

The 1 atm phase equilibria of MORBs are shown in Figure 6.3A, projected from the anorthite apex to the olivine–clinopyroxene–silica plane. This pseudo-liquidus phase diagram can be used for the qualitative description of the 1 atm crystallization and magma mixing processes in MORBs. The equilibrium liquid lines of descent for a picritic and a basaltic magma are shown, as is a simplified model for magma mixing during periodic replenishment of a magma chamber (Figure 6.3B). The fields for primitive MORB glasses (> 9.5% MgO) and more evolved MORBs are shown in Figure 6.3B. It is apparent that almost all MORBs lie near the 1 atm multiple saturation boundaries (or are displaced into the olivine + plagioclase field) and that very few samples project to the right-hand side of the reaction point R. The periodic replenishment of magma chambers will result in the eruption of very few highly evolved or primitive magmas, but most compositions will be buffered near some intermediate value (O'Hara, 1977; Rhodes *et al.*, 1979; Walker *et al.*, 1979). Alternatively it takes about 85–95% crystallization of primitive MORBs for the derivative liquids to reach the reaction point so that, even in the absence of magma mixing, only 5–15% of basalts plot to the right of the reaction point (Grove and Bryan, 1984).

The differentiation of MORB liquids can be further understood from these 1 atm experiments by examination of the compositions of minerals in

Table 6.1 Representative distribution coefficients for 1 atm crystallization of MORBs (in wt% except where stated otherwise)

	Distribution coefficient		
Component	$D^{ol/liq}$	$D^{plag/liq}$	$D^{cpx/liq}$
Na_2O (cation%)	—	0.9–1.5	0.15–0.25
MgO (cation%)	2.8–6.0	0.01–0.05	1.6–2.2
Al_2O_3 (cation%)	0.005–0.03	1.7–2.1	0.2–0.4
CaO (cation%)	0.01–0.04	1.1–1.3	1.2–2.1
Sc	0.25	0.01	2–3
Ti	—	0.01	0.25–0.35
Cr	0.2	0.01	3–8
FeO* (cation%)	0.8–1.2	0.01–0.06	0.4–0.6
Co	4	0.1	0.8–1.2
Ni	4–30	0.01	1–3
Sr	0.016	1.2–3	0.1–0.2
Y	0.02	0.06	0.2–0.6
Zr	0.01	0.05	0.15–0.30
La	0.0001	0.17	0.02–0.06
Ce	0.0002	0.15	0.04–0.10
Nd	0.0004	0.12	0.08–0.35
Sm	0.0007	0.09	0.14–0.45
Eu	0.001	0.38	0.16–0.48
Gd	0.003	0.06	0.19–0.59
Lu	0.022	0.04	0.19–0.53
FeO*/MgO (cation%)	0.26–0.30	—	0.21–0.25

Abbreviations: ol, olivine; plag, plagioclase; cpx, clinopyroxene; and liq, liquid

Principal sources of data: Agee and Walker (1988); Arndt (1977); Frey *et al.* (1978); Fujimaki *et al.* (1984); Grove and Bryan (1983); Hart and Davis (1978); Irving and Frey (1984); McKay (1986)

equilibrium with these liquids. As a result of numerous studies of the partitioning of elements between basaltic liquids and near liquidus minerals, the equilibrium distribution coefficients listed in Table 6.1 are reasonably well known.

The two major uses for these equilibrium distribution coefficients are for crystal–liquid matching tests and to develop computer models for the calculation of liquid lines of descent and more complex magmatic processes. The crystal–liquid matching tests involve dividing the concentration of an element in the crystal of interest by the concentration of the same element in the coexisting glass; if this value is the same as the equilibrium distribution coefficient, it is likely that this crystal is in equilibrium with the surrounding glass. Many crystals in MORBs do not meet this criterion and are interpreted as xenocrysts or as unreacted remnants of evolved and primitive magmas that mixed to form the present sample in which these crystals occur (Watson, 1976; Rhodes *et al.*, 1979; Stakes *et al.*, 1984).

Recent developments in the computer modelling of crystallization processes enable the quantitative modelling of liquid lines of descent using either thermochemical data (Ghiorso, 1985) or empirical distribution coefficients (Nielsen and Dungan, 1983). This approach has several strengths. One is that these computer programs are capable of modelling processes that cannot be easily carried out experimentally. Perfect fractional crystallization and assimilation events are two such processes. Another strength is that these calculations are a relatively quick method for evaluating the effects of variations in the system (e.g. variations in magma composition or oxygen fugacity), which might take weeks or months to fully evaluate experimentally. The validity of these programs ultimately rests on the quality of the experimental data on which they are based, but their accuracy is likely to improve with an increase in the amount and quality of relevant experimental data.

One such application of the programs is to calculate a perfect fractional crystallization liquid line of descent (Figure 6.4). The cation normative An/(An + Ab + Or) versus $Mg/(Mg + Fe^{2+})$ variations for oceanic basalts and gabbroic rocks are shown. The fields for basaltic glasses and minerals from gabbroic rocks from the Mid-Cayman Rise spreading centre (Thompson *et al.*, 1980; Elthon, 1987) are shown. The crystallization of olivine, plagioclase and clinopyroxene from basaltic glasses near the primitive end of the basaltic glasses (labelled P in Figure 6.4) will produce evolved basaltic liquids near the evolved end (labelled E), while producing gabbroic cumulates that lie within the region shown. The same feature is found for the 26°N samples, in which the basalt data are from O'Donnell and Presnall (1980) and the gabbroic data are from Tiezzi and Scott (1980). Also shown are representative trajectories for perfect fractional crystallization liquid lines of descent and cumulate mineral compositions calculated using the program of Ghiorso (1985). The agreement between the slopes inferred for gabbroic rocks and

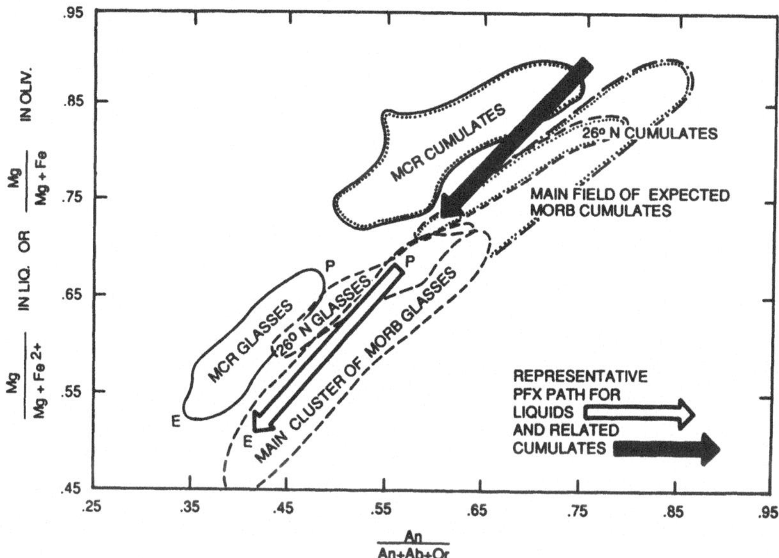

Figure 6.4 The Mg/(Mg + Fe) of olivine versus the An/(An + Ab) of plagioclase from cumulate rocks and the Mg/(Mg + Fe^{2+}) versus An/(An + Ab + Or) (cation) of related basaltic liquids. Fields are from the Mid-Cayman Rise (MCR) Thompson *et al.*, 1980; Elthon, 1987), 26°N on the Mid-Atlantic Ridge (O'Donnell and Presnall, 1980; Tiezzi and Scott, 1980), and MORB glasses (various sources). Also shown, with the large arrows, are representative perfect fractional crystallization (PFX) paths for both the liquids and cumulates (Ghiorso, 1985). For each suite of samples (MCR, 26°N, or MORBs) the trends defined by the cumulate minerals and the basaltic glasses are approximately colinear and it appears that the crystallization of the cumulate minerals will produce the trend of basaltic glasses from the primitive end (P) to the evolved end (E). Fe^{2+} in the basaltic liquid is calculated at the QFM buffer by the method of Kilinc *et al.* (1983).

their spatially related basalts with the slopes calculated for liquid lines of descent and crystal extract paths indicates the utility of these programs in developing a more complete understanding of crystallization processes in mid-ocean ridge magmatic systems.

6.2.3 *Poorly known aspects of low pressure crystallization*

In spite of a general understanding of the low pressure crystallization of MORBs outlined above, there are several aspects of this subject that are poorly known. The most important of these seem to be those described briefly below.

Plagioclase plays an enigmatic role in the differentiation of MORBs. It has approximately the same density as MORB liquids and potentially has a more complex fluid dynamic behaviour during the crystallization of MORB than do olivines or pyroxenes, as suggested by the complex petrographic characteristics of plagioclase crystals (often resorbed or with several apparent intervals of growth). The resorption of plagioclase can have a substantial effect on the compositions of MORBs and, consequently, on liquid lines of descent (Elthon, 1984).

MORBs often contain anomalously calcic plagioclases ($An_{>85}$) and unusually magnesian clinopyroxenes ($Mg/[Mg + Fe] > 0.85$). The anomalously calcic plagioclases in both basalts (Bryan *et al.*, 1976; Hekinian *et al.*, 1976; Flower *et al.*, 1977; Wood *et al.*, 1979b; Stakes *et al.*, 1984) and in abyssal peridotites (Hamlyn and Bonatti, 1980; Tiezzi and Scott, 1980) have been interpreted to suggest that there are basaltic magmas with anomalously high $CaO/Na_2O (>10)$, substantially higher than commonly found in MORBs (<7.0) (Fisk, 1984).

Anomalously magnesian clinopyroxenes occur as megacrysts in some MORBs (Sato *et al.*, 1978; Sinton and Byerly, 1980; Donaldson and Brown, 1977; Flower *et al.*, 1977; Dickey *et al.*, 1977; Wood *et al.*, 1979) and in some oceanic cumulates (Elthon, 1987). These clinopyroxenes have been generally interpreted to have been produced by the moderate to high pressure crystallization of MORBs (Bender *et al.*, 1978; Bence *et al.*, 1979; Elthon *et al.*, 1982).

The compositions of Cr–Al spinels are probably sensitive indicators of processes involved in the early crystallization stages of MORBs, but the only systematic experimental study is that of Fisk and Bence (1980). It appears on the basis of this and other studies (Hill and Roeder, 1974; Nielsen and Dungan, 1983) that liquid composition, pressure and oxygen fugacity influence the compositions of spinels, but the individual effects of these parameters have not been established experimentally for MORBs.

6.2.4 *Extreme differentiation of MORBs at low pressure*

The extreme differentiation (defined here to describe magmatic processes that produce liquids which project to the right-hand side of reaction point R in Figure 6.3B) of MORBs occurs locally in the oceanic crust to produce ferrobasalts, andesites and rhyodacites (Byerly *et al.*, 1976; Christie and Sinton, 1981; Fornari *et al.*, 1983). Petrological studies of these samples indicate that the minerals which crystallize are plagioclase, clinopyroxene, pigeonite, titanomagnetite, ilmenite and apatite (Perfit and Fornari, 1983).

Experimental investigations of the extreme differentiation of MORBs have been reported by Dixon and Rutherford (1979) and Spulber and Rutherford (1983). Dixon and Rutherford (1979) determined the liquid line of descent for a basaltic liquid as it evolved to a ferrobasalt (about 25% FeO*) and then developed liquid immiscibility. The immiscible liquids consisted of low silica (40–46% SiO_2) and high silica (66–70% SiO_2) conjugate liquids (Table 6.2). The high silica liquids are broadly similar to silicic rocks, often called plagiogranites (Coleman and Peterman, 1975), recovered from the oceanic basins and ophiolites (Table 6.2). Dixon and Rutherford (1979),

Table 6.2 Composition of oceanic plagiogranites, extreme differentiates and immiscible liquids

	Sample							
Component	1	2	3	4	5	6	7	8
SiO_2	76.37	78.39	69.67	64.28	70.2	67.6	43.9	68.5
TiO_2	0.42	0.09	0.60	1.22	0.21	0.75	4.61	1.70
Al_2O_3	12.78	12.68	12.30	12.26	12.4	14.2	7.0	11.1
FeO*	0.81	0.75	5.19	11.05	3.62	4.43	23.45	7.86
MnO	0.02	0.01	—	0.23	0.05	0.06	0.55	0.16
MgO	0.87	0.54	0.36	1.04	0.54	0.84	2.32	0.75
CaO	0.84	0.55	2.82	5.14	2.23	3.21	10.18	0.82
Na_2O	7.70	6.66	4.29	4.13	4.15	3.41	1.85	2.86
K_2O	0.07	0.06	1.48	0.58	2.30	0.41	0.42	1.20
P_2O_5	0.02	0.01	0.07	0.21	0.34	0.43	4.87	0.96
Total	99.90	99.74	96.78	100.14	96.04	95.34	99.15	98.91

(1) Aplite dredged from the Argo Fracture Zone, Indian Ocean (Engel and Fisher, 1975)
(2) Aplite dredged from the Mid-Atlantic Ridge (Miyashiro *et al.*, 1970)
(3) Average rhyodacite glass dredged from the Galapagos Spreading Centre (Byerly *et al.*, 1976; Dixon and Rutherford 1979)
(4) Andesite glass from the Galapagos Rift (Perfit *et al.*, 1983)
(5) Glass produced experimentally by the extreme differentiation of Hawaiian basalt (Spulber and Rutherford, 1983)
(6) Glass produced experimentally by the extreme differentiation of MORB (Spulber and Rutherford, 1983)
(7) Average experimental Fe-enriched immiscible conjugate liquid (Dixon and Rutherford, 1979)
(8) Average experimental SiO_2-enriched immiscible conjugate liquid (Dixon and Rutherford, 1979)

therefore, suggested that silicate liquid immiscibility may occur during the extreme differentiation of MORB-type liquids and it is possible that immiscibility plays a role in the formation of some highly silicic rocks within the oceanic crust.

Spulber and Rutherford (1983) studied the crystallization history of a MORB from the Galapagos Spreading Centre and an evolved tholeiitic basalt from Hawaii at 1–3 kbar under hydrothermal conditions. Results from this study indicate that the extreme differentiation of oceanic basalts will lead to andesitic to rhyolitic residual liquid compositions (Table 6.2) similar to glasses reported by Byerly *et al.* (1976) and Fornari *et al.* (1983). This extreme differentiation was the result of the crystallization of olivine, plagioclase and clinopyroxene in the early stages and plagioclase, clinopyroxene, low Ca pyroxene, ilmenite and magnetite from more evolved liquids (Spulber and Rutherford 1983). If the experimental results are viewed from the low temperature direction, it is also possible to produce these silicic magmas by the partial melting of hydrated basalts (amphibolites) within the crust (Helz, 1973).

The experimental studies, therefore, suggest that the three principal mechanisms for producing silica-rich magmas in the oceanic crust are extreme crystal fractionation, liquid immiscibility and the partial melting of amphibolites. These three mechanisms have been individually proposed for various suites of plagiogranites from ophiolites (Saunders *et al.*, 1979; Pedersen and Malpas, 1984). With only a few suites of samples from the oceanic basins studied in any detail, it is not possible to comment on the most common mechanism in the oceanic basins.

6.3 Experimental studies at high pressure

At the time of melt separation from the residual mantle, the primary magma will presumably be in equilibrium with the residual minerals. Petrological studies of abyssal peridotites (Hamlyn and Bonatti, 1980; Dick and Fisher, 1983; Michael and Bonatti, 1985; see also Chapter 15) have shown that olivine, orthopyroxene, an aluminous phase (plagioclase, spinel, or garnet) and (often) clinopyroxene are residual phases that remain after partial melting of the sub-oceanic mantle. Olivine and orthopyroxene, particularly, should be on the liquidus of a primary magma because these minerals persist in the residual mantle until large increments (about 40%) of melting (Mysen and Kushiro, 1977). The experimental approach to determining the nature of primary MORBs, therefore, involves determining the compositions of liquids in equilibrium with olivine + orthopyroxene ± plagioclase/spinel/garnet ± clinopyroxene. If a specific MORB has a composition very similar to a liquid in equilibrium with this mantle assemblage, then it is a strong candidate for a primary magma.

Current discussions of the petrogenesis of primary MORBs almost always focus on the most magnesian primitive MORBs because they are believed to have undergone the smallest amount of crystallization since separation from the mantle. There is not a generally accepted definition of what constitutes a primitive magma, but those basaltic glasses with > 9.5% MgO will be termed 'primitive glasses' here. A tabulation of primitive glasses from the oceanic basins is given in Elthon (1990). This discussion is restricted to MORB glasses and does not include any whole-rock data because of the problems of phenocryst accumulation and alteration that are inherent in many whole-rock analyses.

There are systematic chemical variations in these primitive MORB glasses in which a continuum of compositions is found between two end-members. These two end-member compositions and some primitive MORB glasses are listed in Table 6.3. In light of these substantial chemical variations in primitive MORB glasses, it is likely that substantial chemical variations also occur in primary magmas.

Two major schools of thought on the nature of primary MORBs have developed based on high pressure experimental studies. One school suggests that most MORBs are derived from primary magmas separated from the mantle at about 10 kbar; this group cites the multiple saturation of some

Table 6.3 Compositions of primitive MORB glasses and end-members

Component	End-member glass compositions		Representative primitive MORB glasses				
	H	L	1	2	3	4	5
SiO_2	47.0	50.5	47.83	49.15	49.5	50.3	50.73
TiO_2	1.1	0.5	1.15	0.96	0.66	0.73	0.73
Al_2O_3	~17	~17	16.25	17.58	16.51	16.6	16.88
FeO*	10.0	7.5	9.45	8.01	8.76	7.99	8.22
MnO	0.2	0.2	0.16	—	0.17	0.12	—
MgO	~9.8	~9.8	9.56	9.58	9.86	10.20	9.84
CaO	11.0	13.0	11.80	12.14	12.68	13.20	13.33
Na_2O	2.7	1.6	2.79	2.38	2.08	2.00	1.68
K_2O	0.05	0.05	0.05	0.04	0.04	0.01	0.04
Total			99.04	99.84	100.26	101.15	100.45

Sources of data: (H) Elthon (1989); (L) Elthon (1989); (1) Eaby *et al.* (1984); (2) Natland and Melson (1979); (3) Bryan and Moore (1977); (4) Frey *et al.* (1974); and (5) Melson *et al.* (1979)

MORBs with olivine + orthopyroxene + clinopyroxene + plagioclase or spinel at 7–12 kbar as strong evidence (Fujii and Kushiro, 1977; Presnall *et al.*, 1979; Fujii and Bougault, 1983; Presnall and Hoover, 1987). The other school suggests that most MORBs are instead derived from primary magmas that separated from the mantle at about 20–25 kbar; they cite the absence of orthopyroxene as a liquidus phase at high pressures in other primitive MORBs and note the chemical differences between many primitive MORB glasses and experimentally produced glasses at 10 kbar (O'Hara, 1968; Green *et al.*, 1979; Stolper, 1980; Elthon and Scarfe, 1984). In the following section, high pressure experiments are evaluated in the context of these two competing models and it is concluded that both models have strengths that need to be incorporated in any future comprehensive model.

6.3.1 *Experimental techniques*

Most high pressure experimental studies of MORBs and related compositions have been conducted in a piston-cylinder apparatus similar to that shown in Figure 6.5. The sample is loaded into a capsule that is placed (along with spacers to maintain the correct geometry during compression) inside a graphite heater. The temperature is controlled by an electrical current applied through the graphite heater, whereas pressure is controlled by the force applied to the two opposing pistons (Boyd and England, 1960). The loss of Fe to Pt in contact with the sample can be a significant problem in high pressure experiments (Stern and Wyllie, 1975). Most experiments within the last decade have been performed in graphite capsules inserted within sealed Pt tubes (Figure 6.6). This configuration keeps the basalt from losing Fe

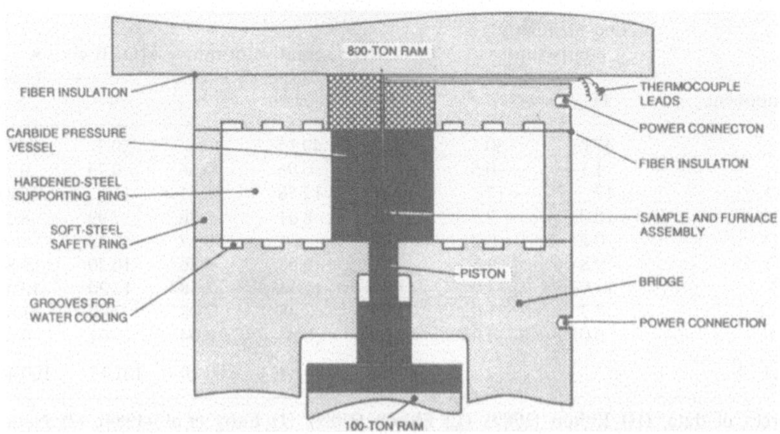

Figure 6.5 Cross-section of high pressure piston cylinder press, modified slightly from Boyd and England (1960). The sample and furnace assembly is enclosed within the pressure vessel and is pressurized by movement of the lower (100 ton) ram.

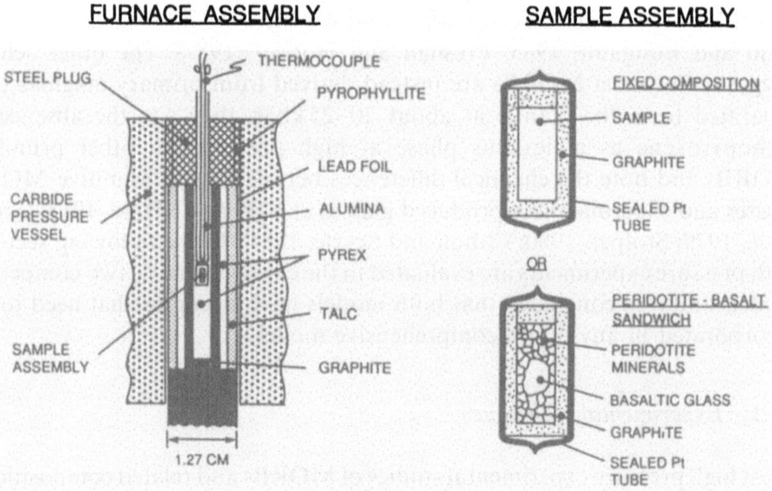

FURNACE ASSEMBLY
STEEL PLUG
THERMOCOUPLE
PYROPHYLLITE
LEAD FOIL
ALUMINA
CARBIDE PRESSURE VESSEL
PYREX
TALC
SAMPLE ASSEMBLY
GRAPHITE
1.27 CM

SAMPLE ASSEMBLY
FIXED COMPOSITION
SAMPLE
GRAPHITE
SEALED Pt TUBE
OR
PERIDOTITE - BASALT SANDWICH
PERIDOTITE MINERALS
BASALTIC GLASS
GRAPHITE
SEALED Pt TUBE

Figure 6.6 Cross-section of a common type (talc–Pyrex) of furnace assembly and two sample assemblies used in high pressure piston-cylinder experimental studies. See text for discussion.

during the experiment unless the basaltic liquid escapes the containment of the graphite capsule and comes into contact with the Pt tube. It is possible to control the oxygen fugacity in piston-cylinder experimental studies using the double-capsule technique (Huebner, 1971). The use of the graphite capsules inside sealed Pt tubes, however, buffers the oxygen fugacity to reducing conditions slightly above the iron–wustite buffer (Thompson and

Kushiro, 1972; Ulmer and Luth, 1988). This oxygen fugacity is similar to that inferred for MORB glasses (Christie *et al.*, 1986).

High pressure experiments are usually conducted for 2–100 h before quenching. Quenching is generally not as rapid in a piston-cylinder apparatus as with 1 atm furnaces and it is common that fibrous quench crystals will grow from the liquid in high pressure experiments; these quench crystals can substantially affect the composition of glasses in these experiments if they are abundant. The experimental run products are generally mounted in epoxy and prepared for electron probe microanalysis of the phases. The publication of analyses of all phases would be advantageous, but this is seldom the case in high pressure studies.

6.3.2 *Results from experimental studies at high pressure*

There are three major types of high pressure experimental studies that have been undertaken to determine the compositions of liquids in equilibrium with mantle assemblages at various depths (pressures). These are: the partial melting of mantle peridotites; basalt–peridotite sandwich experiments; and the determination of the near-liquidus phase equilibria of basalts as a function of pressure.

6.3.2.1 *Partial melting of mantle peridotites.* The experimental melting of peridotite compositions has generally involved the study of the near-solidus conditions where there is a small to moderate amount (< 40%) of liquid present (Mysen and Kushiro, 1977; Jaques and Green, 1979, 1980; Sen, 1982; Takahashi and Kushiro, 1983). Although each of these studies has been a substantial contribution to the understanding of how the mantle melts, their direct application to the origin of primary MORBs is complicated by the experimental difficulties inherent in determining liquid compositions in these types of studies. Both the Mysen and Kushiro (1977) and Jaques and Green (1980) studies had substantial Fe loss from the samples because of their use of Pt-tubes as sample containers, without the graphite inserts generally used today. Jaques and Green (1980), however, estimated the compositions of liquid produced during melting based on the compositions and abundances of the remaining minerals. Sen (1982) partially melted a depleted Hawaiian lherzolite at 9 kbar in graphite containers, which eliminated the Fe loss problems that previous studies had encountered. As quench overgrowth was not a substantial problem in his study, it was possible to directly analyse the experimental glasses, which are similar to some of the silica-rich MORBs (50.6–51.7% SiO_2), except that they have low TiO_2 (0.2–0.4%) and Na_2O (0.2–0.5%).

Takahashi and Kushiro (1983) studied the partial melting of a fertile lherzolite from Hawaii using graphite capsules. They were not able to directly analyse the compositions of glasses because of quench overgrowth problems in small pockets of glass. Instead they used the basalt–peridotite sandwich technique (described in section 6.3.2.2) to estimate liquid compositions in equilibrium with the mantle. A problem with the direct application of the Takahashi and Kushiro (1983) study to the origin of primary MORBs is that most of their liquid compositions are substantially different from primitive MORBs, especially with respect to TiO_2, alkalis and P_2O_5. Although the effects of each of these components on high pressure phase equilibria are not well known, the alkali elements clearly have a major effect (Presnall and Hoover, 1987).

6.3.2.2 *Basalt-peridotite sandwich experiments.* Owing to the technical difficulties involved in determining the compositions of liquid produced by melting of mantle peridotites (e.g. quench crystal overgrowth and the analysis of very small pools of glass), several recent studies have used the basalt–peridotite sandwich technique, which was developed by Stolper (1980) based on earlier studies by Watson (1979) and Walker *et al.* (1979). The basalt–peridotite sandwich is a basalt sample that is surrounded by peridotite minerals (Figure 6.6); the basalt interacts with the surrounding minerals to produce a modified basaltic composition that is, at least in principle, in equilibrium with minerals. The advantages of this technique are that the basalt tends to be concentrated in a large pool that can be easily analysed by electron microprobe, and that quench crystal overgrowths are not a substantial problem (Stolper, 1980).

Rather than simplifying the debate, however, these experimental studies have further complicated it by producing results that contradict each other. Takahashi and Kushiro (1983) and Fujii and Scarfe (1985) have shown that liquid in apparent equilibrium with mantle mineral assemblages at 10 k bar are very similar to some of the most primitive MORB glasses. These studies, in combination with the studies of Fujii and Kushiro (1977) and Fujii and Bougault (1983) which demonstrated that some primitive MORBs are in equilibrium with a lherzolite assemblage at about 10 kbar, provide the strongest evidence that some primitive MORBs are strong candidates for primary magmas.

Stolper (1980) and Falloon and Green (1987), however, have reached the opposite conclusion from similar studies. These workers conclude that there are significant chemical differences between liquids in equilibrium with peridotite minerals at 10 kbar and primitive MORBs. In particular, Falloon and Green (1987) suggest that the Fujii and Scarfe (1985) experiments had significant Fe-loss problems and that the liquids produced by Takahashi and Kushiro (1983) at 10 kbar have much higher FeO, Na_2O and TiO_2 contents and lower CaO contents than primitive MORB glasses.

As seen from the strongly conflicting interpretations of the data, these basalt–peridotite sandwich experiments have not resolved the debate on the origin of primary MORBs. One of the problems is that the composition of the liquid produced in an experiment depends on the compositions and proportions of the peridotite minerals and the basaltic glasses used as starting materials (Elthon, 1990). Another major problem is whether the liquid changes composition to be in equilibrium with the mantle minerals (the intended result) or whether the compositions of the minerals change to be in equilibrium with the basaltic liquid. All phases will change composition in these equilibration experiments, but the amount of change that occurs for a specific phase depends on its abundance and composition, diffusion rates and reaction rates. In those experimental studies for which data are available (e.g. Takahashi and Kushiro, 1983), it is clear that the orthopyroxene rims have changed to be in equilibrium with the basaltic liquid whereas cores of a very different composition are preserved. In this instance, the experiments do not determine what they were designed to test, i.e. the compositions of liquids in equilibrium with mantle mineral compositions. As a result of these problems associated with the compositional effects of the starting materials, the basalt–peridotite sandwich experiments most relevant to the origin of MORBs will have a primitive MORB glass as a starting material and will have mineral compositions, both before and after equilibration, comparable to those in abyssal peridotites. These studies have not yet been reported.

6.3.2.3 *Near-liquidus phase equilibria of basalts.* Numerous studies have shown that the near-liquidus minerals for MORBs and other basalts change with pressure (O'Hara, 1968, BVSP, 1981). For primitive MORBs the 1 atm crystallization sequence is generally olivine + spinel, followed by olivine + plagioclase, then olivine + plagioclase + clinopyroxene, and finally plagioclase + clinopyroxene + low Ca pyroxene (pigeonite). With increasing pressure, olivine is replaced by pyroxene as the liquidus phase, resulting in the multiple saturation of the basalt at some higher pressure.

For MORBs, there are two types of multiple saturation found at 7–12 kbar. In some instances, MORBs are multiply saturated with olivine + clinopyroxene + orthopyroxene ± plagioclase ± spinel (Kushiro and Thompson, 1972; Fujii and Kushiro, 1977; Fujii and Bougault, 1983). Other studies of MORBs and high MgO basalts have not had orthopyroxene as a near-liquidus phase in this pressure interval (Bender *et al.*, 1978; Green *et al.*, 1979; Maaloe and Jakobsson, 1980; Elthon and Scarfe, 1984).

One method for extending the high pressure experimental results on the few samples noted above to the origin of primitive MORB glasses as a group is with pseudo-liquidus phase diagrams such as that shown in Figure 6.7A. The compositions of basalts studied in high pressure experiments are projected in this figure, with the general field for liquids in equilibrium with orthopyroxene at about 10 kbar shown in the shaded region to the lower right of the 10 kbar multiple saturation boundaries. In this field are those MORBs that are saturated with orthopyroxene at about 10 kbar and the liquids produced by the melting of mantle samples at about 10 kbar. The

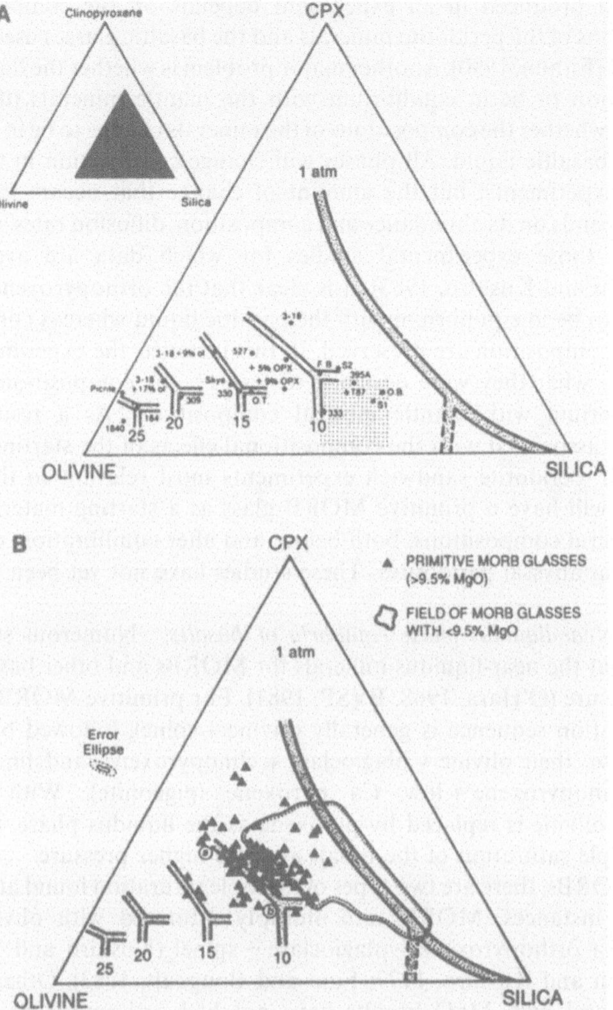

Figure 6.7 Projection of high pressure phase equilibria results onto a portion of the clinopyroxene–olivine–silica plane (see Figure 6.3), with the high pressure boundaries at 10, 15, 20 and 25 kbar (from Elthon, 1989). The 1 atm phase equilibria boundary is from Figure 6.3. Note that liquids in equilibrium with a low calcium pyroxene at about 10 kbar plot in the shaded field to the right of the 10 kbar boundaries. See text for discussion.

approximate locations of multiple saturation boundaries are shown at higher pressures; it is likely that they will show a larger range of compositions than those shown here, but there are insufficient experimental data with which to evaluate the full range of compositions for multiply saturated liquids at pressures >10 kbar.

The compositions of primitive MORB glasses are projected onto this surface in Figure 6.7B. The primitive MORB glasses project over a considerable range of space, but most are scattered along the 10 kbar olivine + clinopyroxene + liquid multiple saturation boundary. Those samples which project near the right-hand end of the basalt cluster (near b in Figure 6.7B) are similar to MORBs that are saturated with orthopyroxene at about 10 kbar and are strong candidates for primary magmas generated at 10 kbar. Those MORB glasses which project near the left-hand end of the basalt cluster (near a in Figure 6.7B) do not appear to be close to saturation with orthopyroxene at 10 kbar or any other pressure. The most likely interpretation for the origin of these MORB glasses is that they are derived by 15–20% crystallization of spinel-bearing dunite or wehrlite from primary picritic magmas produced at 20–25 kbar (O'Hara, 1968; Green *et al.*, 1979; Stolper, 1980; Elthon and Scarfe, 1984; Falloon and Green, 1987).

The primitive MORB glasses which project near the 10 kbar orthopyroxene saturation field in Figure 6.7B are those which are broadly similar in composition to the L end-member in Table 6.3. They generally have lower Na_2O (1.6–2.1%) and incompatible element abundances than the H end-member basalts (2.2–2.8% Na_2O) that lie at the left-hand end of the basalt cluster (in the vicinity of 'a' in Figure 6.7B). Comparisons with Na_2O and incompatible element abundances of possible primary liquids would suggest that melting at 10 kbar to produce L end-member basalts involved a larger extent of melting (about 15–20%) than melting at 20–25 kbar to produce the primary magmas that differentiated to form the H end-member (3–10% melting) (Elthon, 1990).

It is clear from Figure 6.7B that very low pressure (< 5 kbar) melting of the mantle or resorption of orthopyroxene is not a major process in MORB petrogenesis because none of the primitive MORB glasses plot substantially to the right of the 10 kbar multiple saturation boundary. Jaques and Green (1980), for example, have shown that basaltic liquids produced by melting of pyrolite at 5 and 2 kbar have 52.1–53.3% SiO_2, which is higher than primitive MORBs (46–51%). Further, Fisk (1986) has studied this problem experimentally and found that this low pressure interaction produces basalts with substantially higher SiO_2 contents (55–60 wt%) than primitive MORBs.

These phase equilibria studies also provide information on the compositions of crystals in equilibrium with basaltic liquids at various pressures. One of the potentially most important minerals in helping to resolve the question of the pressure of melting for MORBs is orthopyroxene. The covariation of Al_2O_3 and Mg/Mg + Fe in orthopyroxene in abyssal peridotites is shown in Figure 6.8. The Al_2O_3 contents of abyssal peridotite orthopyroxenes range from 3 to 6 wt% (Figure 6.8). The variations in orthopyroxene compositions which result from partial melting are a decrease in Al_2O_3 as Mg/Mg + Fe increases with a larger extent of partial melting (Green *et al.*, 1979: inset to Figure 6.9). For a given extent of partial melting, the Al_2O_3 content of orthopyroxene increases with the pressure of melting (inset to Figure 6.9).

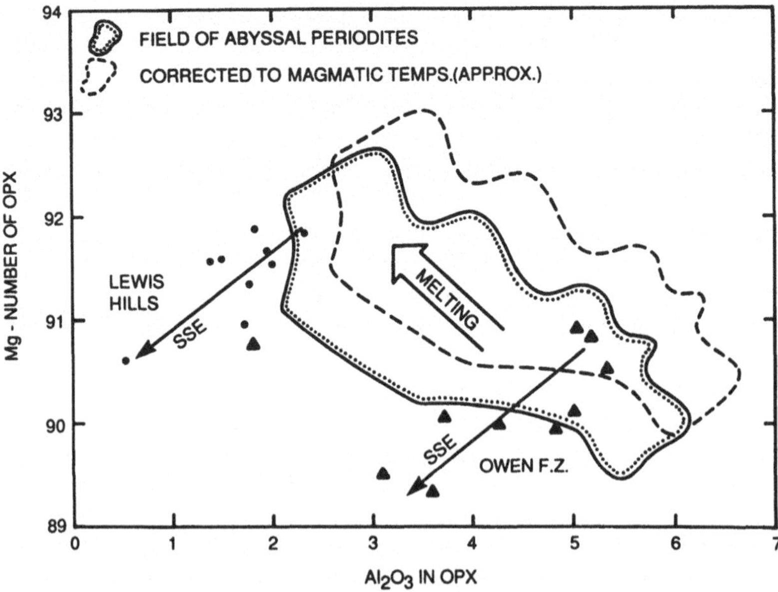

Figure 6.8 Mg number (100 Mg/(Mg + Fe)) versus Al_2O_3 content of orthopyroxene (OPX) from abyssal peridotites (data from Prinz *et al.*, 1976; Nicholls *et al.*, 1981; Dick and Fisher, 1983; Michael and Bonatti, 1985). The large arrow in the centre of the field indicates the trajectory for increasing partial melting. The arrows for the two data sets correspond to estimated sub-solidus equilibration (SSE) trends. The field of abyssal peridotites is shifted slightly to indicate the compositions before SSE. Note that this shift corresponds to an average abyssal peridotite but that individual samples will move different distances along different trajectories; the shift is, therefore, only approximate.

Analyses of liquidus orthopyroxenes (or pigeonites) crystallized from basalts at 7 to 12 kbar have been reported by Kushiro and Thompson (1972), Green *et al.* (1979), and Fujii and Bougault (1983). These pyroxenes contain 1.9, 2.2–2.4, and 4.1% Al_2O_3, respectively. Orthopyroxenes in melting studies of peridotites at about 10 kbar contain 3.0–3.9% Al_2O_3 (Jaques and Green, 1979, 1980; Takahashi and Kushiro (1983). The present data from melting of both MORBs and peridotites suggest that the Al_2O_3 content of orthopyroxenes at about 10 kbar is 2–4 wt%, which is too low to match all but those orthopyroxenes from the most depleted abyssal peridotites (those with the highest Mg/Mg + Fe).

The higher Al_2O_3 contents of many abyssal peridotite orthopyroxenes suggest that melting at higher pressures (20–25 kbar) is required (Elthon, 1989, 1990). The 25 kbar orthopyroxenes in experimental studies of MORB-type compositions have 6.7% (Elthon and Scarfe, 1984) and 5.0% Al_2O_3 (Takahashi and Kushiro, 1983). The 20 kbar orthopyroxenes have 4.6% (Elthon and Scarfe, 1984), 5.5% (Green *et al.*, 1979) and 4.7% Al_2O_3 (Takahashi and Kushiro, 1983). The liquidus orthopyroxenes of high MgO basalts at

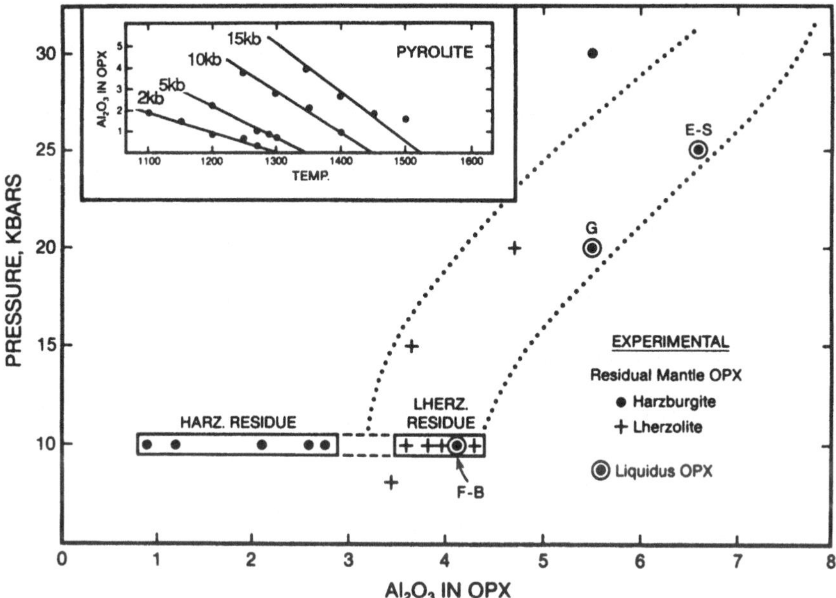

Figure 6.9 The Al_2O_3 contents of orthopyroxene (OPX) versus pressure of equilibration for the experimental melting of peridotites (Jaques and Green, 1980; Takahashi and Kushiro, 1983). Also shown are the compositions of liquidus (or near-liquidus) opxs from high pressure experimental studies of MORB-type liquids (F–B, Fujii and Bougault, 1983; G, Green *et al.*, 1979; E–S, Elthon and Scarfe, 1984). The inset (from Jaques and Green, 1980) shows how Al_2O_3 in opx varies as a function of pressure with increasing melting (temperature).

20–25 kbar match or closely approximate the compositions of the most Al_2O_3-rich orthopyroxenes (5.0–6.5% Al_2O_3) in abyssal peridotites. These results suggest that the most Al_2O_3-rich orthopyroxenes in abyssal peridotites are the result of high pressure (20–25 kbar) melting.

The results from experimental studies of MORBs, the compositions of primitive MORB glasses and the orthopyroxenes in abyssal peridotites suggest that melt separation in the sub-oceanic mantle occurs over a variety of pressures, from about 25–8 kbar. Various investigations have suggested that the melt separation occurred over a relatively small interval; some have advocated melt separation over a small pressure interval at about 7–12 kbar, whereas others have advocated only a high pressure (20–25 kbar) origin. It is apparent that evidence for these restricted pressure intervals is contradictory.

6.3.3 *Poorly known aspects of high pressure equilibria relevant to MORBs*

It may seem surprising for a field that is so important, but there are few high quality experimental studies that are directly relevant to the origin of primary MORB. Much of the debate, consequently, focuses on how to extend these limited results to the interpretation of primary MORB petrogenesis as a whole. It is obvious that more high quality experimental studies are needed. These experiments are needed for both primitive MORB glass compositions and relevant mantle compositions.

High pressure liquid lines of descent for magmas undergoing crystallization in the mantle are poorly established. The substantial gap between the pressure of melt separation from the mantle (10–25 kbar) and the pressure of crustal magma chambers (1–3 kbar) suggests that the opportunity for crystallization at high pressures exists. It has been suggested that the presence of anomalously magnesian clinopyroxene crystals in some MORBs is the result of the crystallization of MORB at high pressures (Bender *et al.*, 1978; Bence *et al.*, 1979; Elthon *et al.*, 1982). It is only possible, at present, to determine liquid lines of descent for high pressure crystallization in the most qualitative manner. These high pressure phase relationships need to be well known in order to better address the roles of high pressure crystallization and the interaction of ascending magmas with the mantle.

6.4 Concluding statements

1. Primitive MORB glasses ($>9.5\%$ MgO) have a substantial range of compositions within which two major end-members are defined. One has high FeO*, Na_2O, TiO_2 and Na_2O/CaO with low SiO_2 and CaO. This is the H end-member. The other has low FeO*, Na_2O, TiO_2 and Na_2O/CaO with high SiO_2 and CaO. This is the L end-member. This substantial range of compositions suggests that primary MORBs also have a substantial range of compositions.
2. The L primitive MORB glasses plot near the 10 kbar multiple saturation region for liquids in equilibrium with a lherzolite. The most likely interpretation of these glasses is that they are close in composition to primary magmas separated from the mantle at about 10 kbar. These primary magmas have probably been produced by relatively large increments of melting (15–20%).
3. The H primitive MORB glasses plot far from the 10 kbar multiple saturation region for liquids in equilibrium with a lherzolite. Their compositions lie on olivine + spinel controlled or olivine + clinopyroxene + spinel controlled liquid lines of descent from primary high MgO basaltic magmas separated from the mantle at 20–25 kbar. These primary magmas were produced by relatively small degrees of partial melting (about 3–10%). The H primitive MORB glasses are not close to primary magma compositions but have undergone substantial crystallization prior to eruption.

4. The compositions of liquidus minerals for primary magmas must match the compositions of minerals in the residual mantle at the time of magma separation. The mineral that offers the best opportunity for constraining the pressure of melt separation in the sub-oceanic mantle is orthopyroxene, which has 2.5–6.5% Al_2O_3 in abyssal peridotites. The liquidus orthopyroxenes in studies of MORBs and the melting of peridotites at 10kbar have about 2–4% Al_2O_3, which matches only a portion of the range found in abyssal peridotites. The high Al_2O_3 contents (5–6.5% Al_2O_3) in orthopyroxenes in abyssal peridotites suggest that melt segregation from the mantle occurs at 20–25 kbar in many localities.
5. The first-order characteristics of the low pressure (1 atm) crystallization of MORBs are well understood as a consequence of experimental and petrological studies of basalts themselves. The high pressure crystallization of MORBs is currently poorly constrained by experimental data.

7 Magmatic processes in oceanic ridge and intraplate settings

MARTIN FLOWER

7.1 Introduction

The global mid-ocean ridge (MOR) system is about 60 000 km in length and represents the single largest supply system of magma to the earth's surface. The oceanic lithosphere also contains islands and archipelagos representing the effects of isolated, but durable, mantle melting anomalies. In attempting to review the processes of magma genesis and evolution in these distinctive environments, this chapter draws on existing syntheses of magmatic processes in the oceanic domain (e.g. BVSP, 1981; Wilkinson, 1982; Hekinian, 1982; Thompson, 1987). Within the oceanic domain, 'magmatic processes' include the partial melting of mantle, the physical extraction of melts from multiphase solid assemblages, and the combined effects of crystallization, mixing and wallrock reaction during the uprise and emplacement of magma. Realistic physical models provide the means to utilize erupted magma as a chemical and thermal probe of the mantle and to better understand the phenomenon of planetary melting. This chapter examines the chemical and petrographic variation of magmas generated at oceanic ridge and intraplate settings in terms of recent experimental results, current knowledge about the physical character of active ridge axes, and real-time studies of active volcanoes.

Since it became evident in the 1960s that the oceanic crust was produced by profuse and continuing magmatism, research has addressed the following key questions. (1) Mantle dynamics and the fundamental causes of melting; is mantle plume activity the dominant factor, or does melting result from passive upwelling in response to lithospheric stretching? (2) The thermal and compositional character of primitive melts; are these of ultramafic picrite or magnesian tholeiite composition and do they represent simple equilibrium batch melts or integrated increments of a polybaric melt column? (3) The configuration of magmatic fractionation processes; are these dominated by closed or open system conditions and are they ubiquitous or localized?

As such questions were clarified in the 1970s and 1980s, models were developed which attempted to satisfy the constraints of phase equilibria,

Table 7.1 Examples of primitive MORB compositions

Composition	Basalt No.[a]									
	1	2	3	4	5	6	7	8	9	10
SiO_2	49.07	48.81	49.19	48.93	49.98	48.2	50.3	49.97	49.57	51.89
TiO_2	0.74	0.73	0.85	0.84	0.77	0.51	0.73	0.82	0.85	0.92
Al_2O_3	16.44	16.13	16.12	16.14	15.24	17.0	16.6	15.23	15.26	14.57
FeO^t	8.86	8.89	8.74	8.75	8.60	8.51	7.99	8.15	8.28	8.53
MnO	0.16	0.16	0.14	0.15	0.13	0.16	0.12	0.14	0.14	—
MgO	10.15	10.15	10.41	10.49	10.11	10.1	10.2	10.66	10.63	10.17
CaO	11.65	11.65	11.91	11.84	12.45	12.7	13.2	12.21	12.11	11.01
Na_2O	2.13	2.13	2.35	2.41	2.23	2.34	2.00	1.94	2.11	2.16
K_2O	0.07	0.07	0.09	0.09	0.07	0.0	0.01	0.16	0.19	0.25
P_2O_5	—	—	—	—	—	—	—	0.10	0.11	—
Cr_2O_3	0.03	0.04	0.08	0.09	—	—	—	0.09	0.09	—
Total	99.30	98.56	99.88	99.73	99.58	99.56	101.15	98.97	99.34	99.60
Mg No.	0.70	0.71	0.71	0.71	0.71	0.71	0.73	0.71	0.72	0.70
CaO/Al_2O_3	0.709	0.717	0.739	0.733	0.817	0.747	0.795	0.801	0.794	0.756
Al_2O_3/TiO_2	22.2	22.1	19.0	19.2	19.8	33.3	18.6	18.6	18.0	15.8
CaO/Na_2O	5.47	5.38	5.07	4.91	5.58	5.43	6.29	6.29	5.74	5.10

[a](1) 519-4-1 (Bryan and Moore, 1977); (2) 519-4-2 (Bryan and Moore, 1977); (3) 525-5-1 (Bryan and Moore, 1977); (4) 525-5-2 (Bryan and Moore, 1977); (5) 530-3-1 (Bryan and Moore, 1977); (6) ARP-74-14-31 (Bryan, 1979); (7) 3-18-7-1, 1 (Frey *et al.*, 1974); (8) ARP-74-10-16 (Fujii and Bougault, 1983); (9) CYP-31-35 (Fujii and Bougault, 1983); (10) Calculated for $T_p = 1280°C$ to produce 7 km oceanic crust (McKenzie and Bickle, 1988). Mg numbers calculated on the basis of $Fe^{2+}/(Fe^{2+} + Fe^{3+}) = 0.9$

chemical mass balances and (more recently) fluid dynamics. In particular, integrated studies of active mid-ocean ridge segments and individual intraplate volcanoes provide remarkably precise information about the longevity and configuration of magma supply systems that complements experimental and theoretical models.

7.2 Compositional diversity of oceanic magmas

There are distinctive differences in the compositional character of magmas processed through MOR and intraplate magma systems, the most obvious being the relative uniformity of mid-ocean ridge basalt (MORB) and the extensive differentiation of oceanic island basalt (OIB) parent liquids. These characteristics appear to reflect fundamental differences in the melting and fractionation regimes in these environments.

7.2.1 *Mid-ocean ridge basalt*

MORB eruptives consist largely of quartz (Qz + Hy)- or olivine (Ol + Hy)-normative tholeiite and, compared to intraplate OIB and continental magmas, are chemically and isotopically homogeneous. Published studies pertaining to ocean drilling programmes such as DSDP, IPOD and ODP, and numerous investigations of dredged basement material, yield a vast geochemical and isotopic database for MOR-generated magma (Table 7.1). Compared to OIB, MORB liquids are richer in CaO and Al_2O_3 and poorer in FeO*, TiO_2, K_2O and P_2O_5 for equivalent values of MgO, and are for the most part confined to MgO values of between about 11 and 5 wt% (Melson *et al.*, 1976). These fundamental characteristics were recognized by Nicholls *et al.* (1964), Nicholls (1965), Engel *et al.* (1965) and others who first established the global significance of MORB magma. The observation of chondrite-normalized depletions in light rare earth elements (LREE) and other incompatible elements in MORB (Gast, 1970) led to the postulate that the oceanic mantle itself had been depleted of elements such as LREE, Rb, U and Th. Isotopic studies corroborated this observation, showing MORB to be depleted in radiogenic Sr and Pb and enriched in radiogenic Nd with respect to single-stage growth from the primordial mantle (Gast, 1968; O'Nions *et al.*, 1978). Such depletions occurred as a result of the time-averaged extraction of melt or a single early differentiation event (O'Nions *et al.*, 1978).

Regional studies of spreading centres (Sigurdsson, 1981; Schilling *et al.*, 1983; Hamelin *et al.*, 1984; Klein and Langmuir, 1987) reveal systematic variations of major and trace elements and isotopic parameters, between normal (i.e. depleted) MORB and enriched magmas resembling those encountered in intraplate settings (see later). In the examples of Iceland, the Azores and the Galapagos archipelagos (Schilling *et al.*, 1983; Hamelin *et al.*, 1984), mantle upwelling is suggested by positive gravity anomalies, enhanced heat flow and thickening of the lithosphere due to excess melting (Kaula, 1973; Watts *et al.*, 1985). These correlate with the observed ridge-longitudinal chemical and isotopic changes from normal (N-) MORB to less saturated magmas enriched in incompatible elements, radiogenic Sr and Pb and normative diopside. Enriched (E-) MORB variants are more widespread than was previously thought and may reflect processes additional to mantle plumes. Several workers noted an association of E-MORB with transform fracture zones and other dislocations of the ridge system (Flower, 1981b; Langmuir and Bender, 1984; Bender *et al.*, 1984).

N-MORB subtypes are recognized from bimodal Ca, Al and incompatible element abundance distributions (Melson *et al.*, 1976; Dmitriev *et al.*, 1984; Viereck *et al.*, 1989). Geochemical studies of basalt from between 30 and 35°N on the Mid-Atlantic Ridge (MAR) show that N-MORB subtypes may occur within single MOR spreading segments (Viereck *et al.*, 1989). Subtle compositional differences are also discerned between fast and slow spreading centres. Nisbet and Pearce (1973) and Scheidegger and Corliss (1981) observed that MORB generated at the East Pacific Rise has higher TiO_2 and lower Al_2O_3 than that formed at the Mid-Atlantic or Mid-Indian Ocean Ridges. Flower (1980, 1981b) attributed this to processes occurring in the magma supply system (e.g. preferred conditions for plagioclase accumulation at the slow spreading axis) rather than to inherent differences in the primitive melts generated at fast and slow spreading ridge axes.

7.2.2 *Oceanic island basalt*

The transition from MORB to ocean ridge hot-spot magmas (e.g. Iceland, the Azores and the Galapagos) clearly reflects changes in partial melting and magma system regimes from intra-plate settings. The chemical variation in OIBs usually reflects a distinctive spatial–temporal association with the developmental stages of eruptive edifices. Parent magmas may range from SiO_2-saturated and oversaturated to strongly undersaturated types of variable isotopic and incompatible element character, reflecting a range of melt segregation depth, melt fraction, and H_2O and CO_2 activities in the source (Kushiro, 1968, 1973; Green *et al.*, 1987). Representative primitive OIB compositions are given in Table 7.2.

The Hawaiian archipelago is probably the best documented example of oceanic islands and involves a sequence of diverse primitive melt types and their respective derivative magmas. At Loihi seamount (Chapter 9), an example of submarine intraplate volcanism adjacent to Hawaii, active 'pre-shield' lavas consist of alkalic undersaturated basalt and basanite. The shield-building stages on Hawaii, Oahu and other islands commenced with

Table 7.2 Examples of primitive OIB (Hawaiian) magma compositions

Composition	Basalt No.[a] 1	2	3	4	5	6	7	8	9	10
SiO_2	44.4	42.4	50.00	47.95	48.93	46.75	44.02	43.02	40.15	39.91
TiO_2	1.68	3.02	1.73	1.37	2.11	1.58	2.05	2.17	1.90	2.76
Al_2O_3	10.6	10.8	11.62	9.29	11.86	9.01	11.82	13.7	12.01	9.13
FeO^t	12.24	13.32	11.12	10.80	11.49	11.13	13.92	12.42	12.42	12.99
MnO	0.19	0.18	0.18	0.19	0.18	0.16	0.20	0.16	0.23	0.18
MgO	16.3	11.9	14.00	21.19	13.02	21.81	13.40	13.76	13.27	15.88
CaO	10.1	11.1	8.83	7.03	9.67	7.30	10.95	10.37	12.73	11.97
Na_2O	1.91	3.13	1.90	1.51	1.96	1.47	2.39	2.74	4.13	3.16
K_2O	0.52	1.2	0.29	0.23	0.41	0.31	0.62	0.74	1.06	1.53
P_2O_5	0.19	0.47	0.175	0.14	0.215	0.16	0.45	0.33	0.98	0.81
Cr_2O_3	n.d.	n.d.	0.10	0.20	0.10	0.21	n.d.	n.d.	n.d.	n.d.
Total	98.13	97.4	99.945	99.90	99.945	99.89	99.82	99.41	98.88	98.32
Mg No.	0.72	0.64	0.71	0.80	0.69	0.80	0.66	0.69	0.68	0.71
CaO/Al_2O_3	0.953	1.03	0.76	0.757	0.815	0.810	0.926	0.757	1.06	1.31
Al_2O_3/TiO_2	6.31	3.56	6.72	6.78	5.62	5.70	5.77	6.31	6.32	3.31
CaO/Na_2O	5.29	3.33	4.65	4.66	4.93	4.97	4.58	3.78	3.08	3.79

[a](1) Alkali basalt from Loihi seamount (Frey and Clague, 1983); (2) Basanite from Loihi seamount (Frey and Clague, 1983); (3) Mauna Loa parental magma, averaged from Wright (1971). Table 9, corrected to 14% MgO by olivine addition (Wright, 1971; Table 15); (4) Calculated composition of Mauna Loa primary magma in equilibrium with olivine ($Fo_{92.5}$) using method of Irvine (1977), from Wright (1984); (5) Kilauea parental magma, averaged from September 1971 eruption (unpublished data cited by Wright and Tilling, 1980), corrected to 13 wt% MgO by olivine addition (Wright, 1971; Table 15); (6) Calculated composition of Kilauea primary magma in equilibrium with olivine $Fo_{92.5}$ using method of Irvine (1977), from Wright (1984); (7) Alkali olivine basalt from Hahaina. West Maui (Macdonald, 1968); (8) Basanite from Kalaupapa, East Molokai (Naughton *et al*., 1980); Nephelinite from the Honolulu Series, Koolau (Clague and Frey, 1982); (10) Nepheline melilitite from the Koloa Series, Kauai (Macdonald, 1968)

massive eruptions of olivine tholeiite, progressing to shield-capping, transitional tholeiite and post-erosional rejuvenescent stages of alkaline olivine basalt, basanite, and/or nephelinite (Wright and Clague, 1989). Silica-undersaturated magmas such as basanite and nephelinite usually appear as rejuvenescent episodes on eroded relict shields (Flower, 1973; Clague and Frey, 1982) but may sometimes form edifices adjacent to earlier, less undersaturated, shields (Strong, 1972; Schmincke and Weibel, 1972; Flower *et al.*, 1976). Alkali basaltic and more undersaturated OIB magmas usually develop extensive low pressure differentiation trends to trachyte and phonolite, respectively (Clague, 1987).

Archipelagos built on the Pacific plate, such as the Hawaiian, Society and Marquesas chains, tend to show simple monotonic age progressions (Shaw and Jackson, 1973; Clague and Dalrymple, 1987; Dupuy *et al.*, 1987) made up of increasingly denuded shields and corresponding phases of rejuvenescent activity. In contrast, island groups built on slow spreading (e.g. Atlantic) lithosphere such as the Canaries, Azores and Cape Verdes, show a more complex compositional variation in time and space, spatial patterns often being better defined than temporal patterns (Schmincke and Weibel, 1972; Flower *et al.*, 1976; Duncan, 1984). Such patterns appear to reflect lithospheric stress distribution, as expressed by rifting and transverse faulting, rather than absolute plate velocities with respect to centres of mantle upwelling (Duncan, 1984).

7.3 Phase equilibrium and fluid dynamic constraints

Phase equilibria, fluid dynamics and the kinetics of magma mixing and wallrock reaction appear to be the principal constraints on magmatic processes.

7.3.1 *Phase equilibria*

On the basis of pioneer experiments in the subsystems CaO–MgO–Al_2O_3–SiO_2–Na_2O (CMASN), Bowen (1928) demonstrated the significance of partial melting and fractional crystallization as fundamental causes of chemical fractionation in magmas. Yoder and Tilley (1962) integrated the existing experimental data for natural and synthetic systems at pressures between 1 atm (101 325 $N\,m^{-2}$) and 30 kbar (1 bar $= 10^5\,N\,m^{-2}$) and established the eutectic-like composition of basalt, the significance of the SiO_2-saturation concept, and the efficacy of thermal barriers in determining fractional crystallization paths. Green and Ringwood (1967) and O'Hara (1968) contributed to this conceptual basis for evaluating the roles of source composition and mineralogy, the degree of partial melting, and polybaric fractionation in producing the range of oceanic magma types. Phase equilibrium constraints are reviewed by Thompson (1987). Following Yoder and Tilley (1962), it was often assumed that mantle melts approximate isobaric invariant compositions provided no phase in the residue is consumed during the melting process. However, it is now recognized that the small number of phases and large number of components theoretically precludes isobaric invariant melting (Takahashi and Kushiro, 1983; Fujii and Scarfe, 1985; Falloon and Green, 1988). Investigations of natural basalt under anhydrous, H_2O-undersaturated and H_2O-saturated conditions have proceeded in parallel with further studies of CMAS and CMASN, aiming to reproduce equilibrium conditions for multiply-saturated primitive melts assuming such conditions to be equivalent to those of partial melting, and establishing the variation of melt composition with degree of melting.

With concern about the validity of isotopic and incompatible element source indicators in erupted magma, the question of whether primary MORB is picritic or resembles the relatively few examples of erupted magnesian tholeiite developed into a major controversy with profound implications for the thermal character of the source (Chapter 6). Presnall *et al.* (1979) reviewed the anhydrous experimental data for CMAS between 1 atm and 10 kbar and argued that the solidus cusp between plagioclase- and spinel-lherzolite represents a likely condition for partial melting, to produce a five-phase saturated tholeiite melt. In contrast, Stolper (1980) proposed that primary MORB is picritic and segregates from a harzburgite residue. This proposal is based on the experimental equilibration of MORB glass at 20 kbar in a 'sandwich' of olivine and orthopyroxene.

The ensuing debate in favour of (O'Hara, 1968; Green *et al.*, 1979; Stolper, 1980; Jaques and Green, 1980; Elthon and Scarfe, 1984; Elthon, 1986) and against (Fujii and Bougault, 1983; Takahashi and Kushiro, 1983; Fujii and Scarfe, 1985; Presnall and Hoover, 1984, 1987) picritic MORB melts further stimulated efforts to reconcile phase equilibrium and geochemical constraints. For example, Presnall and Hoover (1987) projected 40 primitive glass norms, identified from a database of 1700 analyses, into olivine–plagioclase–quartz and olivine–diopside–quartz planes of the olivine–plagioclase–diopside–quartz system, together with the compositions of picrites synthesized between 20 and 35 kbar (Figure 7.1) as a means of testing phase equilibrium models in terms of geochemical mass balances. Two observations are pertinent. Firstly, none of the natural glasses is picritic (erupted picrites are invariably cumulitic), and secondly, primitive glasses form an array between (Ol + Hy) and (Hy + Qz) normative domains which appear to mimic that of non-primitive MORB (Figure 7.2). However, whereas the variation of evolved MORB reflects low pressure fractionation, the primitive glass variation is clearly not constrained by the 1 atm, three phase cotectic defined by studies of evolved MORB (Walker *et al.*, 1979; Fisk *et al.*, 1980; Grove and Bryan, 1983), suggesting that this reflects inherent differences in primary MORB melts. Presnall and Hoover (1987) extended phase equilibrium studies into

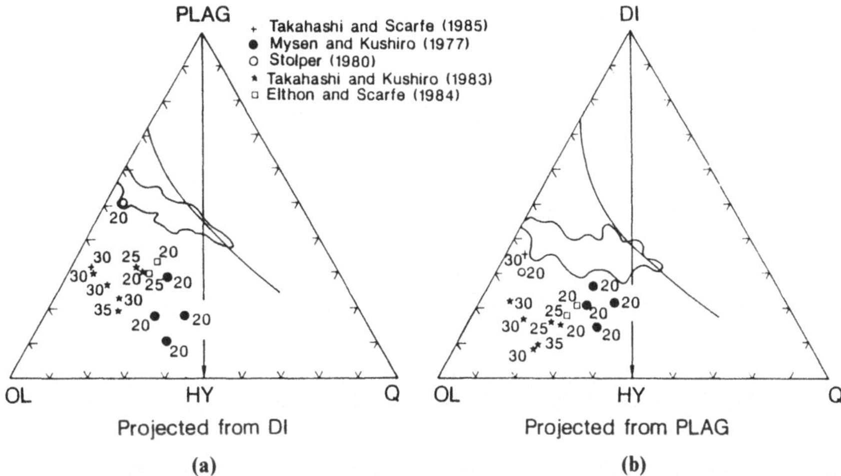

Figure 7.1 Comparison of primitive MORB glasses (outlined) with compositions of melts experimentally synthesized at 20–35 kbar (after Presnall and Hoover, 1987). Compositions are projected into the CIPW normative tetrahedron from (a) diopside (Di) onto the plane olivine (Ol)–plagioclase (Pl)–quartz (Q), and (b) plagioclase onto the plane olivine–diopside–quartz, assuming $Fe^{2+}/(Fe^{2+}+Fe^{3+})$ to be 0.83 (Presnall *et al.*, 1979). Pressures (kbar) at which experimental melts equilibrated are shown. The curved line indicates the path of liquid compositions with decreasing temperature at 1 atm in equilibrium with olivine, plagioclase and clinopyroxene (Walker *et al.*, 1979).

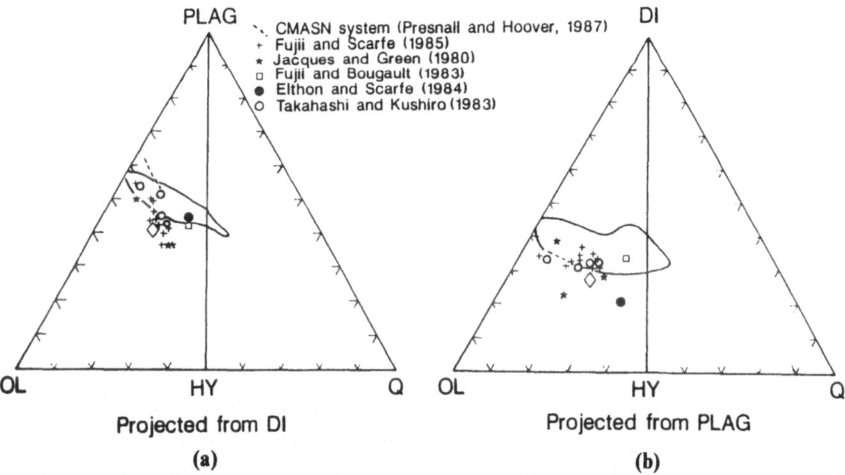

Figure 7.2 Comparison of primitive MORB glasses (outlined) with compositions of melts experimentally synthesized at 8–12 kbar from natural starting materials in equilibrium with olivine, orthopyroxene and clinopyroxene, and (broken line) 10 kbar liquids in the CMASN system (after Presnall and Hoover, 1987) and computed MORB melt for $T_p = 1280°C$ (*) from McKenzie and Bickle (1988). Normative projection method and 1 atm cotectic as in Figure 7.1.

CMASN between 1 atm and 20 kbar and conducted equilibrium melting experiments on a simplified lherzolite consisting of mixes of simulated mantle residue (in the modal proportions reported by Dick and Fisher, 1984) and melt equilibrated with plagioclase, spinel, clinopyroxene, orthopyroxene, and olivine in CMASN. The CMASN melt corresponds well with the primitive MORB glass compositions and the synthetic glass compositions given in Table 7.1 (Figures 7.1 and 7.2) if allowance is made for the absence of Fe, K, Ti and other minor elements.

Two tests for the validity of picrite parent melt models may be applied. The first is a simple graphical demonstration of parent–daughter relationships between synthetic picrite and primitive MORB in terms of the projected phase equilibria. Whereas Figure 7.1 shows that olivine fractionation would satisfy phase equilibrium requirements, which is consistent with observations by O'Hara (1968), Green *et al.* (1979), Stolper (1980), Elthon and Scarfe (1984) and Elthon (1986), Figure 7.2 shows that experimentally produced melts in equilibrium with olivine, orthopyroxene and clinopyroxene at about 10 kbar lie close to the olivine-rich edge of the primitive MORB glass field, such that the latter could represent primary melts or be derived by small amounts of olivine fractionation. So far the role of picrite is unresolved. The second test relies on the geochemical validity of a projected fractionation model, specifically the requirement for a decrease in the magnesium number ($Mg/Mg + Fe^{2+}$) and corresponding increases in the contents of K, Ti and Na resulting from mafic phase fractionation. Presnall and Hoover (1987) applied these tests and observed that the magnesium number of Stolper's (1980) picrite is lower than those of the natural glasses, whereas this and other synthetic picrite candidates (Mysen and Kushiro, 1977; Elthon and Scarfe, 1984; Takahashi and Scarfe, 1985) show K, Ti and Na values too high to be parental to natural MORB glass. Synthetic melt compositions that do satisfy the magnesium number, K, Ti and Na constraints are reported by Fujii and Bougault (1983) and Fujii and Scarfe (1985) (Table 7.1).

The debate continues in the light of the melting experiments by Falloon and Green (1988) on fertile 'MORB pyrolite' (Green *et al.*, 1979) and refractory Tinaquillo lherzolite (Jaques and Green, 1980) in the pressure range 8–35 kbar. Experiments on sandwiches of primitive MORB (DSDP 3-18-7-1 + 17 wt% olivine) in pyrolite yield potential MORB-type melts ranging from Ne- and (Ol + Hy)-normative picrite (at high pressures) to Ol- and Qz-normative tholeiite (at low pressures) between 8 and 25 kbar, reflecting the progressive expansion of the orthopyroxene field with pressure (Takahashi and Scarfe, 1985). Falloon and Green (1988) concur with the views of Bryan *et al.* (1981), Grove and Bryan (1983); Francis (1960) and Klein and Langmuir (1987) that there is a spectrum of primary MORB between tholeiite and picrite, and that most primitive MORB glasses have undergone olivine fractionation (11–25% according to Falloon and Green, 1988). Viereck *et al.* (1989) interpret compositional bimodality to reflect variable degrees of melting under isobaric conditions of 8–12% (N2-MORB) and 13–20% (N1-MORB), and show that buffered Sr and REE contents and major element mass balances are consistent with a plagioclase-bearing residue. This appears to support pressures of at least 20 kbar (Clague and Frey, 1982; Sen, 1982). In contrast to MORB and OIB shield magmas, undersaturated primitive melts are readily identified from lavas entraining mantle xenoliths (examples given in Table 7.2).

7.3.2 *Fluid dynamics*

The generation of magma and its segregation from a convecting mantle may also be considered from a fluid dynamic perspective. Investigations of solid-state creep in olivine led Aherne and Turcotte (1979) to the conclusion that upwelling mantle beneath oceanic ridges begins melting at about 70 km depth. According to these workers the increased permeability due to small amounts of melting and the increasing buoyancy of melt at lower pressures combine to drive the melt upwards until small increments coalesce into magma bodies. Theoretical models of intergranular melt transport (McKenzie, 1984; Richter and McKenzie, 1984; Ribe, 1985) are in excellent agreement with the prediction of Aherne and Turcotte (1979), and indicate a need to re-evaluate models based exclusively on phase equilibria.

McKenzie (1984) estimated that for 'normal' upper mantle temperatures melt will begin to move upwards at porosities of less than 2% within a progressively melting mantle column. Melt compositions generated by continuous decompression melting will differ from those of equilibrium batch melts produced experimentally or simulated by least-squares mass balance calculations. However, assuming that melts continuously re-equilibrate with their surroundings, they will still be constrained to evolve along for example, ol + opx ± cpx(± sp ± pl) + liquid cotectics as the latter shift with decreasing pressure. McKenzie and Bickle (1988) calculated the rate of melt production during mantle decompression and, on the basis of experimentally derived melt compositions (Jaques and Green, 1980), calculated the compositional ranges of melt as a function of decreasing pressure. Recalculated as CIPW norms, these fall within the range of primitive MORB (Presnall and Hoover, 1987; Thompson, 1987), suggesting that, in contrast to the Aherne and Turcotte (1979) and McKenzie (1984) models (which predict continuous mixing of melt increments), compositionally discrete melt batches may be tapped from different depths. Models accommodating differential tapping of this type include the concept of 'magmons' (Scott and Stevenson, 1986), i.e. ascending regions of high porosity that incorporate new liquid from above and relinquish liquid fractions below. These are parameter perturbations leading to the buoyancy-driven penetration of upper partially molten layers by lower layers (Whiteheat *et al.*, 1984; Whitehead, 1986), and fracture propagation through partially molten upwelling mantle (Sleep, 1984; and Spence and Turcotte, 1984).

The fluid dynamic behaviour of ascending magma may also determine its mode of emplacement in the lithosphere. For example, Whitehead *et al.* (1984) attribute the larger scale stability of spreading cells to Rayleigh–Taylor type gravitational instability beneath spreading centres. In this model, the low viscosity and density of partially molten mantle lead to the development of regularly spaced protrusions centred beneath specific ridge segments, from which melt separates to form or feed axial magma chambers and generate new crust. Schouten *et al.* (1985) view this as a cyclic process perpetuated by lithospheric extension, whereby regions of partially molten mantle develop in response to decompression at the base of the depleted asthenosphere beneath the ridge axis. As a result of their lower density and viscosity such regions become gravitationally unstable and develop regularly spaced disturbances. Once established, the disturbed regions yield low viscosity basaltic melt which migrates upwards leaving melt-depleted asthenosphere. The process will then repeat itself as a new region of partially molten mantle develops at the base of freshly depleted mantle. Ridge offsets themselves are probably not sufficient to disrupt the continuity of the partially molten region which, according to the model, is the source of instability.

These processes will be further complicated by the physical effects of chemical fractionation in the magma supply system. Widespread evidence for magma mixing in oceanic magma systems has led to fluid dynamic studies of evolving magma chambers replenished with new magma. The major delimiting factors are the viscosity, density and temperature of melts which are closely dependent on melt composition. Huppert and Sparks (1980) proposed that the mixing of primitive melts with evolved tholeiite will be delayed until their respective physical properties become equilibrated. This process has been verified by experimental studies using aqueous solution analogues (Huppert and Sparks 1984; Sparks *et al.*, 1984). Influxed magnesian melt accumulates at the base of the chamber, maintaining a sharp boundary with overlying cooler, less dense, liquid. While both layers convect vigorously heat is transferred across the boundary layer and precipitated olivine is retained in suspension until thermal equilibration between the layers is achieved and convection ceases. Density differences between primitive, evolved and hybrid melts may effectively produce a ‘density window’ preventing the eruption of high density picrite and evolved iron-rich melts and predisposing the eruption of a narrow compositional spectrum equivalent to ‘average’ MORB (Stolper and Walker, 1980; Sparks *et al.*, 1980).

7.4 Melt generation: active versus passive

Robust physical models for melting in the earth have only recently emerged and implicitly (at least) may involve decompression melting resulting from ‘active’ plume activity (Aherne and Turcotte, 1979; McKenzie, 1984; Courtney and White, 1986; McKenzie and Bickle, 1988), ‘passive’ lithospheric stretching

(Lachenbruch, 1975; Foucher *et al.*, 1982; McKenzie, 1985b; McKenzie and Bickle, 1988), or shear melting resulting from viscous dissipation along zones of lithospheric decoupling (Shaw, 1973; Shaw and Jackson, 1973).

7.4.1 *Passive melting at 'normal' ridge systems*

The acceptance of mantle convection as the 'engine' for seafloor spreading led to the common belief that oceanic ridges are a surface expression of

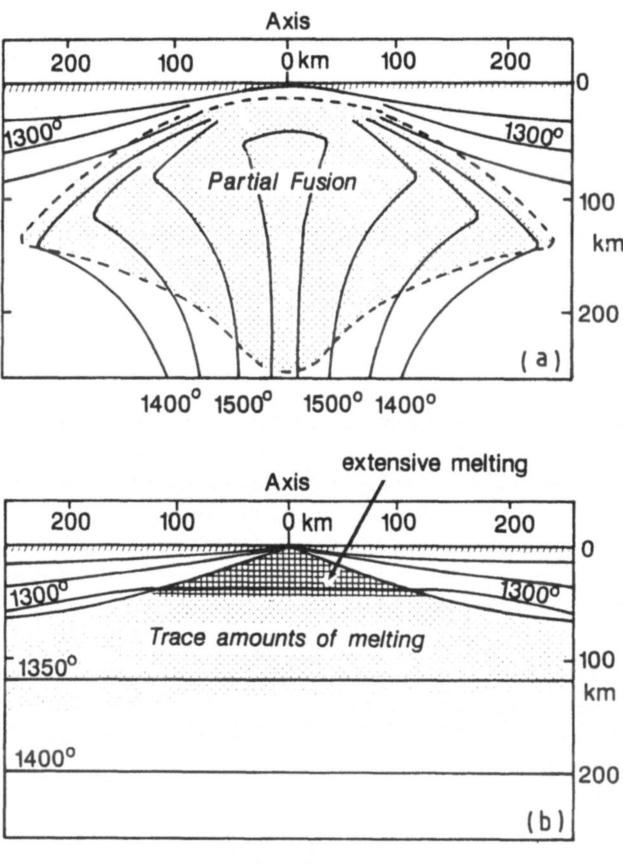

Figure 7.3 (a) Schematic temperature distribution beneath an ocean ridge axis (from Oxburgh, 1980) where ridge spreading coincides with a hot rising mantle jet. A close association between upwelling mantle and spreading ridges presents major conceptual problems for plate tectonic models. At intraplate settings a rising jet may raise mantle potential temperatures (T_p) by >200 C. (McKenzie and Bickle, 1988). (b) Temperature distribution beneath 'passive' ocean ridges where separation of two plates is the sole cause of mantle upwelling. A hot rising plume or sheet is not required to account for high heat flow or shallow bathymetry of ridges. Partial melting will result from decompression alone, given a T_p of 1280°C (from McKenzie and Bickle, 1988).

upwelling mantle (Hess, 1962; Turcotte and Oxburgh, 1967), (Figure 7.3a). However, the assumed association between convective geometry and zones of plate formation produced several conceptual problems concerning the relationship between mantle convection and plate tectonics (McKenzie and Bickle, 1988). If oceanic ridges are viewed as passive features where two plates separate and upwelling mantle fills the resulting gap, the plate tectonic problems are removed. This concept (Figure 7.3b) involves horizontal isotherms at a certain depth and implies that only the uppermost mantle is available for melting. Ridge migration may occur irrespective of mantle convection patterns, as indicated by the gravity anomaly pattern which appears to represent direct evidence for mantle upwelling (Watts *et al.*, 1985).

The volume of partial melt generated through decompression of the asthenosphere depends on the amount of lithospheric stretching and the mantle potential temperature (T_p) (McKenzie, 1984; White *et al.*, 1987). If ridge melting is not connected to upwelling mantle jets, 'passive' melting must be able to generate enough melt to produce the average thickness of oceanic crust (about 7 km) with a T_p of 1280°C. From experimental studies of natural basalt and peridotite, McKenzie and Bickle (1988) parameterized partial melt composition as a function of melt fraction and pressure. The agreement between the calculated compositions of melts produced by adiabatic melting at $T_p = 1280°C$ and natural MORB glass (Table 7.1) is close, suggesting that olivine fractionation was at a minimum in the latter and that primitive melts were buffered by a clinopyroxene-bearing residue. The composition of melt calculated to produce 7 km of oceanic crust at $T_p = 1280°C$ is given in Table 7.3 for comparison with the 10 k bar melt candidates discussed earlier. Significantly, it corresponds to multiply-saturated melts synthesized at 10 k bar (Figure 7.2) (Fujii and Bougault, 1983; Fujii and Scarfe, 1985; Presnall and Hoover, 1987), rather than picrite (Stolper, 1980; Elthon and Scarfe, 1984). Viereck *et al.* (1989) calculated that Mg-rich MORB melts equilibrated with clinopyroxene-poor spinel/plagioclase lherzolite are similar in composition to those sampled from the Mid-Atlantic Ridge (Dick and Fisher, 1984). Agreement between the passive melting model, experimental phase equilibria and mantle–melt mass balances is encouraging (Presnall *et al.*, 1979; McKenzie and Bickle, 1988; Viereck *et al.*, 1989).

7.4.2 *OIB shields: rising mantle jets versus shear melting*

If correlated gravity and geoid anomalies reflect mantle circulation patterns (McKenzie *et al.*, 1980; Parsons and Daly, 1983; Watts *et al.*, 1985), **hot-spot** magmatism may be reasonably attributed to rising mantle jets (e.g. Hawaii, the Galapagos, Iceland, the Azores and Cape Verdes). Upwelling from the lower mantle precludes horizontal isotherms in the upper mantle and necessarily produces isotherm distributions of the type shown in Figure 7.3a.

Table 7.3 Examples of experimentally synthesized melts: 8–12 kbar

Composition	Basalt No.[a]									
	1	2	3	4	5	6	7	8	9	10
SiO_2	49.75	51.09	52.63	49.1	49.3	48.9	49.1	49.5	49.9	49.64
TiO_2	1.33	0.65	0.58	0.49	0.63	0.61	0.68	0.54	0.45	0.59
Al_2O_3	15.97	16.26	14.50	14.3	15.4	17.3	17.0	15.0	14.7	14.34
FeO^t	8.87	7.32	7.39	8.55	7.23	7.02	7.00	6.76	6.03	8.58
MnO	0.18	—	—	0.10	0.11	0.21	0.25	0.25	0.12	—
MgO	9.05	10.25	11.68	12.20	12.2	10.9	11.1	12.7	13.9	13.83
CaO	10.61	12.16	11.14	12.20	12.2	11.8	12.0	12.5	12.4	11.10
Na_2O	2.49	1.91	1.66	1.79	1.84	2.36	2.29	1.86	1.73	1.65
K_2O	0.16	—	—	0.09	0.08	0.07	0.10	0.07	0.05	—
P_2O_5	0.04	—	—	0.14	0.12	0.15	0.16	0.07	0.0	—
Cr_2O_3	—	0.35	0.41	0.28	0.15	0.08	0.08	0.20	0.32	0.59
Total	98.94	99.99	99.99	99.24	99.26	99.40	99.76	99.45	99.68	100.32
Mg No.	0.67	0.73	0.76	0.72	0.75	0.74	0.74	0.77	0.80	0.76
CaO/Al_2O_3	0.664	0.748	0.768	0.853	0.792	0.682	0.706	0.833	0.844	0.774
Al_2O_3/TiO_2	12.0	25.0	25.0	29.2	24.4	28.4	25.0	27.8	32.7	24.3
CaO/Na_2O	4.26	6.37	6.71	6.82	6.63	5.00	5.24	6.72	7.18	6.73
Pressure (kbar)	8	8	8	10	10	10	10	10	10	12
Temperature (°C)	1225	1350	1400	1290	1290	1275	1275	1300	1310	1450
Residue	L	L	H	L	L	L	L	H	H	H

[a](1) Partial melt of lherzolite HK66 (Takahashi and Kushiro, 1983); (2) 27% melt of MPY-87 (Falloon and Green, 1988); (3) 30% melt of MPY-87 (Falloon and Green, 1988); (4) 15% melt of peridotite SM-5 (Fujii and Scarfe, 1985); (5) 15% melt of peridoitite SM-4 (Fujii and Scarfe, 1985); (6) Run no. 111 (SM-2) (Fujii and Scarfe, 1985); (7) Run no. 112 (SM-4) (Fujii and Scarfe, 1985); (8) Run no. 131 (SM-4) (Fujii and Scarfe, 1985); (9) Run no. 100 (SM-4) (Fujii and Scarfe, 1985); (10) 28% melt of MPY-87 (Falloon and Green, 1988)

MPY-87 = MORB pyrolite, L = lherzolite residue, H = harzburgite residue

In such cases the mantle T_p may exceed 1280°C by more than 200°C (McKenzie and Bickle, 1988) and melt fractions and the depth range of melt generation would be markedly different from those at normal oceanic ridges. Whereas there is no *a priori* association between rising jets and spreading ridges (Watts *et al.*, 1985), numerical experiments indicate that plumes may become 'attached' to a migrating ridge axis if the relative movement of the ridge is less than a few millimetres per year (Houseman, 1983). Examples of attached plumes include Iceland, the Azores and the Galapagos, associated with a normal ridge axis, a ridge–transform triple junction, and a ridge–ridge triple junction, respectively. In contrast, Hawaiian magmatism is the surface expression of an intra-plate 'unattached' plume.

The association of hot-spot magmatism and active mantle convection, whether connected to a ridge axis or an intraplate setting, can be modelled if the T_p, lithospheric thickness and magmatic phase equilibria are known (McKenzie and Bickle, 1988). For the case of a plume rising beneath Hawaii, McKenzie (1984) estimated T_p to be about 1550°C and calculated that melting would commence at about 100 km depth. Thompson (1987) observed that the volume of melt generated by such a plume, if segregated at about 70 km depth, would be equivalent to a 15% anhydrous partial melt (similar to the Kilauea parent calculated by Irvine, 1979) of a moderately refractory lherzolite. Chen and Frey (1985) also applied the decompression melting of a rising plume to Hawaiian magmatism and from geochemical mass balances calculated similar melt fractions to produce a Hawaiian-type shield. To explain the decoupling of radiogenic isotopes from their respective parent–daughter trace element parameters, Chen and Frey (1985) proposed that incipient melt contributions from the MORB-like lower lithosphere were incrementally mixed with the rising plume. Feigenson (1986) envisaged that the incipient decompression melting of the uppermost asthenosphere entrained at the edge of a rising plume is enhanced by H_2O released by hydrous phases in the lower lithosphere, and may be tapped as shield edifices migrate from above the plume apex.

McKenzie (1984) calculated that for 'attached' plumes involving high T_p and the additional factor of ridge dilation (Schilling *et al.*, 1983; Hamelin *et al.*, 1984), decompression melting would also begin at about 100 km depth but would yield larger melt fractions, in the case of the Azores equivalent to 35% melting. If the rising mantle is of equivalent fertility to that beneath normal ridges, such melt fractions would require harzburgite residues (Viereck *et al.*, 1989) consistent with the observation of increasingly refractory mantle along-ridge to the Azores (Dick and Fisher, 1984; Michael and Bonatti, 1985). Thus support for mantle upwelling as a cause of intra-plate and ridge-attached hot-spot magmatism rests on the geophysical evidence for mantle circulation, the propensity for mantle melting implied by the calculated isotherm distribution, mantle–melt mass balance and the ridge-longitudinal variation of mantle residues.

In contrast to the concept of decompression melting, shear melting was postulated as the primary cause of intraplate magma production (Shaw, 1973; Shaw and Jackson, 1973) whereby heat is generated by viscous dissipation at the decoupled interface of the asthenosphere and lithosphere. Shear melting was invoked in view of the capacity of a shear couple to allow extensive melting at relatively shallow depths. Shaw and Jackson (1973) envisaged that the propagation of melting through a thermal feedback process is linked to gravitational 'anchors' of refractory peridotite. Lithospheric thinning (Detrick and Crough, 1978), involving the conversion of lithosphere to asthenosphere, is considered to trigger shear melting whereas thermal feedback activates a cyclic runaway melting process.

The concept of gravitational anchors is diametrically opposed to that of plumes and has been disputed by O'Hara (1975) and others. However, Wright (1984), while accepting the mechanism of plumes, favoured shear melting without gravitational anchors to explain Hawaiian shield magmatism in view of its capacity to produce extensive melting at shallow depth and the close temporal association observed between partial melting, magma supply and magmatic replenishment. To produce incompatible element-enriched Kilauea shield magmas, Wright (1984) proposed a model whereby primitive melts are formed from refractory lherzolite residues to MORB formed at the East Pacific Rise, metasomatized by a low fraction nephelinite melt. In this model nephelinite melts (rather than the magmas producing the shield) are formed in response to a thermal plume (Feigenson, 1986). The precise range of compositions produced by the decompression melting of rising mantle awaits further refinement of T_p estimates, available source compositions, and relationships between melt composition and melt fraction.

7.5 Melt transport and storage in the oceanic lithosphere

Most primitive melts are subjected to a range of chemical fractionation processes prior to their emplacement in ridge and intra-plate settings. Erupted MORB and OIB are often saturated with low pressure mineral phase assemblages (Thompson and Tilley, 1969; Thompson and Flower, 1971; Wright and Fiske, 1971; Fisk *et al.*, 1980; Grove and Bryan, 1983), reflecting the extensive fractionation of olivine (O'Hara, 1968). Fluid dynamic experiments confirm the likelihood for the cooling, entrapment and mixing of melts released into the crust (Sparks *et al.*, 1980; Huppert and Sparks, 1980; 1984). The development of SiO_2-rich derivatives from undersaturated OIB parent magmas is ubiquitous, reflecting lengthy crustal residence times for mantle-derived liquids. Seismic evidence indicates complex magma storage and transport in both oceanic ridge and intraplate environments (Nisbet and Fowler, 1978; Macdonald *et al.*, 1984, 1986; Detrick *et al.*, 1987; Ryan *et al.*, 1981; Ryan, 1987). Before examining the tectonic and environmental boundary conditions of magma fractionation, fractionation mechanisms themselves are reviewed.

7.5.1 *Fractionation mechanisms*

7.5.1.1 *Fractional crystallization.* Anhydrous basaltic melts saturated at high pressure with olivine and orthopyroxene ($\pm$ cpx $\pm$ an aluminous phase) will pass through the olivine phase field on rising into the crust (O'Hara, 1968). This predisposition is reflected in both MORB and OIB by the correspondence of experimentally verified low pressure liquidus crystallization assemblages with decreasing MgO and inflected trends of CaO, Al_2O_3, and FeO^t (Fisk *et al.*, 1980; Grove and Bryan, 1983; Thompson and Tilley, 1969; Thompson and Flower, 1971). A typical low pressure crystallization sequence in MORB is: ol(+sp) > ol + pl > (ol) + pl + cpx(+pig), and in OIB: ol(+sp) > ol + cpx > ol + cpx + pl, reflecting the differences in SiO_2 saturation and normative plagioclase/clinopyroxene ratios of their parent melts (Chapter 5). In MORB, olivine appears as euhedral to subhedral phenocrysts, often with included spinel, but is usually absent from the matrix as a result of its reaction with SiO_2-rich melt (Kushiro, 1968, 1973; Bryan, 1983). Plagiocase phenocrysts may have reacted and may show oscillatory zoning, but they also occur as euhedral unzoned (or normally zoned) laths in coprecipitational clusters with olivine. Augitic clinopyroxene is usually ophitic to plagioclase and olivine and (especially in Atlantic MORB) may also appear as subhedral microphenocrysts. Whereas most MORB glasses project at or near the experimentally determined low pressure phase boundaries and reaction points (Bryan, 1983; Grove and Bryan, 1983), several features of MORB variation are not easily explained by low pressure fractional crystallization.

Numerical fractional crystallization models (Flower *et al.*, 1977; Byerly and Wright, 1978; O'Donnell and Presnall, 1980; Bryan *et al.*, 1981; Sigurdsson, 1981) involving olivine, plagioclase and clinopyroxene are required for MORB eruptives which are often devoid of clinopyroxene phenocrysts. The 'phantom clinopyroxene' problem has been interpreted as an effect of high pressure or polybaric fractionation, whereby high pressure clinopyroxene is resorbed while melts re-equilibrate with low pressure assemblages (Walker *et al.*, 1979; O'Donnell and Presnall, 1980; Bryan *et al.*, 1981; Francis, 1986). Rounded clinopyroxene has been cited as evidence of the same problem, although on the basis of low pressure experimental investigations Grove and Bryan (1983) conclude that this need not be the case. A similar problem is presented by plagioclase megacrysts which Flower (1980, 1981b) observed are relatively abundant in Atlantic MORB. Although primitive MORB does not pass through the plagioclase primary phase field during its ascent (O'Hara, 1968), it is reasonable to assume that plagioclase accumulation is facilitated by its buoyancy relative to tholeiite melt and accompanying mafic phases. Fujii and Kushiro (1977) and Kushiro (1980) show that tholeiite melt density exceeds that of plagioclase (An_{90}) at pressures greater than about 6 kbar, suggesting that high pressures favour plagioclase accumulation. Megacrysts of diopside-rich clinopyroxene (Donaldson and Brown, 1977) and Al-rich spinel (Sigurdsson and Schilling, 1976) may be further evidence of high pressure crystallization in MORB (Thompson, 1987), although Wilkinson (1982) has argued that they are mantle xenocrysts.

OIB magmas are constrained by similar phase equilibria, and although experimental studies are less comprehensive, the interpretation of fractional crystallization at least in undersaturated types is facilitated by the presence of xenoliths representing consolidated fractionation products (Jackson and Wright, 1970). Many shield magmas are strongly porphyritic and consist of large volumes of picrite (ol) or ankaramite (ol + cpx). Their compositions and petrographic textures suggest these to be massive cumulate sequences developed within underlying magma chambers (Wright and Fiske, 1971), although phyric shield lithologies may also represent quasi-equilibrium crystal–liquid mixtures (Thompson and Flower, 1971; Cox, 1980). According to the Hawaiian model, magma supply rates increase from the submarine pre-shield (Loihi) stage (Moore *et al.*, 1982; Staudigel *et al.*, 1984) to a maximum during shield construction (Swanson, 1972; Wright and Helz, 1987), then decline during post-caldera collapse and post-erosional rejuvenescent activity. The extent of differentiation of the respective parent magmas and the configuration of magma transport and storage appears to correspond to magma supply rates such that shield magmas are relatively unfractionated whereas subsequent eruptives may be extensively fractionated (Flower, 1973; Clague, 1987).

Silica-saturated and oversaturated shield magmas in the Azores include relatively abundant trachyte and rhyolite differentiation products (Schmincke and Weibel, 1972). The Azorean shields show a spatial variation of SiO_2 saturation character, from saturated and oversaturated types in the north (Graciosa, Terceira), mildly saturated or undersaturated in the centre (Sao Jorge) to strongly undersaturated in the south (Faial, Pico), with the western (Corvo, Flores) and eastern (Santa Maria) extremities exhibiting patterns closer to the Hawaiian model (Flower *et al.*, 1976). In general, undersaturated magmas are rich in xenoliths of lherzolite and harzburgite mantle residues and consolidated cumulate products such as dunite (ol), wehrlite (ol + cpx) and gabbro (cpx + pl), and, rarely, eclogite (gar + cpx), (Jackson and Wright, 1970).

7.5.1.2 *Magma mixing.* The effects of magma mixing in ridge basalt were first described by Donaldson and Brown (1977) and Dungan and Rhodes (1978). Evidence for the mixing of primitive and evolved melts includes the presence of reacted clinopyroxene and plagioclase phenocrysts, magnesian melt inclusions, and the observed linear variation of compatible and incompatible elements. Dungan and Rhodes (1978) proposed that mixing would explain the 'phantom clinopyroxene' problem in MORB. Hybrids of magnesian and

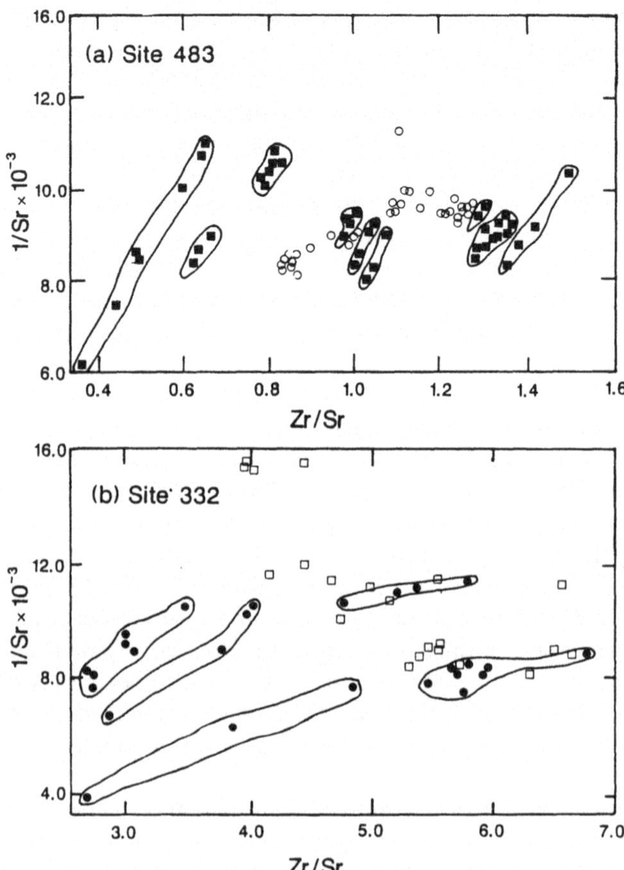

Figure 7.4 Variation of 1/Sr versus Zr/Sr as a function of magma mixing in basalts drilled from DSDP/IPOD Site 483, Gulf of California (EPR 22°N), and Site 332B, Mid-Atlantic Ridge 36–37°N. Massive flow units (outlined) appear to reflect more extensive magma mixing effects than pillowed flows.

evolved melt would show the compositional effects of (cpx-pl) cotectic fractionation while still occupying the primary olivine phase volume. Mixing could also explain the presence of olivine and plagioclase megacrysts more primitive than expected from their host liquid compositions, and the apparent decoupling of incompatible trace elements, as illustrated in Figure 7.4. Although some workers (Rhodes and Dungan, 1979; Stakes *et al.*, 1984) consider that these features invariably signify mixing, others suggest that clinopyroxene, olivine and spinel megacrysts may represent scavenged wallrock material (Wilkinson, 1982; Sparks *et al.*, 1984) and that differences in normative clinopyroxene expressed in least-squares fractionation calculations reflect variable degrees of partial melting or mantle heterogeneity (Byran and Dick, 1982).

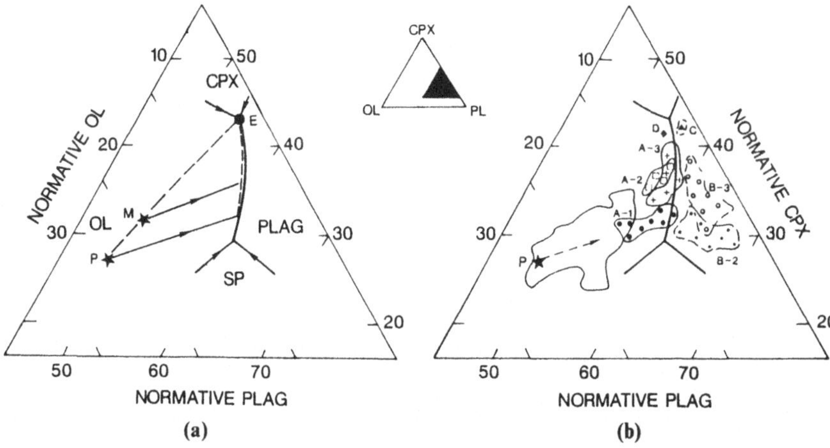

Figure 7.5 Effects of magma mixing on normative MORB composition projected from quartz onto the plane: olivine (Ol)–plagioclase (P)–clinopyroxene (Cpx), (a) Primitive mantle melt (P) fractionates olivine and then coprecipitates Pl and Pl + Cpx at the low pressure cotectic. Mixing of evolved melt (E) with newly influxed melt (P) produces hybrid liquids (M) occupying the low pressure olivine field with a history of (Pl + Cpx) fractionation. (b) Such compositions (e.g. aphyric basalt as shown) do not always conform to low pressure cotectics and may be further deflected if plagioclase accumulates in the melt (e.g. phyric basalts). Data are for basalts drilled from DSDP/IPOD Site 396, MAR 22°N, comprising aphyric groups Al, A1, A2 and A3, and plagioclase-phyric groups B3, B2 and C. Primitive MORB glasses compositions are shown in outline.

Studies of the FAMOUS and AMAR spreading segments (MAR, 36–37°N) (Bryan *et al.*, 1979; Stakes *et al.*, 1984) also support repeated mixing between more and less evolved melt fractions, respectively saturated with (ol + pl) and (ol + pl + cpx). Experimental studies by Walker *et al.* (1979) on basalts from the Oceanographer Fracture Zones (MAR 35°N) show clinopyroxene crystallizing at higher temperatures than olivine such that melts appear to be driven into the clinopyroxene field by mixing (Dungan and Rhodes, 1978; Figure 7.5). In general, the associated evidence of resorbed and 'absent' phenocryst species and discrepancies between the observed phenocrysts, experimental phase equilibria and least-squares fractionation models are indicative that magma mixing is a common process at oceanic ridge. Disagreements persist about the compositions of mixed melts (Walker *et al.*, 1979; Bryan *et al.*, 1981; Perfit and Fornari, 1983), although such differences would be expected given their dependence on the character of mixing systems as determined by spreading rate and magma supply.

Shield-building OIB magmas also show evidence for mixing. Mixing between replenishing primitive magma and stagnating evolved batches has been reported from Kilauea (Wright and Fiske, 1971) and Réunion (Ludden, 1978) and is clearly an important process. Wright (1971), Wright and Fiske (1971), Wright *et al.* (1975) and Wright and Tilling (1980) were able to distinguish hybrid batches from those erupted directly from the mantle or unmixed intermediate storage reservoirs, and observed that after each eruption at Kilauea the shallow storage reservoir is rapidly resupplied. This suggests that eruption and partial melting are closely related, and separated by < 100 years (Wright, 1984).

7.5.1.3 *Metasomatism.* Metasomatism prior to the onset of magma generation is commonly invoked to explain incompatible element enrichment in mantle sources whose isotopic compositions reflect time-integrated depletions in Rb, LREE, U and Th (Menzies and Murthy, 1980; Wright, 1984; Chen and Frey, 1985; and references in Menzies and Hawkesworth, 1987). Studies of mantle xenoliths provide abundant evidence for the operation of metasomatic processes, although in many instances the composition and origin of metasomatizing fluids are poorly constrained. Evidence takes the form of incompatible element enrichment matched by accessory phlogopite, amphibole and apatite, and also major phases such as clinopyroxene with relatively enriched incompatible element contents (Frey and Green, 1974; Hawkesworth *et al.*, 1984, 1987; Roden *et al.*, 1984; Menzies *et al.*, 1987; Kempton, 1987). Several lines of evidence support the hypothesis that metasomatized mantle is a suitable source for oceanic intraplate magmas (Roden *et al.*, 1984).

7.5.1.4 *Dynamic melting.* The idea of dynamic melting was introduced to explain decoupling between incompatible and compatible elements in MORB lavas in 'transitional' ocean ridge segments (e.g. FAMOUS, 36–37°N MAR; Langmuir *et al.*, 1977). This concept rests on the premise that melt extraction is generally not complete and that trapped melt increments may accumulate to give enriched melts leaving residues that yield depleted melts. The appeal of dynamic melting is its versatility in geochemical mass balance modelling, especially its ability to exploit very small differences in K_D between incompatible elements. Wood (1979) invoked dynamic melting to explain the enriched plume magmas on Iceland, depleted suprasubduction lavas from the Troodos ophiolite, and an association of enriched and depleted magmas from Skye in the British Tertiary province from an effectively homogeneous source. In each of these settings the processes which might contribute to source heterogeneity are almost certainly distinct from those along normal ridge sections.

The principal problem concerning dynamic melting is the extent of its influence on melt composition rather than its fluid dynamic precepts. For example, the geochemical expression of dynamic melting at 36–37°N on the MAR would be expected *a priori* along normal ridge segments to the south where element decoupling is actually not apparent (e.g. Viereck *et al.*, 1989). The concept of continuous melting (Aherne and Turcotte, 1979; McKenzie, 1984; Ribe, 1985) is similar to dynamic melting, and the geochemical mass balances involved are those adopted in other open system fractionation models (O'Hara, 1977; O'Hara & Mathews, 1981; O'Hara, 1985; see below). Thompson *et al.* (1985) and McKenzie (1985a) also appeal to models dependent on small differences in solid–liquid K_D values to explain element–isotope decoupling and the observation of radioactive disequilibrium between ^{238}U and its decay product ^{230}Th. Oxburgh (1980) proposed that low degree melts in the peripheral regions of a melt zone will compositionally dominate mantle-derived magma batches, whereas O'Hara (1985) proposed that incompatible element abundance and ratio characteristics in magmas are dominated by vanishingly small melt increments at the edge of partial melt regions and are thus highly susceptible to the shape of molten zones in upwelling mantle.

Albarede (1988) and Langmuir and Planck (1988) recently questioned the compositional significance of variable partial melt shape and related dynamic melting hypotheses, and independently demonstrated that the major and trace element compositions of magmas thus generated are essentially indistinguishable from those predicted for batch equilibrium melting. Whereas the significance of fluid dynamic factors to mantle melting is undisputed and accountable in geochemical models, the predictive aspects of dynamic melting and related concepts of partial melt shape must also take account of the compatible behaviour of major and trace elements that appears to characterize at least some N-MORBs (Viereck *et al.*, 1989).

7.5.2 *Mid-ocean ridge fractionation models*

Early attempts to explain ocean ridge proccesses were based on the lithological and seismic layering of ophiolites (Chapter 4). Cann (1974) developed the concept of a continuously evolving, globally extensive magma chamber through which magma is processed beneath mid-ocean ridges (Chapter 1). Magma fed to the chamber cools and fractionates, crystallization products being accreted laterally to form the plutonic layer 3, and derivative melt is emplaced upwards to form layer 2, consisting of sheeted dykes (the fossile conduit system) and the uppermost eruptive layer. Cann's (1974) model. known as the 'infinite onion' (Figure 7.6a), recognizes that changes in spreading rate profoundly influence the mode of magma supply and the resulting lithospheric structure. Fast spreading rates will attenuate the conduit system so that a laterally stabilized steady-state magma chamber effectively 'bleeds' to the eruptive site. Slow spreading, in contrast, will constrain the width and extend the depth of the magma chamber such that the sheeted dyke layers will be thicker and the consolidated plutonic layers will be more heterogeneous. Calculated thermal conductivities (Sleep, 1975, 1978; Kuznir, 1980) confirm that magma chamber stability is highly dependent on the spreading rate and that at very low rates chambers may be ephemeral or even non-existent. Although most applicable to fast-spreading axes, this concept

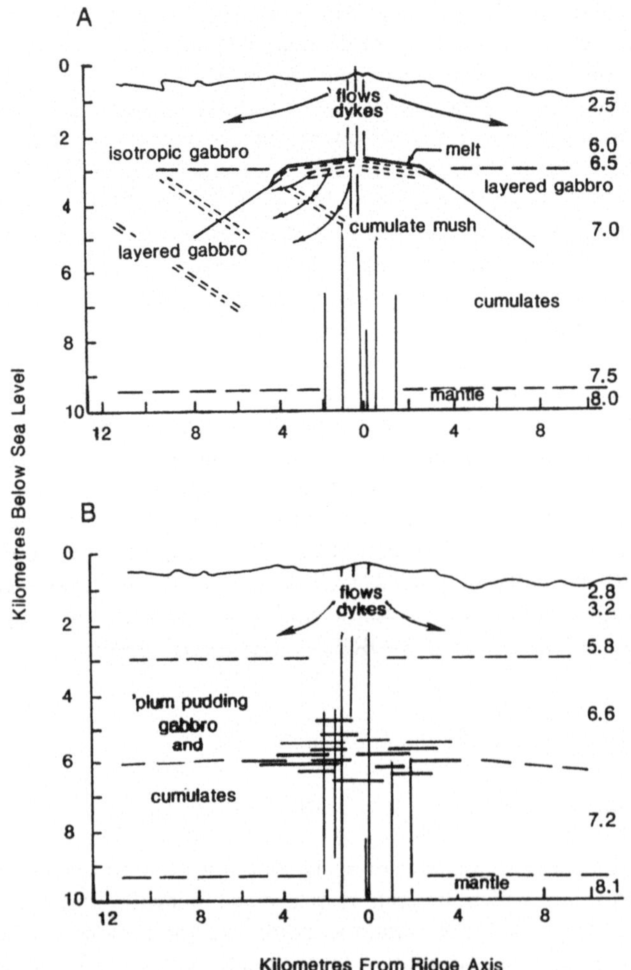

Figure 7.6 Comparison of steady state 'infinite onion' (Cann, 1974) and discontinuous 'infinite leek' (Nisbet and Fowler, 1978) oceanic ridge magma systems, with associated crustal seismic structure (in $km\,s^{-1}$). (a) A large magma chamber continuously replenished by new mantle melts and tapped by eruption was proposed as a steady-state feature stabilized preferentially at fast spreading ridges (Sleep, 1975; Kuznir, 1980). Fractional crystallization produces cumulus phases which accrete to the chamber walls and floor. Compositional steady state is maintained by periodic refilling from below and tapping by eruption (see O'Hara, 1977; O'Hara and Mathews, 1981). Compositional effects would be those of a thoroughly mixed system conditioned by, but not adhering to, low pressure isobaric phase equilibria. (b) A plexus of isolated magma bodies represents melt trapped in the $7.2\,km\,s^{-1}$ layer at the base of elastic crust, from which melt either rises to the surface leaving a cumulate residue, or becomes trapped at depth to form small pockets of gabbro. Expected compositional effects would be those produced by polybaric fractionation in separate coeval transport systems; discrete compositional magma groups, reacted phenocrysts and plagioclase accumulation (Flower, 1980, 1981b; Elthon, 1984). Seismic structures interpolated from (a) Detrick *et al.* (1987) and (b) Nisbet and Fowler (1978).

has served as a paradigm for interpreting data from both slow and fast spreading ocean ridges. Multi-disciplinary investigations of active ridges have involved submersible sampling, gravity, magnetic and heat flow surveys, seismic experiments and (in some instances) coordinated basement drilling.

7.5.2.1 *Slow spreading ridges.* Perhaps the best known of these investigations is the FAMOUS–AMAR programme conducted between 1974 and 1978 by French and American investigators at 36–37°N on the Mid-Atlantic Ridge. The study area consisted of three contiguous spreading segments, one of which (FAMOUS) is currently active and the others (Narrowgate and AMAR) temporarily quiescent (Bryan and Moore, 1977; Stakes *et al.*, 1984). The FAMOUS and AMAR programmes established the volcanic and structural morphology in the three rifted segments. The FAMOUS rift is dominated by axial highs whereas the rift in Narrowgate and AMAR is dominated by normal faults. In FAMOUS, younger median eruptives are mostly pillows of primitive picrite and plagioclase cumulates. Older massive flows form the valley floors and flanks and are (cpx-pl-ol)-saturated and relatively enriched in FeO and incompatible elements (Bryan and Moore, 1977; Hekinian *et al.*, 1976; Le Roex *et al.*, 1981). Bryan and Moore (1977) and Bryan *et al.* (1981) proposed that the chemical zonation of the rift valley floor reflects a zoned steady-state magma chamber (Figure 7.6b), cooler, more evolved parts of which are tapped by flank eruptions whereas hotter, more primitive parts feed magma to the rift axis. The decoupling of incompatible from compatible element abundances (e.g. REE) and ratios was attributed to a combination of volatile transfer (Bryan and Moore, 1977) and differential fractional crystallization and mixing (Bryan *et al.*, 1979).

Flower *et al.* (1977) and Byerly and Wright (1978) observed similar geochemical relationships in basalts drilled on a spreading flow line from FAMOUS but attributed these to multiple fractionation systems tapping discrete sources. Their rejection of onion-type models stemmed from the recognition of compositionally distinct magma groups showing within-group consistency of incompatible and compatible element variation. At upper levels of the deeper sections these groups appeared sequentially, whereas at lower levels they were interlayered and in some cases mixed. Dynamic melting (Langmuir *et al.*, 1977) was a further attempt to explain the geochemical differences without recourse to a heterogeneous source or to processes in the magma supply system (see discussion above). Flower *et al.* (1977), Byerly and Wright (1978) and Le Roex *et al.* (1981) attribute differences in normative and trace element composition to distinct high Ca (low Ni) and low Ca (high Ni) primitive melt batches, derived respectively by melting at high (> 20 kbar) and low (10 kbar) pressures (Le Roex *et al.*, 1981). The FAMOUS and AMAR basalts exhibit many of the features cited as evidence for magma mixing and/or polybaric fractionation including resorbed clinopyroxene, reacted plagioclase megacrysts and between-group clinopyroxene mass balance discrepancies (Flower *et al.*, 1977; Byerly and Wright, 1978; Dungan and Rhodes, 1978). Stakes *et al.* (1984) argued forcibly for open system fractionation processes on the basis of the AMAR data and prefer an onion-like system of the type proposed by Bryan *et al.* (1979) for FAMOUS.

However, despite evidence for low velocity high attenuation regions in the upper mantle (Steinmetz *et al.*, 1977; Fowler, 1976) no studies have convincingly demonstrated the presence of shallow magma chambers beneath the Mid-Atlantic Ridge. Seismic refraction results for the MAR (Fowler and Matthews, 1974) were synthesized by Fowler (1976) in a velocity model that appears to preclude extensive steady-state chambers beneath the axis. The model was supported by the absence of significant shear wave attenuation although small scale (< 2 km wide) pockets of melt could be accommodated. Nisbet and Fowler (1978) accordingly proposed the 'infinite leek' as a limiting case where axial chambers are vanishingly small and rising magma is propagated largely via cracks. Most ocean ridge systems lie somewhere between the onion and leek extremes, the MAR being closer to the latter. The leek model accommodates closed system magma fractionation at the base of the lithosphere and in small subvolcanic chambers such that REE and other incompatible element patterns are established in the source. Flower and Robinson (1979) proposed that the imbrication of lava 'packets', equivalent to discrete eruptive pulses, would preferentially expose the distal early erupted products of each packet at flank sites in the median rift, in contrast to late-stage primitive representatives of packets erupted at the axial active zone. Accordingly the chemical zonation of rift eruptives may be an artefact of the magma emplacement mechanism rather than the surface expression of the magma processing region. It is clear from the evidence for magma mixing in this and other parts of the Atlantic axis that fractionation, even within a leek system, cannot be a simple closed system process.

7.5.2.2 *Fast-spreading ridges.* Since the completion of the FAMOUS–AMAR programme, studies of the active East Pacific Rise (EPR) have thrown further light on magma supply processes beneath ridge axes. The ROSE (Rivera Ocean Seismic Experiment) project was conceived as a means of defining magma chamber stability and crust structure at a fast spreading axis, that of the EPR between the Clipperton Fracture Zone (9°N) and the Rivera Fracture Zone (20°N). Low velocity crustal regions were reported from the EPR at 21°N (Orcutt *et al.*, 1976; Rosendahl *et al.*, 1976; McClain and Lewis, 1982; Reid *et al.*, 1977) and 9°N (Harron *et al.*, 1979). These workers proposed that a magma chamber at 9°N could be 5–10 km wide, although Lewis and Garmany (1982) qualified this with the suggestion that such chambers are probably transient. The analyses of seismograms from 11–13°N on the EPR led Bratt and Solomon (1984) and Bratt and Purdy (1984) to conclude that any axial chamber in this region is confined to narrow

Table 7.4 Examples of experimentally synthesized melts: 20–35 kbar

	Basalt No.[a]									
Composition	1	2	3	4	5	6	7	8	9	10
SiO_2	47.03	46.94	47.52	47.39	48.75	48.11	48.02	47.72	47.84	48.63
TiO_2	1.00	0.70	0.73	0.60	0.53	0.97	0.89	0.86	1.10	1.00
Al_2O_3	16.62	15.44	14.39	14.26	11.97	14.65	13.16	13.38	11.85	10.30
FeO^t	9.67	9.83	9.62	9.44	9.71	9.40	9.41	9.37	10.12	11.20
MnO	—	—	—	—	—	0.16	0.21	0.17	0.21	0.12
MgO	12.99	14.24	14.51	15.46	17.67	14.50	17.06	16.72	18.47	21.41
CaO	9.93	10.53	10.98	10.94	9.56	11.08	9.94	10.06	8.08	6.23
Na_2O	2.63	2.30	1.84	1.60	1.36	1.27	1.04	1.07	1.84	1.64
K_2O	0.12	—	—	—	—	0.03	0.01	0.06	0.55	0.64
P_2O_5	—	—	—	—	—	—	—	—	0.07	0.09
Cr_2O_3	—	0.20	0.38	0.28	0.53	0.21	0.20	—	0.16	0.23
Total	99.99	100.18	99.97	99.97	100.08	100.38	99.94	99.41	100.29	101.49
Mg No.	0.73	0.74	0.75	0.76	0.80	0.75	0.78	0.78	0.78	0.79
CaO/Al_2O_3	0.597	0.682	0.763	0.767	0.799	0.756	0.755	0.752	0.682	0.605
Al_2O_3/TiO_2	16.6	22.1	19.7	23.8	22.6	15.1	14.8	15.6	10.8	10.3
CaO/Na_2O_3	3.78	4.58	5.96	6.84	7.03	8.72	9.56	9.40	4.39	3.80
Pressure (kbar)	20	20	20	20	20	20	25	25	30	35
Temperature (°C)	1420	1430	1450	1475	1500	1400	1475	1470	1540	1600
Residue	L	L	L	H	H	L	—	—	H	H

[a](1) 12% melt of MPY-87 (Falloon and Green, 1988); (2) 17% melt of MPY-87 (Falloon and Green, 1988); (3) 24% melt of MPY-87 (Falloon and Green, 1988); (4) 28% melt of MPY-87 (Falloon and Green, 1988); (5) 36% melt of MPY-87 (Falloon and Green, 1988); (6–8) melt compositions of NT-23 (Elthon and Scarfe, 1984); (9, 10) Equilibrated MORB in peridotite sandwich (Takahashi and Kushiro, 1983)
MPY87 = MORB pyrolite, L = lherzolite residue, H = harzburgite residue

conduits or to small isolated bodies with a vertical thickness of about 1 km, and that along-strike crust structure and accretion are heterogeneous.

Detrick *et al.* (1987) reported multichannel seismic results from between 8°50′N and 13°30′N, and located the top of a crustal chamber at 1.2–2.4 km below the seafloor, which is consistent with the previous estimates of Orcutt *et al.* (1976), Rosendahl *et al.* (1976) and McClain *et al.* (1985). This region had been surveyed by SeaBeam and SeaMarc I (Macdonald *et al.*, 1984; Kastens *et al.*, 1986), dredging (Thompson *et al.*, 1985; Langmuir *et al.*, 1986) and high resolution photography and manned submersibles and contains the Clipperton fracture zone, two large offset spreading centres and several smaller deviations from axial linearity ('devals') (Macdonald *et al.*, 1984; Langmuir *et al.*, 1986). Seismic imaging of the chamber defined a maximum width of 4–6 km (2–3 km from each flank of the rise) whereas a weak discontinuous extension of the chamber's roof probably marks the gabbro–dyke transition (Detrick *et al.*, 1987). The shallowest parts of the rise axis are associated with the thinnest crustal lids whereas the chamber roof is deeper, more discontinuous and, in some instances, completely absent, where the axis is deeper. However, the geophysical evidence strongly suggests that crustal magma chambers are not steady-state features in this part of the EPR. The axial magma chamber reflection deepens towards large offset spreading centres and can be traced into regions where spreading centres overlap, although it disappears before reaching the tips of the overlapping ridges.

7.5.2.3 *Segmented magmatic accretion.* Transform fracture zones segment the ridge system on a scale of up to several hundred km, whereas high resolution geophysical surveys have revealed a cellular structure, defined by smaller offsets and devals, on scales of between 30 and 80 km (Schouten and Klitgord, 1983; Schouten *et al.*, 1985). Spreading cells appear to be a primary feature of oceanic ridges and represent a three-dimensional frame of reference for partial melting and magma fractionation processes. Crust generated within the cells appears to have normal seismic structure, whereas at boundaries between cells thinner anomalous crust is generated irrespective of the scale of the offset (Schouten and Klitgord, 1983). Where no offset is apparent the transition between cells may be complex and unstable on scales of 10 km and 10^5–10^6 years, although it may be stable over longer lengths and timescales (Macdonald *et al.*, 1984).

Segmentation of the MAR is evident from the FAMOUS–AMAR studies (Ramberg *et al.*, 1977) and in the TAG area at 25–27°N (Rona and Gray, 1980), while SeaBeam investigations between 8 and 18°N on the EPR show that axial continuity is disrupted in at least six locations by overlapping (*en echelon*) spreading centres (Langmuir *et al.*, 1986). Segmented magmatic accretion (Whitehead *et al.*, 1984; Schouten *et al.*, 1985) thus provides an alternative to both infinite onions and leeks as a means to transport and process mantle melts at oceanic ridges (Figure 7.7). The discovery of overlapping

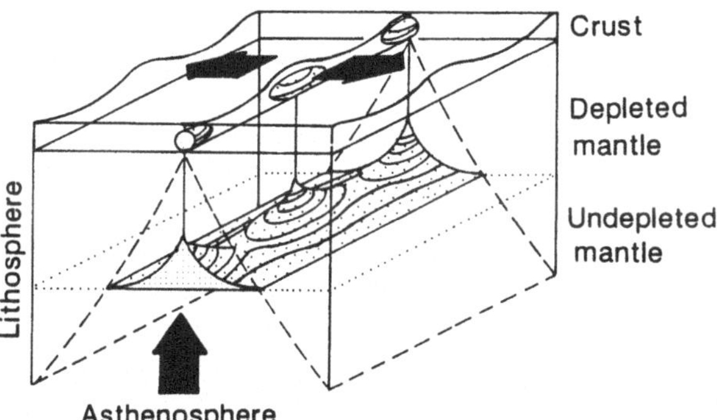

Figure 7.7 Magmatic accretion at a spreading centre (Whitehead *et al.*, 1984). Viscous asthenosphere decompresses in response to separating lithosphere plates. Lithosphere thickens away from the spreading boundary (broken lines) while partial melt accumulates as a continuum below the base of the lithosphere. Owing to its lower viscosity and density the melt zone becomes gravitationally unstable, leading to regularly spaced blobs that rise to form magma chambers. Melt-depleted asthenosphere will continue to rise viscously and, on cooling, become lithosphere. The wavelength of gravitational instability (and offsets causing ridge segmentation) depends on the width of the partial melt zone, hence spreading rate.

spreading centres (OSCs) at the fast spreading EPR also provoked a debate about their significance to magmatic processes. Whereas there is agreement that OSCs are related to the supply of magma from upwelling mantle, unresolved questions concern their relationship to existing axial magma chambers, the required magma budget, the causes of the small offsets giving rise to OSCs and their temporal and spatial evolution. Lonsdale (1983) proposed that the magmatic budget near an OSC is large and that a single steady-state chamber simultaneously feeds the overlapping rifts, and predicts that the lava geochemistry at the two opposing rift tips would be similar. The model also predicts that the magma supply is complex, allowing differential growth of the adjoining ridge tips, despite their sharing a common reservoir.

An alternative model (Macdonald *et al.*, 1984, 1986) appears to better satisfy the requirements suggested by recent imaging of magma chambers along the EPR. Macdonald *et al.* (1986) proposed that overlapping spreading centres are in general overlain by separate, ephemeral reservoirs which may not be synchronous in time. Moreover, these chambers necessarily stem from disparate supply systems and hence accommodate chemically distinct magmas. These views are based largely on the concept that the elevation of the ridge axis is a direct measure of the height reached by upwelling magma along the ridge and is a function of the rate of magma supply relative to spreading rate (Vogt, 1976; Parmentier and Forsyth, 1985, Crane, 1985). Macdonald *et al.* (1986) suggest that the morphology of the EPR results from the temporally and spatially variable emplacement of melt batches injected at discrete points on the rise axis. This model resembles that of Whitehead *et al.* (1984) and Schouten *et al.* (1985) and predicts that melt batches locally swell crustal magma reservoirs to result in a bathymetric high. Magma will migrate 'downhill' from the region of influx towards distal parts of the spreading cell, losing hydraulic head and increasing its distance from the locus of magma replenishment. Such magmas may be tapped where the overlying carapace stretches and fractures. A corollary of this model is that bathymetric lows represent points of zero magma replenishment and reflect the meeting places for magmatic pulses of different provenance. OSCs are interpreted to be a special form of this process, representing the distal ends of magmatic pulses that failed to meet 'head-on' (Macdonald *et al.*, 1986). In general, this model is supported by evidence that the rise axis is underlain by relatively narrow chambers (McClain *et al.*, 1985; Burnett *et al.*, 1985; Detrick *et al.*, 1987) rather than the large chambers predicted by some theoretical thermal models (Sleep, 1975; Kuznir, 1980).

7.5.3 *Intraplate fractionation models*

Intraplate subaerial volcanoes are clearly more amenable to petrological and geophysical study than submerged oceanic ridges. Two lines of research have been particularly rewarding in developing models for OIB fractionation. The first is coordinated real-time studies of active volcanoes, which have yielded information on eruptive chronology, properties of the mantle source and the configuration of magma transport and reservoir systems (Wright and Swanson, 1987). The second line is petrological and geochemical investigations of older eroded intraplate edifices. These may be partially enveloped by rejuvenescent activity and may provide a means of identifying intraplate primitive melts and constructing and testing magma fractionation models.

The Kilauea volcano is one of five shields constituting the island of Hawaii, perhaps the best and most studied example of oceanic intraplate magmatism. The Kilauea volcanic system has been studied intensively by personnel and associates of the US Geological Survey Hawaii Volcano Observatory, which has allowed the repeated testing of models for magma uprise and fractionation. Fundamental properties of the Kilauea system have been established which are clearly applicable to other intraplate and ridge-related hot-spot systems. A primary region of magma storage was located by three-dimensional seismic methods between 2 and 4 km below the summit area, extending laterally, and tapped by distributary rifts of several kilometres extent (Ellsworth and Koyanagi, 1977; Ryan, 1988; Ryan *et al.*, 1981; Thurber, 1984). Studies of the relationship between seismic activity and ground deformation yield information on magmatic intrusion and indicate that magma transport rates through the system are rapid (Klein *et al.*, 1987). Secondary storage reservoirs were located below the flank rift zones (Wright and Fiske, 1971; Swanson *et al.*, 1976b; Ryan *et al.*, 1981) and the dynamic relationship of these to the primary reservoir and conduit system was defined in three-dimensional space (Swanson

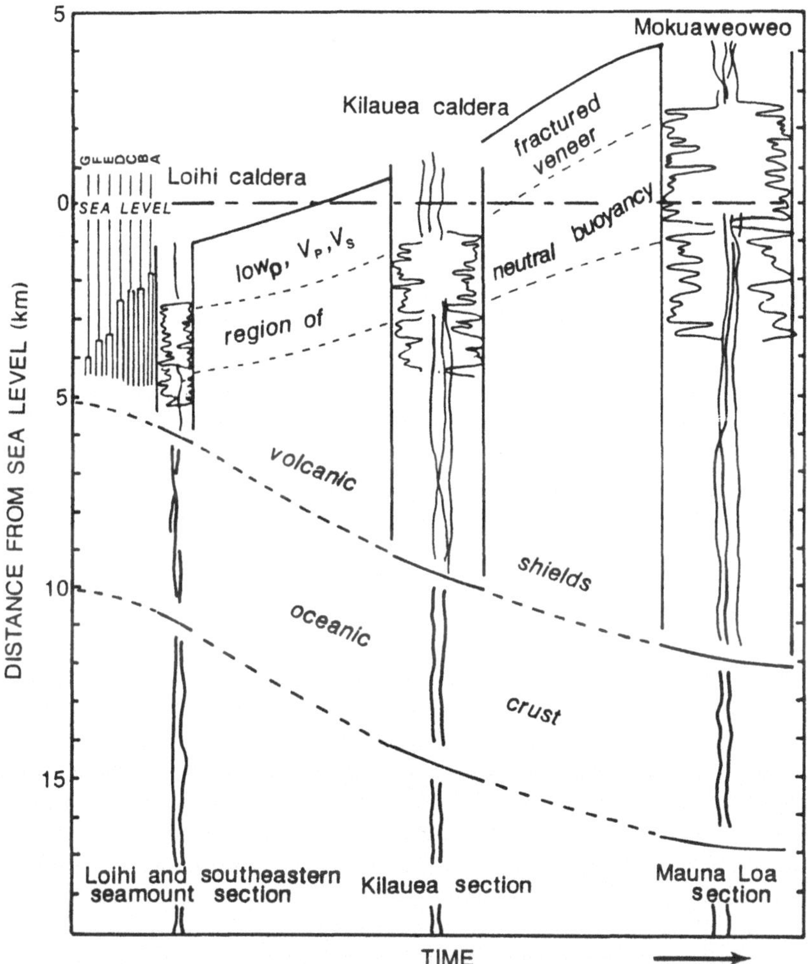

Figure 7.8 Schematic model for the development of magma systems in intraplate oceanic volcanoes, based on Hawaii (from Ryan, 1987). As a volcanic shield develops from an infant submarine state (e.g. Lohi seamount) to maturity (e.g. Mauna Loa) it adjusts isostatically beneath its eruptive load. The zone of neutral buoyancy (determining the size and depth of magma chambers) migrates upwards with progressive elevation of the fractured volcano surface. Crust structure and transitions between the oceanic crust and volcanic shield have been constrained by geophysical studies cited by Ryan (1987). Active magma reservoirs and their dyke systems are shown in outline. Seamounts adjacent to Hawaii are ranked according to their summit depth below sea level: A – Apuupuu; B – Palmer; C – Dana; D – Day; E – Indianapolis; F – Hohonu; G – Green.

et al., 1976a; Ryan *et al.*, 1981; Dvorak *et al.*, 1986). Studies of the petrographic and compositional variation for all eruptions since 1952 yield quantitative models of fractional crystallization and magma mixing within rift zones, serving as a prototype for similar models adopted for oceanic ridges and ridge-related hot-spot loci (Wright and Fiske, 1971; Wright *et al.*, 1975; Bjornsson *et al.*, 1978; Wright and Tilling, 1980; Garcia and Wolfe, 1988).

One of the most significant observations at Kilauea is the close connection between shallow and deep processes. The rate of magma supply to the shallow storage region has been estimated at 0.1 km^3 per year (Swanson, 1972; Dzurisin *et al.*, 1984). Estimates of eruptive output and growth by intrusion have been correlated with isostatic subsidence of the pile. Magma drained from shallow reservoirs during summit or rift activity is rapidly replenished from the mantle (Jackson *et al.*, 1975; Swanson *et al.*, 1976a), apparently without prior entrapment in deeper parts of the system.

Ryan (1987) examined the implications of density differences between magma and the surroundings of magma reservoirs and rift systems. A zone of neutral buoyancy exists above which magma will descend under the influence of negative buoyancy and below which magma will tend to ascend. In Hawaii the region of neutral buoyancy coincides with the location of summit reservoirs at Kilauea and Mauna Loa, and provides the long term stability of reservoirs and their rift systems. These observations led Ryan (1987, 1988) to develop a generalized model for the Hawaiian shields. As shield volcanoes mature and increase in elevation their magma storage reservoirs undergo a systematic increase in elevation, as illustrated in the progression from an infant submarine stage (Loihi seamount) to subaerial maturity (Mauna Loa) (Figure 7.8). Mechanically, this process is marked by the simultaneous elevation of the low density fractured shield surface region, the level of neutral buoyancy and the deeper region separating the field of fracture and pore fluid compression (above 9 km local depth) from the field of solid phase compression (below 10 km). The rising reservoirs leave consolidated remnants of pre-existing reservoirs in their wake, now characterized by high elastic wave velocities. At oceanic ridge systems the process of lateral magma injection along the axis of evolving ridge segments is expected to follow the horizon of neutral buoyancy (Ryan, 1987), a corollary of which is that the injection of magma as dykes is maximized along the strike of an active segment.

7.6 Concluding statements

1. The combined effects of partial melting conditions and the dynamics of mantle overturn and lithospheric plate spreading are crucial determinants of magmatic properties, as measured at the earth's surface. It is important to understand such processes as they provide the means to interpret

thermal and compositional attributes of the mantle and, in turn, its dynamic history.

2. Recent research has contributed quantitative and, in many instances, experimentally verifiable models in four principal areas: (1) Production of magma within the earth depends largely on the ambient temperature of the asthenosphere–lithosphere boundary, as determined by factors such as lithospheric extension and mantle upwelling. (2) The compositions of primary (unfractionated) melts, once established for oceanic ridge or intra-plate tectonic settings, are sensitive to thermal and compositional variations in the mantle, providing a basis for further interpretation of mantle characteristic trace element and isotopic parameters. (3) In both oceanic ridge and intra-plate systems the chemical fractionation of magma depends on the extent of lithospheric dilation, magma supply rates and the combined constraints of isostasy and fluid dynamics. (4) Despite the complexities of open system partial melting and magma fractionation processes, valid interpretations of geochemical data are possible given a rigorous combination of phase equilibrium, mass balance and fluid dynamic factors.
3. Much work remains to reconcile the predictions of dynamic melting concepts with simple geochemical mass balance models and to develop such models in a geophysical context. However, it is clear that definitive models of oceanic magmatism as a planetary phenomenon now exist based on diverse lines of geological and geophysical research.

8 Metamorphic and hydrothermal processes: basalt–seawater interactions

GEOFFREY THOMPSON

8.1 Introduction

The volcanic seafloor, once formed, undergoes changes in its physical, chemical and mineralogical properties as a result of interactions with seawater. Part II of this book discusses the importance of many of these initial basalt properties and how their measurement can be interpreted to infer the source, melting history, magmatic evolution and volcanic processes which formed the oceanic basalts. This chapter discusses how these properties can be changed, and the resulting metamorphosis and evolution of the oceanic crust as it ages and passes from its place of origin at the accreting plate margin to its eventual return to the mantle at subduction zones.

8.1.1 *Importance of seawater–rock interactions*

Seawater–rock interactions are important in a number of ways and the most significant of these are described in the following sections.

8.1.1.1 *Controlling the chemistry of seawater.* Fluxes of ions between the reactants, basaltic rocks and seawater, help to buffer and maintain the steady-state composition of the oceans. How the oceans maintain a steady-state composition has long been of interest to geochemists and chemical oceanographers. The importance of the role of seawater–basaltic crust interactions in this process has only recently been recognized. Discussion of this phenomenon and estimates of some of the fluxes involved can be found in papers by Deffeyes (1970), Hart (1970), Corliss (1971), Hart (1973), Thompson (1973), Spooner and Fyfe (1973), Wolery and Sleep (1976), Humphris and Thompson (1978a), Mottl and Holland (1978), Holland (1978), Edmond *et al.* (1979a, b), Hart and Staudigel (1982), Thompson (1983, 1984), Alt *et al.* (1986a) and Wolery and Sleep (1989).

8.1.1.2 *Changing the chemistry of the oceanic crust.* The result of this process is that the crust returned to the mantle at subduction zones is different in composition from that formed at accreting plate margins. This subducted crust affects the eventual composition of island arc magmas, and even the composition of oceanic island basalts (Kay, 1980; Perfit *et al.*, 1980; Chase, 1981; Hofmann and White, 1982; Morris and Hart, 1983; Hart and Staudigel, 1989).

8.1.1.3 *Formation of hot springs and ore deposits on the seafloor.* The discovery of iron-rich sediments associated with spreading centres was one of the first observations which suggested that hydrothermal circulation might be extensive in the oceanic crust (Bostrom and Peterson, 1966; Bischoff, 1969). In recent years hot springs and the associated ore deposits of polymetallic sulphides, iron oxides, iron silicates and manganese oxides have been reported from all the major oceanic ridges and back-arc basins. Reviews of these findings are discussed in reports such as Edmond *et al.* (1981), Thompson (1983), Rona *et al.* (1984), Rona (1988), Scott (1985) and Thompson *et al.*, 1988.

8.1.2 *Controls of seawater–rock interactions*

The extent and rate of seawater–rock interactions are controlled primarily by the following processes.

8.1.2.1 *Solution circulation.* The amount of seawater passing through the rock (the water: rock ratio) is a function of the permeability. The permeability is controlled by the morphology of the particular rock, the type of geological formation and the tectonic setting, which controls crack formation.

8.1.2.2 *Temperature of reaction.* In general, the higher the temperature the faster and greater the extent of chemical reaction. The minerals formed and precipitated from seawater–rock interactions vary as a function of temperature and are the key to identifying the different metamorphic facies. Chemical exchange between the basalt and seawater also varies with temperature, not only with respect to rate but also to the direction of exchange. Some elements which are added to the rock from seawater under low temperature conditions may be leached at higher temperatures.

8.1.3 *Effects of seawater–rock interactions*

Seawater–rock interactions influence many of the properties of oceanic basalts.

8.1.3.1 *Acoustic properties.* Compressional wave velocities are a function of the density of the substrate. Seawater–rock interactions generally hydrate the rock and reduce the density (Christensen and Salisbury, 1972; Fox *et al.*,

1973). Changes in the compressional wave velocity as a result of low temperature reactions and hydration of the oceanic crust with distance from the ridge axis have been used to deduce the time and rate of chemical reactions (Hart, 1973).

8.1.3.2 *Magnetic properties.* Unaltered oceanic basalts typically have a high magnetic susceptibility (up to 30×10^4 emu cm^{-3}) and a high remanent magnetization (up to 113×10^4 emu cm^{-3}). These values markedly decrease with distance from the spreading axis due to the oxidation and breakdown of the iron–titanium minerals, which control the magnetic properties, by reaction with seawater at low temperatures (Irving *et al.*, 1970; Watkins and Paster, 1971; Marshall and Cox, 1971; Fox and Opdyke, 1973). Basalts that have been metamorphosed by high temperature reactions also show a low magnetic susceptibility and very low remanent magnetization, in addition to a change in the Curie temperature (Luyendyk and Melson, 1967). Areas of the seafloor which have been altered by hydrothermal circulation show distinct lows in the residual magnetic intensity (Rona, 1978a, b; Tivey *et al.*, 1989, Woolridge *et al.*, 1990), a property which can be used in prospecting for ore deposits on the seafloor.

8.1.3.3 *Porosity.* Cracks, pores and voids in the fresh basalts are progressively infilled by the precipitation of new minerals produced as a result of seawater–rock interactions. Hydrothermal circulation in the crust requires a permeable medium in addition to a temperature gradient, and porosity is a critical property in modelling circulation. The porosity also affects the bulk or integrated compressional wave velocity of the geological formation. Near spreading centres, where fresh basalts are common, the porosity is very high as a result of plentiful voids in and between the individual lava formations. Although the compressional wave velocities of the fresh, dense basalts as measured in hand specimens are high, the bulk compression velocities of the geological formations are very low. However, as the voids and spaces become infilled with secondary minerals resulting from water–rock interactions, the bulk compressional wave velocities of the geological units actually increase, even though the hand specimen may be less dense than the fresh basalt precursor (Purdy, 1987).

8.1.3.4 *Heat flow.* Both the observation and measurement of heat flow are directly affected by the circulation of seawater in the volcanic crust, which efficiently extracts heat from the lavas. The deviation from model values of heat flow (Figure 8.1), based on the conductive cooling of the oceanic basaltic crust near the accreting plate margin, is due to convective cooling of the crust by hydrothermal circulation (Langseth and von Herzen, 1970; Talwani *et al.*, 1971; Sclater *et al.*, 1974; Andersen and Hobart, 1976). This discrepancy between the theoretical and actual observations can be used to estimate the

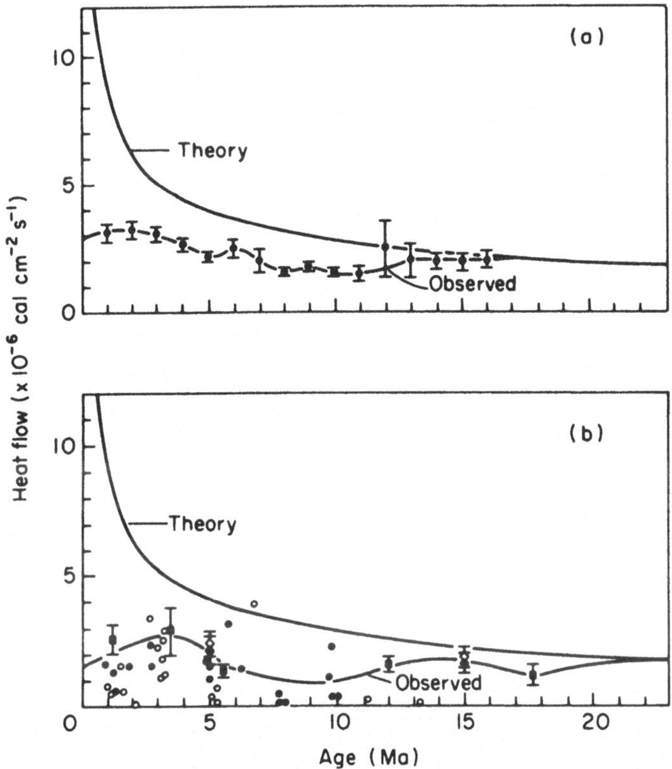

Figure 8.1 Theoretical versus observed heat flow profiles as a function of crustal age for (a) fast spreading and (b) slow spreading mid-ocean ridges. For data sources and discussion, see Wolery and Sleep (1976).

magnitude of heat transfer (Williams and von Herzen, 1974; Andersen and Hobart, 1976). It can also be used to calculate the volume and flow-rate of oceanic hydrothermal systems (Wolery and Sleep, 1976). Downflow and upflow zones of individual hydrothermal cells may also be recognized from heat flow measurements and their relative distributions (Langseth *et al*., 1983).

8.1.3.5 *Chemical composition.* The exchange of ions between seawater and basaltic rock results in chemical fluxes that may influence and control the composition of seawater and produce basalts which are very different in composition from the precursor. Changes in the composition of basalt must be considered if the altered samples are used to deduce petrogenetic processes (Hart, 1971; Thompson, 1973, 1983, 1984; Andrews, 1977, 1978; Floyd and Winchester, 1978; Ludden and Thompson, 1978, 1979; Honnorez, 1981; Alt *et al*., 1986a; Bienvenu *et al*., 1990).

8.1.3.6 *Mineralogy.* Seawater–rock interactions produce a variety of new minerals. These are generally hydrated, less dense and of a different chemical composition to the precursors. The assemblages of minerals produced are a function of the physical and chemical conditions of the reaction. In general these can be grouped into metamorphic facies, although the metamorphic textures often seen in the classic continental facies are not commonly observed (Miyashiro *et al.*, 1971; Humphris and Thompson, 1978a; Cann, 1979; Elthon, 1981; Thompson, 1984).

The effects of seawater–rock interactions can generally be considered under the term metamorphism as, whatever the temperature of reaction, the basalt is fundamentally changed in composition and mineralogy. In practice, however, metamorphism is often considered as the result of two processes. The reactions at low temperatures, $< 70°C$, often ambient seawater temperatures, are pervasive, ubiquitous and continue for long periods of time ($> 10^6$ years). These processes are often referred to as weathering, halmyrolysis, or low temperature alteration. The second process occurs at elevated temperatures of the order of 70–400°C or higher and is generally restricted to regions near accreting plate margins or where localized mid-plate volcanism may be present, i.e. the heat source is magma or hot rocks, and the time of reaction is relatively short (10^2–10^4 years). These higher temperature reactions are often referred to as hydrothermal alteration and result from thermally driven convecting seawater. In this chapter metamorphism is considered under these two thermal regimes.

8.2 Low temperature alteration

8.2.1 *Dredged basalts*

Early studies of basalts recovered from the seafloor by dredging recognized that many of the samples had undergone some reaction with seawater (Wiseman, 1937; Nicholls and Bowen, 1961; Bonatti, 1965; Moore, 1966; Honnorez, 1967, 1972; Hay and Ijiima, 1968a, b; Paster, 1968; Hart 1969, 1970, 1971, 1973; Philpotts *et al.*, 1969; Miyashiro *et al.*, 1969; Hart and Nalwalk, 1970; Miyashiro *et al.*, 1969, 1971; Matthews, 1971; Hekinian, 1971; Jakobssen 1972, 1978; Muehlenbachs and Clayton, 1972, 1976; Thompson, 1973; Melson and Thompson, 1973; Shido *et al.*, 1974; Frey *et al.*, 1974; Scott and Hajash, 1976; Ludden and Thompson, 1978, 1979; Stakes and Scheidegger, 1981; Bienvenu, 1990. A good review of many of the early studies can be found in Honnorez (1981). In general, these studies noted that the reactions with seawater at low temperatures led to the palagonitization of the glass rims of pillows (the commonest form of dredged basalts), oxidation of Fe^{2+} to Fe^{3+}, hydration and the alteration of groundmass minerals to smectite clays.

The alteration of dredged basalts typically takes place under oxidative conditions and at high water to rock ratios. Chemically the basalts gain in elements such as K, Cs, Rb, B, P, U, Th and Li, and lose Ca, Mg and Si, a process which is to some extent dependent on the degree of alteration. Elements such as Fe, Mn, Na, Cu, Ba, Sr and the LREE vary in mobility depending on the local conditions within a given pillow. Isotopically the basalts show increased $^{18}O/^{16}O$ and $^{87}Sr/^{86}Sr$ ratios. Elements such as Al, Ti, Zr, Ta, Hf, Nb and the HREE remain unaffected. Mineralogically the glass is palagonitized, plagioclase is often altered to K-feldspar and smectite, olivine to smectite, and titanomagnetite to titanomaghemite. Vugs and cracks are often infilled with hydrated iron oxides and smectites.

Honnorez (1981), based on previously reported work, concluded that the submarine alteration of basaltic glass at low temperatures resulted in the formation of phillipsite, smectites and Fe–Mn oxides. The alteration occurs in three stages. The initial stage is characterized by Na > K phillipsite and K–Mg-rich smectite, and the bulk glass is hydrated and oxidized with a concomitant increase in the K, Na and Mg contents and a loss of Ca. The second, or mature, stage has a similar mineralogy but the phillipsite typically has a K > Na content, and the smectite is Fe–Mg saponite or nontronite with a high K concentration. The final stage has Ca-poor, K > Na phillipsite, K–Fe–Mg smectite and hydrated Fe–Mn oxides. The uptake from seawater is K > Mn > Na with Ca, Mg and Si being lost from the glass.

Some workers (Hart, 1970, 1973; Thompson, 1973, 1983, 1984; Honnorez, 1981; Wolery and Sleep, 1989) have attempted to calculate the fluxes of elements exchanged during the alteration of dredged basalts and the duration of such reactions. Hart (1976) showed that basalts recovered by dredging, in addition to those from the upper parts of drilled sections of oceanic crust, showed increasing water contents with distance from the ridge crest out to 80 Ma (Figure 8.2). Thompson (1983) showed that dredged basalts had increasing alteration out to 57 Ma, although glass alteration appeared to have been completed within 5–10 Ma (Figure 8.3).

8.2.2 *Drilled basalts*

Studies of basalts recovered by drilling have typically shown that alteration at depth within the crust is not always the same as that in dredged basalts or the drilled basalts from the tops of the cores. Lower water to rock ratios and less oxidative, or even anoxic, conditions are more typical (Bass *et al.*, 1973; Bass, 1975, 1976; Andrews, 1977, 1978, 1980; Robinson *et al.*, 1977; Honnorez *et al.*, 1978, 1983; Hart and Staudigel, 1978; Floyd and Tarney, 1979; Donnelly *et al.*, 1979a, b, c; Staudigel *et al.*, 1979; Mevel, 1979; Richardson *et al.*, 1980; Bohlke *et al.*, 1980, 1984; Ailin-Pyzik and Sommer, 1981; Thompson, 1983; Alt and Honnorez, 1984; Alt *et al.*, 1986a, b; Wolery and Sleep, 1989).

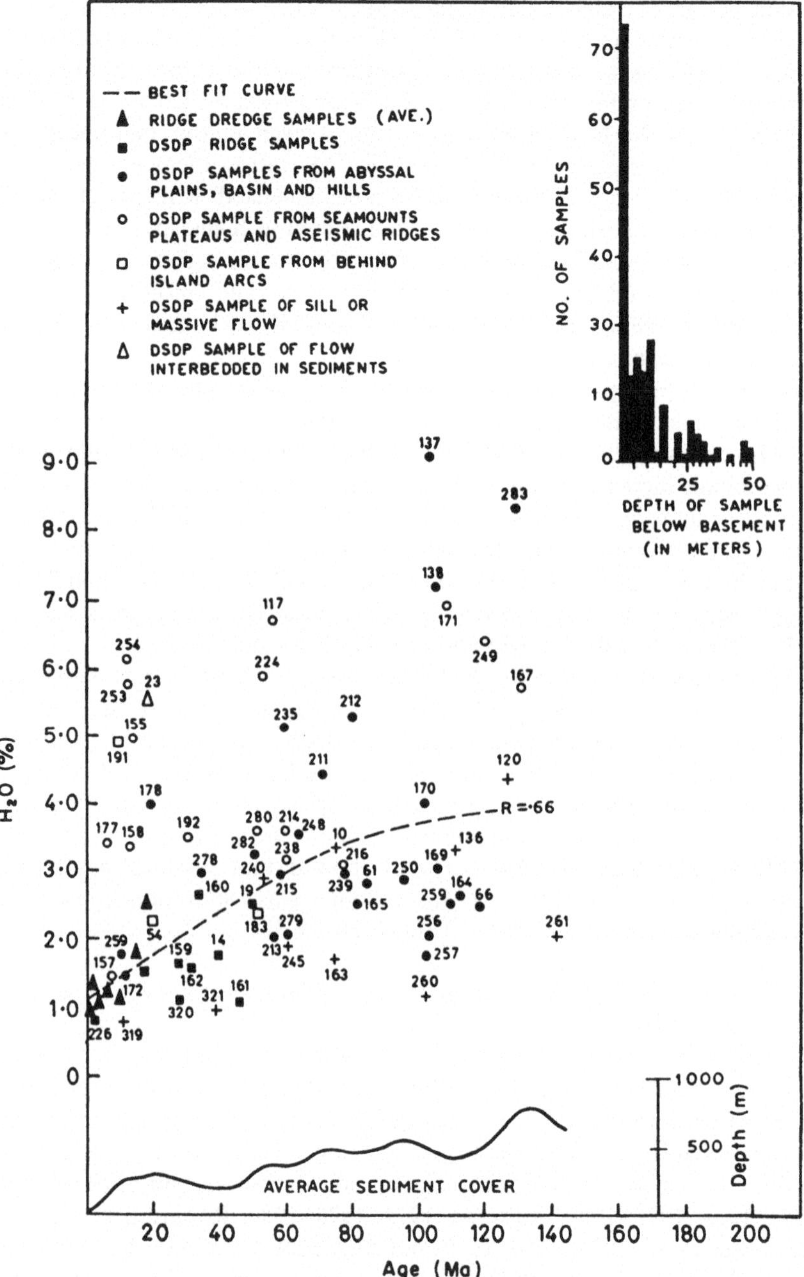

Figure 8.2 Average H_2O contents of basalts from DSDP basement sites as a function of age. Most of these sites are shallow (less than 10 m penetration) and show water uptake for at least 80 Ma (from Hart, 1976).

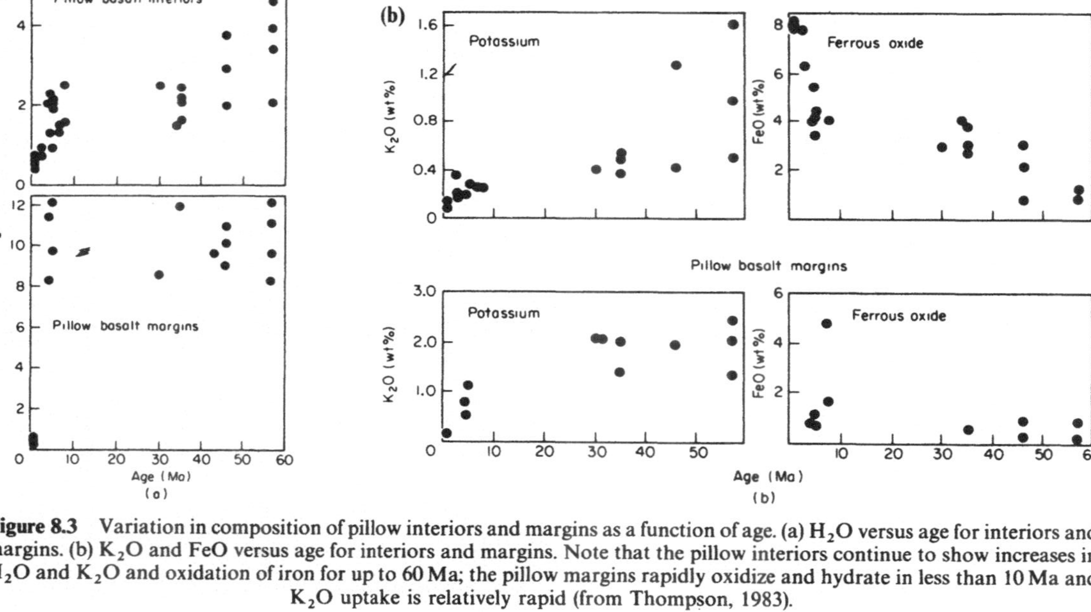

Figure 8.3 Variation in composition of pillow interiors and margins as a function of age. (a) H_2O versus age for interiors and margins. (b) K_2O and FeO versus age for interiors and margins. Note that the pillow interiors continue to show increases in H_2O and K_2O and oxidation of iron for up to 60 Ma; the pillow margins rapidly oxidize and hydrate in less than 10 Ma and K_2O uptake is relatively rapid (from Thompson, 1983).

Table 8.1 Generalized mineralogy of the alteration zones in pillow and massive basalts of DSDP hole 396B (from Bohlke *et al.*, 1981)

	Grey interior	Black halo	Brown zone
Pillow basalt:			
Plagioclase	Not altered	Not altered	Not altered
Pyroxene	Not altered	Not altered	Not altered
Olivine	Not altered	Not altered	Hydrous Fe oxides
Titanomagnetite	Titanomaghemite	Titanomaghemite	Titanomaghemite and secondary Fe and Ti oxides
Early authigenic pyrite	Hydrous Fe oxides	Hydrous Fe oxides	Hydrous Fe oxides
Vugs	A. saponite/calcite (± Fe, Mn oxides) B. 'amorphous Fe silicate' (± Fe, Mn oxides or phillipsite)	Glauconite-smectite/saponite or Fe, Mn oxides	'Amorphous Fe silicate'/Fe, Mn oxides, phillipsite, or calcite
Veins			Smectites/Fe, Mn oxides/phillipsite/calcite
Massive basalt:			
Plagioclase	Not altered		Not altered
Pyroxene	Not altered		Not altered
Olivine	Not altered		Saponite and hydrous Fe oxides
Titanomagnetite	Titanomaghemite		Titanomaghemite
Early authigenic pyrite	Hydrous Fe oxides		Hydrous Fe oxides
Vugs	Saponite/calcite		Saponite/Fe, Mn oxides/calcite, aragonite
Veins			Smectites/Fe, Mn oxides/calcite

For detailed descriptions, see Böhlke *et al.* (1980)

Table 8.2 Summary of alteration stages and associated mineralogical and chemical changes in DSDP hole 417 basalts (from Alt and Honnorez, 1984)

Alteration	Conditions	Resultant alteration zones	Mineralogical changes	Chemical changes
1. Local high temperature	High temperature reaction of seawater with cooling basalt	Restricted to coarse grained portions of massive units	Cpx + pl + ol partly replaced by sap + chl	Undetermined
2. Hydrothermal	Mixing of cooled hydrothermal fluids with large volumes of oxygenated seawater	Black halos around exposed surfaces	Voids filled and ol partly replaced by celad-nont + Fe-hydroxide	$+Fe^T$, Fe^{3+}/Fe^T, K, Rb H_2O^+, $\delta^{18}O$ −Ca, Mg
3a. Oxygenated seawater alteration	Circulation of oxygenated seawater	Brown–light grey zone pairs, brown zone assoc. with dark grey zone, brown + earthy pillow rim zone, breakdown of glass	Voids filled + glass + pl + ol replaced by beid, Fe-beid, Al-sap, K-spar, Fe-hydrox.; sulphides + mt oxidized	$+Fe^{3+}/Fe^T$, K, Rb, P. H_2O^+, $\delta^{18}O$ −Ca, Mg, Na, Si, Al_3Mn
3b. Sub- to anoxic seawater alteration	Reaction with seawater at low water-rock ratios, or with a large basaltic component	Dark grey zones, further alteration of associated brown zone	Voids filled + glass + pl + ol replaced by sap + py	$+Fe^{3+}/Fe^T$, H_2O^+, $\delta^{18}O$ −Ca, Mg, Si, Al
4. Zeolite and calcite formation	Restricted circulation of seawater-derived fluids with lowered Mg/Ca and higher pH	Zeolite zones, calcite veins	Voids filled + pl replaced by an + nat.; calcite in veins	+Ca, Na, H_2O^+, CO_2, $\delta^{18}O$
5. Late anoxic	Restricted circulation(?) sub- to anoxic		Local ppt of Mn-calcite + py in oxidized rocks	Undetermined

(cpx) clinopyroxene; (pl) plagioclase; (ol) olivine; (mt) magnetite; (sap) saponite; (chl) chlorite; (celad-nont) celadonite-nontronite; (beid) beidellite; (k-spar) K-feldspar; (an) analcite; (nat) natrolite; (py) pyrite

As with the studies of dredged basalts, the drilled basalts indicate that the alteration is path dependent, rarely is equilibrium reached, and it is also time dependent so that effects on a given rock may be additive. In general, the alteration is diffusion controlled with a variety of fronts, leading to zonation in individual samples. It is thus difficult to use any given quantitative factor, such as the amount of H_2O added, the exchange of ^{18}O, or the addition of K, as a definitive marker of the degree of alteration. However, some general agreements on the kinds and stages of alteration have been reached, primarily based on studies of the longer drill cores so far recovered, e.g. DSDP drill holes 396, 397 (Table 8.1) (Andrews, 1977, 1978; Bohlke *et al.*, 1980, 1984; Robinson *et al.*, 1977), DSDP drill holes 417A, 417D (Table 8.2) (Donnelly *et al.*, 1979a, b; Alt and Honnorez, 1984; Bohlke *et al.*, 1984; Staudigel *et al.*, 1981, 1989) and DSDP drill hole 504B (Honnorez *et al.*, 1983; Alt *et al.*, 1986a).

Bohlke *et al.* (1984) and Alt and Honnorez (1984) suggest that the simple model of low temperature alteration as a basalt–smectite mixing line expressed by $\delta^{18}O$ versus H_2O (Muehlenbachs and Clayton, 1972) is not always valid, and deviations from linearity in this plot are due to changes in the water to rock ratios, oxidative versus anoxic conditions and other factors. Following the observations of workers such as Honnorez (1981) and Thompson (1983, 1984) that there are differences between dredged and drilled basalts, they conclude that there are three main types of low temperature alterations.

8.2.2.1 *Oxidative alterations.* These occur under conditions of open, permeable, high water to rock ratios, typical of dredged basalts and the upper parts of drill cores. These often show the following two major alteration zones.

- Brown zones where the plagioclase is altered to beidellite (an Al-rich smectite) and K-feldspar, the olivine to celadonite or nontronite and hydrated iron oxides, and titanomagnetite and sulphides are oxidized. The clinopyroxene is unaltered. Vugs and cracks have hydrated iron oxides and celadonite and the basaltic glass is palagonitized.
- Light grey zones where the plagioclase is altered to K-feldspar, beidellite and calcite, olivine to beidellite and calcite, and the titanomagnetite and sulphides are oxidized. Vugs and cracks have celadonites and calcite and any glass is palagonitized.

Chemically, the basalts are marked by the addition of K, Rb, Cs, H_2O, P, Fe^{3+}, B and U, and the loss of Ca, Mg, and Si.

8.2.2.2 *Anoxic alterations.* These occur under restricted flow, low permeability and low water to rock ratios and are typically found in the deeper parts of drill cores. Two major zones are often noted.

- Dark grey interior zones where the plagioclase is altered to saponite

(Mg-rich smectite), olivine to saponite and calcite, and clinopyoxene, titanomagnetite and sulphides are unaltered. Vugs and cracks typically have saponite, pyrite and minor calcite. Glass often remains unaltered.

- Light brown peripheral zones where the plagioclase is altered to saponite, olivine to saponite, celadonite and hydrated iron oxides. Vugs and cracks have calcite and hydrated iron oxides. Glass is palagonitized.

Chemically these basalts show very little change, although the elements may be redistributed. The water contents are increased but still low; oxidation is restricted to only the light brown zones where the alkali metals K, Rb, Cs and B are slightly increased. Berndt and Seyfried (1986) noted that under anoxic conditions the Li contents may also be increased.

8.2.2.3 *Late stage alteration.* As the crust is sealed off from continuing circulation, the last stages of alteration are characterized by the presence of zeolites (such as analcite and natrolite) and calcite. Chemically the basalts show the addition of Ca, Na, H_2O and CO_2.

8.2.2.4 *Discussion of alteration.* Richardson *et al.* (1980), using the $^{87}Sr/^{86}Sr$ contents of the vein minerals as indicators of the time of deposition (the $^{87}Sr/^{86}Sr$ ratio reflects that of seawater, and the changing Sr isotopic composition of seawater is known as a function of time) suggest that in the deeper crust, oxidative and anoxic alterations take place within the first 3 Ma although palagonitization may take place within 1 Ma. Late stage alteration may be completed within 10 Ma. After this period, no further alteration occurs in the deep crust, except possibly some compaction and dehydration.

Low temperature alteration is ubiquitous and can lead to major effects on the chemical composition and mineralogy. Honnorez (1981) suggests that the term 'low temperature alteration' covers all reactions up to the presence of clear metamorphic minerals such as lawsonite, wairikite and chlorite. Cann (1979) suggests that low temperature alteration is considered as a distinct facies (the brownstone facies) in the metamorphism of the oceanic crust, and that it covers the temperature range 0 to approximately 50–70°C. Seyfried and Bischoff (1979) noted experimentally that at temperatures up to 70°C, alkali metals were taken up by the rock from seawater, as observed in dredged basalts, but at higher temperatures they were leached.

As noted in the previous discussion, the extent and kind of alteration may vary within a single pillow or morphological unit, depending on the porosity, diffusion front and the pathway of the water. DSDP holes 417A and 417D show a marked example of this variation and heterogeneity of alteration (Figure 8.4). Drilled only 400 m apart, 417D was located on a basement high and was extremely altered, mainly under oxidative conditions. Drill hole 417A was located on a basement low and was much less altered, with anoxic, low water to rock ratio alteration dominating (Donnelly *et al.*, 1979a, b, c).

In this instance it is thought that the differences are not solely due to topographic effects; the alteration of 417D was probably effected by warm water (about 30°C) (Lawrence, 1979) circulating upwards as part of an off-axis hydrothermal cell. The topography effect may have led to funnelling of the flow to that site, whereas the nearby basement low saw only cold, downward circulating seawater. Topographic effects on the extent of low temperature seawater alteration were noted for the Troodos ophiolite (Gillis and Robinson, 1985, 1988, 1990). Beneath basement lows, apparently quickly sealed to seawater circulation by sediment and umber infill, the depth of low tempera-

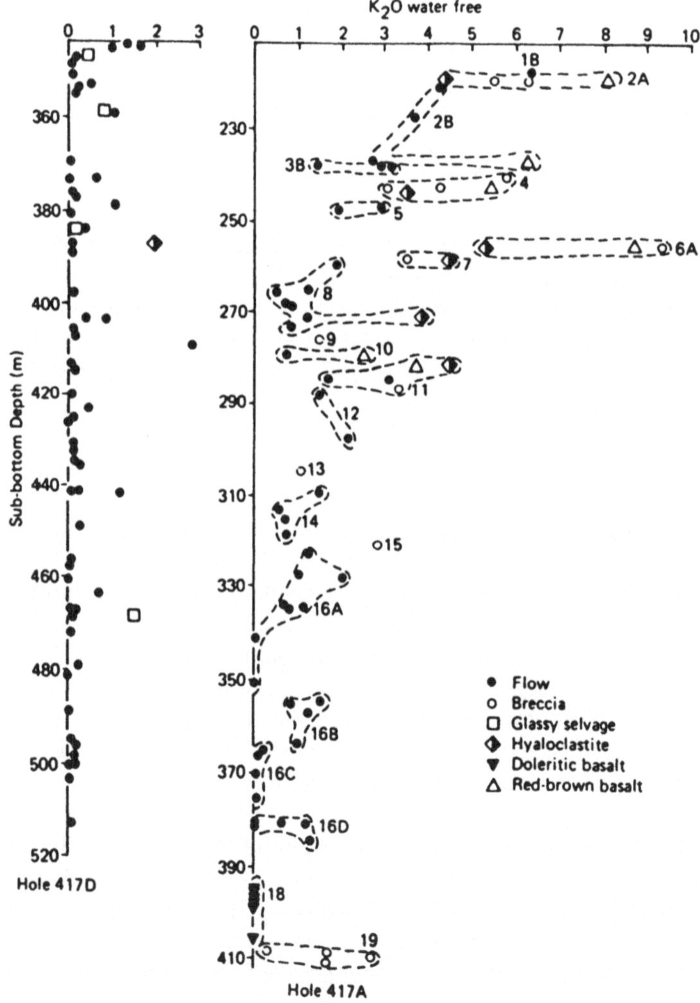

Figure 8.4 K_2O contents (calculated on a water-free basis) versus depth in DSDP drill holes 417A and 417D (from Donnelly *et al.*, 1979c).

ture alteration into the crust is much less than on the basement highs which remained sediment free and open to circulation for some considerable time. Troodos, as at hole 504B (Alt *et al.*, 1986a), has a low temperature altered crust overlying a high temperature metamorphosed basaltic crust. Clearly, in estimating the total effects of seawater alteration, both low and high temperature alteration have to be considered.

8.3 High temperature reactions

8.3.1 *Dredged basalts*

The recovery of metamorphosed basalts ranging from zeolite to amphibolite facies is confirmation that hydrothermal circulation is widespread in the oceanic crust. The metamorphosis of the basalts results in major chemical and mineralogical changes, although the classic schistose textures of continental metabasalts are not often found. The first recovery of metabasalts was during the *Challenger* expedition (Murray and Renard, 1891). Other reported recoveries include Shand (1949), Quon and Ehrlers (1963), Melson *et al.* (1966), Melson and van Andel (1966), Bogdanov and Ploshko (1968), Melson *et al.* (1968), Cann (1969), Aumento *et al.* (1971), Miyashiro *et al.* (1971), Melson and Thompson (1971), Cann (1971), Chernysheva (1971), Thompson and Melson (1972) and Hekinan and Aumento (1973).

Humphris and Thompson (1978a,b) have reported extensively on metabasalts recovered from the ocean floor. They noted that these rocks are predominantly found in the axial valley of slow spreading ridges or from large fault scarps in transform faults. They are rarely found in deep drill holes

Table 8.3 Metabasalts dredged from active oceanic ridges and their reported mineralogies (from Humphris and Thompson, 1978a)

Metamorphic facies	Reported minerals	References[a]
Zeolite	Analcite, stilbite, heulandite, natrolite–mesolite–scolectite series, mixed-layer chlorite–smectite	1
Prehnite–pumpellyite	Prehnite, chlorite, calcite, epidote	2
Greenschist	Albite, actinolite, chlorite, epidote, quartz, sphene, hornblende, tremolite, talc, magnetite, nontronite	3
Amphibolite	Hornblende, plagioclase, actinolite, leucoxene, quartz, chlorite, apatite, biotite, epidote, magnetite, sphene	4

[a](1)Aumento *et al.* (1971); Miyashiro *et al.* (1971); Shido *et al.* (1974); (2) Hekinian and Aumento (1973); Mevel (1981); (3) Melson *et al.* (1968); Miyashiro *et al.* (1971); Aumento *et al.* (1971); Shido *et al.* (1974); Aumento and Loncarevic (1969); Melson *et al.* (1966); Melson and Thompson (1971); Bonatti *et al.* (1975); Thompson and Melson (1972); Cann (1971); Hekinian (1968); Chernyseva (1971); (4) Cann and Funnell (1967); Cann (1971); Aumento *et al.* (1971); Bonatti *et al.* (1971); Bonatti *et al.* (1975); Bogdanov and Ploshko (1968); Rozanova and Baturin (1971)

(Cann, 1979), with the exception of DSDP hole 504B (Alt *et al.*, 1986). The predominant types of rock recovered by dredging are greenstones, that is, metabasalts of the greenschist facies. Table 8.3 shows the mineral assemblages reported in recovered metamorphosed rocks from the seafloor.

With regard to the development of metamorphic facies, Humphris and Thompson (1978a) suggested the inclusion of the pumpellyite–prehnite facies between the zeolite and greenschist facies, although reported occurrences are very rare. However, neither Elthon (1981) nor Cann (1979) included this in their classifications. Elthon (1981) and Miyashiro (1973) noted that the products of ocean floor metamorphism differ from those of continental metamorphism, particularly in the hornblende-bearing facies. Based on observations from the seafloor (Miyashiro *et al.* 1971), from ophiolites (Elthon and Stern, 1978) and from geothermal regions (Kristmannsdottir, 1976; Brown and Ellis, 1970), Elthon (1981) suggested a subdivision of the greenschist facies, and renaming of the hornblende to actinolite facies (Table 8.4; Figure 8.5). In basalts recovered from the seafloor, the amphibolite facies is rare, although it is often reported in recovered metagabbros (Miyashiro *et al.*, 1971; Miyashiro *et al.*, 1979; Ito and Anderson, 1983; Tiezzi and Scott, 1980; Stakes and O'Neill, 1982; Mevel, 1987).

Table 8.3 shows reported mineralogies in dredged metabasalts. Figure 8.6 shows the congruent mineral reactions that take place over a range of temperatures (Cann, 1979). Greenschist facies rocks predominate in the recovered dredge hauls. Table 8.5 shows the different sub-assemblages that have been reported in recovered seafloor rocks of the greenschist facies, and the alteration of the primary minerals (Gillis and Thompson, 1990).

Chemically the major change noted in altered metabasalts is that Mg is taken up by the basalt and Ca leached, on an approximately 1:1 molar basis. K, Si, B, Rb and Li are generally leached from the rock, although retrograde low temperature reactions may result in the later uptake of alkali metals. Other elements such as Na, Fe, Mn, Sr, Ba, Co and Ni show various effects which are very dependent on the conditions of reaction such as temperature,

Table 8.4 Equilibrium metamorphic assemblages and proposed facies classification for ocean floor metamorphism (from Elthon, 1981)

Type of assemblage	Classification	Temperature range (°C)	Mineral assemblage[a]
I	Lower greenschist facies	230–320	Epi + chl + alb + sph + calc ± qtz ± sulph
II	Upper greenschist facies	320–525	Actin + epi + chl + alb + sph ± qtz ± sulph
III	Lower actinolite facies	525–600	Actin + calc plag + sph
IV	Upper actinolite facies	> 600	Actin + calc plag + mag + ilm

[a](epi) epidote; (chl) chlorite; (alb) albite; (sph) sphene; (calc) calcite; (qtz) quartz; (sulph) sulphide; (actin) actinolite; (calc plag) calcic plagioclase; (mag) magnetite; (ilm) ilmenite

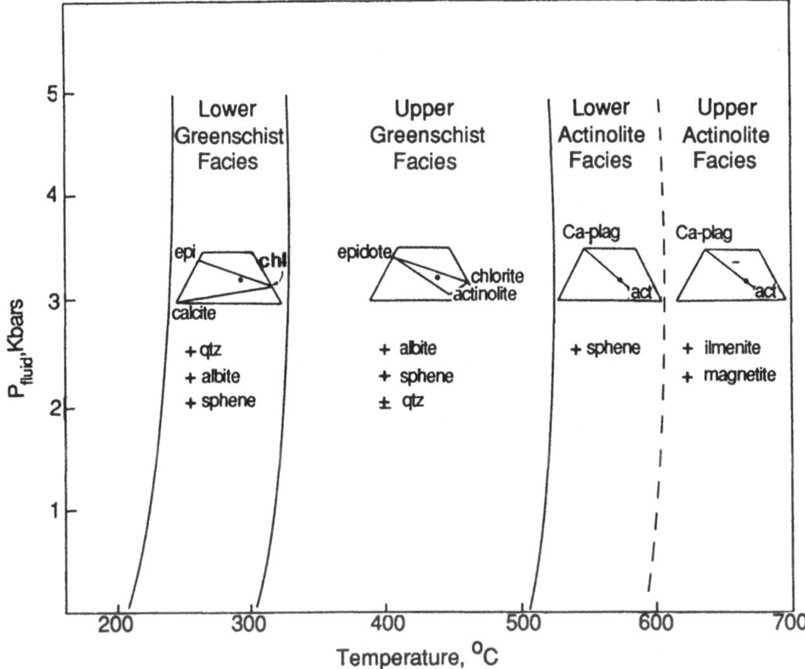

Figure 8.5 Facies classification for hydrothermal metamorphism in the oceanic crust. Metamorphic assemblages are projected into an ACF diagram, where A = moles of (Al_2O_3 + Fe_2O_3 − Na_2O − K_2O), C = moles of CaO and F = moles of FeO + MgO + MnO. The black dot is an average basalt composition. Temperatures are deduced from experimental data and from metamorphic assemblages observed in geothermal regions (from Elthon, 1981).

water to rock ratio and the mineralogy (Humphris and Thompson, 1978a, b).

Mottl (1983), based on experimental results for seawater–basalt reactions at high temperatures, and the observed mineralogy of dredged metabasalt, suggested that the mineral assemblage is a direct result of the water to rock ratio. Figure 8.7 shows the predicted relationship and the principal mineral assemblages. In this model it is assumed that any anhydrite formed has been redissolved on cooling of the system. The model is for reactions occurring in the 300–450°C range. It predicts that, at low water to rock ratios, the typical greenstone assemblage of chlorite + albite + epidote + actinolite predominates. At high water to rock ratios (> 50) chlorite + quartz is the principal assemblage. As noted by Humphris and Thompson (1978a) and Cann (1969), examples of the two extreme assemblages can often be found in a single sample, in which alteration presumably occurred at different seawater to rock ratios at the rim compared to the core. The result is shown in Table 8.6 with the respective mineralogies.

Mottl (1983) also noted that the composition of minerals such as chlorite and actinolite should vary with respect to Mg and Fe content as a function

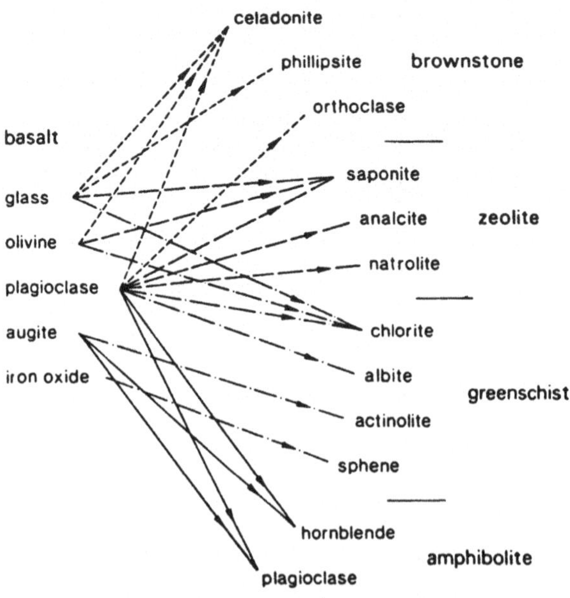

Figure 8.6 Metamorphic facies and congruent mineral reactions for basalt–seawater reactions over a range of temperatures (from Cann, 1979).

Table 8.5 Alteration assemblages and primary mineral alteration in oceanic greenstones (from Gillis and Thompson, 1990)

Type of assemblage	Primary mineral alteration[a]	Secondary assemblage[b]	Degree of alteration (vol%)	References[b]
I	Cpx, amphi (S) Plag, albite (M–T) Oliv, talc ± chl (T) Ti-mag, sphene (S) Mesos, chl + qtz + epi (S–T)	Alb + chl + qtz ± epi	10–100	1–3, 4, 6
II	Cpx, amphi (S–T) Plag, epi (S) Oliv, talc ± chl (T) Ti-mag, sphene (S–T) Mesos, amphi + chl	Calcic plag + amphi + chl ± trace epi	10–100	1–4
III	Chlorite–quartz breccias	Chl + qtz ± sulph	100	5, 7
IV	Epidosites	Epi + qtz + mag + chl	100	8

[a](Cpx) Clinopyroxene; (plag) plagioclase; (oliv) olivine; (Ti-mag) titanomagnetite; (mesos) mesostasis; (amphi) amphibole; (chl) chlorite; (qtz) quartz; (epi) epidote; (sulph) sulphides; (mag) magnetite; (s) slight; (M) moderate; (T) total

[b](1) Melson and Van Andel, 1966; (2) Melson *et al.*, 1968; (3) Miyashiro *et al.*, 1971; (4) Humphris and Thompson, 1978; (5) Mottl, 1983; (6) Alt *et al.*, 1986; (7) Delaney *et al.*, 1988; (8) Quon and Ehlers, 1963

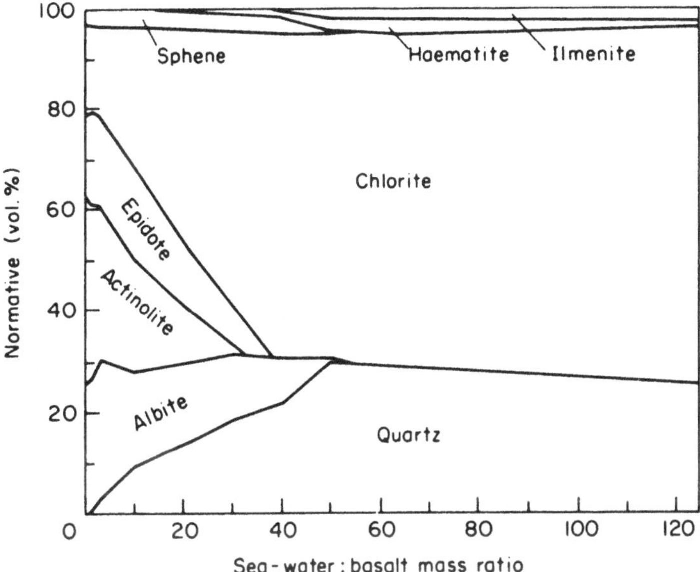

Figure 8.7 Model predicting mineral assemblages and proportions produced when basalt reacts with seawater in different water to rock mass ratios. The model is based on experimental data and observed assemblages in dredged metabasalts (from Mottl, 1983).

Table 8.6 Gains (+) and losses (−) of chemical components in core (orthospilite) and glassy rind (hyalospilite) of metamorphosed basalt pillow compared to the fresh precursor, and changes observed in basalt glass altered at water to rock ratios of 10:1 and 62:1 (from Cann, 1969, and Seyfried *et al.*, 1978)

Component	Orthospilites	Basalt glass, (water to rock 10.1)	Hyalospilites	Basalt glass, (water to rock 62:1)
SiO_2	+7.02	−1.63	−7.77	−3.39
Al_2O_3	0	0	0	0
Fe_2O_3	+0.19	−0.10	+10.78	+0.46
MgO	+1.15	+1.97	+4.98	+9.46
CaO	−2.16	−2.92	−8.93	−9.32
Na_2O	+1.88	+0.13	−1.14	−1.61
K_2O	−0.06	−0.10	−0.08	−0.08
H_2O^+	+2.83	+6.18	+7.76	+11.38
Mineralogy	Albite, chlorite, actinolite, pyrite, quartz, sphene	Albite, saponite, chlorite–saponite, actinolite, anhydrite, quartz, pyrite	Chlorite	Chlorite–smectite, anhydrite

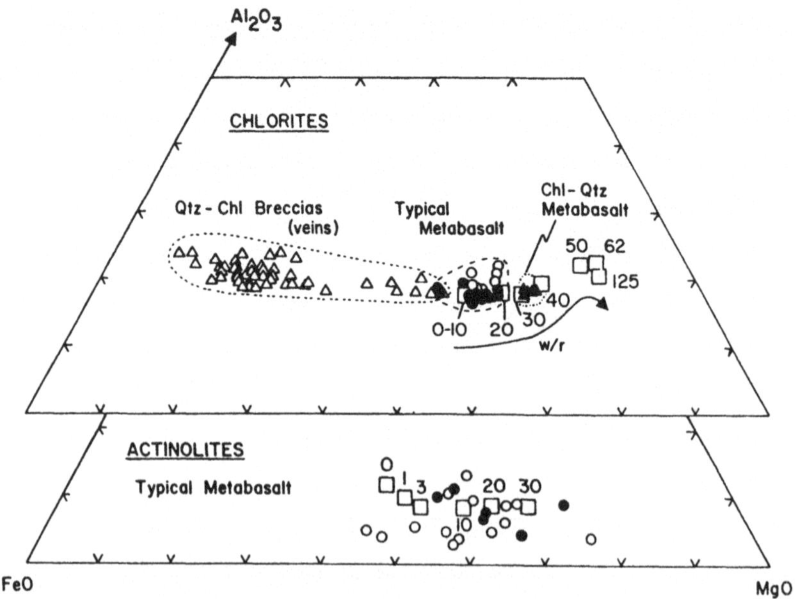

Figure 8.8 Compositions of chlorites and actinolites from seafloor greenstones compared with those from a normative model based on experimental data. Recovered metabasalts are depicted by open and closed circles and closed triangles. Chlorites from quartz–chlorite veins and breccias are iron-rich and depicted as open triangles. The open squares are model predictions for various water:rock ratios (from Mottl, 1983).

of the composition of the hydrothermal solution and the water to rock ratio. Figure 8.8 shows the predicted model data for mineral compositions compared to observed values. Mottl (1983) suggested that the iron-rich chlorites probably formed in a hydrothermal upflow zone where the water to rock ratios are high but the solutions are magnesium-poor due to previous reactions with basalt.

The predominance of metabasalts of the greenschist facies and the paucity of rocks of the zeolite facies suggests that the rapid heating of seawater occurred to temperatures greater than 200°C and that the transition zone must be relatively sharp.

8.3.2 *Experimental evidence*

Reactions of basalts with seawater at temperatures from 70 to 500°C, and pressures up to 1 kbar, have provided insight into the resulting modification of the basalt and the reacting fluid (Bischoff and Dickson, 1975; Hajash, 1975; Seyfried and Bischoff, 1977, 1979, 1981; Mottl and Holland, 1978; Seyfried *et al.*, 1978; Mottl *et al.*, 1979; Mottl and Seyfried, 1980; Seyfried, 1987; Berndt *et al.*, 1988, 1989). Summaries of the results can be found in Mottl (1983), Thompson (1983) and Seyfried (1987), and clearly show that the principal variables controlling the mineral assemblages are temperature, water to rock ratio and type of rock. For example, glassy basalt alteration is characterized by the formation of albite, actinolite–tremolite and chlorite, whereas crystalline dolerite alteration results in albite, clinozoisitic epidote and chlorite (Seyfried 1987; Berndt *et al.*, 1988, 1989).

Within the past decade studies of vent fluids from hot springs on the seafloor (Edmond *et al.*, 1979a, b, c; Von Damm, 1988; Campbell *et al.*, 1988), combined with experimental data, have suggested that vent fluids are controlled by rock buffering and probably by equilibrium controls with respect to a greenschist-type mineral assemblage at depth. Thermodynamic calculations suggest that the fluid compositions are close to or at saturation with respect to quartz, albite, muscovite, mixed-layer smectite–chlorite, epidote and pyrrhotite at temperatures of 350°C (Bowers and Taylor, 1985; Bowers *et al.*, 1988).

Although the mineral assemblages observed in rocks so far recovered from the seafloor differ from the experimentally and theoretically deduced assemblages, they are in agreement with an origin from reaction with heated seawater in hydrothermally driven circulating cells close to volcanic heat centres.

8.3.3 *Ophiolite evidence*

Steinmann (1927) was the first to describe the so-called trinity assemblage of serpentine, spilite (metabasalt) and chert, later recognized as ophiolite assemblages and fragments of oceanic crust (Gass, 1968; Thayer, 1969; Coleman, 1977). Studies of ophiolites have given important information on the nature and extent of hydrothermal reactions in the crust, e.g. Troodos (Moores and Vine, 1971; Spooner *et al.*, 1977; Baragar *et al.*, 1987, Schiffman *et al.*, 1987; Schiffman and Smith, 1988; Richardson *et al.*, 1987; Gillis and Robinson, 1990), Newfoundland (Coish, 1977; Casey *et al.*, 1985), Taiwan (Liou and Ernst, 1979), Chile (Elthon and Stern, 1978), Oman (Hopson *et al.*, 1981; McCullough *et al.*, 1980) and Josephine (Harper *et al.*, 1988). In general, except for spatially restricted discharge zones where epidosites predominate, the alteration of the extrusive sequence in ophiolites is similar to that observed in the oceanic crust (Gillis and Robinson, 1988). However, alteration in the deeper parts, such as the sheeted dyke complex of ophiolites, appears to be more uniformly pervasive and epidote-rich than metamorphic rocks recovered from the seafloor (Baragar *et al.*, 1987; Gillis and Robinson, 1990).

8.3.4 *Drilled basalts*

DSDP hole 504B on the Costa Rica rift is the only deep basement site to recover significant amounts of metamorphosed basalt. Alt *et al.* (1986) have described in detail the mineralogical and chemical effects of the high

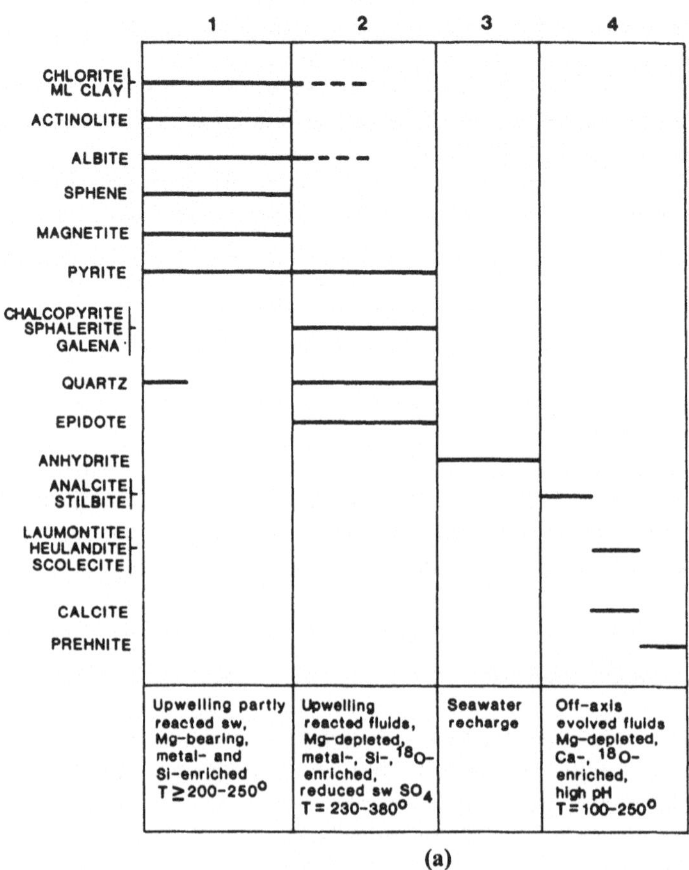

Figure 8.9 Summary diagram for mineralogy in the alteration stages in (a) the lithological transition zone and dyke sections and (b) the pillow section of DSDP drill hole 504B (from Alt *et al.*, 1986a).

temperature alteration (200–380°C) of basalts and sheeted dykes in a 1 km section of the crust (Figure 8.9).

Zeolite facies alteration is noted, principally in the lithological transition zone from pillow lavas to dykes, but is a fairly restricted zone. Saponite, celadonite, natrolite, mesolite, thompsonite, analcite and apophyllite, with aragonite and calcite, predominate. Minor scolecite, heulandite, laumontite, chabazite and stilbite were also reported.

Greenchist facies alteration is the dominant form of metamorphism both in the lower pillow lavas and the transition zone, and in the dykes. Reactions with upwelling hot hydrothermal fluids are deduced to have taken place. One episode resulted in the deposition of a mineralized stockwork with

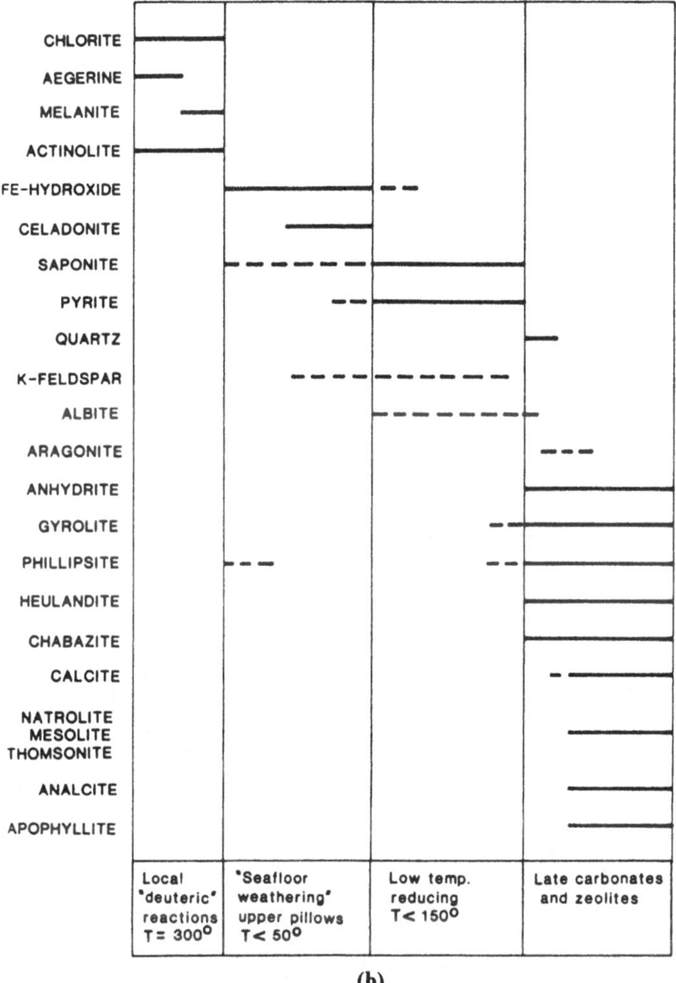

(b)

Figure 8.9 *(Continued)*

massive sulphides such as pyrite, chalcopyrite, sphalerite and minor galena. The principal alteration of massive and pillowed basalts is consistent with observations of dredged metabasalts from the ocean floor. Olivine is replaced by chlorite, mixed-layer clay minerals and minor pyrite. Plagioclase is replaced by albite, heulandite, scolecite and minor prehnite and chlorite; only rarely are epidote and calcite noted. Clinopyroxene is often overgrown by, or has reaction rims of, actinolite. Titanomagnetite is replaced by sphene, and glass is altered to clay minerals, chlorite and sphene. Vein and crack infilling is

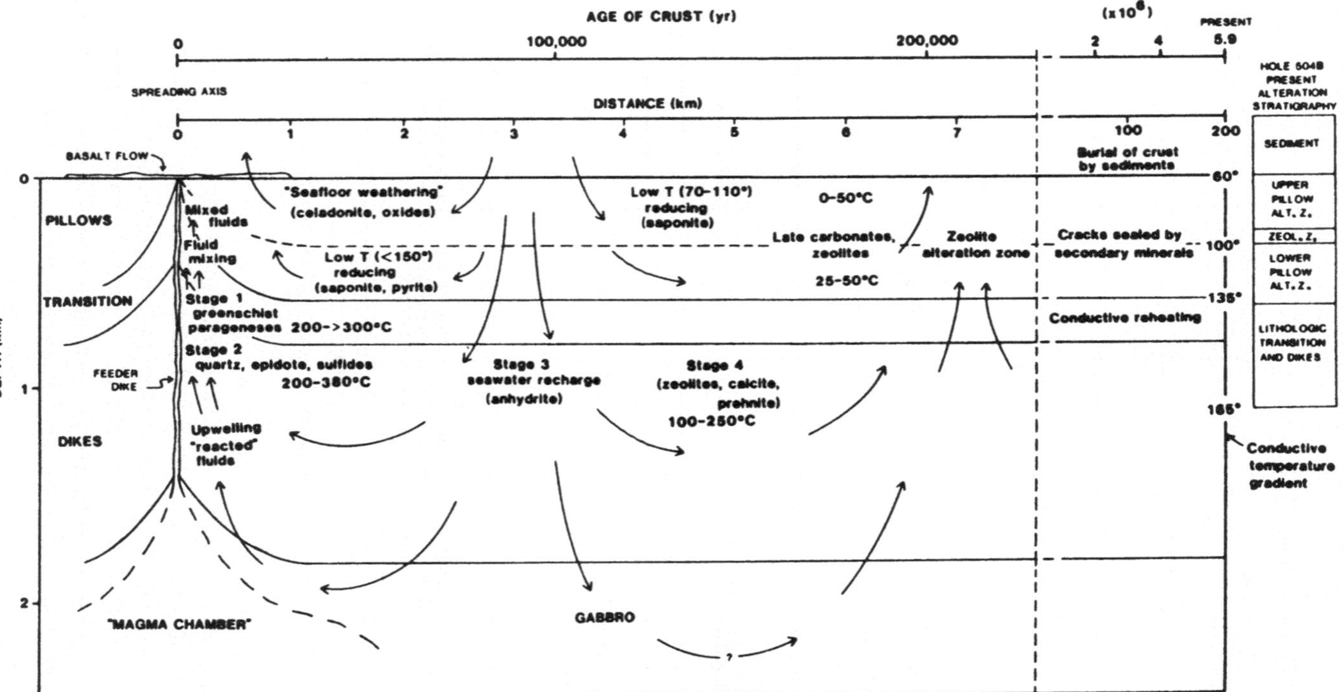

Figure 8.10 Summary diagram for the structure and evolution of the hydrothermal system at DSDP drill hole 504. The schematic cross-section shows the structure of the crust, the distribution of circulation cells, and the various alteration processes with distance from the spreading centre (from Alt *et al.*, 1986a).

typically by zeolites, calcite, epidote, quartz and sulphides. Chemically the high temperature altered zones gain in Mg and lose Si, Ca, and K.

Alt *et al.* (1986) have shown that at DSDP hole 504B, the 5.9 Ma crust has undergone at least four stages of alteration at low and high temperature (Figure 8.10). These stages are as follows.

Firstly, the dykes reacted with seawater (200–300°C) in the upwelling zone of an axial convection cell at the spreading axis, resulting in the formation of greenschist facies paragenesis in veins and host rocks. Mixing of upwelling hydrothermal fluids with seawater circulating in the upper pillow lava section occurred in the lithological transition zone and caused a steep temperature gradient at the base of the pillowed section. The initial effects of low temperature (< 50°C) seafloor weathering began in the upper pillow lavas under oxidizing conditions. Following refracturing of the dykes, a second stage of axial upwelling occurred and hydrothermal fluids (200–300°C) reacted with the dykes. Again, mixing of these fluids with cold seawater in the lithological zone resulted in the deposition of quartz, epidote and sulphides. Low temperature seafloor weathering continued in the upper zone progressing downwards into the crust. Seawater recharge penetrated to depths of 1 km into the crust and anhydrite was locally precipitated.

Off-axis alteration of the dykes was characterized by the formation of zeolites, calcite and prehnite precipitated in veins from fluids of 100–250°C. Low temperature alteration in the upper pillow lava zone evolved into anoxic, low water to rock ratio conditions as the basement was covered with sediment and pathways in the rock became clogged with secondary minerals.

As noted in dredged metabasalts, hydrothermal alteration of the crust in DSDP hole 504B differs from that observed in ophiolites. The transition from low to high temperature is marked and the gradual transition from zeolite to greenschist facies, seen in ophiolites, is not observed. Extensive alteration and development of epidosites, as noted in ophiolites, is not seen in DSDP drill hole 504B. It is not certain whether this is a function of some post-seafloor metamorphic effects in the ophiolites, or variable metamorphic effects in the oceanic crust with the very limited sampling not having recovered all types.

8.4 Concluding statements

1. Basalts on the seafloor undergo reaction with seawater over a range of temperatures. High temperature alteration results from the hydrothermal circulation of seawater associated with volcanism and the formation of new crust at spreading centres. Reactions typically occur in the range 200–400°C and result in predominantly greenschist facies assemblages. The transition from low to high temperature is apparently rapid and zeolite facies assemblages are restricted. High temperature amphibolite facies are

found in deeper levels of the crust, usually in plutonic sequences such as gabbros or peridotite intrusions.

2. Low temperature alteration (3–70°C) is ubiquitous. The upper part of the seafloor basement may undergo reactions for up to 80 Ma. The lower portion of the crust generally reacts under oxidative and/or reducing conditions but is relatively quickly sealed from continued reaction after about 10 Ma. Clay minerals and calcite predominate in the low temperature facies.
3. The oceanic crust as a whole shows the effects of both high and low temperature alteration. The mineralogical effects are summarized in Table 8.7. In Table 8.8 the observed chemical changes are summarized, together with the net effect on the total chemical budget when the two kinds of alteration are considered. Table 8.9 summarizes the estimated fluxes compared to river inputs. The latter estimates indicate that seawater–basalt reactions are an important process in the geochemical cycles of the earth.

Table 8.7 Summary tabulation of mineralogies reported for various metamorphic facies in oceanic basalts

Facies	Mineralogy
Halmyrolysis	Celadonite, phillipsite, palagonite, saponite, montmorillonite, nontronite, Fe–Mn hydroxide, orthoclase, smectite, beidellite, titanomaghemite, calcite
Zeolite	Analcite, stilbite, heulandite, natrolite–mesolite–scolecite, chlorite–smectite, saponite, chabazite, thompsonite, apophyllite
Prehnite–pumpellyite	Prehnite, chlorite, calcite, laumontite, epidote, heulandite
Greenschist	Albite, actinolite, chlorite, epidote, quartz, sphene, hornblende, tremolite, talc, magnetite, nontronite, sulphides
Amphibolite	Hornblende, plagioclase, actinolite, leucoxene, quartz, chlorite, apatite, biotite, epidote, magnetite, sphene

Table 8.8 Summary of elemental rock gains and losses at low and high temperature, together with a summary of the net exchange over the full temperature range

Conditions	Rock gains	Rock losses
Low temperature (< 100°C)	H_2O, K, P, Mn, (Fe), B, Li, Rb, Cs, U, Cu, Zn, LREE, (Ba)	Si, Ca, Mg, (Na), (Sr)
High temperature (> 100°C)	H_2O, Mg, (S)	Sr, Ca, K, Mn, (Fe), B, Li, Rb, Cs, Ba, Sr, (Cu), (Ni), (Zn), (U)
Net exchange over full range of temperatures	Mg, K, B, Rb, H_2O, Cs, U, C	Si, Ca, Ba, Li, Fe, Mn, Cu, Ni, Zn

Table 8.9 Estimates of hydrothermal fluxes between oceanic basaltic basement and seawater (from Thompson, 1983)

	Mass flux (10^{14} g y^{-1})				Mass flux (10^{10} g y^{-1})			
	Si	Ca	Mg	K	B	Li	Rb	Ba
Case A:								
Surface	−0.006	−0.045	−0.03	+0.013	+0.45	+0.44	+0.14	+0.43
Basins	−0.52	−0.082	−0.26	+0.09	+2.69	+2.42	+1.37	+2.73
Flanks	−0.2	−0.47	−0.11	+0.22	+5.12	+3.7	+4.23	+1.1
Axis	−0.87	−1.3	+1.87	−0.49	(−)0	−111	−20.5	−46
Total	−1.60	−1.90	+1.47	−0.17	+8.26	−104.54	−14.76	−41.74
River	−1.99	−4.88	−1.33	−0.74	−47.0	−9.4	−3.2	−137.3
Basement flux as % of river flux	80.4	38.9	110.5	23.0	17.6	1112	461	30.4
Case B:								
Surface	−0.006	−0.045	−0.03	+0.013	+0.45	+0.44	+0.14	+0.43
Basins	−0.52	−0.082	−0.26	+0.09	+2.69	+2.42	+1.37	+2.73
Flanks	−0.2	−0.47	−0.11	+0.22	+5.12	+3.7	+4.23	+1.1
Axis	−0.087	−0.13	+1.0	−0.049	(−)0	−11.1	−2.05	−4.6
Total	−0.82	−0.73	+0.6	+0.27	+8.26	−4.54	+3.69	−0.34
River	−1.99	−4.88	−1.33	−0.74	−47.0	−9.4	−3.2	−137.3
Basement flux as % of river flux	41.2	14.9	45.1	36.5	17.6	48.3	115.3	0.2

PART III ENVIRONMENTS

9 Oceanic islands and seamounts

PETER FLOYD

9.1 Introduction

In the oceanic environment the majority of volcanic activity is concentrated at plate margins, especially at ocean ridges or spreading centres where the oceanic crust is generated. However, the interiors of oceanic plates are invariably pockmarked by numerous basaltic submarine volcanoes (seamounts and guyots) and emergent oceanic islands that testify to extensive, but often localized intraplate volcanism. Oceanic islands within 30° north and south of the equator may be encircled with reefs, or form atolls if the volcanic pedestal is submerged and capped by reefal limestone. The Hawaiian islands are probably the best known example of intraplate volcanism and were considered by Wilson (1963a, b) to be generated by a local upper mantle thermal perturbance that was the source of anomalous mid-plate or hot-spot volcanism. Many of the characteristic linear chains and island groups related to hot-spot activity are sited on large-scale topographic swells (1000–2000 km in width) that elevate some volcanic domains 1–2 km above the adjacent seafloor (Monnereau and Cazenave, 1988; White, 1989). The geophysical properties of swells (e.g. geoid, gravity and heat flow anomalies) suggest that they are probably supported by deep convective thermal perturbances derived from below the lithosphere (Crough; 1983; Watts *et al.*, 1985).

The study of oceanic islands and seamounts is important for a variety of reasons.

- They provide comparative chemical and morphological information on hot-spot relative to other forms of volcanism, and help constrain the nature of mantle processes and compositional domains.
- Young seamounts developed on or near ridge axes provide constraints on models of spreading centre plumbing.
- The thermal and mechanical properties of the lithosphere can be evaluated from the loading generated by seamounts and islands.

- They enable a better overall composition of the oceanic crust to be determined, especially the volcanic portion that is returned to the mantle via subduction.

Uncontaminated by passage through thick continental crust, the chemically heterogeneous basalts of oceanic intraplate volcanism provide a unique window into the composition and structure of the mantle and its kinematics relative to that supplying the oceanic ridges. Although such volcanism has tectonic significance globally, little is known about seamounts in particular, relative to spreading centre volcanics, and many questions concerning their origin, distribution and structural development are only now being fully addressed (Watts, 1984; Keating *et al.*, 1987).

The extent of submarine intraplate volcanism (compared to oceanic islands) has only recently been realized with the development of high resolution sonar techniques that can map volcanic structures of only 50–100 m in height (Chapter 2). The Pacific ocean floor alone is estimated to be covered by about 55 000 basaltic seamounts (Batiza, 1982) or well over 1×10^6 seamounts, including those < 1 km high and assuming a (questionable) uniform distribution (Fornari *et al.*, 1987; Smith and Jordan, 1988). Although the Atlantic ocean floor has not been as extensively studied as the Pacific, it appears less well populated, with estimates in the region of 9000 for seamounts > 1 km high (Litvin and Rudenko, 1973; Kharin *et al.*, 1976). In both oceans the majority of small to medium sized seamounts (1–2 km high) appear to cluster on or around the active axial ridges (Emery and Uchupi, 1984; Fornari *et al.*, 1987; Batiza *et al.*, 1989).

Although estimates vary, intraplate volcanism may consist of up to 25% of the oceanic crust (Batiza, 1982; Jordan *et al.*, 1983), representing a major contribution to the oceanic lithosphere. However, although individual oceanic volcanoes produce vast volumes of lava at high eruption rates ($0.018\,km^3\,y^{-1}$ for the Hawaiian islands; Bargar and Jackson, 1974), the magma production rate for oceanic intraplate volcanism is estimated as about $2.4\,km^3\,y^{-1}$ for intrusive and extrusive products combined or only about 11% of the spreading ridge rates (Fisher and Schmincke, 1984).

This chapter is concerned with the structure and development of oceanic volcanoes, in addition to the compositional variation displayed by selected examples in different eruptive environments. Oceanic intraplate volcanism can be conveniently divided into the following categories: mid-plate, linear chains of seamounts and islands (e.g. Hawaiian-Emperor, Samoan, Tasmantid); linear aseismic ridges (e.g. Walvis Ridge, Ninetyeast Ridge); island groups adjacent to spreading axes (e.g. Azores, Tristan da Cunha); ridge flank and near-axis young seamounts (e.g. Lamont seamounts on East Pacific Rise).

As Iceland represents the product of a hot-spot centred on an active ridge it is considered as an elevated segment of the Mid-Atlantic Ridge and discussed in Chapter 13.

9.2 Hypotheses of intraplate volcanism

In both the oceanic and continental environments intraplate volcanism is generally concentrated in narrow, highly active, linear zones. Oceanic islands and seamounts often form chains, whereas continental regions are scarred by major rift zones, both of which can be thousands of kilometres long. As the chemical nature of intraplate basaltic magmas differs considerably from those generated by shallow melting at spreading ridges, they are often assumed to have been derived from a generally deeper and more enriched source, although the character and depth of the source often depends on the mantle model invoked.

Two main hypotheses have been advanced to explain the origin of intraplate hot-spot volcanism: (1) mantle plumes and (2) propagating fractures.

9.2.1 *Mantle plume model*

In this model hot-spot volcanism in the result of the pressure-release melting of uprising convecting mantle (a thermal plume), which does not usually generate a plate boundary. Laboratory experiments and theoretical modelling indicates that plumes have a mushroom-like cross-section with a long, narrow stalk set deep within the mantle and a broad head with down-curling edges that spreads out under the lithosphere (Whitehead and Luther, 1975; Olson and Singer, 1985; Griffiths, 1986; White and McKenzie 1989a; Griffiths and Campbell, 1990). Plume stalks are generally narrow (about 100 km across), whereas the heads may have diameters up to 2000 km and are composed of very hot mantle material 100–200°C above normal potential mantle temperatures of about 1280°C (White and McKenzie, 1989a). The head diameter is predicted to relate to the depth of origin, with plumes derived from the core–mantle boundary producing large heads (1000–2000 km in diameter) and those from the 650 km boundary layer much smaller heads < 300 km diameter (Griffiths and Campbell, 1990). Laboratory experiments indicate that large volumes of buoyant source fluid must accumulate at an unstable boundary layer before the plume can penetrate the overlying medium (Oslon and Singer, 1985). Generally, little interaction is envisaged between the plume and surrounding mantle such that a continuing supply of deep-seated, extra hot source fluid being carried up the plume tail progressively enlarges the head. However, analyses by Griffiths and Campbell (1990) have indicated that large volumes of relatively cool surrounding mantle can be stirred into and entrained by plumes (derived from the core–mantle boundary) by the time the head is deflected by the lithosphere. This results in the cooling and enlargement of the head and may account for some of the heterogeneity displayed by plume-generated picritic and basaltic magmas (Campbell and Griffiths, 1990).

Plumes cause uplift, fracture, extensive associated volcanism and, in some instances, eventual rifting of extensionally-stressed continental lithosphere and the production of new oceans (White and McKenzie, 1989b). Plumes with large diameter heads are an effective means of generating the sudden appearance of the large volume continental flood basalts that have eruptive provinces about 2500 km across (White *et al.*, 1987; White and McKenzie, 1989a, b; Richards *et al.*, 1989; Hooper, 1990; Campbell and Griffiths, 1990). In the oceanic regime, long before the upsurge of interest in the relationship of plumes and continental flood basalts, the early observations by Wilson (1963a, b) within the Hawaiian chain produced the idea of hot-spot volcanism being caused by a fixed mantle source over which the oceanic plate passed. The line of volcanic islands produced represent the visible trace on the migrating ocean floor of the stationary plume below. Burke and Wilson (1976) have identified about 122 plume-related hot-spots that have been active during the last 10 Ma with about 43% in the oceanic area (Figure 9.1). Morgan (1971) developed the notion of deep-seated and fixed mantle plumes (derived from the mantle–core boundary) that could be used to define a frame of reference for recording the absolute motion of past and present lithospheric plates by back-tracking along the volcanic trace (Morgan, 1972, 1983; Wilson, 1983; McDougall and Duncan, 1980; Duncan and Clague, 1985).

Fundamental assumptions of the plume model in its simplest form include: all plumes are fixed in the mantle relative to each other; the plume is continuous in its activity and is long-lived; there is little mechanical

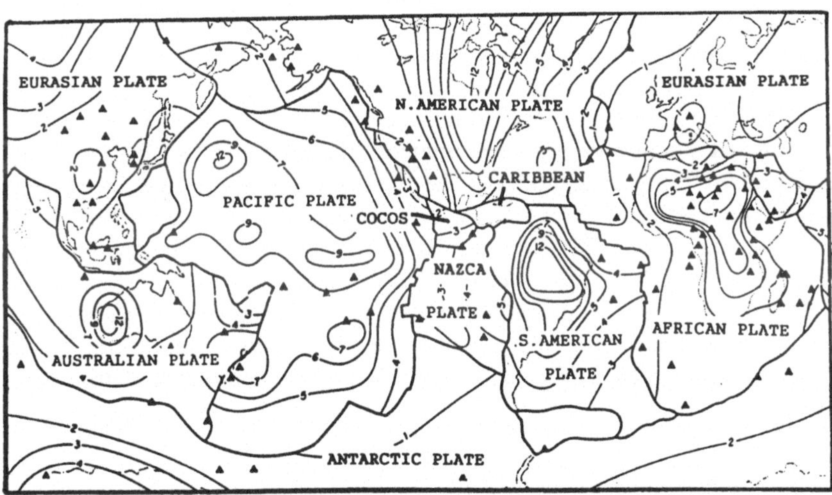

Figure 9.1 Distribution of global hot-spots (triangles) related to mantle plumes (Burke and Wilson, 1976) and superimposed lithosphere 'vulnerability parameter' contours (Pollack *et al.*, 1981). Low values (< 7) characterize areas vulnerable to penetrative magmatism.

interference between the relative motions of the plume (vertical) and the lithospheric plate (horizontal).

Many of the characteristics of Pacific island and seamount chains (Wilson, 1963a, b) can be explained by the presence of a plume. The characteristics include their linearity and parallelism to other chains within the same plate, systematic age progression along the chain away from the current activity situated at one end and the constant rate of inferred progression along the chain which matches that measured for seafloor spreading within any one plate. However, there are various features within linear chains that do not fit this simple version of the model (reviewed by Okal and Batiza, 1987). For example, linearity offsets related to fractures, *en echelon* island systems suggesting adjacent, multiple plumes, variable and intermittent volcanism over a long period in one volcanic group and no-age progression along the chain. If the plume model is used to explain these additional features of oceanic volcanoes it requires modification, whereby intraplate volcanism represents the products of a continuum from strong, continuous deep-seated plumes to intermittent, upwelling blobs of variable size, strength, longevity and depth of origin (Schilling and Noe-Nygaard, 1974; Okal and Batiza, 1987). However, a number of perturbations to plumes, such as erratic plate motion, effects of local transforms, ridge–plume interactions and differential magma storage (Epp, 1984) could produce some of the apparently anomalous volcanic effects.

There are various aspects of intraplate oceanic volcanism that cannot be satisfactorily explained on a modified, waxing and waning, or even perturbed plume model. The development of volcanism at different times and places along a chain, such that there is no age progression, suggests the activity is related to 'hot-lines' rather than hot-spots (Bonatti and Harrison, 1976) and reflect variable upwelling mantle convection spouts aligned to the direction of plate motion. Many small, individual and isolated seamounts developed near or on the flanks of spreading axes cannot readily be generated by deep-seated plumes in this environment (Batiza, 1982). A further major problem for the plume model is the development of massive, large-scale, synchronous volcanic events throughout the Pacific basin during the Cretaceous (Schlanger *et al.*, 1981; Rea and Vallier, 1983) and their limited reactivation in the Tertiary (Haggerty *et al.*, 1982; Schlanger *et al.*, 1987).

Mantle plumes are currently popular not only for generating hot-spot volcanism within oceanic plates, but the enormous continental flood basalt provinces (e.g. Karoo, Deccan) associated with continental rifting and break-up.

9.2.2 *Propagating fracture model*

Even allowing for a number of problems, the plume model suggests that much intraplate volcanism could be related to deep, convective mantle processes. However, although shallow plumes are a theoretical possibility (Parmentier *et al.*, 1975), strong plumes, if they exist, should at least produce seismically detectable, anomalous, low velocity mantle under active centres. Unfortunately, no unequivocal anomalous mantle has been detected below Hawaii (Ellsworth and Koyanagi, 1977), which is generally considered the prime example of plume-generated oceanic volcanism. This leaves open the possibility of other mechanisms to explain the distribution and genesis of intraplate volcanism.

One characteristic feature of continental intraplate volcanism is its association with major crustal fracture zones (e.g. the East African Rift) such that the location of activity appears to be controlled by the (often ancient) structure of the local lithosphere. The propagating fracture model envisages the development of tensional fractures, produced during lithospheric migration, which funnel magmas derived from the asthenosphere to penetrate the lithosphere above (Oxburgh and Turcotte, 1974; Turcotte and Oxburgh, 1978). Heat generated by shearing at the lithosphere–asthenosphere boundary would cause melting and the production of magma at relatively shallow depths (Shaw and Jackson, 1973). In this model intraplate volcanism occurs where the structure of the lithosphere allows and its distribution will depend on a combination of factors such as lithospheric age, thickness and speed of movement (Bailey, 1977; Gass *et al.*, 1978). Thin, slow moving lithospheric plates are more prone to fracturing than older, thicker segments and are therefore more vulnerable to penetration by uprising melts. Although Vogt (1981) has questioned the general application of some of these factors to intraplate volcanism, there does appear to be a crude relationship on the global scale between such lithospheric 'vulnerability' and hot-spot volcanism (Figure 9.1) (Pollack *et al.*, 1981).

Each model provides a reasonable explanation of a variety of different features of intraplate volcanism, whether oceanic or continental. The part played by tensional fracture zones is particularly important, especially in funnelling and focusing magmas through the lithosphere. For example, it has been suggested that Hawaiian magmas may penetrate the lithosphere via a set of south-east propagating fractures that account for the two parallel lines of volcanic loci observed (Jackson and Wright, 1970).

9.3 Seamount distribution and morphology

The subaerial portions of oceanic basalt volcanoes are readily available for observation, although it is only recently that good quality bathymetric maps have allowed the distribution and shape characteristics of seamounts to be documented. Abundance estimates are generally based on counting and measuring all volcanic features within defined regions or along specific seafloor tracks (using sonar equipment such as SeaMarc; see Chapter 2).

From these measurements statistical estimates of areal distributions can be made.

A number of general features emerge for seamount surveys, although the following comments are mainly based on Pacific ocean data.

- Estimates of the abundance per 10^6 km^2 of ocean floor for all Pacific seamounts (with heights > 300 m) range from 1600 to 2400 (Jordan *et al.*, 1983; Abers *et al.*, 1988). However, seamount density on very young crust near the East Pacific Rise increases to > 9000 seamounts per 10^6 km^2 (including small seamounts, 50 m high) and suggests a non-uniform distribution with the highest density closely related to possible conduits such as fracture zones, transforms and offset spreading axes (Fornari *et al.*, 1987; Smith and Jordan, 1988). The very young oceanic crust of the ridge flanks appears to be a major location for the development of small volcanoes not related to hot-spot processes.
- The abundance of seamounts per unit area generally increases with lithospheric age (Batiza, 1982; Smith and Jordan, 1988) and implies that not all non hot-spot volcanoes are produced near the ridge axes, but are developed as the crust ages. Batiza (1982) inferred that the production rate (volcanoes per unit area per unit time) for Pacific off-axis seamounts was inversely proportional to the square root of the lithospheric age with the highest production near the ridge crest on crust younger than about 1 Ma (Fornari *et al.*, 1987). The actual number of seamounts per unit area (especially for those of small size) increases markedly between the ridge axis and 5 Ma crust and then, within error limits, remains relatively constant on crust up to 40 Ma (Smith and Johnson, 1988; Abers *et al.*, 1988).
- Variations in the size of volcanoes indicate that seamount populations show an approximately exponential distribution with numerous small volume structures and relatively few large edifices (Batiza, 1982; Aber *et al.*, 1988; Smith and Jordan, 1988). Relative seamount sizes are a function of lithospheric age and increase with the square root of the basement age (Vogt, 1974a). As the plate thickens the length of the column of magma feeding the volcano also increases such that the height is isostatically controlled. The proportion of large volcanoes appears to increase more rapidly with crustal age than smaller seamounts, which may actually decrease. This is partly due to the older, thicker lithosphere being able to support larger structures, as well as the smaller seamounts becoming buried by sediment and thus being less readily detectable.

Apart from the hot-spot related islands and associated seamounts developed away from actively spreading centres on crust of variable age, many of the numerous non hot-spot seamounts that pepper the ocean floor were initially generated near to, or on, ridge flanks. Virtually all small seamounts had their origin on young, thin crust (< 10 Ma) tapping distinct magma batches via fractures related to the ridge system (Fornari *et al.*, 1987). As the crust moves away from the axis small batches of melt would not be able to penetrate the cooler and thicker crust and small seamount production would virtually cease (Abers *et al.*, 1988). In contrast, larger magma volumes are more likely to penetrate thicker crust in the vicinity of fracture zones and this enables larger seamounts to grow further away for the ridge (Vogt, 1974b; Lowrie *et al.*, 1986). Continued growth depends on the nature and size of the off-axis magma source and local fracture patterns, in addition to the thickness of the lithosphere required to support larger structures.

Both subaerial and submarine volcanic edifices often display a symmetrical cone-like structure with a slightly concave upwards slope on the flanks that steepens towards the summit. This characteristic shape suggests a universal mechanism for growth with the edifice surface approximating to a surface of constant hydraulic potential (Lacey *et al.*, 1981; Augevine *et al.*, 1984). Essentially the shape is determined by the hydraulic resistance of the (assumed porous) edifice to magma flow, such that it seeks the path of least resistance to the surface. This general model can be applied to both large subaerial volcanoes and small submarine seamounts, although in both instances the cone-shape can be modified by parasitic cones, active axial fissures and erosional features.

Recent investigations of seamount bathymetry and observation from submersibles have enabled their overall morphology to be determined in terms of structural development, in addition to providing data for shape statistics (the interrelationship of height, basal diameter, flatness of summit area, crater size, slope). Analysis of the shape of Pacific seamounts (Taylor *et al.*, 1980; Batiza and Vanko, 1983; Jordan *et al.*, 1983; Smith and Jordan, 1988; Smith, 1988) indicates that seamounts range from about 50 m to 4 km in height with the majority being below about 1.5 km, the seamount summit height is about one fifth of its basal radius, the average slope angle is $18 \pm 6°$ and the summit area may show various degrees of flatness (summit/basal radius ratio) such that a truncated cone-shape is commonly produced. Statistically most of the shape variation of seamounts can be expressed in terms of summit height and flatness with large seamounts being preferentially cone-shaped, whereas smaller seamounts often display flat summit areas (Smith, 1988). Flat-topped submarine volcanoes are generally referred to as guyots, and in many instances can be shown to be large drowned volcanoes truncated by erosion at or near sea level (Menard, 1984). However, some seamounts with summit plateaux were constructed in deep water by normal volcanic processes involving caldera collapse (Simkin, 1972; Batiza and Vanko, 1983; Fornari *et al.*, 1984).

Morphological studies of basaltic seamounts, covering a range of sizes, indicate that they often display constructional features similar to their subaerial counterparts. Small, young seamounts near the East Pacific Rise (EPR) spreading axis range from conical domes to truncated cones with

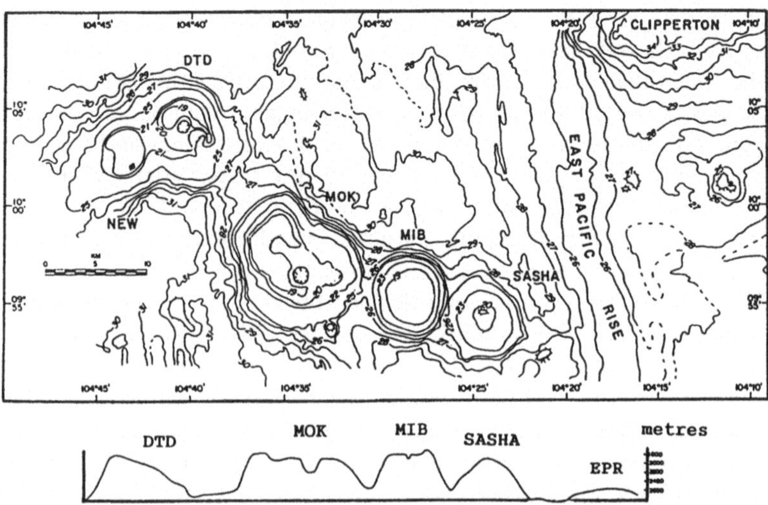

Figure 9.2 SeaBeam bathymetric map (contours × 100 m) and cross-section of the Lamont seamounts adjacent to the East Pacific Rise at 10°N (Fornari *et al.*, 1984, 1988a).

summit plateaux or craters (Batiza and Vanko, 1983; Fornari *et al.*, 1984), in addition to more irregular plan-forms controlled by local fractures (Fornari *et al.*, 1987). The Lamont seamounts (ranging in age from about 200–830 Ka) near the EPR at 10°N (Figure 9.2) have profiles which suggest an evolutionary constructural sequence as the seamounts age and move away from the axis (Fornari *et al.*, 1984). Initially small conical volcanoes (< 1 km high) are built up on the seafloor via summit eruptions, which, as a result of subsequent flank activity, cause central collapse and the development of a summit crater or larger caldera. Continued growth may be effected by eruption of lavas from localized vents fed by ring fractures in the summit area and the development of a summit plateau. Explosive, phreatomagmatic eruptions are common with the development of bedded (and slumped) hyaloclastites within craters and associated with waning, late-stage summit activity (Lonsdale and Batiza, 1980; Batiza *et al.*, 1984).

Together with their distribution, the external shape of volcanoes provides evidence both for their temporal growth and for the influence of local tectonics on magma generation and mode of emplacement. In the case of young seamounts their shape may be strongly influenced throughout growth by local tectonic factors such as active fracture zones both near ridges and in intraplate regions. For example, the most recent activity in the Hawaiian chain at Loihi seamount (about 25 km south-east of Hawaii shows a summit plateau with nested craters sitting astride a cresent-shaped rift zone (Figure 9.3) which is sub-parallel to similar features through Mauna Loa and Kilauea (Fornari *et al.*, 1988b).

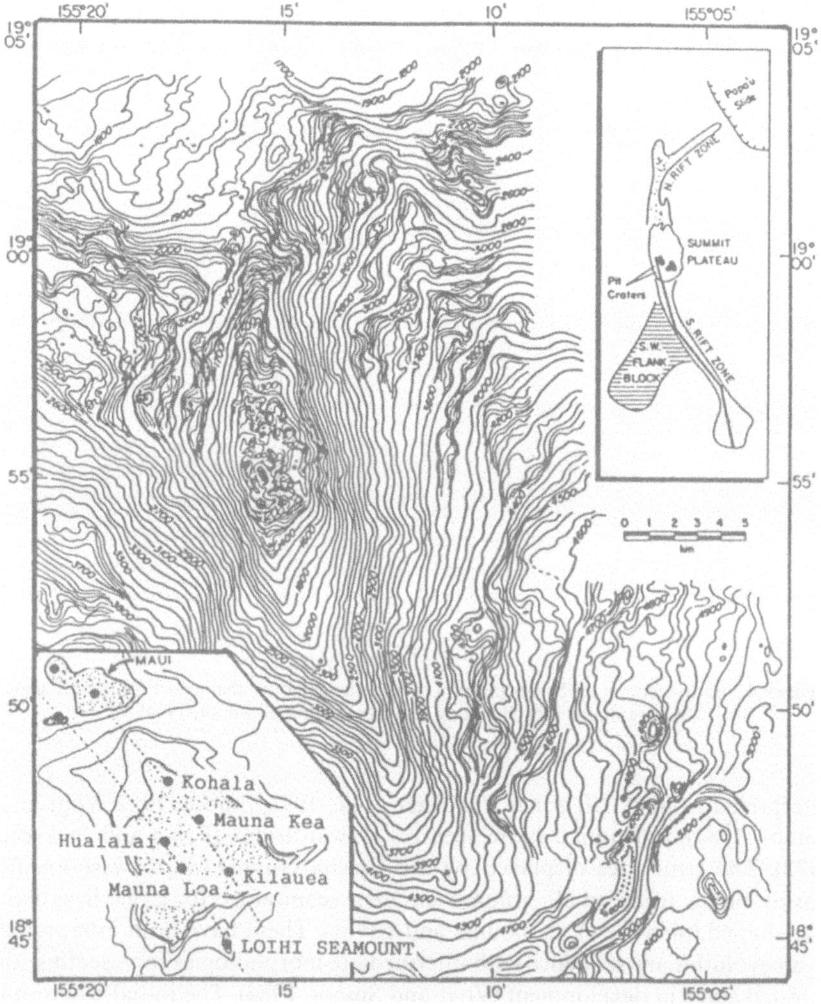

Figure 9.3 Bathymetry of Loihi seamount off Hawaii (Fornari *et al.*, 1988c). Upper inset shows the major morphological features of the seamount. Location of Loihi seamount off Hawaii shown in lower inset.

Here, as near the tectonic zone of ridges, seamount growth is often governed by conduit geometry which is influenced to various degrees by the local tectonic stress pattern, Larger (and generally older) seamounts developed mid-plate show both similar and more complex morphologies which continue to grow as the volcano emerges above sea level. Characteristic of western Pacific seamounts is the development of multiple, nested summit craters and large calderas, extensive summit plateaux and radial ridges (Figure 9.4),

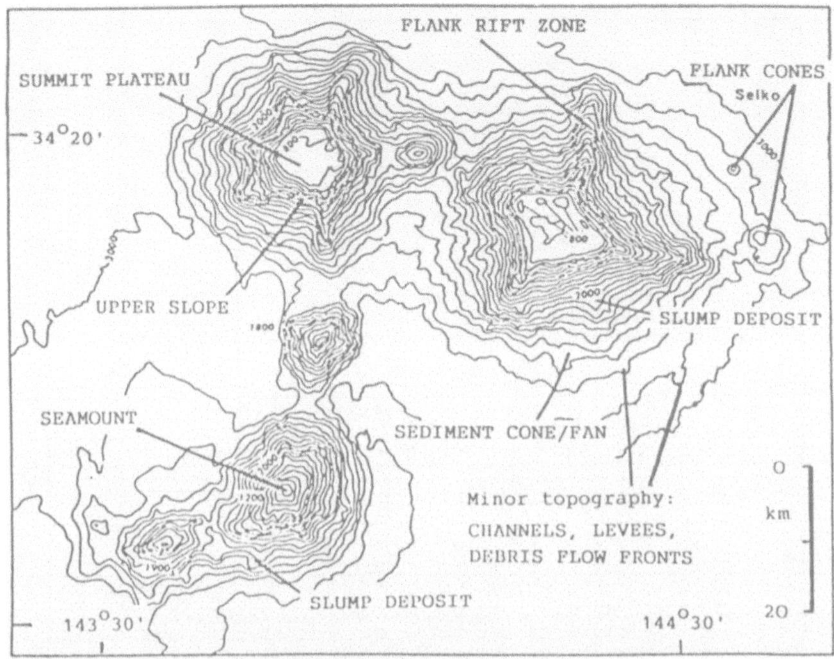

Figure 9.4 Bathymetry of the Seiko cluster in the Geisha seamount chain, north-western Pacific, showing major morphological features (Vogt and Smoot, 1984).

interpreted as flank rift zones (Hollister *et al.*, 1978; Smoot, 1982; Vogt and Smoot, 1984), not unlike those seen on Hawaii today (Fiske and Jackson, 1972). The structures displayed by the Geisha Guyots (north-west Pacific Basin) range from small volume circular seamounts to large irregular, star-shaped (plan view) seamounts and guyots. These have been interpreted as an evolutionary sequence with intermediate morphologies representing an arrested stage in development (Vogt and Smoot, 1984). The initial seamount growth is governed by central magma conduits (producing seamounts with a circular base), whereas subsequent growth modifies this shape with eruptions fed via lateral dykes supplying flank rift zones (producing a star-shaped base). Although construction rates are very different, this interpretation of the magma plumbing system for the development of large seamounts is similar to models for subaerial edifices, such as Kilauea on Hawaii (Ryan *et al.*, 1981; Ryan, 1988).

Seamount morphology is not only governed by the chemical and physical properties of the constituent lavas and their rate of eruption, but, as indicated above, the local tectonic environment exerts a control over the geometry of the conduits that direct magma flow during progressive growth.

9.4 Internal structure and composition

Although the overall structure and lithologies of the subaerial portion of oceanic volcanoes are well known, the submarine edifice which comprises the vast bulk of intraplate volcanoes is poorly documented and largely inferred. Drilling into seamounts rarely penetrates more than a few tens of metres into the basaltic structure below the sediment capping and gives little information about structural development.

Models for the internal structure and composition of seamounts and oceanic islands are derived from a number of sources including: seismic, gravity and magnetic surveys (Harrison and Brisbin, 1959; Menard, 1964; Batiza and Watts, 1986); general submarine morphology and eruption styles via bathymetry, actual observation and sampling (Moore and Fiske, 1969; Batiza and Vanko, 1983; Bonatti and Harrison, 1988; Fornari *et al.*, 1988b); and, cross-sections of tectonically uplifted ancient seamounts (MacPhearson, 1983; Staudigel and Schmincke, 1984).

Many of our ideas concerning the structure and composition of oceanic volcanoes have been based on the so-called Hawaiian model, although there is no reason to suppose that all edifices will show the same evolutionary and compositional development. The gross volcanic structure of large mid-plate

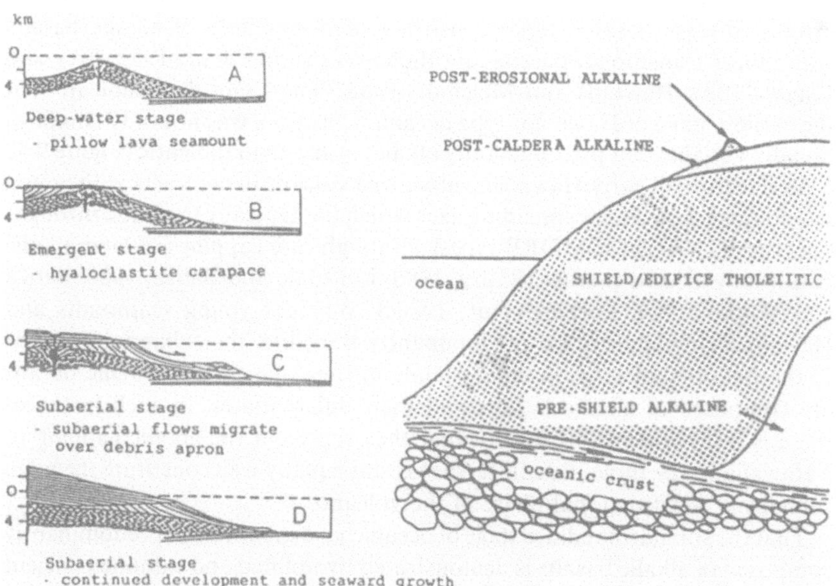

Figure 9.5 Inferred structure and growth (parts A to D) of a Hawaiian-type edifice (Moore and Fiske, 1969), together with a diagrammatic cross-section showing the relationship between the four main Hawaiian eruptive stages (Clague, 1987).

volcanoes such as on Hawaii (Moore and Fiske, 1969), are inferred to be constructed of a massive submarine edifice composed predominantly of pillow lavas, which are replaced upwards by a shallow water carapace of pillow breccias and hyaloclastites on which subaerial flows develop (Figure 9.5). Loading caused by the growth of large edifices flexes the adjacent oceanic lithosphere to produce a moat which is filled by high level volcaniclastics derived by the mass-wasting and gravitational failure of shallow submarine slopes (Fornari and Campbell, 1987; Lipman *et al.*, 1988). The crustal and subcrustal structure immediately below the volcano (inferred from seismic and gravity data) probably consists of massive intrusives, sill and dyke complexes, variably metamorphosed by injections of new magma that are stored at higher and higher levels as the structure grows (Batiza and Watts, 1986). It has been known for some time that the subaerial portions of Hawaiian volcanoes evolve through three main stages, each of which erupts basaltic lavas of a distinct chemical composition (Macdonald and Katsura, 1964; Macdonald, 1968). These stages are an early, tholeiitic shield-building stage, a post-caldera alkalic stage and finally, after several millions of years of non-activity and erosion, the post-erosional, strongly alkalic stage. It is generally assumed that the main submarine edifice is composed of tholeiitic basalts similar to the subaerial shield lavas and was built in < 1 Ma (Macdonald, 1968). However, the most recent Hawaiian activity at Loihi seamount, which respresents an early submarine and thus pre-shield stage, exhibits diverse basaltic types consisting predominantly of alkalic basalts with minor transitional basalts and tholeiites (Moore *et al.*, 1982; Frey and Clague, 1983; Hawkins and Melchior, 1983). The Loihi data indicate that the earliest stage of Hawaiian-type oceanic volcanoes was not only compositionally variable, but predominantly alkalic rather than tholeiitic (Figure 9.5).

This feature of early Hawaiian submarine volcanism contrasts with young seamounts located near spreading axes which are typically tholeiitic throughout (variably depleted MORB-types), although rare cappings of later alkalic basalt may also be present (Batiza, 1980; Lonsdale and Batiza, 1980; Batiza and Vanko, 1984; Fornari *et al.*, 1980c). Whereas young seamounts and Hawaiian-type edifices are predominantly tholeiitic, the subaerial portions of the majority of large oceanic islands in the Atlantic and Pacific oceans are composed of alkalic basalts and their differentiates. In such instances there is little evidence to suggest that they represent the alkalic capping to a Hawaiian-type tholeiitic edifice; they could equally well constitute the main submarine constructional phase of the volcano.

That the submarine edifice stage of oceanic volcanoes can be predominantly composed of alkalic basalts is demonstrated by uplifted and exhumed ancient seamounts, such as the La Palma seamount series, Canary Islands (Staudigel and Schmincke, 1984) and the Snow Mountain volcanic complex, California (MacPherson, 1983). The thick volcanic (and intrusive) sequences exposed provide evidence for the internal structure of seamounts and the processes

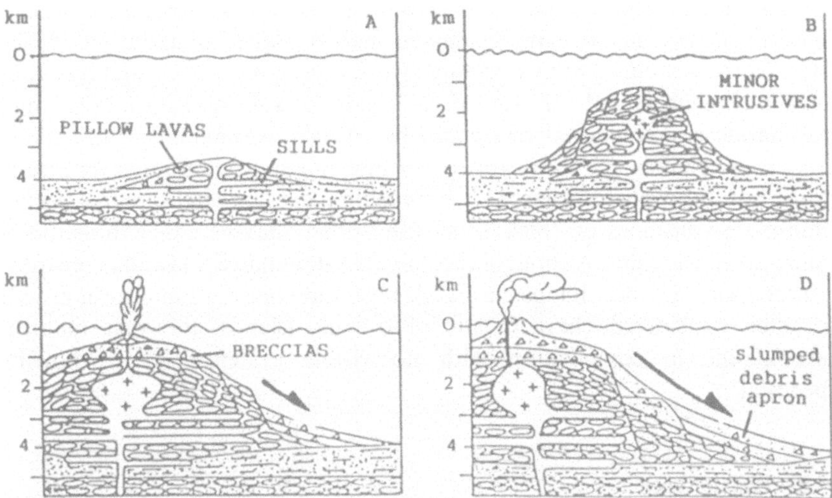

Figure 9.6 Evolution (parts A to D) of a seamount from the early submarine stage to emergent stage based on the La Palma seamount series, Canary Islands (Staudigel and Schmincke, 1984). The ocean crust basement (stippled) is composed of pillow lavas (ovals) and deep-sea sediments (short lines).

involved in their development from small submarine volcanoes to emergent islands.

The La Palma seamount is divided into three magmatostratigraphic units (Staudigel and Schmincke, 1984) consisting of a basal plutonic series (alkali gabbros, ultramafic cumulates) cut by dykes, a massive sill–dyke complex with locally developed sheeted dykes that feed higher level volcanics and an upper submarine series of mildly differentiated alkali basalt pillow lavas, breccias and hyaloclastites. One significant feature is the importance of intrusives (mainly sills and small volume plutons) relative to volcanics, not only in the early stages of development, but throughout the submarine constructional stage (Figure 9.6). The nature and progressive development of the extrusive products which make up a large proportion of the growing seamount are mainly governed by water depth, together with eruptive rate and lava composition. The initial deep water stage is characterized by lenticular, massive flow and pillowed basalts which build up steep slopes surrounded by minor pillow breccias on the flanks. As the seamount builds upwards, a critical water depth is reached (about 750 m for the volatile-rich La Palma basalts) when gases in the lava are explosively released. At this stage the products of explosive volcanism (breccias and hyaloclastites) increase markedly relative to the proportion of lava flows. This shallow water stage essentially produces a thick tephra cap to the deep water lavas below; the slumping of such material produces an apron of volcaniclastic debris far beyond the summit area (Figure 9.6).

In summary, the submarine growth stage of oceanic volcanoes involves both intrusive and extrusive processes, with the latter changing in effusive style as the seamount builds upwards from a deep water to a shallow water eruptive environment. In terms of basalt lava composition, the main submarine constructional phase can be either predominantly tholeiitic or alkalic. Tholeiites are characteristic of non hot-spot seamounts on young, thin, hot lithosphere adjacent to spreading axes, whereas plume-related mid-plate volcanoes on thick, older lithosphere may be either tholeiitic or, more likely, alkalic. Although the tholeiite-dominated Hawaiian eruptive pattern with distinct changes in basalt type and composition during growth is often considered characteristic of oceanic islands, the vast majority are alkalic in composition with no visible evidence for a tholeiitic pedestal.

9.5 Basalt types

Basalts from oceanic islands and seamounts show a much greater diversity of petrographic and chemical composition than the olivine tholeiites of the mid-ocean ridges. In general, they exhibit a range of basaltic types, including tholeiites, alkali basalts, transitional basalts and highly undersaturated basanites and nephelinites, the proportions of which may vary from one island or island group to another.

The commonly hypersthene-normative tholeiites are variably olivine phyric along with Cr spinel, two pyroxenes (diopside and hypersthene or pigeonite), basic plagioclase and Fe–Ti oxides. The nepheline-normative alkali basalts contain abundant phenocrystic and matrix olivine, titaniferous augite, plagioclase and titanomagnetite; clinopyroxene and plagioclase may occur as glomerocrysts. In more evolved basaltic compositions alkali feldspar, together with magmatic brown amphibole (kaersutite) and biotite, may also be present. Hydrous phases are absent from tholeiitic oceanic island basalts. As the proportion of normative (and modal) nepheline increases, the alkali basalts merge into undersaturated basanites and nephelinites. Between the common tholeiitic and alkalic compositions are the hypersthene-normative transitional basalts with chemically intermediate features or mildly alkaline characteristics (in terms of total alkali content). Oceanic island basalts may show a considerable range in total alkali content, such that Baker (1973) grouped the islands of the southern Atlantic into mildly alkaline (Ascension, Bouvet), moderately alkaline (Gough, Tristan da Cunha, St. Helena) and strongly alkaline or potassic (Cape Verde Islands) types. The alkali basalt series is also chemically subdivided into a potassic suite (basalt–trachybasalt–trachyandesite) and a sodic suite (basalt–hawaiite–mugearite–benmoreite) on the relative proportions of the alkali elements (Middlemost, 1985; Le Bas *et al.*, 1986). Unlike tholeiite-dominated islands, alkalic suites are often typified by more evolved basaltic compositions (trachybasalt, hawaiite) than true basalt (Baker, 1973).

Alkali basalts may be associated with minor intermediate and acid differentiates developed mainly via the low pressure fractional crystallization of initially olivine + plagioclase or olivine + clinopyroxene, for example, Gough Island (LeMaitre, 1962; Zielinski and Frey, 1970), St. Helena (Baker, 1969), Réunion (Upton and Wadsworth, 1966; Zielinski, 1975) and the Eastern Caroline Islands (Mattey, 1982). The acidic end-products of the alkali suite may be represented by undersaturated nepheline-bearing trachytes and phonolites or oversaturated quartz-bearing alkali rhyolites. Differentiation within tholeiitic Hawaiian lavas is largely restricted to the basaltic compositional spectrum and is related to the gravity settling of olivine and Cr spinel, together with flow concentration of pyroxene and plagioclase (Wright and Fiske, 1971; Wright *et al.*, 1976). Studies of the Kilauea Iki lava lake have, however, demonstrated the existence of more fractionated dacitic-rhyolitic melts segregated from the main basalt host by filter pressing (Wright and Helz, 1987).

The petrography and phase compositions of associated oceanic island tholeiites and alkali basalts are often distinctive and reflect the composition and crystallization history of the basaltic melts. For example, clinopyroxenes from alkali basalts tend to be more calcic (diopside), are often highly titaniferous and have a higher proportion of non-pyroxene quadrilateral components than tholeiitic clinopyroxenes. Also, spinel compositions may be distinctive, with tholeiitic spinels generally being richer in Cr_2O_3 than those from alkali basalts, although the Cr content is sensitive to the reduction state (fO_2) of the melt during crystallization. An example of the composition and relationships between phenocrysts and matrix phases in associated Hawaiian tholeiites and alkali basalts is documented in the Basaltic Volcanism Study Project (BVSP, 1981).

9.6 Geochemical features

The initial studies on oceanic basalts, which revealed distinct chemical differences between those erupted at mid-ocean ridges (MORB) and in oceanic islands (OIB) (Engel *et al.*, 1965; Tatsumoto *et al.*, 1965; Gast, 1968), have been amply confirmed by subsequent data and emphasize subtle, in addition to gross, variations within both these environments. The major oxides and compatible trace element compositions generally overlap, although OIB tend to exhibit greater ranges in element abundances than MORB rocks or glasses, both overall and at the same fractionation level (BVSP, 1981). The most significant differences in composition between OIB and MORB are shown by incompatible trace elements and isotopes, and provide evidence for a heterogeneous mantle (Chapter 15).

Table 9.1 Comparison of minor and trace element composition of average MORB and OIB (samples 1–3), together with alkali basalts (4–6) and tholeiites (7–9) from oceanic islands and seamounts

Composition	Sample No. (location)								
	1 (N-MORB)	2 (E-MORB)	3 (OIB)	4 (Bouvet)	5 (St. Helena)	6 (Gough)	7 (EPR)	8 (Kilauea)	9 (Loihi)
Minor oxides (wt.%):									
TiO_2	1.269	1.002	2.872	2.28	3.40	3.51	1.19	2.61	2.20
K_2O	0.072	0.252	1.44	0.87	1.17	2.77	0.07	0.48	0.32
P_2O_5	0.117	0.143	0.621	0.38	0.58	0.81	0.11	0.27	0.20
Trace elements (ppm):									
Ba	6.3	57	350	175	335	921	16	125	88
Cr	290	46	150	26	177	291	354	630	764
Cs	0.007	0.063	0.387	0.07	0.11	0.3	0.07	—	—
Hf	2.05	2.03	7.8	4.4	5.7	7.6	1.89	4.47	2.92
Nb	2.33	8.3	48	26	55	44	—	24	10
Ni	138	32	100	21	92	212	66	—	270
Rb	0.56	5.04	31	13	20	54	5	8.4	5.8
Sc	44	39	—	23	28	27	36	31	30
Sr	90	155	660	548	658	918	62	368	300
Ta	0.132	0.47	2.7	1.92	3.96	3.09	—	—	0.73
Th	0.12	0.60	4	1.98	4.02	5.16	0.03	1.14	0.40
U	0.047	0.18	1.02	0.51	0.56	1.02	0.02	—	—
V	—	—	—	212	281	202	—	—	270
Y	28	22	29	27	29	27	—	22	18
Zn	—	—	—	75	106	107	—	—	—
Zr	74	73	280	186	254	352	47	141	112

Rare earth elements (ppm):									
La	2.5	6.3	37	19.2	36.7	51.7	1.4	14.6	9.6
Ce	7.5	15	80	45	85	109	5.81	36.5	24.5
Pr	1.32	2.05	9.7	—	—	—	—	—	—
Nd	7.3	9.0	38.5	24	39	49	—	—	15.5
Sm	2.262	2.6	10	—	—	—	2.7	5.55	4.06
Eu	1.02	0.91	3	2.11	2.78	3.36	1.00	1.92	1.53
Gd	3.68	2.97	7.62	—	—	—	—	—	—
Dy	4.55	3.55	5.60	—	—	—	—	—	—
Ho	1.01	0.79	1.06	—	—	—	—	—	—
Er	2.97	2.31	2.62	—	—	—	—	—	—
Yb	3.05	2.37	2.16	—	—	—	2.51	1.90	1.66
Lu	0.455	0.354	0.300	—	—	—	0.38	0.27	0.23
Ratios:									
Ba/Nb	2.7	6.9	7.3	6.7	6.1	20.9	—	5.2	8.8
La/Nb	1.1	0.8	0.8	0.7	0.7	1.2	—	0.6	1.0
Zr/Hf	36	36	36	42	45	46	25	32	38
Ti/Zr	103	82	61	74	80	60	152	111	118
Th/Ta	0.9	1.3	1.5	1.0	1.0	1.7	—	—	0.5
La/Sm	0.95	2.42	3.70	—	—	—	—	—	—
La/Yb	0.82	2.66	17.13	—	—	—	0.56	7.68	5.78
K/Rb	1067	415	386	555	486	426	116	474	458
P/Ce	69	42	34	37	30	33	83	33	36

[a] 1–3: MORB and OIB averages (mainly Sun and McDonough, 1989); 4–6: S. Atlantic Islands (Weaver *et al.*, 1987); 7: EPR seamount (Batiza and Vanko, 1984); 8: Kilauea, Hawaii (BVSP, 1981); 9: Loihi seamount, Hawaii (Frey and Clague, 1983)

The following sections outline some of the main chemical features of OIB relative to the depleted characteristics of normal-type MORB (Chapters 11–13).

9.6.1 *Incompatible element abundances*

Both tholeiites and alkali basalts are invariably enriched in the most incompatible trace elements (Cs, Ba, Rb, Th, U, K, Nb, Ta, La, Ce, Sr, Nd, P, Sm; listed in order of decreasing incompatibility) and exhibit a far greater range of absolute abundance than MORB. Selected incompatible element contents and ratios of OIB relative to normal-type (N-) and enriched-type (E-) MORB show distinctive features (Table 9.1), although the wide range of OIB compositions actually encompasses E-MORB such that they may be difficult to distinguish.

The Zr/Nb ratios for N-MORB are typically high (>30; Table 9.1 and Erlank and Kable, 1976), whereas E-MORB and OIB (both tholeiites and alkali basalts) overlap with ratios varying between 4 and 15 (BVSP, 1981; Weaver *et al.*, 1987). Similarly, the La/Ta ratios for N-MORB are 18–20, whereas E-MORB and OIB have ratios of 10–14 (Saunders, 1984; Sun and McDonough, 1989). Chondrite-normalized multi-element plots for different oceanic islands (Figure 9.7) show typically humped patterns with variable degrees of incompatible element enrichment at approximately the same level of chemical fractionation. With increasing incompatibility the patterns peak around Nb–Ta and may then decrease in an irregular fashion with or without a secondary Ba peak.

The depletion of the highly incompatible elements (relative to K) indicates that some of these basalts have been derived from an OIB mantle source that was already partially depleted in these particular elements (via melting out of a small basaltic fraction) rather than from a 'primitive' or 'pristine' reservoir which would exhibit a progressive enrichment pattern for all incompatible elements.

9.6.2 *Rare earth elements*

The rare earth elements (REE) are strongly fractionated with chondrite-normalized diagrams showing variable light REE enrichment (relative to the heavy REE) that increases, together with the total REE content, from tholeiitic to alkaline compositions (Figure 9.8). For example, the $(La/Yb)_N$ ratios broadly increase from Hawaiian tholeiites (3–5), alkali basalts (5–7) to

Figure 9.7 Chondrite-normalized multi-element patterns for (A) average N-MORB, E-MORB and OIB, (B) OIB tholeiites and (C) OIB alkali basalts from various oceanic localities (data and references in Table 9.1). Note the particularly depleted characteristic of the EPR seamount tholeiite relative to the enriched patterns for most OIB.

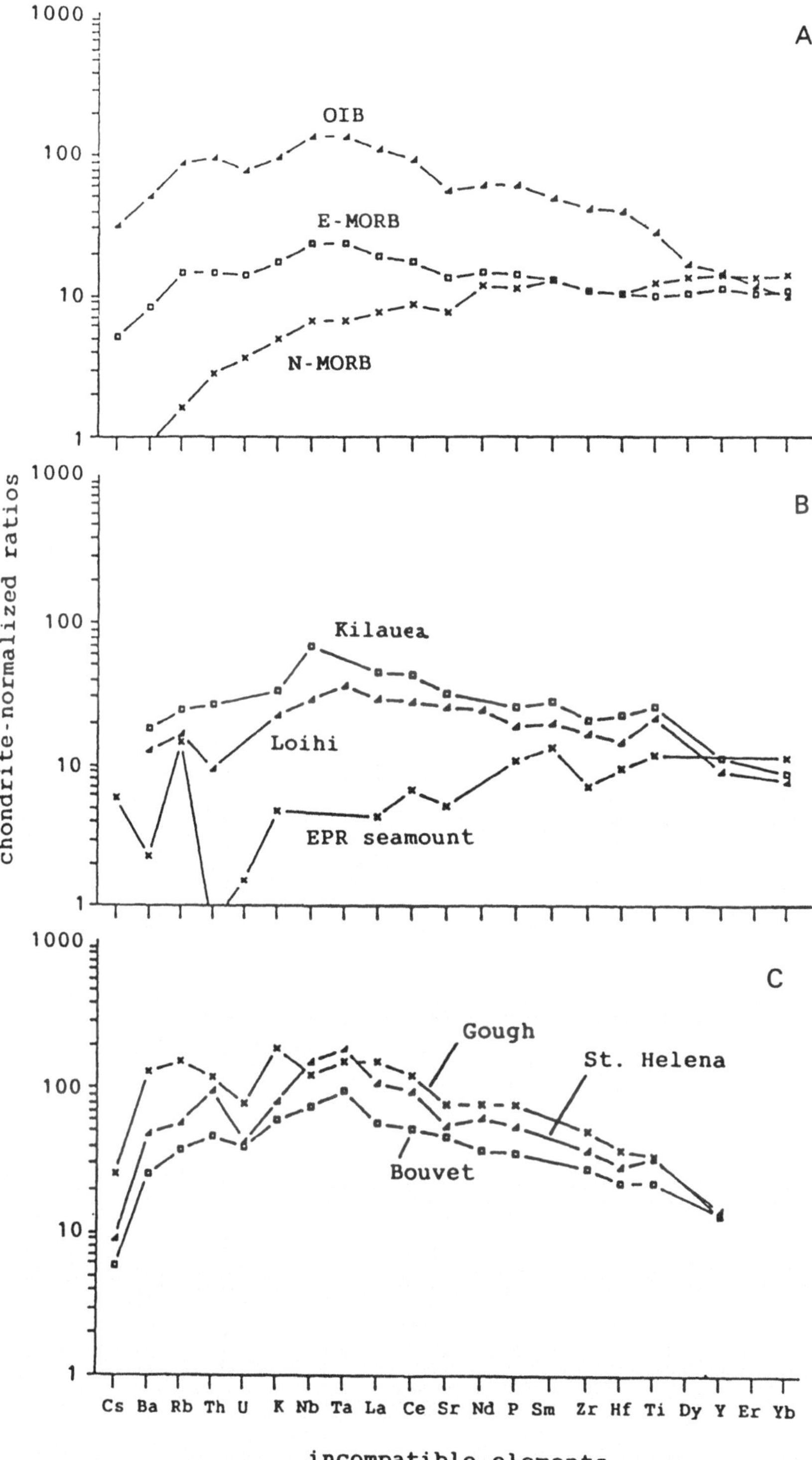

A
OIB
E-MORB
N-MORB
B
Kilauea
Loihi
EPR seamount
C
Gough
St. Helena
Bouvet
chondrite-normalized ratios
1000
100
10
1
Cs Ba Rb Th U K Nb Ta La Ce Sr Nd P Sm Zr Hf Ti Dy Y Er Yb
incompatible elements

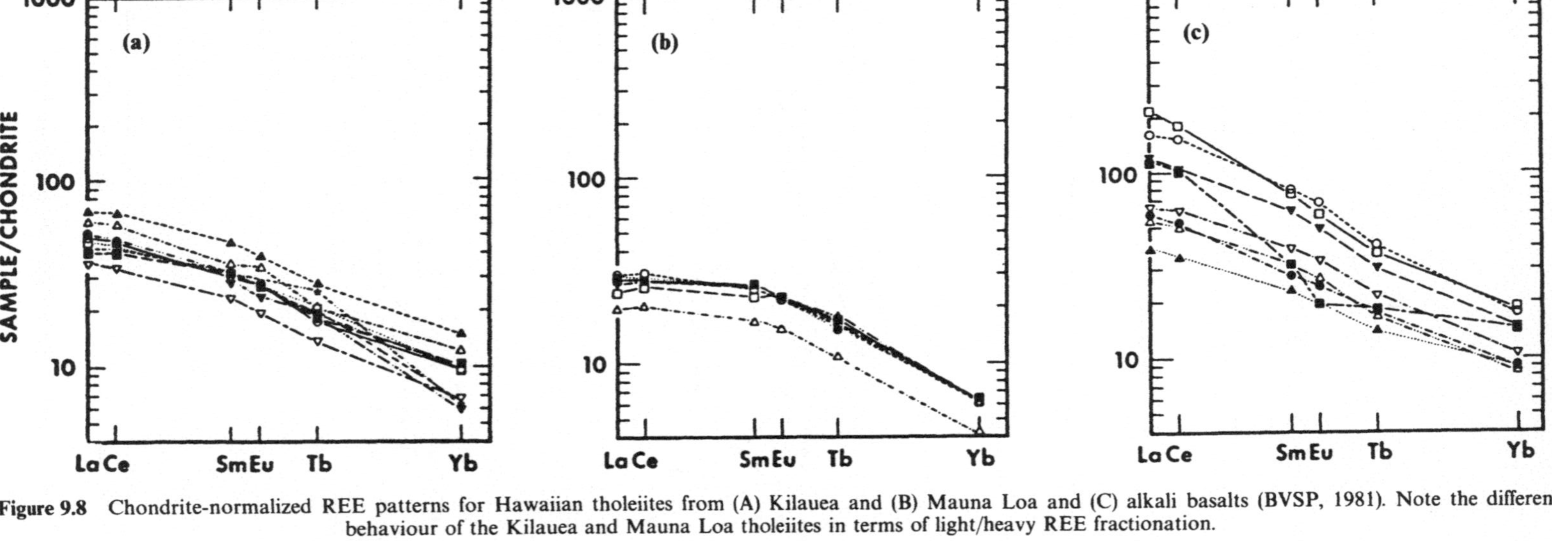

Figure 9.8 Chondrite-normalized REE patterns for Hawaiian tholeiites from (A) Kilauea and (B) Mauna Loa and (C) alkali basalts (BVSP, 1981). Note the different behaviour of the Kilauea and Mauna Loa tholeiites in terms of light/heavy REE fractionation.

basanites/nephelinites (> 10) (Schilling and Winchester, 1969; Leeman *et al.*, 1980; Clague and Frey, 1982), a feature also exhibited by the recent basalts of Loihi seamount off Hawaii (Frey and Clague, 1983). The characteristic light REE enrichment of OIB relative to MORB indicates derivation from an enriched source (with light/heavy REE ratio > chondritic) and for highly undersaturated rocks, with very steep normalized patterns, the presence of residual garnet in the source (Kay and Gast, 1973; Shimizu and Arculus, 1975; Clague and Frey, 1982).

9.6.3 *Highly incompatible element ratios*

The ratios of highly incompatible elements with small distribution coefficients (e.g. K/Rb, Ba/Nb, La/Nb, Th/Ta, Th/La, Ce/P) can vary between islands and also within basalts from the same island. As these ratios are not changed by fractional crystallization or partial melting (except very low degrees of melting, < 1%), they reflect the composition of the mantle and indicate that the OIB 'source' is heterogeneous.

Figure 9.9 shows the wide range of Ta/La and Th/La ratios in various OIB relative to the confined N-MORB distribution. More specifically, islands in the southern Atlantic Ocean, such as Tristan da Cunha and Gough Island,

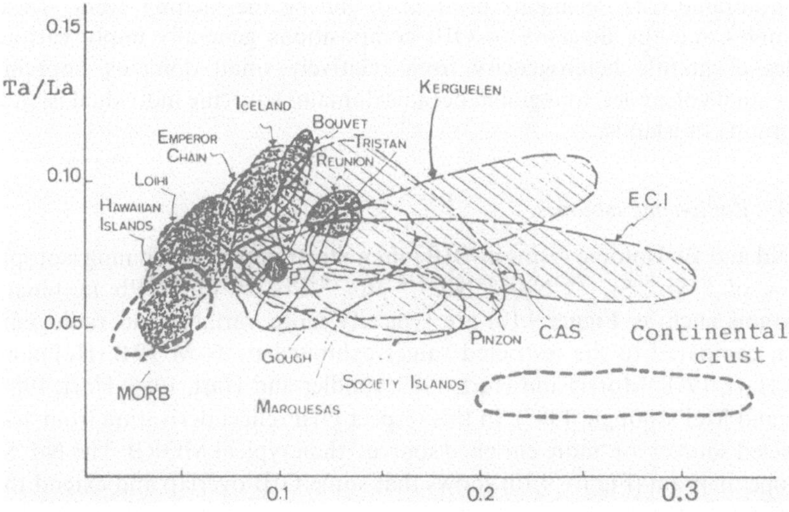

Figure 9.9 Variation in Ta/La and Th/La ratios in various OIB relative to the restricted MORB field (Loubert *et al.*, 1988). The two OIB fields are distinguished on the basis of different Sr–Nd isotopic relationships, with the dark fields representing less radiogenic OIB and the lined fields representing more radiogenic OIB. CAS = Cook–Austral–Samoa Islands and ECI = Eastern Caroline Islands.

have higher La/Nb, Th/Ta and Ba/La ratios than Ascension, Bouvet and St. Helena (Weaver *et al.*, 1987), which are also matched by isotopic differences (Sun and McDonough, 1989) and reflect the incorporation of crustal-derived materials (rich in La, Th, Ba) in the OIB source. In the same ocean, the islands of the Azores plateau show significant differences in REE fractionation, incompatible element ratios (La/Sm, U/Th, Ba/Zr, Zr/Nb) and Sr–Nd–Pb isotopes (Flower *et al.*, 1976; White *et al.*, 1979; Hawkesworth *et al.*, 1979; Dupre *et al.*, 1982; Marriner *et al.*, 1982) that suggest considerable variation in the mantle below this island group. On a smaller scale, systematic variations across individual islands may be observed that are probably produced by the mixing of sources or magmas (see section 9.8.3) (Hawkesworth *et al.*, 1979; Dupre *et al.*, 1982).

In addition to chemical differences within island groups, temporal variations may also be significant. In a number of instances tholeiites and alkali basalts erupted from the same Hawaiian volcano have similar K/Ba and P/Zr ratios (thus implying derivation from a similar source), but differ from the wide range of ratios exhibited by the later post-erosional basalts (Clague *et al.*, 1980; Frey and Clague, 1983). Although the strongly alkaline, late basalts are rich in incompatible elements with distinct incompatible element ratios, they are less radiogenic than most Hawaiian edifice basalts (Chen and Frey, 1983). These two geochemical characteristics indicate that they were generated from an isotopically-depleted source that was subsequently enriched with large ion-lithophile (LIL) elements prior to or during the melting event. These examples and the diversity of OIB compositions generally imply various scales of mantle heterogeneity from relatively small domains, sourcing individual volcanoes, to regional oceanic domains sourcing individual islands or groups of islands.

9.6.4 *Radiogenic isotopes*

Sr, Nd and Pb isotopic ratios in OIB (often illustrated by combining isotopic ratios of $^{87}Sr/^{86}Sr$, $^{143}Nd/^{144}Nd$, $^{207}Pb/^{204}Pb$, $^{206}Pb/^{204}Pb$ in binary diagrams such as Figure 9.10) are typically more variable and radiogenic when compared to the restricted values exhibited by N-MORB (Hofmann and Hart, 1978; Morris and Hart, 1983; Zindler and Hart, 1986; Hart, 1988; Sun and McDonough, 1989). In this respect OIB reflect derivation from 'less depleted sources' or 'more enriched sources' than typical MORB. The Nd–Sr isotope diagram (Figure 9.10) shows that some OIB overlap and extend the MORB mantle array towards the bulk earth composition, implying that many OIB have been derived from a source that has been depleted in Rb relative to Sr, and Nd relative to Sm for a long time (O'Nions *et al.*, 1977). That is, the source of OIB must have had low Rb/Sr and high Sm/Nd ratios, but not to the same degree as the depleted source of MORB. However, because such a depleted source of OIB (on the basis of the isotopic data) is

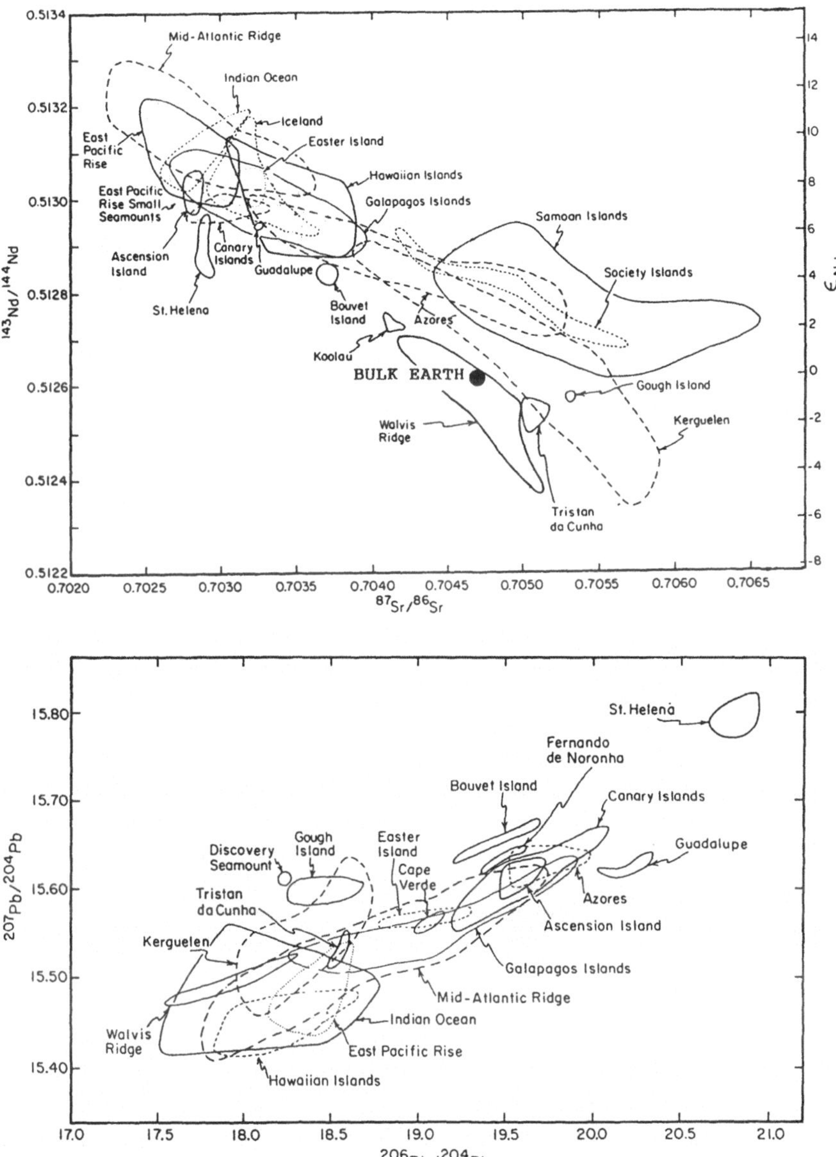

Figure 9.10 Variation in Sr–Nd–Pb isotopic ratios for various oceanic islands and seamounts relative to MORB (Staudigel *et al.*, 1984).

at variance with the incompatible element enriched ratios (e.g. high Rb/Sr) often observed in OIB, a recent enrichment of the source prior to partial melting is required (Norry and Fitton, 1983). This feature has led to the suggestion that mantle enrichment of some form (possibly via metasomatism) is a necessary prerequisite or trigger for the production of OIB (especially alkali basalt) generally (Menzies and Murthy, 1980; Bailey, 1987).

Around the bulk earth composition and towards highly radiogenic values (lower Nd coupled with higher Sr isotopic ratios, Figure 9.10), the negative linear relationship between the isotopic ratios becomes more diffuse with a considerable spread of data. Some of the OIB that fall in this region have been derived from sources variably contaminated by high $^{87}Sr/^{86}Sr$ crustal or lithospheric material. For example, the Samoan Islands source may have been contaminated by pelagic sediments (White and Hofmann, 1982), whereas the Kerguelen array represents mixing between an enriched plume and lithosphere-derived melts (Storey *et al.*, 1988).

Pb isotopic data, in a similar manner to Sr and Nd, overlaps some MORB values, but extends the data arrays to much higher isotopic ratios (e.g. $^{206}Pb/^{204}Pb = 17.5–21.0$ and $^{207}Pb/^{204}Pb = 15.4–15.8$; Sun, 1980; Staudigel *et al.*, 1984). The Pb–Pb isotope diagram (Figure 9.10) highlights isotopic differences between oceanic islands and, in particular, the highly radiogenic nature of St. Helena (Sun, 1980; Chaffey *et al.*, 1989), which is also anomalous in that it falls below the mantle array in the Sr–Nd isotope diagram. Not only is there an overall linear relationship between the Pb isotopic ratios (Figure 9.10) for OIB, but each island or island group (e.g. Bouvet, Galapagos) also shows a similar feature. This isotopic relationship is generally considered to reflect mixing between a depleted MORB-type source and enriched mantle source, such as a plume incorporating various U-rich components that have been stored in the mantle. The Pb content of MORB melted from the depleted asthenosphere is very low and is in strict contrast to the enriched values of plume-generated oceanic islands. Pb isotopic values thus provide a good fingerprint for U-rich oceanic crust and continental lithosphere-derived materials in OIB source regions.

9.6.5 *Gaseous isotopes*

Rare gas isotopic ratios, such as $^{3}He/^{4}He$ and $^{40}Ar/^{36}Ar$, are markedly different in MORB and OIB providing that atmospheric contamination can be ruled out (Fisher, 1986, 1989).

He isotopic ratios are 6–10 times atmospheric values in MORB, whereas in some OIB (e.g. Hawaii, Samoa, Réunion, MacDonald Seamount) the ratios are particularly high and may vary between 20 and 50 times atmospheric. These high OIB values are interpreted as indicating the presence of primordial ^{3}He derived from a primitive (or less depleted), deep mantle source (Lupton and Craig, 1975; Craig and Lupton, 1976; Kurz *et al.*, 1982). However, other oceanic islands (e.g. Gough, Tristan da Cunha, Canary Islands, Kerguelen) have similar or lower $^3He/^4He$ ratios than MORB (Kurz *et al.*, 1987; Vance *et al.*, 1989). Relative to other isotopic data, which do not identify a primitive mantle composition (see earlier), a variable, but high, primitive gas component in plume sources (Hart *et al.*, 1983; Kurz *et al.*, 1987) suggests decoupling from other, less volatile, isotopes. Vance *et al.* (1989) have suggested that volatiles, such as He, are introduced into the lithosphere below oceanic islands via CO_2- and H_2O-rich fluids, independently of incompatible element enrichment or metasomatism.

$^{40}Ar/^{36}Ar$ ratios are typically high in MORB (average about 15 000; Fisher, 1986) and are indicative of radiogenic ^{40}Ar growth in the mantle, whereas OIB ratios are considerably lower and suggest the presence of an atmospheric component (Fisher, 1989).

Correlations between He and Ar isotopes and other isotopic systems are not always developed, such that end-member mantle components are not well constrained and simple source mixing is inadequate to explain the variation observed (see Chapter 14).

9.7 The mantle and OIB

Both incompatible element and isotopic ratios emphasize the variability of OIB (relative to the more constant composition of N-MORB) and by implication the heterogeneous nature of the OIB mantle source or sources and their apparent isolation from the MORB source. Many OIB, exhibiting wide ranges of composition, show a coupled relationship between highly incompatible element ratios and radiogenic isotopes (Palacz and Saunders, 1986, Loubet *et al.*, 1988; Saunders *et al.*, 1988; Sun and McDonough, 1989), which can be used to identify different components in the source. These studies indicate that simple binary mixing between a depleted N-MORB source and a primitive OIB-type source cannot produce the full range of compositions exhibited by oceanic basalts and that a number of 'end-member' source components are required (Zindler *et al.*, 1982; White, 1985; Zindler and Hart, 1986; Hart, 1988). The origin of the chemical diversity of OIB is related to the mixing of variable proportions of at least three mantle end-member components (Zindler and Hart, 1986; Hart, 1988): depleted mantle, DM (the MORB source); enriched mantle, probably two types, EMI and EMII, which may obtain their particular characteristics from long-term, mantle-stored, recycled oceanic crust, sediments and subcontinental lithosphere; and material with a high $^{238}U/^{204}Pb$ ratio or μ (designated as HIMU component) such as the St. Helena source.

The geochemical fingerprinting of oceanic basalts via both MORB and OIB provides data that constrain the size and nature of mantle sources and their relative isolation within the framework of a convecting mantle. For example, the presence of strong plumes with a coherent chemical signature over a long time period suggests the isolation of a source from the rapidly convecting upper mantle that might be expected to mix potential contaminants or heterogeneities.

In this context some mantle models suggest that enriched hot-spot basalts have been derived from deep plumes arising from a separate (primitive) or undifferentiated lower mantle source (Allegre, 1982; Morris and Hart, 1983). However, there appears to be no unequivocal isotope (except possibly He isotopes) or trace element data which suggest a deep-seated, primitive source for OIB (White, 1985; Hofmann, 1986; Hofmann *et al.*, 1986). It seems more likely that the variable 'enriched' characteristics of OIB have been derived from dense oceanic crust and subcontinental lithosphere that, via subduction, have sunk and are stored near the 670 km seismic discontinuity that marks the upper–lower mantle convecting regimes (Ringwood, 1982; Hofmann and White, 1982; McKenzie and O'Nions, 1983; Sekine *et al.*, 1986). After a suitable time period (about 1–2 Ga) thermal activation eventually recycles this material as buoyant diapirs (plumes and blobs) into the upper mantle and provides a source for OIB (Figure 9.11A). Some of the recycled material will also become entrained as small-scale heterogeneous streaks and patches within the upper mantle due to rapid convection (McKenzie, 1979). These isolated domains (sometimes referred to as 'plums' within the mantle 'cake') are scattered and mixed throughout a predominantly depleted mantle matrix and could provide local sources for OIB (Figure 9.11B). In both the mantle models illustrated in Figure 9.11, the highly variable character of OIB is a reflection of the heterogeneous nature of these mantle contaminants (plumes or plums) together with their reaction and mixing with depleted asthenospheric components during melting events (Sun, 1985).

On a larger scale than OIB plumes or plums, apparently long-lived heterogeneities of global extent will also have consequences for our perception of the convective regime in the mantle that allows their preservation. For example, the large-scale isotopic DUPAL anomaly (a major enriched mantle feature), originally recognized in the Indian Ocean (Dupre and Allegre, 1983; and named after these workers), was found to be global in extent and situated between the equator and 50°S latitude (Hart, 1984). This Sr–Nd–Pb isotopic anomaly is found in both OIB and MORB, as well as some southern hemisphere continental basalts, and has been traced back in time to 120 Ma basalts from the Indian Ocean and south Pacific (Weis *et al.*, 1989; Smith *et al.*, 1989). The DUPAL anomaly has been related to geophysical anomalies in the deep mantle where enriched plumes derived from instabilities on the core–mantle boundary layer are swept upwards by low degree equatorial

Figure 9.11 Mantle models and the possible sources for OIB magmas. (A) Plume-type model with storage of OIB components at the lower–upper mantle boundary (largely after Davies *et al.*, 1989). (B) Plum-type model with OIB components scattered throughout upper mantle, as well as a thermal plume (after Zindler *et al.*, 1984).

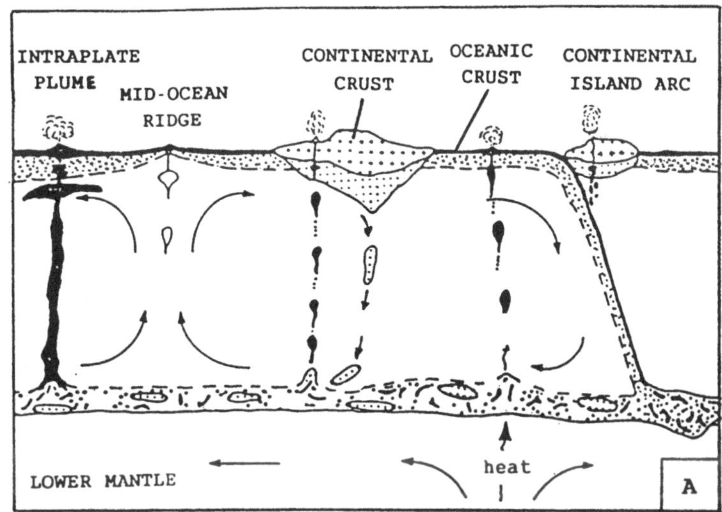

Subducted oceanic lithosphere

Subcontinental lithosphere & delaminated fragments

Rising plumes & blobs

Depleted upper mantle

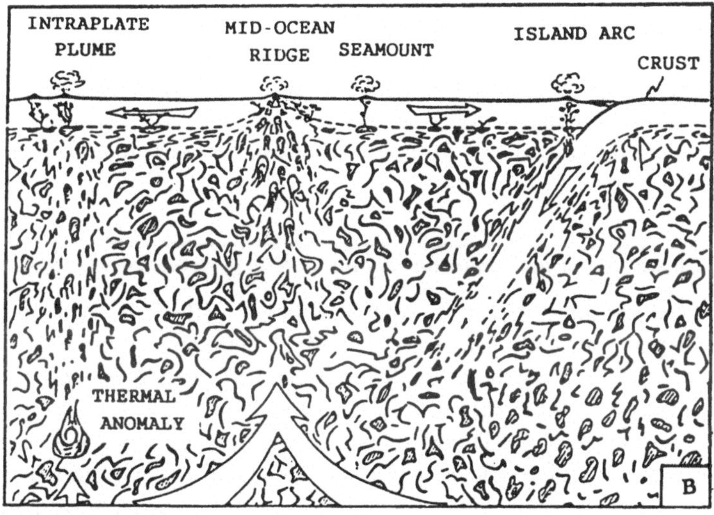

Subducted oceanic crust

Depleted peridotite

Metasomatized peridotite

Partially molten material

convective upwelling (Hart, 1988; Castillo, 1988). Although this apparent correlation between geochemical and geophysical anomalies is speculative, it emphasizes the differences of scale and depth of 'origin' considered for mantle heterogeneities that can produce OIB; either the deep mantle–core and the upper–lower mantle seismic boundaries (for the plume models) or throughout the upper mantle (for the plum model).

The recognition of long-lived isotopic heterogeneity and discrete mantle sources has provided support for a chemically stratified, two-layer convecting mantle with a thermal boundary layer at about 670 km. However, some geophysical evidence has indicated that whole-mantle convection may also be possible (Silver *et al.*, 1988; Olson *et al.*, 1990) and leaves open the question of the degree of involvement of a possible primitive lower mantle and the dispersion of lithosphere-derived heterogeneities via plate tectonic mechanisms. Galer and O'Nions (1985) have suggested that the large-scale transfer of undifferentiated lower mantle into the upper mantle is required to balance bulk earth Th/U ratios. Davies (1990) considers the lower mantle to be sufficiently old and heterogeneous to supply the chemical variation displayed by OIB, together with possible enhancement via the gravitational settling of subducted oceanic crust. However, at the moment it is generally considered that there is little evidence for the involvement of deep primitive mantle in basalt source regions (White, 1985).

9.8 Chemical variation and tectonic setting

This section outlines some of the chemical variations shown by oceanic island and seamount basalts in different tectonic settings, such as linear chains developed mid-plate, island groups and seamounts situated near spreading axes.

9.8.1 *Linear island and seamount chains*

The classic example of linear volcanic chains within the oceans is the elbow-shaped Hawaiian–Emperor Seamounts chain that stretches for about 6000 km across the north Pacific from Hawaii to the Meiji guyot near the Aleutian trench in the north-west. The volcanic activity covers about 75–80 Ma (Dalrymple *et al.*, 1981) and shows an age progression along the chain with current activity restricted to Hawaii (Kilauea volcano) and the Loihi seamount in the extreme south-east (Figure 9.12). The progressive nature of volcanic age with subsequent erosion and subsidence along the chain from Hawaii to Meiji relates to the movement of the Pacific plate over a stationary sub-lithospheric plume (see Section 9.2.1)

As recent accounts of Hawaiian basaltic volcanism (BVSP, 1981; Decker *et al.*, 1987) contain good summaries of general petrology and geochemistry,

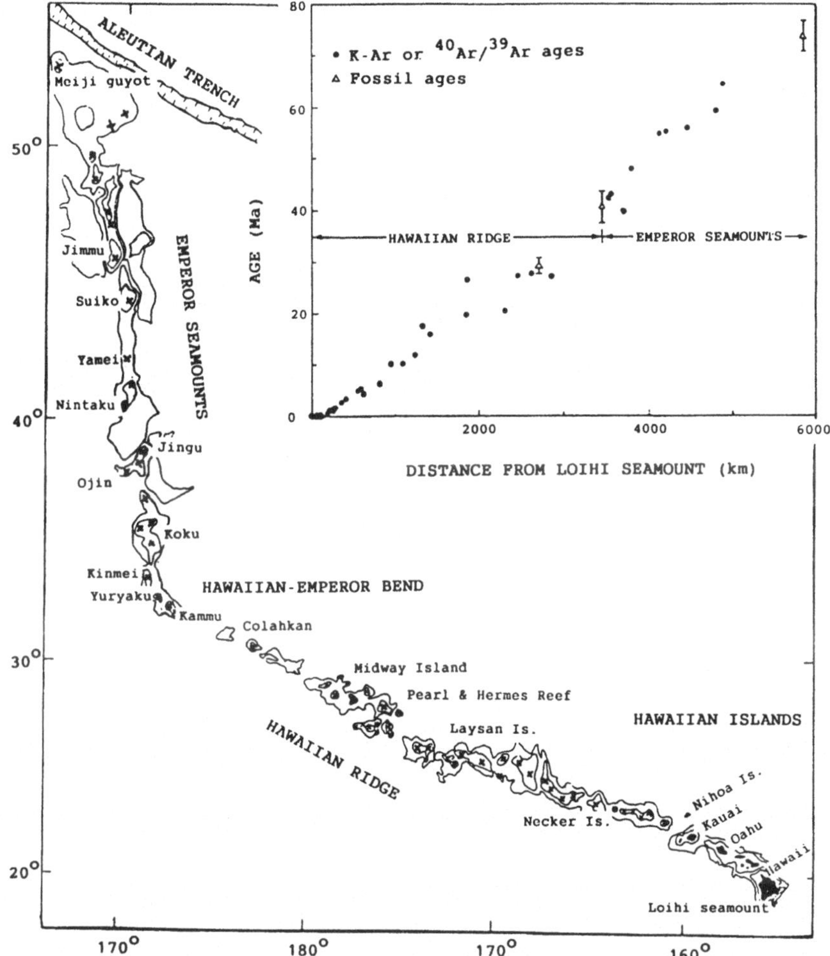

Figure 9.12 Hawaiian–Emperor volcanic chain and age progression data (inset) plotted from Loihi seamount (Clague, 1987).

two specific features of Hawaiian volcanism will be considered here: juvenile activity and the chemical signature of the Hawaiian plume with time.

9.8.1.1 *Juvenile mid-plate volcanic activity.* Hawaiian volcanism is characterized by four sequential eruptive stages (see Section 9.4) which produce basalts of distinctive chemical composition ranging from early submarine alkalic to predominantly shield-building tholeiites, post-caldera alkalic and finally post-erosional highly alkalic types (Macdonald and Katsura, 1964; Clague, 1987; Clague and Dalrymple, 1987). The pre-shield submarine stage has only recently been recognized from studies of the Loihi seamount off south-east Hawaii and illustrates the highly variable nature of the basaltic volcanism of very young mid-plate volcanoes.

Loihi, which is hydrothermally active, represents the youngest volcano (age range of dredged samples, 0.5–1.6 Ka) of the Hawaiian–Emperor chain and features a complex range of vesicular basaltic types consisting of older alkali basalts and basanites (some of which contain ultamafic xenoliths), and younger tholeiites and transitional basalts (Moore *et al.*, 1982; Malahoff *et al.*, 1982). Some of the geochemical features of the Loihi basalts are shown in Figure 9.13 and illustrate their overall similarity to Hawaiian shield tholeiites and post-caldera alkali basalts; alkalic Loihi samples are distinct from post-erosional subaerial alkalic types. The normalized REE patterns for Loihi tholeiites (Figure 9.13A) mirror recent basalts from Kilauea [$(La/Sm)_N > 1$] rather than Mauna Loa, which exhibits flat light REE patterns (BVSP, 1981). The Sr–Nd isotopic relationships (Figure 9.13) for Loihi basalts overlap

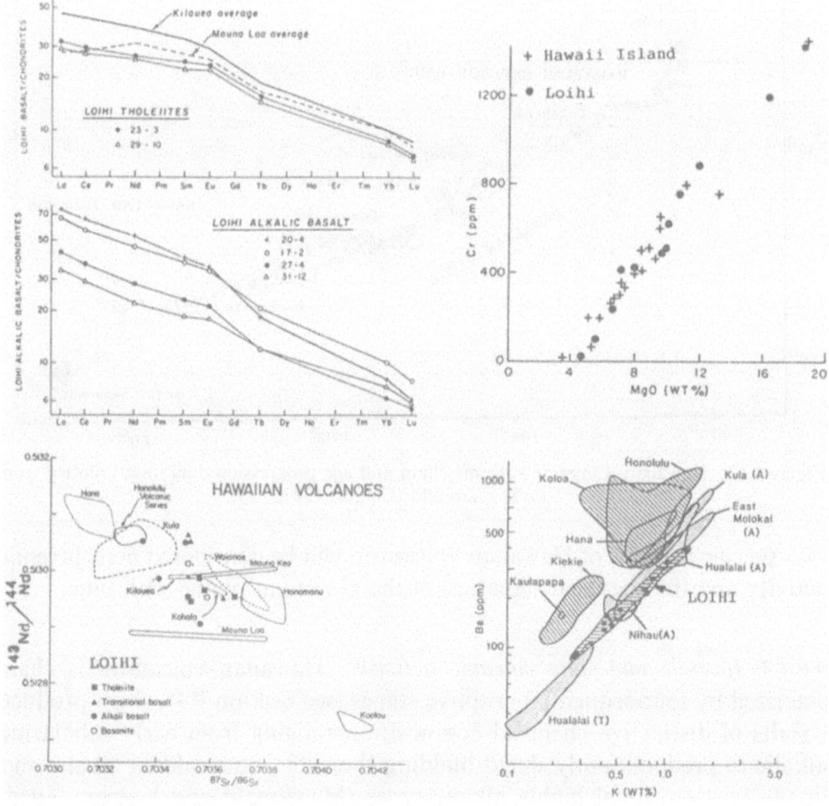

Figure 9.13 Chemical features of Loihi seamount basalts, off Hawaii. T = tholeiites and A = alkali basalts in Ba–K plot. Data from Frey and Clague (1983) and Staudigel *et al.* (1984); comparative Cr and MgO data for Hawaii from BVSP (1981).

those of the subaerial tholeiites and alkalic basalts, all of which are more radiogenic than the post-erosional alkalic types and N-MORB.

Whereas the broad range of chemical composition for all Loihi basalts is continuous (Figure 9.13), low pressure fractional crystallization is inadequate to explain the variation, but can be partly satisfied by variable partial melting of a (postulated) homogeneous source (Hawkins and Melchior, 1983). However, relationships between highly incompatible element ratios (La/P, Nb/P, Ba/P) and radiogenic isotope variation suggest that mixing of source components is necessary to account for some of the range. The Sr–Nd isotopic relationships indicate that the sources are not as depleted as N-MORB, although the end-member components of the Loihi source are not readily defined. He isotopes and inert gas studies, for example, suggest that one of the sources is a primitive undegassed component (Rison and Clague, 1983; Kurz *et al.*, 1983). Mixing between a MORB source and a less depleted plume or even a primitive source are two possibilities (Chen, 1987), although three or more distinct sources are generally considered necessary (Staudigel *et al.*, 1984). However, as pointed out by Wright and Helz (1987) there is no real consensus as to the number, mantle depth or composition of sources based on the isotope data for Hawaiian basalts generally (Chen and Frey, 1983, 1985; Hofmann *et al.*, 1984; Staudigel *et al.*, 1984; Roden *et al.*, 1984; Stille *et al.*, 1986; Frey and Roden, 1987).

In summary, although it is generally recognized that the main edifice building stage of Hawaiian shield volcanoes is tholeiitic, the Loihi data suggest that the initial submarine stage features small volume, compositionally diverse, low percentage partial melts derived from an (isotopically) variably depleted, heterogeneous source (Frey and Clague, 1983; Lanphere, 1983; Staudigel *et al.*, 1984; Clague, 1987).

9.8.1.2 *Chemical coherence of plume activity with time.* The mid-plate Hawaiian–Emperor chain provides the opportunity to study plume-generated hot-spot volcanism over a considerable time period, unaffected by the influence of spreading centres and continental crust. In particular, was the chemical–magmatic signature of the Hawaiian plume uniform over the 70 Ma history of the chain? Six seamounts of the Emperor chain (Koko, Ojin, Nintoku, Yomei, Suiko and Meiji, Figure 9.12) have been drilled by the DSDP (Legs 19, 32 and 55), and provide data on the earlier products of the hot-spot relative to the recent Hawaiian volcanoes, although not all reached volcanic basement. The best data set (Leg 55) indicated that similar basaltic types, eruption sequences and relative volumes were present to those exhibited by the subaerial lavas of Hawaii. Although small chemical variations were noted between the seamounts (Kirkpatrick *et al.*, 1980), the overall basaltic compositions were comparable to and within the range of recent Hawaiian basalts (Bence *et al.*, 1980; Clague and Frey, 1980).

Normalized REE patterns for Emperor Seamount tholeiites show minor

REE fractionation [$(La/Yb)_N$ about 2–4], with generally flat light REE distributions resembling the REE patterns of Mauna Loa (Clague and Frey, 1980). The ratios of highly incompatible trace elements (La/Ta, Ta/Th, Nb/Th, Nb/U, Ba/Rb) in both tholeiites and alkali basalts are virtually constant throughout the whole chain (Cambon *et al.*, 1980; Hofmann, 1986) and suggest that the Hawaiian source has remained grossly constant in composition throughout time (Clague, 1981). However, the $^{87}Sr/^{86}Sr$ ratios are variable, with low values typical of the 50–60 Ma period (0.7033 at Suiko seamount) before increasing (to about 0.7037) at the Emperor–Hawaiian bend (42–44 Ma) and then remaining relatively constant within a wider range of values (Lanphere *et al.*, 1980). The lower Sr isotopic values correspond to seamounts built on younger (and thinner) oceanic crust and indicate the initial involvement of depleted asthenosphere in their generation relative to the isotopically enriched seamounts and islands on old crust. The apparent homogeneity of the Hawaiian plume with time can only be considered on the large scale, as many studies on the Hawaiian islands have demonstrated the highly heterogeneous nature of the source on the small scale.

9.8.2 *Linear aseismic ridges*

These ridges, found throughout the major oceans, are linear, non-spreading structural highs, free of seismic activity and composed of a volcanic base with a cover of volcanogenic and semipelagic sediments. The early development of the ridges indicates very shallow water conditions followed by subsidence due to cooling and contraction away from a heat source such as a hot-spot (Detrick *et al.*, 1977). The linear nature of the (now) totally submerged volcanic ridge is generally considered to have been the result of plate migration over a stationary plume. Early comparative data (Hekinian and Thompson, 1976) showed that aseismic ridges are distinct from N-MORB, being chemically evolved with higher K, Ti, Ba and Zr coupled with lower Mg, Ni and Cr.

Two examples of aseismic ridges are considered below: the Walvis Ridge in the southern Atlantic Ocean and the Ninetyeast Ridge in the eastern Indian Ocean.

9.8.2.1 *Walvis Ridge.* The Walvis Ridge represents the trace of the present Tristan da Cunha plume on the Africa plate over a period of about 120 Ma (Morgan, 1983; Duncan, 1984; White and McKenzie, 1989b). As this plume is situated under the flank of the Mid-Atlantic Ridge, the symmetrical opening of the southern Atlantic produced a counterpart to the Walvis Ridge, called the Rio Grande Rise, on the American plate (Figure 9.14). Both dredging and drilling (DSDP Legs 39, 72, 74 and 75) have verified the Walvis Ridge basement age progression away from the Tristan da Cunha hot-spot towards the African continent.

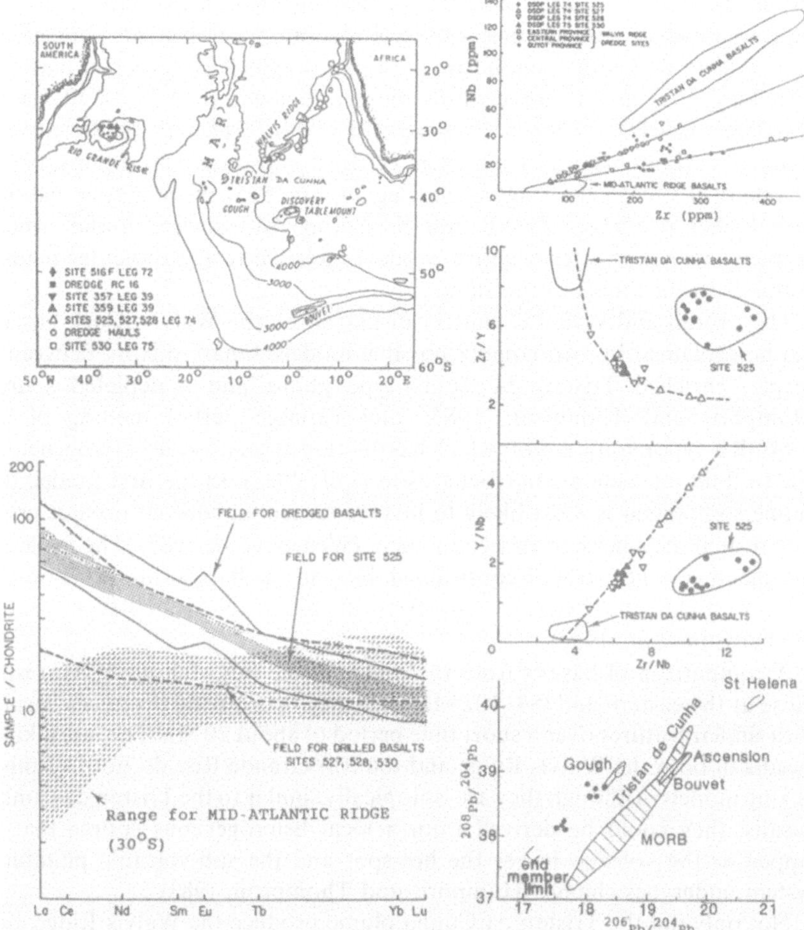

Figure 9.14 Chemical features of Walvis Ridge and Tristan da Cunha basalts. Data from Richardson *et al.* (1982) and Humphris and Thompson (1983). Map shows the location of Walvis Ridge and Rio Grande Rise in the southern Atlantic.on either side of the Mid-Atlantic Ridge (MAR).

Tristan da Cunha is composed of a nepheline-normative, differentiated K-rich alkali basalt suite, dominated by trachybasalts, and exhibits enriched incompatible element and radiogenic isotope features typical of OIB (Figure 9.14) (Baker *et al.*, 1964; O'Nions *et al.*, 1977; Sun, 1980; Weaver *et al.*, 1987). Samples obtained from the Rio Grande Rise consist of enriched tholeiites (with E-MORB-type features) as well as later (87 Ma) alkali basalts similar to the Tristan da Cunha suite (Fodor *et al.*, 1977; Thompson *et al.*, 1983). DSDP drilled and dredged basement samples from the Walvis Ridge (Figure 9.14), however, are predominantly aphyric and variably plagioclase phyric, enriched tholeiites (pillow lavas and massive flows) with relatively high abundances and variable ratios of incompatible elements (but lower than Tristan da Cunha), chondrite-normalized light REE enriched patterns [$(La/Yb)_N$ 3–8] and isotopic ratios more radiogenic than N-MORB with $^{87}Sr/^{86}Sr = 0.70417$, $^{143}Nd/^{144}Nd = 0.51270$, $^{206}Pb/^{204}Pb = 18.32$ (Richardson *et al.*, 1982; Humphris and Thompson, 1983; Thompson and Humphris, 1984). At one drill site on the ridge crest (Leg 74, Site 525), more alkalic basalts, with higher $(Ce/Yb)_N$, Zr/Nb and Sr isotope ratios were found, which probably represent later eruptive products relative to the tholeiites mainly sampled on the flanks of the ridge.

The overall and varied chemistry displayed by the Walvis Ridge basalts can be explained by two equally possible models: binary mixing between a deeper, enriched Tristan da Cunha-type plume and a depleted source (Humphris and Thompson, 1983), and variable partial melting of an E-MORB-type mantle source which has developed small-scale heterogeneities due to fluid metasomatism (Richardson *et al.*, 1982). In the first model, the plume component is most likely to involve ancient subducted oceanic crust with a small pelagic sediment component (Weaver *et al.*, 1987). The depleted end-member is not so well constrained, but, on the basis of the Pb isotopic data, excludes the extensive involvement of an N-MORB-type source (Richardson *et al.*, 1982).

A comparison of basalts from the centre of the ridge (about 68 Ma) with those at the eastern end (84–102 Ma) indicates that mixing produces basalts with similar features over a short time period of about 20 Ma. The late alkalic basalts of both the Walvis Ridge and the Rio Grande Rise do not fit simple mixing models. Although they are isotopically similar to the Tristan da Cunha basalts, they could be derived from a local heterogeneous source that is tapped as the volcano leaves the hot-spot and the sub-volcanic plumbing system undergoes change (Humphris and Thompson, 1983).

Not only did the Tristan da Cunha plume produce the Walvis Ridge, but it interacted with and influenced the composition of the adjacent Mid-Atlantic Ridge (MAR) segment. Relative to normal segments, the Tristan area of the spreading axis features anomalously high [$(La/Sm)_N > 1$] and low Zr/Nb (< 20) ratios relative to N-MORB (commonly < 1 and > 25, respectively). These features are characteristic of incompatible element-rich mantle domains and indicate that the plume-generated Tristan anomaly is large enough to influence the source of the nearby axis basalts (Schilling *et al.*, 1985). The enriched MAR basalts can be shown to be derived by mixing between an N-MORB source and the Tristan da Cunha plume source (Humphris *et al.*, 1985) and this suggests that a sub-lithospheric channel connects the plume and ridge axis.

Interaction between the active ridge and nearby plume systems via pipe flow and the dispersion of enriched asthenosphere into depleted mantle beneath the ridge (Vogt, 1976; Morgan, 1978; Schilling *et al.*, 1985) is also a feature of the MAR adjacent to the Azores plume (Schilling, 1975) and the American–Antarctic Ridge near the Bouvet plume (Le Roex *et al.*, 1985).

9.8.2.2 *Ninetyeast Ridge.* The north–south trending Ninetyeast Ridge (Figure 9.15) is the longest aseismic ridge in the world (about 4500 km) and represents the volcanic trace of the Kerguelen plume on the Indian plate as it moves northwards (Luyendyk, 1977; Duncan, 1978). The volcanic

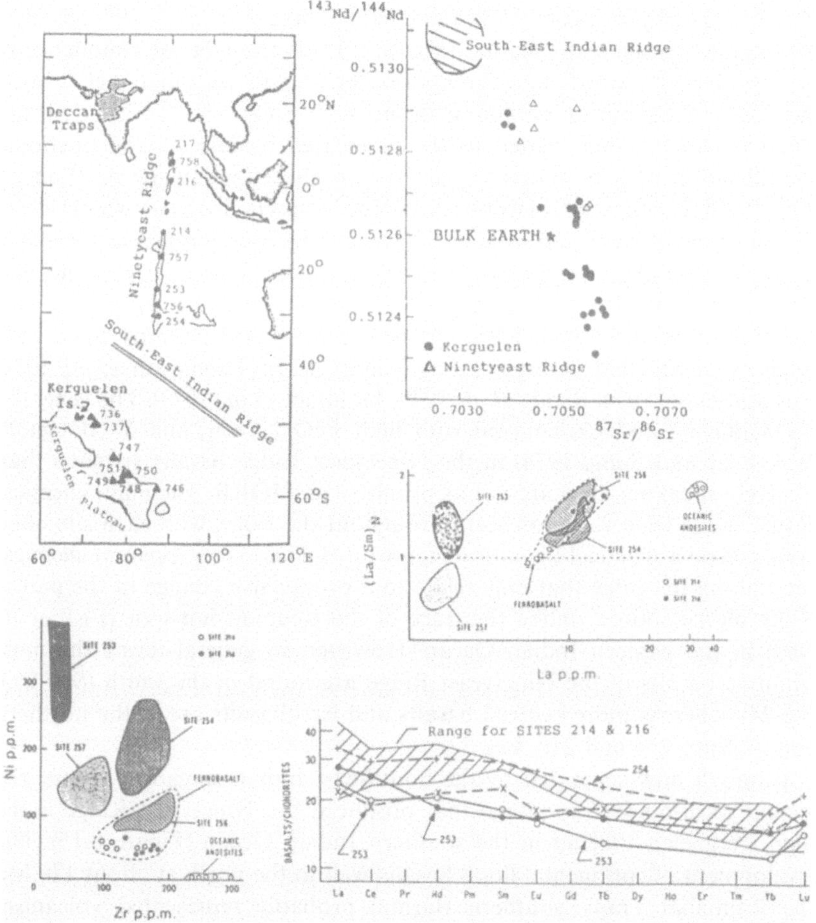

Figure 9.15 Chemical features of Ninetyeast Ridge basalts and evolved iron-rich differentiates. Data from Frey and Sung (1974), Ludden *et al.* (1980), Mahoney *et al.* (1983), Storey *et al.* (1988). Map shows the location of Ninetyeast Ridge and Kerguelen Plateau in the eastern Indian Ocean, together with relevant DSDP and ODP drill sites. DSDP (dots): Leg 22 (Sites 214, 216, 217); Leg 26 (Sites 253, 254). ODP (triangles): Leg 119 (Sites 736, 737, 746); Leg 120 (Sites 747, 748, 749, 750, 751); Leg 121 (Sites 756, 757, 758).

basement and volcanogenic sediments have been drilled by the DSDP (Legs 22 and 26) and more recently by the ODP (Leg 121) which demonstrated that ages generally increase from the south (about 40 Ma) towards the north (about 80 Ma). The highly vesicular nature of many of the basalts and the presence of *in situ* hyaloclastites and air-fall tuff layers indicate that parts of the ridge were subaerial or produced in a very shallow water environment (Luyendyk, 1977; Fleet and McKelvey, 1978).

Ninetyeast Ridge basement is mainly composed of variably vesicular, quartz- and hypersthene-normative aphyric and olivine–plagioclase phyric tholeiites and rare picrites that are often highly altered with the development of secondary phyllosilicates (Hekinian, 1974; Thompson *et al.*, 1974; Kempe, 1974). One characteristic feature (Figure 9.15) is the relative abundance of more evolved intermediate differentiates (ferrobasalt with high FeO*/MgO and TiO_2, tholeiitic or oceanic andesite with $SiO_2 > 55\%$; Sites 214 and 216, Leg 22) that are related to the associated basalts via the fractional crystallization of clinopyroxene, plagioclase and titanomagnetite (Ludden *et al.*, 1980). The presence of iron-rich, evolved rocks is also a persistent feature of plume-generated oceanic islands adjacent to spreading ridges (e.g. Iceland, Galapagos, Azores; Byeryl *et al.*, 1976; and Chapter 5). Relative to N-MORB most of the basalts are incompatible element enriched tholeiites, with chondrite-normalized light REE enriched and depleted patterns and a wide range of Sr and Nd isotopic ratios (Figure 9.15) (Thompson *et al.*, 1974; Frey and Sung, 1974; Reddy *et al.*, 1978; Mahoney *et al.*, 1983). The generally low MgO, Ni and Cr, coupled with high FeO*, TiO_2, and V (Bougault, 1974; Frey and Sung, 1974) in the Ninetyeast Ridge basalts suggests that, relatively speaking, they are not as primitive as MORB. The main chemical feature is the wide variation seen throughout the ridge, with each site often having its own chemical signature (Figure 9.15). There is no apparent chemical trend along the ridge that may relate to a progressive change in the nature of the plume source, unlike the trace of the Réunion hot-spot (Fisk *et al.*, 1989) in the western Indian Ocean. However, in general terms the most primitive basalts of the Ninetyeast Ridge are found in the south (Site 253, Leg 24), whereas more evolved basalts and ferrobasalts are in the northern section (Sites 214 and 216, Leg 22).

Although now separated by the South-east Indian spreading centre, the current location for the plume that produced the Ninetyeast Ridge is the huge Kerguelen Plateau in the southern Indian Ocean (Figure 9.15). The development of continental flood basalts well to the north at about 120 Ma (the Rajmahal Traps, southern Burma) probably represented volcanism associated with the initial plume head, whereas the Ninetyeast Ridge itself may be a reflection of the volcanic effects of the tail of the plume (Richards *et al.*, 1989). However, the size of the Kerguelen plateau suggests either reactivation of the plume and the development of a new head or a slowing down of the northwards migration of the Indian plate. The Kerguelen Island

and Heard Island tholeiitic shield volcanoes, capped by minor, late alkali basalts, represent the most recent activity of this plume source. The majority of the Kerguelen Plateau basement and subaerial edifice-building stage is composed of about 40 Ma tholeiites with low incompatible element abundances and ratios transitional between N-MORB and E-MORB, and a wide range of OIB-type isotopic ratios (Dosso *et al.*, 1979; Dosso and Murthy, 1980; Mahoney *et al.*, 1983; Storey *et al.*, 1988). The Kerguelen Plateau has some chemical affinities to similar, large oceanic plateaux of the western Pacific (Floyd, 1989) and in this context may represent a substantial thickening of the local Cretaceous oceanic crust. The chemical similarity and overlap between the incompatible element and isotopic ratios of the Kerguelen and Ninetyeast Ridge basalts (Figure 9.15) indicates that they have been derived from the same complex source, possibly involving variable interactions between the Kerguelen plume and two other components—depleted asthenosphere (White and Hofmann, 1982) and old oceanic lithosphere (Storey *et al.*, 1988). Kerguelen shows one of the widest ranges of isotopic values for any plume-related OIB with a DUPAL signature (see Figure 9.10) and with the Ninetyeast Ridge data underlines the continuing heterogeneity of this mantle source over some 120 Ma.

9.8.3 *Island groups adjacent to spreading axes*

As well as single islands on or near the flanks of the Mid-Atlantic Ridge, which represent the present loci of mantle plumes and are related to hot-spot traces, there are also various tight groups of young islands, such as the Cape Verdes, Canaries and Azores, for which a similar plume origin may be postulated (Morgan, 1983). The Azores are one of the most interesting island groups in that the wide chemical variation seen between and within individual islands implies a highly heterogeneous mantle source, and also, local tectonic controls play an important role in determining the site of volcanic activity.

The Azores consist of nine volcanic islands developed on the 700 km long, NW–SE trending, submarine Azores Platform (average water depth 2 km) that sits astride the MAR between 37 and 40°N (Figure 9.16). The Azores Platform marks the site of a triple junction between the American, Eurasian and African plates (Figure 9.16), with the islands aligned along the Eurasian–African junction which is interpreted as a complex transform fracture zone (Laughton and Whitmarsh, 1975). The elevation of the whole triple junction area and the development of the oceanic islands is generally related to an underlying thermal plume (Schilling, 1975), although the string of islands does not show any marked age progression away from the MAR towards Africa. Activity occurs all along the chain with most of the islands composed of young lavas (from historic eruptions back to about 1 Ma), although the

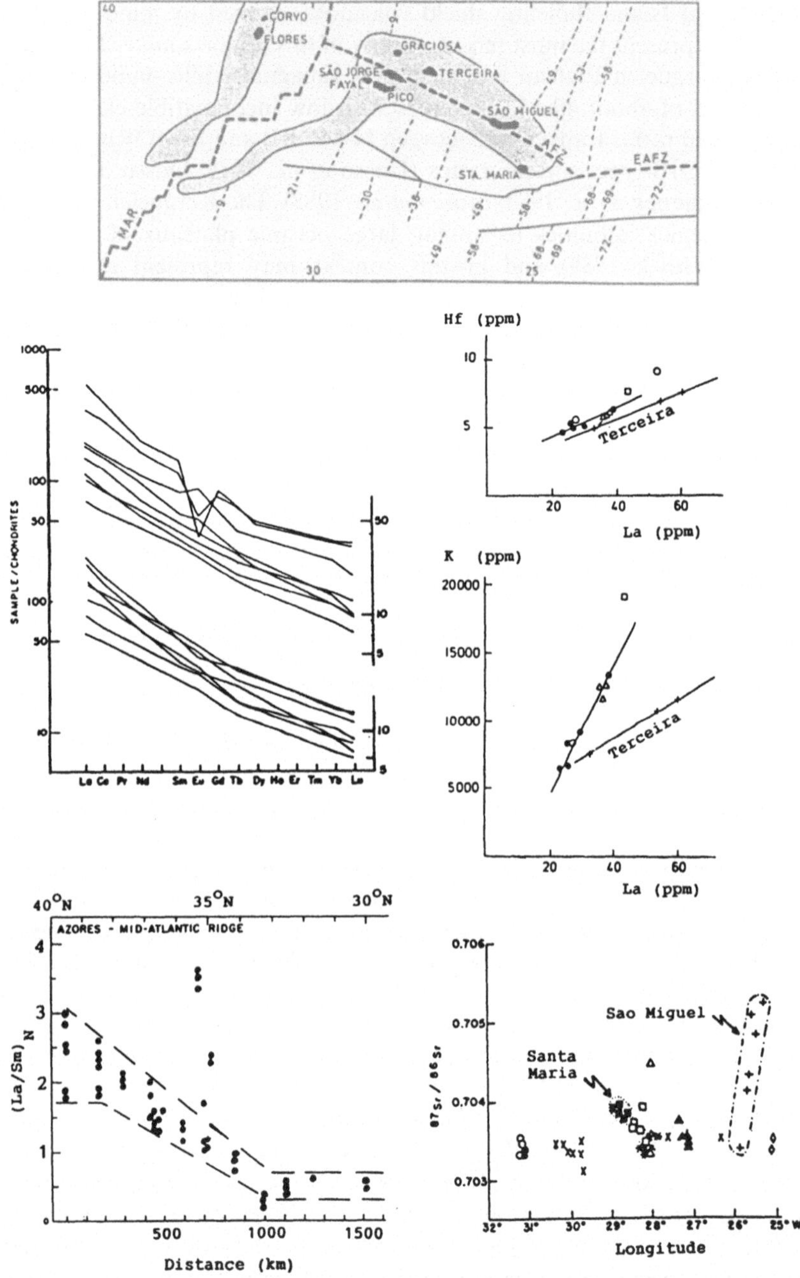

Figure 9.16 Chemical features of the subaerial Azores basalts and submarine plateau adjacent to the Mid-Atlantic Ridge (MAR). Data from Schilling (1975), Flower *et al.* (1976), White *et al.* (1979). Note the wide range of Sr isotope ratios for Sao Miguel and Santa Maria relative to the other islands and submarine basalts from the Azores Plateau and Terceira Trough (crosses). Map shows the location of the Azores Plateau and Islands astride part of the Mid-Atlantic Ridge.

oldest volcanics are recorded on the most easterly islands of Sao Miguel (4 Ma) and Santa Maria (5.5 Ma) (Feraud *et al.*, 1980). The site of the activity and its composition appears to be related to the local lithospheric stress pattern that allows the mixing of melts from the MAR rift and Azores sources to migrate upwards through leaky fracture zones parallel to the Azores Platform axis (Feraud *et al.*, 1980; Flower, 1981a).

The Azores lavas are predominantly phyric alkali olivine basalts with intermediate trachytic differentiates, together with minor strongly nepheline-normative basanites and hypersthene-normative transitional basalts (Flower *et al.*, 1976; Self and Gunn, 1976; White *et al.*, 1979). Oversaturated peralkaline rocks (comendites and pantellerites) are found on Terceira and Sao Miguel (Schmincke, 1973). Although fractional crystallization involving olivine, clinopyroxene (and later, plagioclase, minor amphibole) are important in the evolution of hawaiite and mugearite differentiates (Flower *et al.*, 1976; White *et al.*, 1979), many small volume intermediate lavas are mixed-magma hybrids between basalt and alkali feldspar phyric trachytes (Storey *et al.*, 1989). In general, however, smooth trends of major and trace element variation within suites are indicative of progressive fractionation involving the observed phenocryst phases.

Azores basalts are characteristically enriched in incompatible trace elements with typical OIB chondrite-normalized humped patterns showing positive Ba and Nb anormalies, strongly fractionated REE patterns [$(La/Yb)_N$ 9–14], but higher Ba/Nb, Ba/La and lower Pb/Ce ratios relative to other OIB (Flower *et al.*, 1976; White *et al.*, 1979; Davies *et al.*, 1989). Isotopically the basalts are more radiogenic ($^{87}Sr/^{86}Sr = 0.70332–0.70514$; $^{206}Pb/^{204}Pb = 19.33–20.02$; $^{207}Pb/^{204}Pb = 15.57–15.75$) than N-MORB (White *et al.*, 1979; Dupre *et al.*, 1982; Davies *et al.*, 1989), and deviate from the linear mantle array on the Nd–Sr isotope diagram towards higher Sr isotope values (Hawkesworth *et al.*, 1979). Some of these features are shown in Figures 9.10 and 9.16.

Geochemically the most interesting aspect of the Azores basalts is the wide range of highly incompatible trace element ratios and isotopic values, which not only vary between islands, but within individual islands. Some features persist throughout time, typifying differentiated suites from specific islands, and reflect considerable heterogeneity in the mantle source regions below the islands. For example, Terceira basalts have distinct K/La, Th/La, U/La and Hf/La ratios relative to all the other islands, whereas Sao Miguel basalts have the highest LIL element contents, high K/Na and the greatest light REE enrichment (Flower *et al.*, 1976). Faial and Pico have generally lower Sr isotope ratios, whereas Sao Miguel exhibits the highest and widest range (White *et al.*, 1979). All the islands have different and separate trends on Pb–Pb isotope diagrams (Davies *et al.*, 1989). Within-island variation may be illustrated by Sao Miguel, which shows a marked and systematic increase in Sr isotope ratios from MORB-type values in the west

(0.703) to higher values (> 0.705) in the east. Together with Nd isotope data, this suggests magma or source mixing below the island (Hawkesworth *et al.*, 1979). The low $^{87}Sr/^{86}Sr$ component lies within the MORB mantle array for the MAR on the Sr–Nd isotope diagram (Figure 9.10). Similar mixing is implied by the variation in Pb isotopes within another island, Terceira, where post–caldera lavas have more MORB-like signatures than earlier lavas (Dupre *et al.*, 1982).

Geochemically the MAR segment adjacent to the Azores Platform is incompatible element enriched relative to N-MORB (DSDP Legs 37 and 82) and has similar Sr isotopic ratios, light REE enrichment, and $(La/Sm)_N$ ratios to the Azores Plateau and islands (Schilling, 1975; Flower *et al.*, 1976). Schilling (1975) interpretes this enrichment as a result of the influence of the enriched Azores plume on a normally depleted MORB source, producing a chemically anomalous ridge segment (similar to Tristan da Cunha and Bouvet). As to the nature of the Azores plume itself, a detailed isotopic and trace element interpretation of the Azores basalts indicates its highly heterogeneous nature, involving mixing between a depleted MORB source and possibly three other enriched sources consisting of recycled oceanic lithosphere and subcontinental lithospheric mantle components (Davies *et al.*, 1989).

9.8.4 *Ridge flank young seamounts*

Recent investigations of seamounts situated on < 7 Ma oceanic crust adjacent to the East Pacific Rise between 9 and 14°N and at 21°N (Figure 9.17) have provided an insight into both the morphological and chemical evolution of very young ridge flank volcanoes (Batiza, 1980; Batiza and Vanko, 1984; Batiza *et al.*, 1984; Zindler *et al.*, 1984; Fornari *et al.*, 1988a, b; Graham *et al.*, 1988). The geochemical diversity shown by some of these young seamounts (Figure 9.17) has important implications for mantle structure and the nature of magmatic processes at spreading centres (Chapter 7). For example, although the seamount basalts generally exhibit similar depleted chemical features to the adjacent EPR (which has typical N-MORB characteristics), a wide range of chemical compositions is also displayed that implies the existence of small-scale source heterogeneities in the vicinity of the ridge (Fornari *et al.*, 1988a, b; Graham *et al.*, 1988).

Two aspects of the seamount basalts will be considered: the implications of seamount chemical variability and their relationship to the magmatic plumbing system of the EPR.

Morphological features of the Lamont group of seamounts (at about 10°N; Figure 9.17), developed normal to the strike of the EPR on progressively older oceanic crust, have been outlined in section 9.3. The majority of the seamount basalts are poorly vesicular (a few percentage vesicles only), aphyric

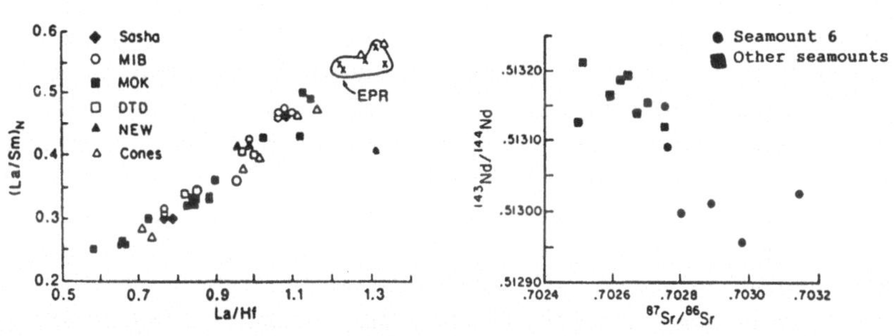

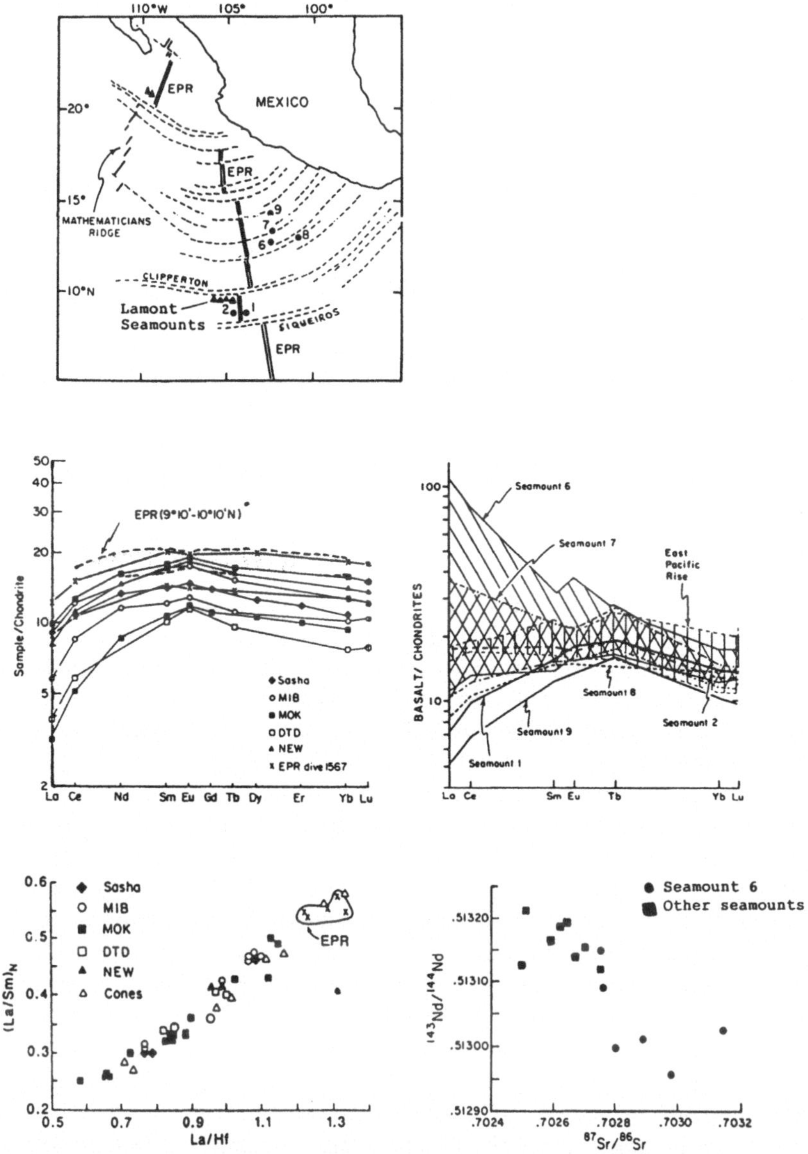

Figure 9.17 Chemical features of young seamounts on the east and west flanks of the East Pacific Rise. Data from Zindler *et al.* (1984), and Fornari *et al.* (1988a, b). Note the often highly depleted character of the seamounts (relative to EPR) and also the wide range of chemical features displayed by seamounts 6 and 7. Map shows the location of the Lamont seamount group (Sasha, MIB, MOK, DTD, NEW) and various numbered seamounts.

or sparsely olivine–plagioclase phyric tholeiites not dissimilar to typical EPR basalts (Batiza and Vanko, 1984; Zindler *et al.*, 1984). Clinopyroxene phenocrysts are, however, typically absent from basaltic glasses of the Lamont seamounts and many compositions display the high MgO, Ni and Cr contents indicative of primitive melts. Some compositions approach primary mantle melts with magnesium numbers commonly between 65 and 75 (Batiza and Vanko, 1984). Although element trends characteristic of low pressure fractionation genetically link the lavas of some seamounts, the degree of chemical evolution exhibited is relatively minor and may terminate before the appearance of any phenocrystic clinopyroxene. Limited chemical evolution, and the primitive nature of the basalts and glasses, suggest that melts have been rapidly transported upwards from the mantle via independent channelways and not ponded in crustal magma chambers (Zindler *et al.*, 1984). This picture is different to the EPR plumbing system with large, fractionating, sub-axial magma chambers that store and periodically erupt melts (Langmuir *et al.*, 1986; Detrick *et al.*, 1987; Macdonald, 1989) and indicates that there is little connection between seamount and ridge magma systems.

Geochemically (Figure 9.17) most of the basalts are incompatible element depleted tholeiites similar to the adjacent EPR basalts and exhibit incompatible element ratios, REE patterns and Sr–Nd–Pb isotopic systematics typical of N-MORB generally (Zindler *et al.*, 1984; Allan *et al.*, 1987; Fornari *et al.*, 1988a,b; Graham *et al.*, 1988). However, the seamount basalts show the following differences to EPR lavas. They are often significantly more primitive with strongly depleted light REE patterns and lower $(La/Sm)_N$ ratios (Fornari *et al.*, 1988a), and they are enriched in volatiles (H_2O, F, Cl, S) at the same level of chemical fractionation (Aggrey *et al.*, 1988). The most interesting chemical feature, however, is the presence of incompatible element enriched tholeiites and minor alkali basalts, which cannot be genetically related to the depleted tholeiites by variable melting of a common MORB source. The wide spectrum of patterns of REE fractionation in the basalts, from depleted to enriched, is illustrated by seamounts mainly on the east flank of the EPR, especially seamounts 6 and 7 (Figure 9.17). The alkali basalts appear to be minor eruptions associated with later crater development or capping the tholeiitic edifices and are confined to the morphologically-evolved volcanoes on crust older than 3 Ma (Batiza and Vanko, 1984). This temporal relationship is similar to the subaerial eruptive stages typified by mid-plate Hawaiian-type volcanoes.

The overall chemical diversity seen between the seamounts, such as variation in light REE fractionation and highly incompatible element ratios (Rb/Sr, La/Hf, Th/Hf), together with a coupled range of Sr–Nd–Pb isotopic ratios (Figure 9.17), suggests derivation from strongly (but variably) depleted, highly heterogeneous sources (Zindler *et al.*, 1984; Fornari *et al.*, 1988a, b). Some seamounts, showing the full range of depleted to enriched compositions (Figure 9.17), can be generated by simple two-component mixing of a depleted melt (derived from a MORB source) with a more enriched melt (derived from ancient subducted oceanic crust or metasomatized mantle) (Zindler *et al.*, 1984). The individual chemical nature of the seamounts indicates that the heterogeneities present in the depleted mantle below ridges are present on a very small scale (< 5 km, Fornari *et al.*, 1988a) with relatively enriched 'plums' residing in a MORB source matrix (Figure 9.11 B). The generation of small volume melts away from the ridge axis involving variable proportions of entrained plums and MORB matrix produce the heterogeneities observed in the seamounts, whereas under the EPR, large volume melting, mixing and subsequent ponding eliminates any original source differences.

9.9 Concluding statements

1. Intraplate volcanism is represented by the numerous oceanic islands and seamounts that pepper the ocean floor and may in some areas constitute up to 25% of the oceanic crust. In terms of the locus of original activity a distinction may be made between those volcanoes that are generated mid-plate on relatively old, cold crust and related to localized mantle hot-spots, and young volcanoes generated on hot, young crust adjacent to spreading axes. Although the latter may eventually migrate away from the ridge into more mid-plate situations, their continued growth and development depends on the efficiency of the heat source and melt extraction, together with the thickness of the lithosphere to support larger structures. The non-uniform distribution of seamounts (per 10^6 km^2 of ocean floor) is related to crustal age (generally increasing with age on young crust, then becoming relatively constant) and the location of fracture zones and transforms (which provide magma channelways).
2. The mantle plume model and the propagating fracture model are two hypotheses that explain many of the features of intraplate (oceanic) volcanism. In the former, activity is related to upwelling, thermochemical plumes derived from the deep mantle, whereas the latter envisages activity being funnelled via tensional fractures generated in mobile, stressed lithosphere of suitable vulnerability. The plume model is alive and well, remaining popular for linear volcanic chains with age progression, although its manifestation is seen more as a continuum from strong, long-lived, continuous plumes to waxing and waning blobs of variable strength and depth of origin. This model is not applicable to young ridge-flank seamounts, the location and morphological development of which are governed by fractures that focus melts from shallow, local source domains.
3. The initial deep water stages of volcano growth feature steep structures constructed of both extrusive pillow lavas and intrusives. These are replaced, as the volcano builds towards sea level, by lavas and increasing proportions of hyaloclastites produced by the explosive release of trapped

volatiles at a critical water depth. Downslope slumping and mass wasting of hyaloclastites and lavas produces widespread debris aprons on which subaerial flows eventually develop. The compositional variation of oceanic islands is often based on the Hawaiian model, with four eruptive growth stages: early submarine (alkalic), main submarine-to-subaerial edifice, (tholeiitic), post-caldera (alkalic) and post-erosional (highly alkalic). However, a tholeiite-dominated main constructional phase represents only one expression of intraplate volcanism. Plume-related mid-plate volcanism, generally developed on old crust, can have either a predominantly tholeiitic or alkaline edifice, although seamounts on young hot crust are typically tholeiitic. When observed, the replacement of tholeiites by later, minor highly alkaline basalts, may be a consequence of a change in the sub-volcanic plumbing system and melting regime as the volcano moves away from its primary heat source.

4. Relative to the chemically-depleted features of mid-ocean ridge basalts, oceanic island basalts, both tholeiites and alkali basalts, are generally enriched in incompatible elements at the same fractionation level and are also more radiogenic. Another distinctive features of OIB is the very wide range of incompatible element abundances and isotopic ratios displayed, such that different basaltic suites exhibit various degrees of enrichment. OIB chemical variation is a reflection of the heterogeneous nature of mantle sources and the interaction of variably enriched OIB melts with more depleted asthenospheric- and lithospheric-derived melts. OIB (together with MORB) data enables a number of specific mantle end-member compositions to be identified. Possible original sources for OIB include subducted oceanic lithosphere and subcontinental lithosphere now residing at the upper–lower mantle boundary or mantle-core boundary (eventually rising as 'plumes') and/or entrained in the convecting upper mantle (as isolated, smeared out 'plums') before being involved in melting.

10 Back-arc basins

ANDREW SAUNDERS and JOHN TARNEY

10.1 Introduction

As illustrated in this book, oceanic basalts are compositionally diverse. This diversity arises not only from low pressure fractionation in sub-volcanic magma chambers, but also during the solid–liquid fractionation that accompanies melting of the mantle. Isotopic and chemical studies also reveal gross heterogeneity in the composition of the sub-oceanic mantle (Chapter 15) reflecting major differentiation and recycling (e.g. subduction of oceanic crust) throughout the past 2.5 Ga, or even longer. To fully understand, and thus evaluate, all of these factors, it is necessary to study basalt genesis in many tectonic settings. The study of magmatism at destructive plate boundaries provides important information about the scale of crustal recycling, which may in time influence our understanding of the evolution of the crust and mantle.

This chapter deals with a special type of oceanic basalt, namely that erupted within extensional basins at destructive plate boundaries. These basins, commonly termed back-arc basins, are often floored by mafic oceanic crust, and are formed by extensional and magmatic processes akin to those occurring in the major ocean basins. In general terms, extension and separation of the lithosphere above a subduction zone allows mantle peridotite to upwell and decompress. Other things being equal, most silicates have lower melting points at lower pressures. Thus, mantle may ascend to a sufficiently high level that its melting point is reached, despite there being no extra input of thermal energy. This will lead to the formation of basaltic (or basalt-like) melts, in much the same way that is envisaged for the formation of mid-ocean ridge basalts (MORB).

Consequently, the compositions of back-arc basin basalts are similar to MORB, but we use the description 'similar' advisedly. These basalts frequently, but not always, show slight, systematic compositional differences from MORB. In particular, many back-arc basalts are transitional in their composition between MORB and island-arc basalts, possibly because the source of the back-arc basalts has been contaminated, or metasomatized, by fluids from a subduction zone. These data offer the potential to evaluate the role of subducted fluids beneath island arcs because, unlike most arc magmas, back-arc magmas are erupted at abyssal depths, and closely preserve their original volatile contents.

Studies of back-arc basin basalts began in the early 1970s following a spate of interest in the origin of marginal basins in general, but despite almost two decades of study, including five legs of the Deep Sea Drilling Project devoted to back-arc and arc objectives, several problems remain outstanding. (1) Not all back-arc basalts have compositions transitional between arc and mid-ocean ridge basalts; why? (2) There is strong evidence that back-arc basalts have sampled several mantle components, and not just a subduction-related component: what do these components represent? (3) Many back-arc basalts have high volatile contents: how do these volatiles affect the melting processes of the back-arc mantle, and subsequent magmatic fractionation? (4) How do these processes differ from those affecting MORB genesis?

Back-arc basins and their associated igneous rocks are not merely of interest to investigations of basalt petrogenesis. It is widely believed that many ophiolites are the remnants of back-arc basins, or at least some form of basin associated with subduction systems, obducted during plate collision. It is therefore essential to fully characterize modern marginal basins, from structural, petrological and sedimentological standpoints, in order to fully understand the provenance of ancient ophiolites. Looking even further back in geological time, marginal basins have been proposed as modern analogues for Archaean greenstone belts (Tarney *et al.*, 1976; Tarney and Windley, 1981), enigmatic assemblages of mafic and ultramafic rocks formed when mantle temperatures were higher than the present day.

10.2 Formation of marginal basins

Back-arc basins are a variety of marginal basin, the majority of which are located within and behind the island arc festoons of the western Pacific (Figure 10.1). There are a few exceptions, namely the Gulf of California and the Bransfield Strait on the eastern rim of the Pacific Ocean, the Grenada Trough and the Scotia Sea in the west-central and west-southern Atlantic Ocean, and the Aegean and Tyrrhenian Seas in the Mediterranean. Nevertheless, over 80% of the basins rim the Australasian–Eurasian plates, and it is noteworthy that Wegener (1929) recognized this association and suggested that the basins opened by extension at the trailing edges of moving continents. Other observations and interpretations may, however, be equally valid, but before discussing which of these mechanisms is responsible for marginal basin formation, it is necessary to emphasize that not all marginal basins are produced by back-arc extension. Marginal basins were defined by Karig (1971) as 'small, semi-isolated basins or series of basins lying behind the volcanic chains of island arc systems', and they correspond to Keunen's (1950) marginal seas. Within this broad definition, three major, distinct types of

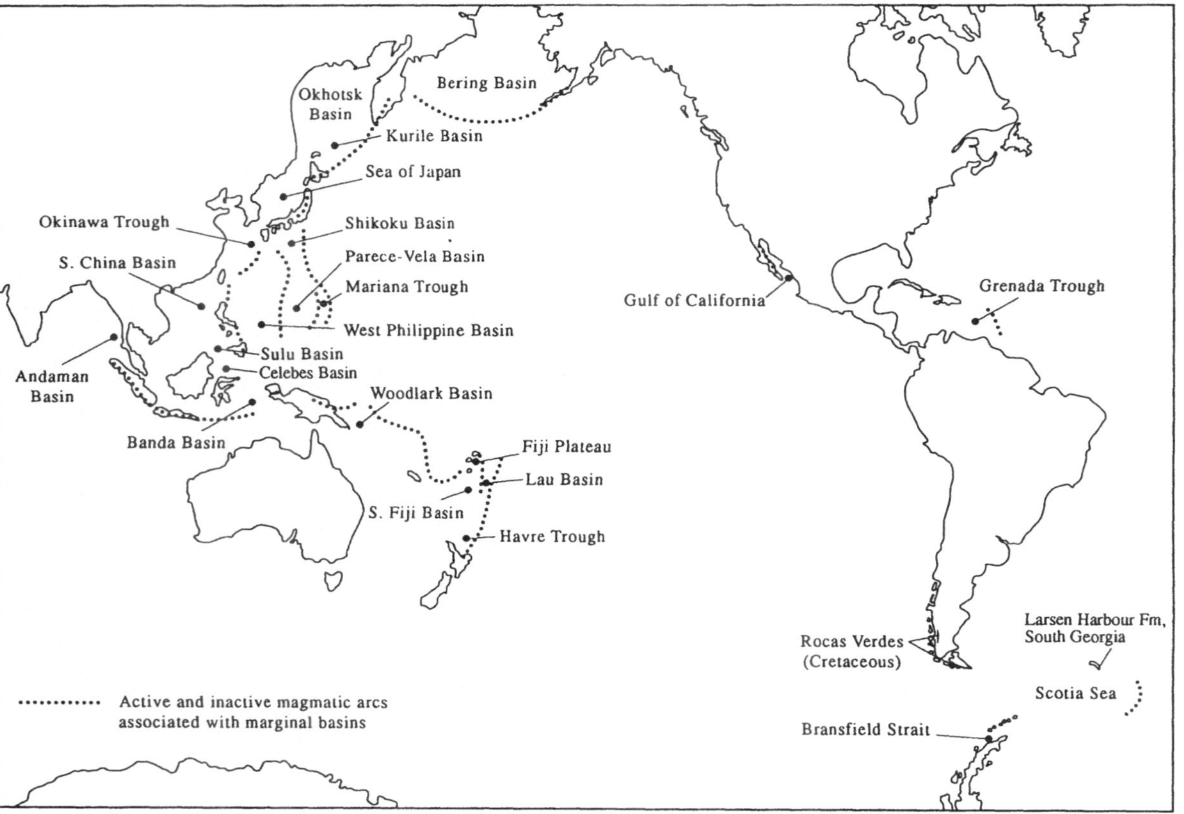

Figure 10.1 Map of the eastern hemisphere showing the association of marginal basins (extensional or back-arc basins, pull-apart basins, and trapped ocean basins) with destructive plate margins. The majority of the earth's marginal basins, including at least one fossil basin (the *rocas verdes* in southern Chile) are found in the Pacific Basin.

marginal basin may be recognized: back-arc basins, pull-apart basins, and trapped marginal basins.

Back-arc (or inter-arc) basins form by extension and seafloor spreading behind, or within, an island arc, and their formation is thus associated with contemporaneous subduction-zone activity. Back-arc basins may be either active or inactive, depending on whether or not the process of extension has ceased. The life-span of an active, extending basin is about 15 Ma, after which extension ceases, the heat flow diminishes, and the basin floor is buried beneath a thickening carapace of sediment. Active basins are characterized by a rough seafloor topography, and although the seafloor is often deeper than normal oceanic crust of equivalent age, the depths of the active spreading centres (about 2000–4000 m) are within the depth range of oceanic ridges (0–4800 m). Several back-arc basins have poorly developed magnetic lineations, which may be a function of less coherent spreading processes (possibly with many ridge jumps) than those operating in normal ocean basins (Lawver and Hawkins, 1978). This is only a broad observation; the East Scotia Sea back-arc basin, for example, has excellent magnetic lineations (Barker, 1972; Barker and Hill, 1981), and Weissel (1981) has demonstrated that many basins have well defined lineations.

A second category of marginal basin is the pull-apart or 'leaky' transform fault basin, (e.g. the Gulf of Calfornia and the Andaman Sea) (Wilson, 1965). These basins have developed at plate margins subjected to highly oblique convergence and, in the case of the Gulf, ridge-trench collision (Moore, 1973; Curray *et. al.*, 1979). This type of basin is uncommon, but they are likely to form at any oblique collision zone. Pull-apart basins are not, of course, restricted to the marginal basin environment; they may form at any strike-slip boundary. In this chapter, we shall include only those pull-apart basins specifically associated with subduction processes.

The third major category of marginal basin, the trapped basin, (e.g. the Aleutian Basin: Cooper *et al.*, 1977) forms when subduction begins within an oceanic plate, resulting in the entrapment and isolation of oceanic lithosphere beneath and behind a new arc system. The magnetic lineations within the basin may be oblique to the arc, and the basin crust does not owe its origin directly to subduction processes. The lithosphere is therefore the same as that found beneath the normal ocean basins, and will not be considered further in this account.

Why do some convergent plate margins develop back-arc and pull-apart basins, while others do not? The existence of a strongly extensional regime at such margins appears at first sight to be paradoxical, but it is clear that subduction processes in some way control the extension and formation of back-arc basins (Karig, 1971; Uyeda and Kanamori, 1979; Molnar and Atwater, 1978). In the case of pull-apart basins, such as the Gulf of California, these processes are readily understood; the conservative plate boundary, with its strong dextral shear couple, has a slight angular divergence, with the result

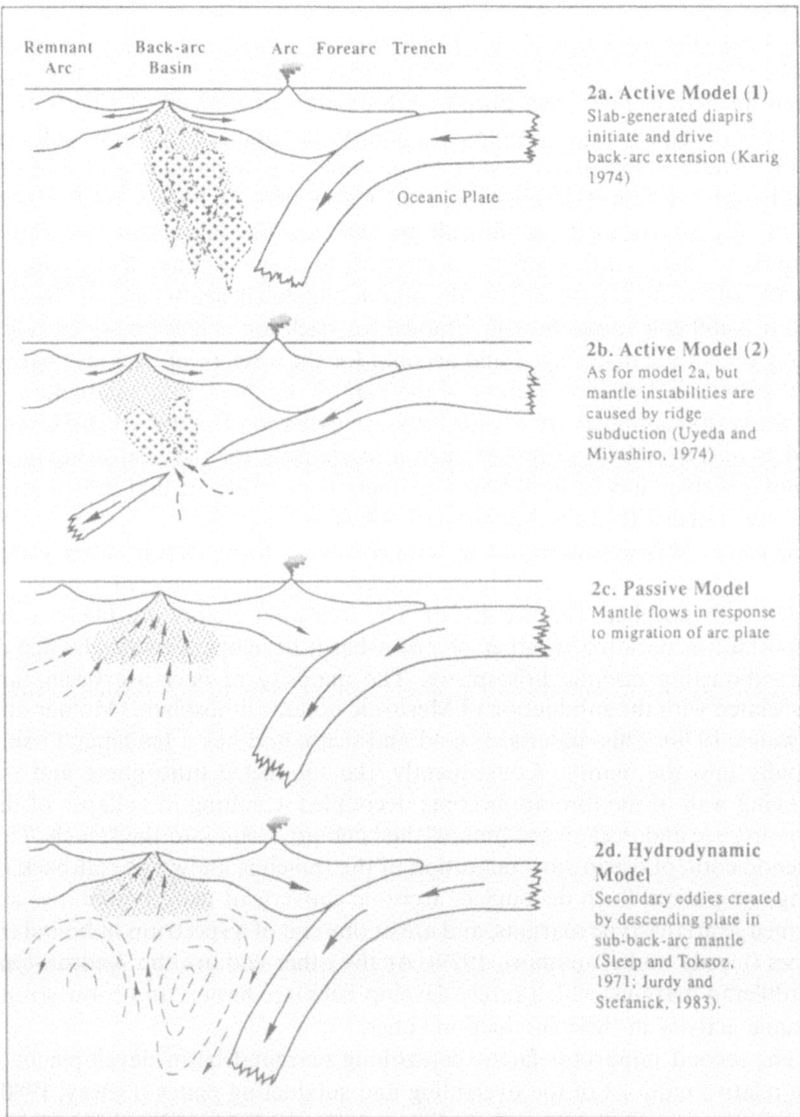

Figure 10.2 Schematic diagrams of various models that have been proposed for back-arc basin formation. (a) Active model of Karig (1974), where diapirism resulting from water and/or heat from the subduction zone causes active spreading in the back-arc or inter-arc region. (b) The active model of Uyeda and Miyashiro (1974) involving subduction of a spreading axis, and invoked to explain the formation of the Sea of Japan. (c) Passive upwelling in response to lithosphere extension (Packham and Falvey, 1971). (d) Hydrodynamic model, with secondary mantle convection being induced by drag along the Benioff Zone (McKenzie, 1969; Sleep and Toksoz, 1971; Jurdy, 1979; Jurdy and Stephanick, 1983).

that the Pacific plate is separating slowly from the North American plate.

In those arc-basin systems where subduction is approximately orthogonal to the trench, the mechanisms are less apparent. Several models of back-arc basin formation have been proposed; four are summarized in Figure 10.2. In the first broad category, plumes of mantle, destabilized by fluids and heat in the mantle wedge above a subduction zone, rise diapirically and split the overlying lithosphere (Hasebe *et al.*, 1970; Karig, 1974; Oxburgh and Turcotte, 1970) (Figure 10.2a). It is difficult to evaluate the importance of active diapirism. There is little geophysical evidence for large volumes of low velocity mantle ascending as plumes from the subducting slab beneath back-arc basins, and it is difficult to see how any model for back-arc extension which relies solely on mantle diapirism could account for the absence of back-arc basins in the eastern Pacific. A variant of the diapiric model is that subduction of an active spreading centre causes back-arc extension (Figure 10.2b); Uyeda and Miyashiro (1974) proposed such a mechanism for the Sea of Japan. It is not a viable model for all basins as there is no evidence that most basins are associated with the subduction of ridges.

The second broad category of back-arc basin formation involves either passive upwelling of mantle (Figure 10.2c), or subduction-induced convection in the asthenosphere (Figure 10.2d). The following factors are likely to be important in deciding whether or not a basin develops. Firstly, the age of the subducting oceanic lithosphere. The majority of back-arc basins are associated with the subduction of Mesozoic oceanic lithosphere (Molnar and Atwater, 1978). This material is cold and dense and has a tendency to sink rapidly into the mantle. Consequently, the subducted lithosphere and the hanging wall of the fore-arc become decoupled, resulting in collapse of the fore-arc, arc and oceanward limb of the back-arc basin into the trench. This phenomenon of oceanward migration of the trench is known as roll-back or hinge migration. Such decoupled, aseismic convergent plate boundaries are termed Mariana-type margins, and are at one end of a spectrum of boundary types (Uyeda and Kanamori, 1979). At the other end are the Andean-type cordilleran margins, which rarely develop back-arc basins, and show strong seismic activity in their subduction zones.

The second important factor controlling marginal basin development is the relative motions of the overriding and subducting plates (Dewey, 1980). This is illustrated schematically in Figure 10.3, which is simplified to consider the case where convergence is perfectly orthogonal. The critical variables are the relative velocities of the hinge migration or roll-back (V_r) and the overriding plate (V_o). If we assume that tectonic erosion and accretion at the trench are negligible, then V_a, the velocity of the arc plate, is the same as the rate of hinge migration. If V_o is less than V_a, extension can occur. The connection with age of subducting lithosphere thus becomes apparent; old, cold lithosphere will show faster roll-back (V_r), which will increase V_a.

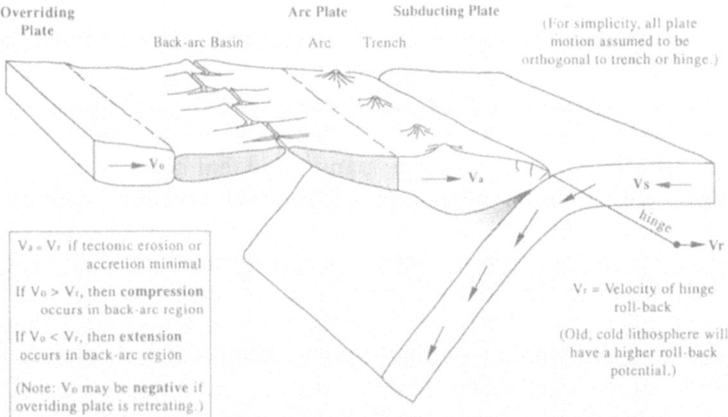

Figure 10.3 Schematic diagram, modified after Dewey (1980), illustrating the importance of plate vectors in back-arc extension. For simplicity, all plate motions are assumed to be at right angles to the trench. Essentially, if the velocity of roll-back (V_r), or hinge migration, is greater than that of the overriding plate (V_o), a gap will open between the overriding plate and the arc plate. Mantle passively upwelling into the gap between the separating plates will undergo decompressive melting similar to that occurring beneath mid-ocean ridges.

Similarly, if the overriding plate moves away from the trench, V_o becomes negative, and extension will again occur. It thus seems that subduction of old oceanic lithosphere facilitates back-arc extension, but it is not a prerequisite condition. Studies of small basins such as the Bransfield Strait at the northern end of the Antarctic Peninsula illustrate this point. Young occeanic lithosphere was subducted at the South Shetland trench until about 4 Ma when spreading at the adjacent Drake Passage spreading centre dramatically slowed (Barker and Burrell, 1977). Soon after, between 1 and 2 Ma ago, extension occurred behind the South Shetland Arc, and formed the narrow Bransfield Strait (Weaver *et al.*, 1979). It appears that extension within the overriding plate was a direct response to the roll-back of the adjacent segment of oceanic plate.

Jurdy and Stefanick (1983) have suggested that the limited life span of back-arc spreading, and the lag time (6–10 Ma) before extension re-commences, is a function of secondary, subduction-induced flow in the mantle wedge (Figure 10.2d). McKenzie (1969), Sleep and Toksoz (1971) and Toksoz and Bird (1977) were among the first to propose such an induction model, but it is not clear to what extent back-arc spreading is actually caused by secondary eddies, or merely accompanies them.

Which of the various models actually operates in any back-arc system will potentially have serious implications for models of basalt generation. The uprise of buoyant mantle will provide a different mantle thermal profile, and hence possibly different conditions of melting, than the situation where the mantle flows passively in response to plate extension. Some back-arc basins have thin crust, less than the average thickness in the major ocean basins (e.g. Mariana Trough, 5 km; LaTraille and Hussong, 1980), which suggests that the volume of melt produced at the ridge crest is smaller than that of the average oceanic crust. It is important to stress that oceanic crust exhibits considerable variability in its thickness, possibly because of the thermal conditions in the underlying mantle, and that back-arc basin crustal thicknesses fall within the oceanic range. These observations, coupled with the observation that back-arc spreading centres are among the deepest spreading axes, suggest, however, that the potential temperature in the back-arc mantle is low. (The potential temperature is the temperature that a given volume of mantle would have if it was brought to the surface, along the adiabatic gradient, without melting. Mantle associated with plumes has a high potential temperature; conversely, mantle beneath the mid-ocean ridge system has a lower temperature; see McKenzie and Bickle, 1988.) This in turn would suggest that the active diapirism model is not applicable, and that passive, or slab-induced flow, is more likely in back-arc basins. Slab-induced convection would allow replenishment of the basalt source region by mantle material of different composition, originating within regions remote from the subduction zone. We shall return to this aspect in the section on processes.

10.3 Back-arc extension and magmatic activity: an overview

The majority of the world's back-arc basins are strictly oceanic: the basin is separating two fragments of oceanic plate. The active volcanic arc is apparently sitting on oceanic lithosphere, although this is often difficult to prove, and the fore-arc lithosphere is also oceanic (Figure 10.4). To the rear of the basin is an inactive, often submerged remnant arc, which was abandoned as the basin extension transported the active arc trenchwards. Such basins probably begin life by rifting of the volcanic arc, a potential line of weakness, as shown by the rifting of the southern Bonin Island arc by the northward-propagating Sumisu Rift. Rifting of continental lithosphere also produces back-arc basins, such as the Sea of Japan, the Bransfield Strait and the Cretaceous *rocas verdes* basin, the latter now preserved as ophiolite complexes in southern Chile. In these basins, termed ensialic back-arc basins, the adjacent active and remnant (if present) arcs are rooted on continental crust.

The earliest geochemical study of back-arc basalts, from the Mariana Trough, was published by Hart *et al.* (1972). They recognized that although these tholeiitic basalts resemble MORB, there are some important differences, particularly in the greater abundances of large ion lithophile (LIL) trace elements such as Ba, relative to high field strength (HFS) elements such as Ti or Zr. Studies of basalts from the Lau Basin (Gill, 1976; Hawkins, 1977; Volpe *et al.*, 1988), the Scotia Sea (Tarney *et al.*, 1977; Saunders and Tarney, 1979) and the Mariana Trough (Natland and Tarney, 1982; Sinton and Fryer,

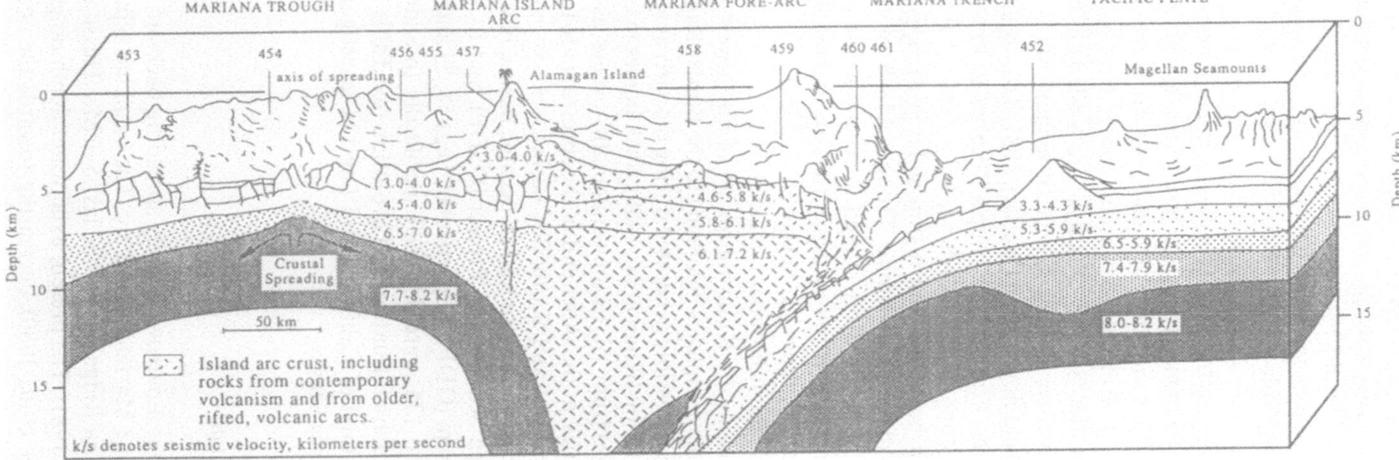

Figure 10.4 Schematic diagram showing an interpretation of the crustal structure of the Mariana trough, arc,,and forearc system, and the location of the DSDP Leg 60 drill sites. Note the absence of any appreciable accretionary prism on this diagram, confirmed by drilling at Sites 460 and 461, despite the half kilometre or so of sediment present on the Pacific Plate. Redrawn from Fryer and Hussong (1981).

1987; Volpe *et al.*, 1987) confirmed these suggestions, and show that many back-arc basalts have a geochemical character transitional between MORB and island-are basalts. Not all back-arc basalts have transitional compositions, however; some basalts from the Lau Basin are indistinguishable from MORB (Hawkins 1976; Volpe *et al.*, 1988).

Basalts from ensialic basins often show strong arc-like characteristics; true MORB are not found unless the basin is very wide. Islands in ensialic back-arc basins erupt a range of magma types from basalt through to rhyolite, but the available data suggest that the floors of such basins are made predominantly of basaltic material. Not all ensialic basins are floored by oceanic crust. Both the Aegean Sea and the Sea of Ohkotsk are underlain by attenuated continental crust, and there is no evidence of magmatism in these basins, although the crustal heat flow is high (Weissel, 1981; Toksoz and Bird, 1977). Crustal thinning has occurred in the Taupo-Rotorua Depression, a part of the Taupo Volcanic Zone of North Island, New Zealand. The depression, characterized by high heat flow and a thin sialic crust, appears to be the landward extension of the Lau-Havre Trough. Much of the associated magmatism is rhyolitic and possibly related to crustal fusion (Cole,1984). In North Wales, UK, there is strong evidence that an ensialic basin partially opened during Ordovician times (Kokelaar *et al.*, 1984), but again probably failed to fully separate the lithospheric plate.

The Chilas Complex in the Kohistan region of north-west Pakistan, and similar mafic-ultramafic complexes belonging to the Border Ranges of Alaska, are large magma bodies possibly generated during the early stages of intra-arc rifting and basin formation (Khan *et al.*, 1988). Such bodies may represent an end-member of a spectrum of tectono-magmatic events associated with extension in arc terranes, at the other end of the spectrum are fully developed basins such as the Shikoku and Scotia Sea basins.

Studies of back-arc basins and indeed all oceanic basins, have been greatly enhanced by the Deep Sea Drilling Project (DSDP), and its successor, the Ocean Drilling Program (ODP). In particular, Legs 58, 59 and 60 and, more recently, Legs 126 and 135, were designed specifically to address problems of back-arc and island arc formation and evolution (see Table 10.1). Legs 64 and 65 drilled in the Gulf of California and Leg 124 investigated the origin of the Sulu and Celebes Seas.

10.3.1 *Mariana Trough*

The Mariana Trough is a narrow, actively spreading back-arc basin located behind the calc-alkaline Mariana island arc. The basin has separated the arc from the West Mariana Ridge during the last 5–6 Ma (Hussong *et al.*, 1981; Hussong and Uyeda, 1981). The present axial water depth ranges from 3.5 to 4.5 km at 18°N (Bibee *et al.*, 1980), decreasing towards the north where the

Table 10.1 Marginal basins: major basalt-recovery sites drilled during the Deep Sea Drilling Project and Ocean Drilling Program

Leg	Site	Location	Basin	Basement penetration (m)	Age of oldest sediment (Ma)	Basement characteristics	Selected references[a]
58	442	28° 59.04′N 136° 03.43′E	Shikoku	160	Early Miocene (18–21)	Drilled *ca.* 50 km west of the axial zone; massive and pillowed vesicular basalts. Compositionally similar to N-type MORB	1–5
	443	29° 19.65′N 137° 26.43′E	Shikoku	116	Early–Middle Miocene (14–15)	Drilled *ca.* 95 km east of the axial zone; massive and pillowed vesicular basalt flows or sills, compositionally similar to N-type MORB	1–5
	444	28° 38.25′N 137° 41.03′E	Shikoku	*ca.* 35	Early–Middle Miocene (14–15)	Drilled *ca.* 90 km east of the axial zone, two basalt units with characteristics similar to E-type MORB (kaersutite-bearing intrusion and N-type MORB (lower unit)	1–5
	446	24° 42.04′N 132° 46.49′E	Daito	Probably not reached	Early Eocene	Kaersutite-bearing and kaersutite-free basalts with E-type MORB characteristics, emplaced as 23 post-Early Eocene sills	1–5
59	447	18° 00.88′N 133° 17.37′E	West Philippine	*ca.* 180	Mid-Oligocene	Massive and pillowed basalts compositionally indistinguishable from N-type MORB	6–8
	449	18° 01.84′N 136° 32.19′E	Parece-Vela	*ca.* 40	Late Oligocene	Drilling recovered approximately 7 m of basalt with N-type MORB chemistry	6–8
	450	18° 00.02′N 140° 47.34′E	Parece-Vela	Not reached (intrusion?)	Middle Miocene	Approximately 4 m of recovered basalt with N-type MORB chemistry	6–8

Table 10.1 (*Continued*)

Leg	Site	Location	Basin	Basement penetration (m)	Age of oldest sediment (Ma)	Basement characteristics	Selected references[a]
60	453	17° 54.42′N 143° 40.95′E	Mariana Trough	*ca.* 150 (not true basement)	Early Pliocene	120 km west of the central graben; gabbroic breccias recovered.	9–11
	454	18° 00.78′N 144° 31.92′E	Mariana Trough	*ca.* 100	Early Pleistocene (0.9–1.6)	28 km west of the central graben; highly vesicular basalts with calc-alkaline affinities	9–11
	456	17° 54.68′N 145° 10.77′E	Mariana Trough	*ca.* 40	Early Pleistocene (1.6–1.8)	37 km east of the central graben. Pillow basalt flows with interbedding of N-type MORB and calc-alkaline compositional affinities	9–11
64	474	52° 57.56′N 108° 58.68′W	Gulf of California (Gulf mouth)	54	Late Pliocene (3.2)	Uppermost units appear to be off-axis sills; pillow basalt and massive basalt flows form bulk of recovered basement. Chemically the basalts resemble East Pacific Rise MORB	12, 13
	477	27° 01.85″N 111° 24.02′W	Gulf of California (Guaymas Basin)	(34) (True basement probably not recovered)	Pleistocene to Recent	This site was in the southern active rift of the Guaymas Basin. Basalts appear to have been emplaced as sills. Compositionally similar to MORB tholeiites, but with enhanced LILE concentrations (especially Sr)	12–14
	478	27° 15.18′N 111° 30.46′W	Gulf of California (Guaymas Basin)	(117) (True basement probably not recovered)	Pleistocene to Recent	12 km NW of the southern active rift of the Guaymas Basin. Massive basaltic or doleritic units, the lowermost in excess of 100 m thick; probably sills. Compositionally similar to the basalts from Site 477	12–14
	481	27° 15.18′N 111° 30.46′W	Gulf of California (Guaymas Basin)	Basement not reached	Pleistocene to Recent	SW end of the northern active rift of the Guaymas Basin. Massive and sheet-like basalt units which appear to have been intruded as sills. Compositionally similar to the basalts from Sites 477 and 478	12–14

65	482	22° 47.34′N 107° 59.57′W	Gulf of California (Gulf Mouth)	*ca.* 90	Pleistocene (approx 0.5)	The youngest site from a transect across the EPR. Massive and pillowed basalt basalt flows with N-type MORB compositions	15, 16
	483	22° 53.0′N 108° 44.9′W	Gulf of California (Gulf Mouth)	*ca.* 160	*ca.* 2	The third oldest site of a transect across the EPR. Massive and pillowed basalt flows with N-type MORB compositions. Upper units may represent off-axis activity	15, 16
	485	22° 44.9′N 107° 54.2′W	Gulf of California (Gulf Mouth)	178	Pleistocene (approx 1)	A site from a transect across the EPR. Massive and pillowed basalt flows with N-type MORB compositions, interbedded with pelagic sediments	15, 16
124	767	4°47.5′N 123°30.2′E	Celebes Sea	Slight	Middle to Upper Eocene	Only 47 cm of plagioclase-olivine basalt recovered. Shipboard analysis indicates affinities with N-type MORB	17
	768	8° 00.04′N 121° 13.18′E	Sulu Sea	222	Lower Miocene	Pillow basalts, basalt sheet flows, and massive sills. Shipboard analysis suggests that the basalts are transitional between MORB and arc tholeiites	17
	770	45 km NE of site 767	Celebes Sea	106	Middle to Upper Eocene	Massive and pillowed basalt. No chemical data available	17
126	791	30° 54.91′N 139° 50.66′E	Sumisu Rift (Bonin Islands)	*ca.* 200	Pleistocene	Drilled in the centre of the Sumisu Rift. Hole 791B recovered a sequence of basalt flows and a unit of microvesicular basalt ('mousse')	18

[a]References (1) DeVries Klein *et al.*, 1980; (2) Dick *et al.*, 1980; (3) Dick, 1982; (4) Marsh *et al.*, 1980; (5) Wood *et al.*, 1980a; (6) Kroenke *et al.*, 1980; (7) Mattey *et al.*, 1980; (8) Wood *et al.*, 1980b; (9) Hussong *et al.*, 1981; (10) Fryer *et al.*, 1981; (11); Wood *et al.*, 1981; (12) Curray *et al.*, 1982; (13) Saunders *et al.*, 1982a, b; (14) Einsele *et al.*, 1980; (15) Lewis *et al.*, 1983; (16) Saunders, 1983; (17) Leg 124 Shipboard Party (1989); (18) Leg 126 Shipboard Party (1989); Leg 135 of the ODP has successfully recovered basalt sections from the Lau Basin.

Trough shallows into the narrow extensional zone presently sundering the Volcano Arc (Stern *et al.*, 1984). The Mariana Trough and adjacent ridges and basins have been studied extensively via dredging, submersible and drilling operations, and in particular the transects across the Mariana Arc and Trough-Parece Vela system (Legs 59 and 60: see Figure 10.4, and Table 10.1), have confirmed the suggestions by Karig (1971) and Karig *et al.* (1978) that back-arc rifting separated active and remnant arcs. Geochemical studies of the West Mariana Ridge and Palau–Kyushu Ridge remnant arcs show an island arc, subduction-related signature whereas drilled sequences from the Mariana Trough contain both tholeiitic and calc-alkaline basalts from the same drill hole (Wood *et al.*, 1980b, 1981; Mattey *et al.*, 1980; Tarney *et al.*, 1981).

The Mariana Trough basaltic rocks are nepheline- to quartz-normative tholeiites, with a higher volatile and LIL element content than MORB (Hart *et al.*, 1972; Garcia *et al.*, 1979; Fryer, 1981; Fryer and Hussong, 1981; Fryer *et al.*, 1981; Wood *et al.*, 1981; Sinton and Fryer, 1987; Volpe *et al.*, 1987). The Mariana Trough does not extend far beyond the northern or southern limits of the arc itself; in the north, it narrows into a northward-propagating rift zone. The island of Iwo Jima, one of the Volcano Islands located at the northern end of this propagating rift, contains evolved and enriched trachyandesites, distinct from normal arc lavas. Stern *et al.* (1984) have interpreted the unusual compositions of these lavas as a manifestation of the earliest stages of inter-arc rifting. Literally, the Volcano Arc is being unzipped by the northward propagating Mariana Trough.

The West Mariana Ridge, the remnant arc to the west of the Mariana Trough, was active volcanic arc during Miocene times. Behind this earlier arc, back-arc spreading formed the Parece-Vela Basin, which is now an inactive back-arc basin. Basalts recovered from this basin during Leg 59 of the DSDP are tholeiites similar to MORB. The Parece-Vela Basin is in turn backed by the Palau-Kyushu Ridge, which consists of primitive island arc tholeiites of late Eocene to mid-Oligocene age (Hussong *et al.*, 1981). The Parece-Vela Basin continues northwards into the now inactive Miocene Shikoku back-arc basin. This basin was drilled during Leg 58 of the DSDP (deVries Klein *et al.*, 1980), and the majority of the samples are vesicular tholeiites (Marsh *et al.*, 1980; Dick *et al.*, 1980; Dick, 1982). However 'enriched' kaersutite-bearing basalts were recovered at Site 444 in the Basin, and at Site 446 in the Oki-Daito Basin (Marsh *et al.*, 1980).

Although the Mariana Trough is a very youthful feature, it is apparent that subduction-related arc and extensional activity has been underway in this region of the western Pacific throughout much of the Tertiary probably from the time, some 43 Ma, when the Pacific plate changed its direction of motion from WNW to NNW (Jackson *et al.*, 1972).

10.3.2 *Sumisu Rift*

The Sumisu Rift, located immediately to the west of the Izu-Bonin Island arc, is a very youthful almost nascent back-arc basin. It is some 40 km wide, 110 km long, and its floor is 2000–2275 m deep. Drilling (during Leg 126) within the Rift recovered basaltic rocks, the most spectacular of which is a basaltic froth, or 'mousse' (Leg 126 Scientific Drilling Party, 1989). At the time of writing, no data are available for these samples.

10.3.3 *Lau Basin*

The Lau Basin is bounded in the east by the Tonga Arc, with the southern continuation as the Kermadec Arc, both of which are erupting primitive island-arc tholeiites (Ewart *et al.*, 1973; Ewart and Hawkesworth, 1987). The Lau Basin has a complex spreading history, involving several easterly-directed ridge jumps, although the record of back-arc activity in this region of the south-west Pacific is less prolonged than in the Mariana region. The Lau Basin narrows southwards into the Havre Trough, the landward continuation of which is the Taupo Volcanic Zone in North Island, New Zealand.

Lavas from the Lau Basin are predominantly tholeiites, but their compositions vary from being indistinguishable from MORB, to having strong arc-like characteristics (high volatile contents, high LIL element contents) (Hawkins, 1976, 1977; Hawkins and Melchior, 1985; Volpe *et al.*, 1988; Sinton *et al.* in press; Jonhson and Sinton, in press). It is apparent that the strongest arc-like signatures are found in basalts from the more southerly parts of the basin, where the spreading axis is closest to the Tonga Arc (J.A. Pearce, personal communication.) The picture is further complicated by the presence of a hot-spot component in basalts from the northern part of the basin. Some of this hot-spot, or oceanic island basalt component probably reflects the influence of the nearby Samoan plume (Volpe *et al.*, 1988), but could also represent a more regional effect.

10.3.4 *East Scotia Sea*

The East Scotia Sea, a rapidly extending basin in the South Atlantic (Barker, 1972), is unusual among back-arc basins in that it has well developed magnetic lineations. The basin has developed during the last 8 Ma in response to subduction at the South Sandwich Trench, although it is probably the latest of a series of basins which opened at various times in the middle to late Tertiary (Barker and Hill, 1981). These basins now form a complicated collage of essentially Pacific oceanic crust beneath the Scotia Sea, which protrudes into the western Atlantic basin. The basalts from the East Scotia Sea are all tholeiites; two of the four available dredges consist of highly vesicular quartz-normative basalts and basaltic andesites (Tarney *et al.*, 1977, 1981;

Saunders and Tarney, 1979, 1984). All show variable enrichment of LIL elements.

10.3.5 *Bransfield Strait*

Bransfield Strait is a narrow ensialic basin which separated the continent-based South Shetland Arc from the Antarctic Peninsula (Barker and Griffiths, 1972). Strictly speaking, this is not a back-arc basin, because the predominantly Mesozoic–middle Cenozoic arc-related magmatism on the Antarctic Peninsula had ceased by the time Bransfield Strait opened around 1 or 2 Ma. The basin is only 50 km wide, and its opening is probably a direct response to the slowing down of oceanic ridge spreading at the nearby Drake Passage spreading centre (Barker and Burrell, 1977). Magmatism in the Strait is seen on three active or recently active volcanic island: Deception, Bridgeman, and Penguin Islands, and on small seamounts located along the axis of the trough (Weaver *et al.*, 1979). Apart from Penguin Island, which has erupted only nepheline-normative, alkaline basalts, the Bransfield lavas range from basalt to basaltic andesite and, on Deception Island, which has a protracted history of magmatism, they have evolved to rhyodacite. All of the lavas of these islands, including Penguin Island, have a calc-alkaline trace element signature. Basaltic andesites dredged from axial seamounts also have a subduction signature (Fisk *et al.*, in press), but there are no samples from the floor of this basin.

10.3.6 *Japan Sea*

The Japan Sea is interpreted as an inactive back-arc basin, having opened around 15 Ma. Palaeomagnetic evidence from Japan suggests that the basin opened very rapidly, perhaps within 1 Ma, implying very high speading rates (Otofuji and Matsuda, 1983, 1984). At the time of writing, no published data are available for basement rocks from the Japan Sea, although the Ocean Drilling Program completed two Legs in this region in 1990.

10.3.7 *Sulu, Banda and Celebes Seas*

These small seas are marginal basins of doubtful origin; some workers believe them to be trapped oceanic crust, whereas others believe them to have formed by back-arc extension. The sedimentary and igneous record in the Celebes Sea suggests that it formed in an open ocean setting during middle Eocene times (42 Ma) (Leg 124 Shipboard Party, 1989). The Sulu Sea, however, appears to have formed by back-arc or intra-arc extension in the early to early–middle Miocene; basalts from the Sulu Sea crust are transitional between MORB and island-arc tholeiites. No basement rocks are available from the Banda Sea, although it is likely that the crust consists of a collage of trapped and back-arc crustal types (M. Audley-Charles, personal communication).

10.3.8 *Gulf of California*

The Gulf represents the classic leaky transform of Wilson (1965). It opened in response to oblique dextral motion between the Pacific and North American plates at about 5 Ma (Moore, 1973). Seafloor spreading is presently under way at short centres, in deep basins along the Gulf, that are offset by long transform faults. In 1978, DSDP Legs 64 and 65 recovered extensive basement sequences from the mouth of the Gulf, and from a spreading axis half-way along the Gulf, in the Guaymas Basin (Curray *et al.*, 1982; Lewis *et al.*, 1983). An unusual aspect of the Guaymas Basin sequences is that the basalts were emplaced into poorly consolidated sediments as massive sills; few or no flows were recovered. Chemically, the basalts resemble MORB, although again a slight enrichment in LIL elements is seen in the basalts from the Guaymas Basin (Saunders *et al.*, 1982a, b). This enrichment is not as great as that seen in ensialic back-arc basins, such as the Bransfield Strait.

10.3.9 *Rocas verdes ophiolite complex, Chile*

Several of the world's major ophiolite complexes have been ascribed to back-arc basin formation (e.g. Oman, Pearce *et al.*, 1981; Zimbales Range of Luzon in the Philippines, Hawkins, 1980). Where ophiolites are parautochthonous or autochthonous, their origin, back-arc basin or major ocean basin, is more easily deduced. An example of such a parautochthonous basin is the *rocas verdes* in southern Chile. Back-arc extension during the Late Jurassic and Early Cretaceous formed a narrow basin, widening towards the southern end of South Chile, behind the volcanic arc of the Patagonian batholith (Dalziel *et al.*, 1974; Dalziel, 1981). This basin, or series of *en echelon* basins (cf. Gulf of California) was closed during Middle Cretaceous times, and preserved as a series of discontinuous ophiolite lenses. Similar basins have been interpreted from ophiolitic or basaltic material further north in Chile (Bartholomew and Tarney, 1984) and Peru (Atherton *et al.*, 1983), and along strike on South Georgia Island (the Larsen Harbour Complex: Storey and Mair, 1982; Alabaster and Storey, 1990).

Compositionally, the *rocas verdes* mafic rocks from Sarmiento, at the northern end of the basin, are LIL enriched, light REE enriched basalts, dolerites, gabbros, silicic plagiogranites and trondhjemites. Basalts from the southern end of the *rocas verdes* in the Isla Tortuga area are MORB-like, with light REE depleted characteristics (Stern, 1979, 1980). The basalts and diabases of the Larsen Harbour Complex exhibit a diversity of compositions, ranging from early basalts with high LIL/HFS ratios (and ε_{Nd} +2 to +4), to later basalts with strongly MORB-like characteristics (ε_{Nd} about +8 (Albabaster and Storey 1990)). These relationships are similar to those noted by Stern (1979), namely that the early magmatism is sampling a relatively enriched source, whereas the later magmatism is tapping a depleted MORB source (most clearly seen in the wider, more developed, southern part of the basin). Bransfield Strait, Antarctica (Saunders *et al.*, 1979; Tarney *et al.*, 1981) may be a modern analogue of the *rocas verdes* basin, although Dalziel (1981) and Alabaster and Storey (1990) have suggested that the Gulf of California may be a more appropriate chemical and tectonic analogue. It is perhaps noteworthy that the highest degree of LIL element enrichment is found in the narrow, or early, parts of the complex, which is consistent with melting of subduction-contaminated mantle. Nevertheless, it is worth emphasizing that further work is required on the *rocas verdes* and the rocks from the floor (rather than islands) of ensialic back-arc basins such as Bransfield Strait, before definitive statements on tectonic analogues of this ophiolite can be made, but it is unlikely that chemical parameters alone will allow us to discriminate precisely between a Gulf of California or Bransfield Strait type setting.

10.4 Compositional diversity of back-arc basin basalts

The majority of back-arc igneous rocks are tholeiitic basalts, mineralogically similar to MORB, although some nepheline-normative basalts occur sporadically. A major difference from MORB is the proportion of quartz-normative tholeiites, basaltic andesites and more differentiated rocks recovered from islands and from some dredges; these rocks are part of tholeiitic or calc-alkaline differentiation series. Higher water contents in the magmas and source regions of some back-arc basins may be responsible for these compositional differences, but depth and extent of mantle melting may also be important. Trace elements and isotope ratios tell a different story. Many back-arc basalts have higher abundances of Ba, K, Rb, Th and light REE compared with normal MORB, which strongly supports the notion that the source of back-arc basalts is preferentially enriched in these elements.

10.4.1 *Textures and mineralogy*

Texturally and mineralogically, many back-arc basin basalts are indistinguishable from MORB. Being erupted in contact with water, textures range from intergranular to intersertal dolerites, and intergranular to glassy basalts. Both massive and thin sheet flows, in addition to pillow basalts, occur. One textural difference between MORB and back-arc basalts is the large content of vesicles in the latter, indicative of a high volatile content in the magma (Saunders and Tarney, 1979; Marsh *et al.*, 1980; Dick, 1980, 1982; Leg 126 Shipboard Party, 1989).

The main phenocryst phases in back-arc basalts are a combination of olivine, plagioclase, chrome spinel $\pm$ clinopyroxene. The range of assemblages is similar to those found in MORB. The frequent occurrence of clinopyroxene is attributable to the higher fractionation state of back-arc magmas (see later). The compositions of the mineral phases are within the range of MORB. From his detailed study of olivines in the Shikoku Basin basalts, Dick (1982) interpreted the low Fe–Mg distribution coefficients in coexisting olivine–glass pairs as being due to a high proportion of Fe^{3+} in the magma. Such increased oxidation state is consistent with the high volatile contents (high fO_2 and fH_2O) predicted by the vesicularity. Reported plagioclase compositions are within the range of MORB. High-Ca clinopyroxenes are found in evolved basalts from the Mariana Trough (Hart *et al.*, 1972; Sinton and Fryer, 1987) and the East Scotia Sea (Saunders and Tarney, 1979). Dick and Bullen (1984) found that spinels from back-arc basin basalts and MORB are similar in composition. Some back-arc basalt spinels have lower Mg numbers (100 Mg/ Mg + Fe) at a given Cr number (100 Cr/Cr + Al) than those in MORB, but these differences do not appear to be sufficient to warrant their use as mineralogical discriminants.

The alkaline units recovered from the Shikoku and Oki-Daito Basins contain both sodic and more potassic feldspars. The intrusive basalts recovered at Site 446 in the Oki-Daito Basin also contain kaersutite amphibole and titaniferous calcic pyroxenes.

10.4.2 *Major elements*

The high content of volatiles, particularly water, in many back-arc basin magmas appears to have affected their crystallization and differentiation histories. This is best shown by comparing back-arc glass compositions with MORB glasses; the major oxide contents provide important information about the types of mineral phases extracted from (or added to) the liquid. It is important to use analyses of glass, rather than of whole rock, because this eliminates the possibility of analytical bias as a result of the presence of cumulus crystals. However, few analyses of basaltic glass also have full trace element and isotopic data, so the representative data in Table 10.2 include whole-rock analyses also. It is important to restate our earlier comments about the paucity of high quality data (Saunders and Tarney, 1984). There are to our knowledge no complete elemental and isotopic analyses of back-arc basalts in the literature. Table 10.2 is a compilation of data, and combines data from rocks of similar composition where necessary. This is an unsatisfactory method but is the best available. In addition, data for key elements (e.g. Th, Ta and Nb) are still not available for the majority of published analyses, especially fresh glass.

Table 10.2 contains major and trace element data for selected back-arc basalts and, for comparison, an analysis of a mid-ocean ridge basalt. Several features of back-arc basalts become apparent when scanning these data. The silica content varies from less than 48 to over 53%, compositions that encompass basaltic andesite. Consequently, the normative mineral compositions

Table 10.2 Published major, (wt.%), trace element (ppm) and isotope data for selected back-arc basin basalts

Basin	East Scotia Sea			Mariana Trough				Lau and North Fiji Basins			Bransfield Strait		
sample	Dredge 20	Dredge 23[a]	Dredge 24[b]	Mara 39-1[c]	Mara 39-8[c]	454A[d]	456A[d]	K5-14	S164/1	K19-7	B.138.2	P.640.1b	N-type MORB[b]
Type[e]	wr & gl	wr & gl	wr & gl	glass	glass	wr	wr	glass	glass	glass	wr	wr	wr
SiO_2	50.7	50.4	53.8	50.95	52.02	49.3	50.7	49.53	53.80	49.25	51.89	52.88	49.54
TiO_2	1.29	1.46	0.61	0.75	1.48	0.86	1.18	2.11	1.75	2.04	1.49	0.64	1.24
Al_2O_3	16.6	16.4	14.5	17.52	16.16	14.4	16.6	15.48	14.69	15.85	16.20	17.68	16.03
Fe_2O_3	1.01	1.09	1.09	1.0	1.2	1.12	1.17	1.3	1.4	1.3	1.1	0.9	1.2
FeO	6.68	7.18	7.34	6.6	8.0	7.47	7.82	9.0	9.0	8.8	7.5	5.9	7.7
MnO	0.16	0.16	0.17	0.17	0.17	0.12	0.16	0.19	0.20	0.16	0.18	0.13	0.16
MgO	7.7	7.4	7.7	7.60	5.85	13.3	5.23	6.46	4.19	6.41	6.11	6.14	7.6
CaO	11.12	10.84	10.8	11.64	10.47	10.97	11.23	10.94	8.31	10.93	10.07	10.30	12.17
Na_2O	3.2	3.39	1.79	2.37	3.58	2.36	3.27	3.01	3.61	3.00	4.07	3.53	2.60
K_2O	0.34	0.43	0.24	0.26	0.29	0.18	0.62	0.20	0.57	0.68	0.28	0.47	0.05
P_2O_5	0.19	0.20	0.08	0.12	0.19	0.09	0.11	0.24	0.29	0.21	0.21	0.06	0.12
H_2O^+	0.733	0.945	2.042	1.18	1.48	n.d.	n.d.	0.720	1.352	0.800	n.d.	n.d.	(0.120)
CO_2	0.378	0.167	0.187	0.05	0.04	n.d	n.d.	0.089	0.110	0.051	n.d.	n.d.	(0.163)
CIPW Norms[f]													
Qz	0.0	0.0	8.6	0.0	0.0	0.0	0.0	0.0	5.2	0.0	0.0	0.0	0.0
Or	2.0	2.0	1.4	1.5	1.7	1.0	3.7	1.2	3.4	4.0	1.7	2.78	0.3
Ab	26.5	28.9	13.7	20.1	30.3	19.8	28.0	25.5	30.6	25.4	34.4	29.9	22.0
An	30.7	28.5	32.3	36.4	27.2	27.8	29.0	28.2	22.2	27.8	25.1	31.0	31.9
Ne	0.0	0.0	0.0	0.0	0.0	0.0	0.0	0.0	0.0	0.0	0.0	0.0	0.0
Di	19.3	20.0	17.7	16.7	19.3	20.5	22.0	20.1	14.3	20.6	19.2	16.0	22.5
Hy	10.1	4.3	22.9	21.0	13.9	6.6	6.7	13.2	16.3	5.8	6.3	15.0	11.8
Ol	6.8	10.6	0.0	0.2	2.9	20.0	5.5	3.9	0.0	8.9	7.5	1.3	5.5
Mt	1.5	1.6	1.6	1.4	1.7	1.6	1.7	2.0	2.0	1.9	1.6	1.3	1.7
Il	2.5	2.8	1.2	1.4	2.8	1.6	2.3	4.0	3.3	3.9	2.8	1.2	2.4
Ap	0.5	0.5	0.2	0.3	0.4	0.2	0.3	0.6	0.7	0.5	0.5	0.1	0.3
Trace elements:													
Ni	63	64	42	127	72	382	51	144	59	159	35	40	53
Cr	270	270	295	260	200	532	106	278	110	430	141	130	306

Zr	107	130	40	60	109	58	84	146	161	162	144	58	65
Nb[g]	5	8	1	5	7	< 1	< 1	5	10	24	2	1	2
Ta[g]	0.45	0.62	0.19	n.d.	n.d.	0.11	0.16	n.d.	n.d.	n.d.	n.d.	n.d.	0.10
La[g]	5.0	7.8	n.d.	n.d.	n.d.	1.98	5.15	2	11.27	18.3	8	2	1.9
Ce	16.1	19.0	6.5	n.d.	n.d.	6.6.	13.1	7	23.51	37.6	21.9	7.92	8.7
Nd	11.1	13.1	4.6	6.76	12.21	n.d.	n.d.	11.21	16.52	20.5	14.4	5.47	8.1
Sm	3.29	3.94	1.46	2.12	3.89	n.d.	n.d.	4.76	4.57	4.97	2.12	1.56	2.85
Eu	1.25	1.44	0.56	n.d.	n.d.	0.86	1.25	n.d.	1.44	1.54	1.46	0.65	1.3
Dy	4.69	5.24	1.99	n.d.	n.d.	n.d.	n.d.	n.d.	7.04	5.89	4.92	2.02	n.d.
Yb	2.87	3.02	1.59	n.d.	n.d.	n.d.	n.d.	n.d.	4.23	2.78	2.99	1.15	3.14
Y	29	30	14	21	33	18	26	41	33	32	26	10	27
Ba	49	55	77	48	47	49	81	30	103	199	88	70	17
Sr	193	212	123	177	164	143	251	140	183	216	177	332	118
Rb	5.77	6.06	3.73	3.39	3.61	2	8	3.53	8.1	15.4	3	11	0.1
Th[g]	0.45	0.56	0.30	n.d.	n.d.	0.24	0.63	bdl	n.d.	2	2	2	0.12
S (wt%)	0.057	0.038	0.002	n.d.	n.d.	n.d.	n.d.	0.086	0.077	0.104	n.d.	n.d.	(0.069)
F (wt%)	0.035	0.028	0.069	n.d.	n.d.	n.d.	n.d.	0.089	0.007	0.009	n.d.	n.d.	(0.005)
Cl (wt%)	0.013	0.014	0.052	n.d.	n.d.	n.d.	n.d.	0.040	0.100	0.037	n.d.	n.d.	(0.005)
Isotope ratios:													
$^{87}Sr/^{86}Sr$	0.70282	0.70297	0.70325	0.702993	0.702823	n.d.	n.d.	0.703482	0.703617	0.703589	0.70336	0.70349	(0.7028)
$^{143}Nd/^{144}Nd$	0.51313	0.51306	0.51305	0.513075	0.513122	n.d.	n.d.	0.513051	0.513051	0.512849	n.d.	n.d.	(0.5131)
$^{206}Pb/^{204}Pb$	18.09	17.96	18.31	n.d.	n.d.	n.d.	n.d.	n.d.	n.d.	n.d.	n.d	n.d.	(18.0)
$^{207}Pb/^{204}Pb$	15.50	15.46	15.55	n.d.	n.d.	n.d.	n.d.	n.d.	n.d.	n.d.	n.d.	n.d.	(15.45)
$^{208}Pb/^{204}Pb$	37.68	37.49	37.99	n.d.	n.d.	n.d.	n.d.	n.d.	n.d.	n.d.	n.d	n.d.	(37.5)
$\delta^{13}C$‰	−13.2	−9.0	n.d.	n.d.	n.d.	n.d.	n.d.	n.d.	n.d.	n.d.	n.d	n.d.	−4.2 to −7.5
References	1–4	1–5	1–5	6	6	7	7	8,9	9,10	8,9	11	11	4,12–14

Notes: [a]**Combined data for dredged samples 23.4 (volatile data) and 23.3 and 23.5 (major, trace and isotope data)**

[b]**Combined data for dredge samples D24.11 (volatile data) and D24.14 (major, trace and isotope data)**

[c]Dredge samples, axial valley

[d]DSDP samples

[e]Wr, whole-rock or gl, glass samples

[f]**Fe_2O_3, FeO and CIPW norms calculated using Fe_2O_3/FeO ratio of 0.15**

[g]Nb, Ta, La and Th data for dredge 20 taken from replicate analyses of sample D20.35 (ADS, unpublished data; and Saunders, 1983)

[h]Combined data for N-type MORB from DSDP sample 483/15/2, 1-6, EPR (major and trace element data); average EPR MORB (H_2O, CO_2, SO_2, F, and Cl data); average **MORB (Sr, Nd, Pb, and C isotopes).**

(n.d.) Not determined; (b.d.l.) Below detection limit

[i]**References: (1) Saunders and Tarney, 1979; (2) Muenow *et al.*, 1980; (3) Cohen and O'Nions, 1982; (4) Mattey *et al.*, 1984; (5) Hawkesworth *et al.*, 1977; (6) Volpe *et al.*, 1987; (7) Wood *et al.*, 1981; (8) Sinton *et al.*, in press; (9) Aggrey *et al.*, 1988; (10) Johnson and Sinton, in press; (11) Weaver *et al.*, 1979; (12) Saunders, 1983; (13) Byers *et al.*, 1986; (14) Ito *et al.*, 1987**

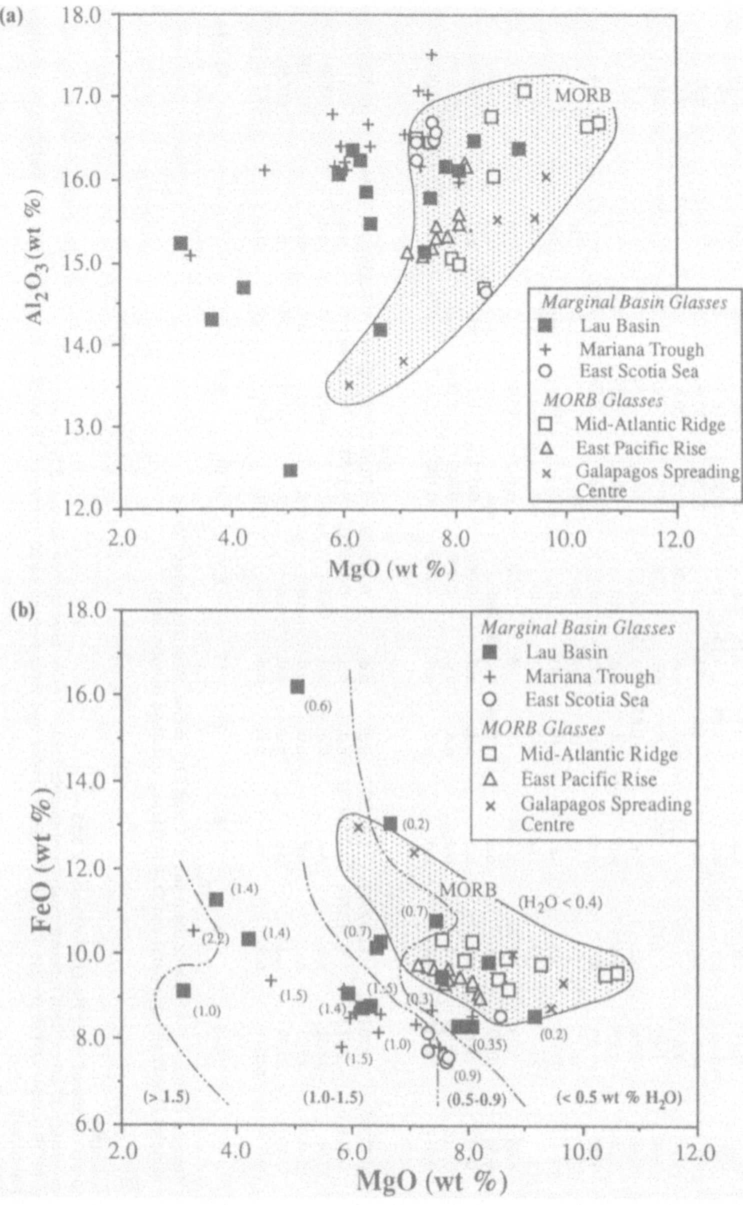

Figure 10.5 (a) MgO (wt%) versus Al_2O_3 (wt%) and (b) total iron as FeO (wt%) in back-arc basalts and basaltic glasses (Lau, Scotia Sea and Mariana Trough), compared with mid-ocean ridge basalt glasses (East Pacific Rise, Mid-Atlantic Ridge and Galapagos Spreading Centre). Note that the back-arc basalts are generally more fractionated (lower MgO) than MORB, but that the Al_2O_3 contents are higher and the FeO contents are lower, at equivalent fractionation states. Part (a) also indicates H_2O content of the analysed glass (in parentheses adjacent to data points). Data sources: East Scotia Sea, Muenow *et al.* (1980); Lau Basin, Aggrey *et al.* (1988b); Mariana Trough, Garcia *et al.* (1979); Volpe *et al.* (1987); Galapagos Spreading Centre, Byers *et al.* (1984); Mid-Atlantic Ridge, Delaney *et al.* (1978); East Pacific Rise, Byers *et al.* (1986).

also vary, from mildly alkaline nepheline-normative basalts through to quartz-normative tholeiites. Although MORB show a similar range, such strongly quartz-normative basalts are rare. Quartz-normative compositions are often associated with basalts with high Mg numbers. This was demonstrated by Sinton and Fryer (1987) in their comprehensive study of dredged and drilled basalts from the Mariana Trough. Back-arc basalts also tend to contain more Na_2O and Al_2O_3 and less total iron than MORB of equivalent fractionation state, although this is less apparent from Table 10.2.

To illustrate these differences, Al_2O_3 and FeO are plotted against MgO as an index of fractionation (Figure 10.5). Most of the plotted data are published glass analyses, for which volatile data are also available (see later), and we have contoured the diagrams for water contents. The MORB data are from 'normal' ridge segments, avoiding those segments clearly affected by a plume component (e.g. FAMOUS), but the data set is not comprehensive. The diagrams confirm previous observations (Sinton and Fryer, 1987) that many back-arc basalts contain less Fe and Al than most MORB, despite often being more fractionated (lower MgO). Note, however, that some Lau Basin basalts (a group of low volatile, MORB-like basalts discussed later) overlap with the MORB fields. Furthermore, some MORB from deep spreading axes, such as the Antarctic–Australia Discordance on the South-east Indian Ridge (SEIR), have low FeO contents (Klein *et al.*, 1988). Why should these variations exist?

There are three important factors to consider. First, the effects of high level, low pressure fractionation. Sinton and Fryer (1987) (among others) have argued that the high Al_2O_3 contents of the Mariana Trough basalts (Figure 10.5a) result from the delayed crystallization of plagioclase. This delay may not be large, perhaps as little as 5 units of Mg number, from Mg number 65 (in MORB) to 60 (in back-arc basalts). The effect on Figure 10.5a would be to produce a more extended, flatter, or even negative fractionation slope as MgO, but not Al_2O_3, is removed during olivine crystallization. Without the crystallization of plagioclase, the main Al-bearing phase at low pressures, the Al_2O_3 content of the liquid increases as olivine is removed. What could cause such a delay? The most likely cause is the increased water content of the back-arc magmas (Figures 10.5 and 10.6), which has the effect of expanding the olivine and pyroxene stability fields at the expense of plagioclase (Green and Ringwood, 1968). However, the higher Al_2O_3 content of back-arc magmas may also result from smaller degrees of melting of the mantle source.

The explanation for the generally lower Fe contents of back-arc basalts is less readily explained by low pressure fractionation. It appears that the parental magmas already have low Fe contents. This brings us to the two other important factors which control magma composition, particularly Fe/Mg ratio, namely conditions of melting of the mantle source, and the composition of the source itself.

Sinton and Fryer (1987) have evaluated the FeO and MgO contents of

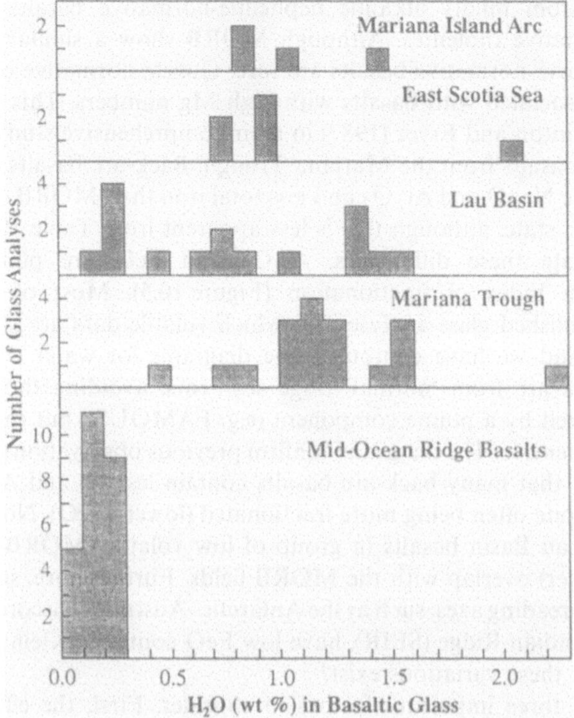

Figure 10.6 Published H_2O abundance data for basaltic glasses from mid-ocean ridges and back-arc basins. Data sources as in Figure 10.5. All data determined by high temperature mass spectrometry (for details see Delaney *et al.*, 1978) on glasses, except for Mariana Trough analyses of Volpe *et al.* (1987) which were determined by coulometric methods.

basaltic glasses from the Mariana Trough in terms of source composition and conditions of melting. They attribute the lower Fe contents of the Mariana Trough basalts to low temperature and pressure melting of mantle diapirs rising along cooler adiabatic gradients than those beneath normal mid-ocean ridges. However, it is necessary to stress that several interconnected variables may be at work here. The FeO content of a primary melt may be decreased by: (1) higher degree of melting, (2) lower pressure of melting, (3) lower temperature of melting, or (4) an increase in the forsterite content of the source (Langmuir and Hanson, 1980); it is very difficult to identify the role played by each of the variables in the formation of a given suite of basalts.

Klein and Langmuir (1987) have suggested that the deepest oceanic ridges (including those in back-arc basins) are associated with the smallest degrees of melting (hence producing the thinnest crust), and that the basalts are formed at the lowest pressures and temperatures. Extending this line of reasoning to their study. Sinton and Fryer (1987) argue that the extent of melting involved in the formation of the Mariana Trough basalts is towards the lower end of the range estimated for MORB (about 13%); this is consistent with the depth of the Mariana spreading axis (about 4000 m), and the thickness of Mariana crust (5 km) (LaTraille and Hussong, 1989). Dick (1980, 1982) came to similar conclusions for the Shikoku and Daito back-arc basins, arguing that the greater depth of these basins compared with oceanic lithosphere of a similar age, is caused by cooler convecting mantle within the back-arc environment. Thus, it would appear that basalts from the Mariana Trough and Shikoku Basin are produced by *shallow, small degree* melting of *cooler* mantle.

The lower temperature of melting may be facilitated by the depression of the mantle solidus by enhanced water contents, but some deep mid-oceanic ridges (e.g. the SEIR) also erupt low FeO, high Na basalts via shallow level melting of mantle. No published volatile data are available for these SEIR samples, but there is no reason (from trace element data, Klein *et al.*, 1988) to predict that their volatile content is as high as in Mariana Trough samples. Thus the role of water in controlling the composition of the primary back-arc magmas *via* the degree of melting may be less than anticipated. As the presence of water has the effect of reducing the mantle solidus, the hydrous nature of many back-arc basalt glasses leads us to predict that they result from *high* degrees of melting of their source. This prediction is in conflict with the comments just made regarding the high Al_2O_3 and Na_2O contents of the Mariana Trough basalts (and other back-arc basalts), unless their source is compositionally distinct, in terms of major elements, from MORB mantle. Bearing in mind that many back-arc basalts have enhanced contents of K, Ba and other LIL elements, enrichment of Na in the source may be thought likely. It is important to stress that disentangling the relative effects of source composition from conditions of melting is still fraught with uncertainties. Nonetheless, Sinton and Fryer (1987) have produced an internally consistent model whereby the Mariana Trough basalts are produced by degrees of melting comparable to those predicted by Klein and Langmuir (1987) for the deepest oceanic ridges. It is not apparent that the high volatile contents have substantially affected the Mariana melt compositions but they may have helped melt extraction by lowering the melt viscosity.

10.4.3 *Volatiles*

Some dredged back-arc basalts, even those erupted at abyssal depths, are highly vesicular, pointing to a greater dissolved gas content in the magma than in MORB magmas. Perhaps the most extreme example of this is the recent recovery of microvesicular basaltic ‘mousse’, a type of basalt froth erupted at water depths in excess of 2000 m, from the Sumisu Rift, an incipient back-arc basin behind the Bonin Island arc (Leg 126 Shipboard Party, 1989). The obvious conclusion is that back-arc basin basalts contain a hydrous component from the adjacent subduction zone. To corroborate this, comparative studies have been made of the volatile content of back-arc basin and MORB glasses (Garcia *et al.*, 1979; Muenow *et al.*, 1980; Hawkins and Melchior, 1985; Volpe *et al.*, 1987; Aggrey *et al.*, 1988b). High temperature mass spectrometry allows the accurate and precise measurement of low concentrations of volatile species (H_2O, CO_2, CO, S, Cl and F) in basaltic glasses and glass inclusions in phenocrysts. If these data are to provide important information, the magmas must not have undergone excessive degassing. Unfortunately, most subaerially erupted lavas lose volatiles during eruption, but magmas extruded on the ocean floor may more closely preserve their volatile contents because of the high confining pressure during eruption.

Published H_2O^+ data for basaltic glasses and inclusions from ocean ridges (EPR and MAR), back-arc basins (Lau, Scotia Sea and Mariana), and from a small seamount in the northern Mariana arc have been compiled (Figure 10.6). There is a progressive increase in H_2O content from MORB, through back-arc basin basalt, to island-arc basalts and andesites. In a similar manner to trace elements, the abundance of water is dependent on the fractionation state of the magma; an evolved ferrobasalt or andesite is likely to contain more volatile material than a primitive basalt, if the volatile species behave incompatibly (evidence from the Galapagos Rift suggests that water acts highly incompatibly). It is therefore more useful to consider water to element ratios.

H_2O^+ content is plotted against various fractionation indices (TiO_2, K_2O and MgO) in Figure 10.7a–c. From these diagrams, it is apparent that: (1) the water content of many back-arc basalt glasses is higher than that of MORB glasses, at equivalent TiO_2 or MgO content. (2) The absolute contents of both K_2O and H_2O are higher in most back-arc glasses, although the H_2O/K_2O ratios of MORB and back-arc basalts are similar. This suggests that both K_2O and H_2O are enriched in the back-arc basalt source. (3) There are weakly positive correlations between K_2O, TiO_2 and H_2O, and all three compounds correlate negatively with MgO. This observation is consistent with H_2O behaving incompatibly in basaltic magma. (4) The Lau basin basalts fall into two main groups (high and low water contents), which suggests that a variety of sources is being tapped by the Lau basin magmas; again, this is consistent with the trace element and isotopic data. Similar conclusions can be drawn from the Mariana and Scotia Sea data; one highly vesicular sample from the latter has over 2 wt% H_2O^+ (Muenow *et al.*, 1980).

CO_2 and H_2O show no obvious correlation (Figure 10.7d), although MORB have consistently higher CO_2/H_2O ratios. The lack of correlation within either group may be a result of the preferential degassing of CO_2 from the cooling magma, even at high confining pressures. From the limited Cl and F data that are available, it appears that back-arc basalt glasses also have higher Cl and F contents than MORB (Aggrey *et al.*, 1988). This is illustrated by the summary diagram, Figure 10.8, where we have normalized back-arc basalt contents with an average MORB from the East Pacific Rise

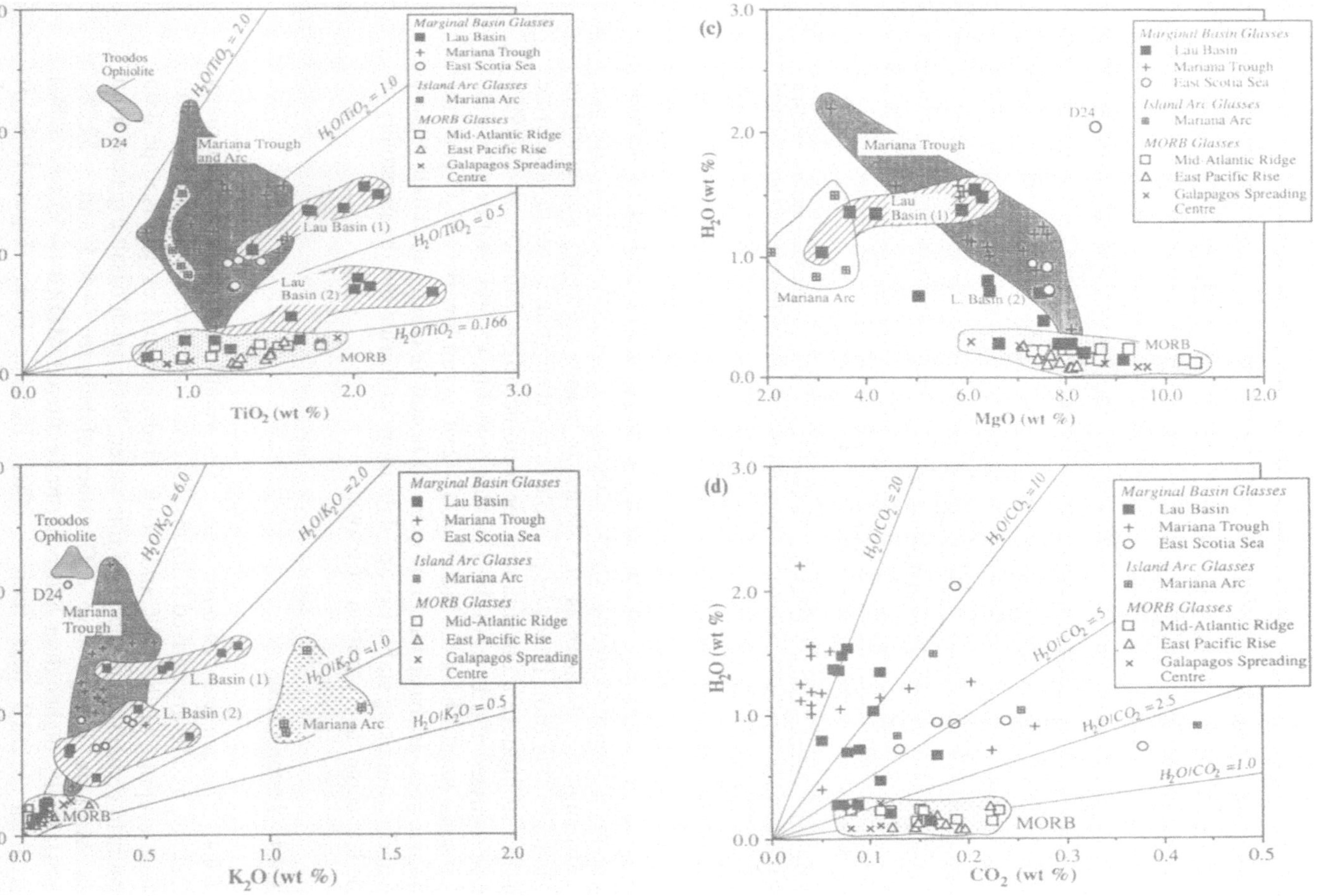

Figure 10.7 H_2O versus (a) TiO_2, (b) K_2O, (c) MgO and (d) CO_2 in basaltic glasses from mid-ocean ridges and back-arc basins. Note that the Lau Basin data fall into at least two groups, one with elevated $H_2O:TiO_2$, $H_2O:CO_2$ and $H_2O:K_2O$ ratios, the other with more MORB-like ratios. Data sources as in Figure 10.5.

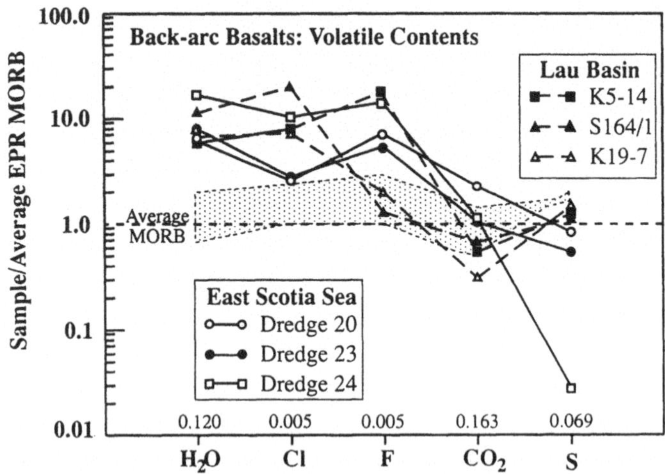

Figure 10.8 Volatile data in MORB and back-arc basalt glasses normalized against average MORB from the East Pacific Rise (values from Byers *et al.*, 1986). The relative enrichment of H_2O, Cl and F in back-arc basalts is clearly illustrated by this diagram. Data sources as for Figure 10.5.

(Byers *et al.*, 1986). For comparison, the range of EPR and MAR volatile contents is indicated by the stippled region.

Although the abundances of volatiles in basalts are highly vulnerable to secondary alteration, it may be possible to use their abundances in submarine glasses to produce a discriminant for tectonic origin. Muenow *et al.* (1990) have analysed glasses from the Troodos ophiolite, Cyprus, and demonstrate that some basalts have the high H_2O (> 2 wt%) and low H_2O/K_2O contents of some back-arc basin basalts (Figure 10.7a and b). Ideally, such data require confirmation using stable isotope data in order to rule out the possibility of the late-stage incorporation of seawater.

10.4.4 *Isotope data*

10.4.4.1 *Oxygen and hydrogen data.* The volatile data indicate that many back-arc basalts are derived from magmas with a higher volatile content than those erupting at the EPR and MAR. Hydrogen and oxygen isotope data can help constrain the source of this water. Water-rich basaltic glasses from the Mariana Trough have δ^2D values of between -46.5 and -32.2‰, substantially higher than hydrothermally altered MORB (-50 to -80‰) or fresh MORB (less than -74‰) (Poreda, 1985) (Figure 10.9). It is therefore unlikely that the high water concentrations in the Mariana Trough basalts represent shallow level processes such as hydrothermal circulation, or entrainment of blocks of hydrothermally altered basalt in the sub-ridge

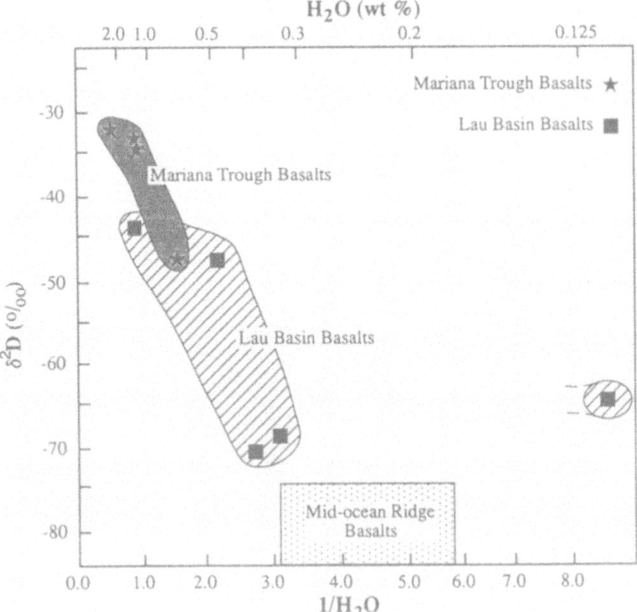

Figure 10.9 δ^2D versus H_2O content in basaltic glasses from the Mariana Trough and MOR. From Poreda (1985).

magma chamber. It is more probable that the water is intrinsic. Other data for the Mariana Trough basalts also rule out the involvement of near-surface seawater: the δ^{18}O value of Trough basalts is the same as that of MORB (+5.8 to +6.2‰, Ito and Stern, 1981) and analysed chromites embedded in olivine have oxidation states, indicating a high oxygen fugacity in the magma (Melchior, 1981). Having ruled out these high level processes, Poreda (1985) concluded that the water originates from the mantle source, and ultimately from the subducting lithosphere.

The wide variation of volatile contents in the Lau Basin basalt glasses suggests that the underlying mantle is heterogeneous. This is also borne out the stable isotope data. Lau Basin basalts with low H_2O contents also have low δ^2D values (less than −60‰), similar to MORB.

10.4.4.2 *Strontium and neodymium data.* Sr and Nd isotope data for back-arc basalts (Mariana Trough, Lau Basin and East Scotia Sea), and their associated arc basalts, are plotted in Figure 10.10. The Mariana Trough basalts have low $^{87}Sr/^{86}Sr$ ratios (mean $^{87}Sr/^{86}Sr = 0.7029$, Volpe *et al.*, 1987). This is higher than the most depleted MORB from the EPR and MAR (< 0.7026), but overlaps with the range of MORB compositions. Other back-arc basalts have higher $^{87}Sr/^{6}Sr$ ratios: East Scotia Sea, 0.7029–0.7034; and Lau Basin: 0.7030–0.7042 (Saunders and Tarney, 1979; Volpe *et al.*, 1988). Unlike the

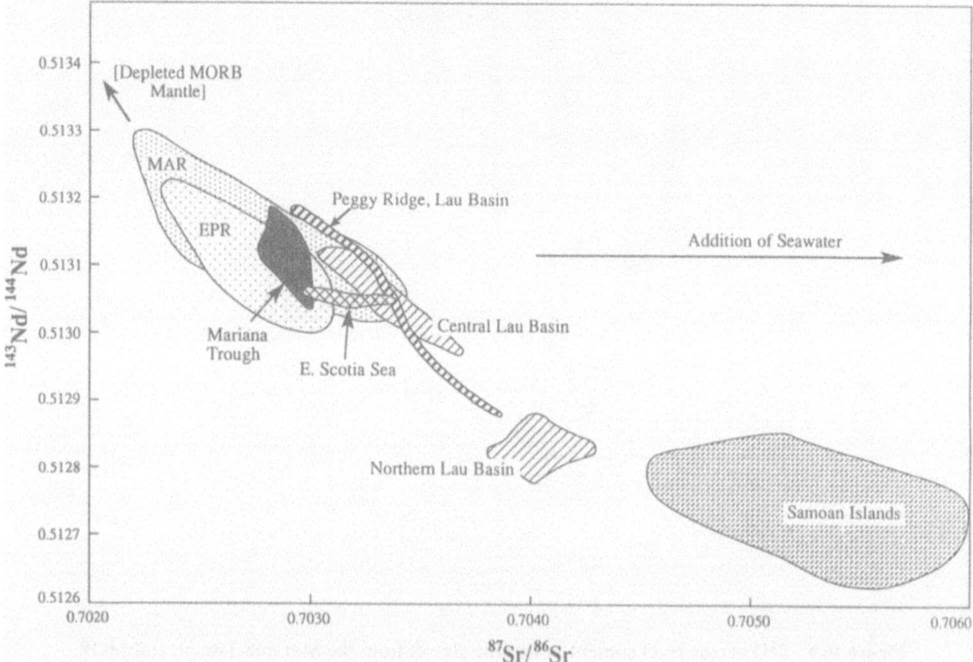

Figure 10.10 $^{87}Sr/^{86}Sr$ versus $^{143}Nd/^{144}Nd$ for basalts from the Mariana Trough, Lau Basin, and Scotia Sea. Data sources: Volpe *et al.* (1987, 1988); Sinton *et al.* (in press); Saunders and Tarney (1979). Note that back-arc basin basalts do not show strong enrichment in $^{87}Sr/^{86}Sr$, unlike some arc suites, but they do show a dispersion to low $^{143}Nd/^{144}Nd$ values. This has been taken to suggest that mantle components in the back-arc source are not simply depleted MORB mantle (high $^{143}Nd/^{144}Nd$) plus a high $^{87}Sr/^{86}Sr$ subduction component (see Volpe *et al.*, 1987). MORB and Samoan oceanic island basalt data from a variety of sources summarized in Palacz and Saunders (1986).

basalts from the East Scotia Sea, the Lau Basin and Mariana Trough basalts span a wide range of compositions in Sr–Nd isotope space with a strong inverse correlation. This is an important observation, because it indicates that slab-derived seawater, carrying only volatiles and alkali elements, is not the only enrichment mechanism in these arc – back-arc regions. Metasomatism by subducted seawater alone would produce a strong dispersion to high $^{87}Sr/^{86}Sr$ ratios, without affecting Nd. There must therefore be involvement of a mantle component with low $^{143}Nd/^{144}Nd$ ratio.

10.4.4.3 *Lead data.* Limited Pb isotope data are available for Mariana Trough, Parece-Vela Basin and East Scotia Sea basalts. The Mariana Trough basalts have unradiogenic Pb ratios, lying at the lower end of the Pacific MORB spectrum ($^{206}Pb/^{204}Pb$ 18.190–18.194; $^{207}Pb/^{204}Pb$ 15.421–15.487; $^{208}Pb/^{204}Pb$ 37.777–37.805: Meijer, 1976). The Scotia Sea basalts extend to slightly more radiogenic compositions, with $^{206}Pb/^{204}Pb$ ranging up to 18.3 (Table 10.2; Cohen and O'Nions 1982), but these values are still well within the range of N-type MORB. Unfortunately, there are presently no published Pb isotope data for the Lau Basin basalts. Such data could confirm suggestions made by Volpe *et al.* (1987) that isotopic and chemical variations in Lau Basin basalts result from mixing between different compositions in the underlying mantle.

10.4.4.4 *Helium data.* Basaltic glasses from the Mariana Trough and from the Peggy Ridge in the Lau Basin have MORB-like helium ratios (R/R_A about 8), which is consistent with derivation from a mixture between MORB-like mantle and alkali- and water-rich fluids from the subducting slab (Poreda, 1985). MORB-like helium ratios are found in most arc volcanics from the circum-Pacific belt, indicating that the He ratio in arc basalts is controlled by the mantle wedge component (Poreda and Craig, 1989). Samples from the Rochambeau Bank in the northern Lau Basin have higher $^3He/^4He$ ratios (R/R_A about 11), providing evidence for a mantle source beneath the northern Lau which is perhaps related to the high ^3He Samoan hot-spot (Poreda, 1985).

10.4.4.5 *Carbon data.* Carbon isotopes provide further evidence of the cycling of subducted material into the source of back-arc basalts. The $\delta^{13}C$ of magmatic CO_2 varies from −4.2 to −7.5‰ in mid-ocean ridge basalts, and from −7.7 to −16.3‰ in glasses from the Scotia Sea and Mariana Trough (Figure 10.11, Mattey *et al.*, 1984). Thus magmatic CO_2 from back-arc basalts is, on average, 5‰ lighter than equivalent CO_2 in MORB. This could be explained by dilution of the MORB CO_2 component with isotopically lighter carbon, perhaps derived from subducted, oceanic-rich pelagic sediment. Oceanic crust may contain carbon in several forms: intrinsic magmatic CO_2, most of which will have been lost during magmatic degassing and hydrothermal alteration; minor, secondary carbonates in veined and altered basalts; pelagic carbonates; and authigenic carbonates and kerogens. Of these possible sources of carbon, kerogen is most likely to provide a suitable light carbon component, because its $\delta^{13}C$ values ranges from − 20 to − 30‰. Furthermore, thermal cracking of the kerogen in the subduction zone system could produce very light carbon ($\delta^{13}C$ about 50‰) (Mattey *et al.*, 1984).

10.4.4.6 *Summary.* Sr, Nd, Pb, He, O, H and C isotope data are presently available for a limited range of back-arc basalt glasses, but no individual samples have published data for all of these isotopes. Back-arc basin basalts show a greater range of Sr, Nd, H and C isotope values than MORB. Overlap of back-arc basalt and MORB isotope data is partial for Sr and Nd, total for Pb and O, but there is no overlap of the carbon isotope data. For some

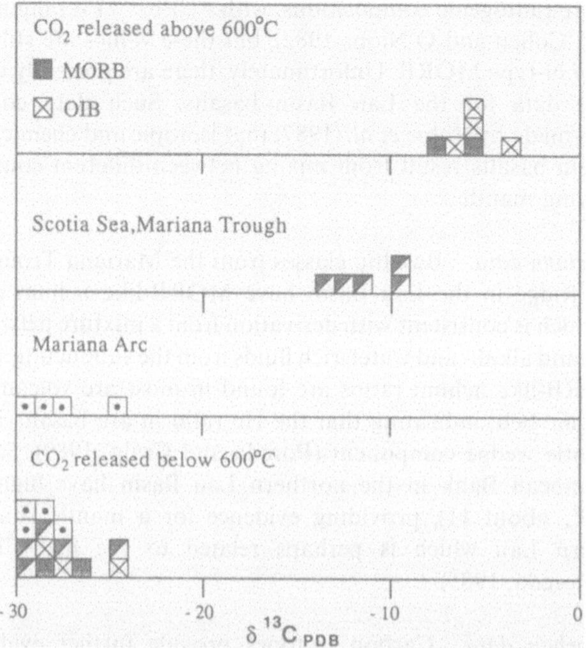

Figure 10.11 Distribution of carbon isotopes in submarine basalt glass, modified from Mattey *et al.* (1984). Values of $\delta^{13}C$ determined at temperatures in excess of 600°C are considered to be more representative of magmatic carbon values.

back-arc basalts, especially those from the Mariana and Scotia Sea Basins, the isotopic variations are consistent with derivation from a MORB mantle source variably contaminated by a slab-derived component. The slab-derived component appears to contain a light carbon component from oceanic material. Basalts from the Lau Basin exibit a wide range of He, Nd and Sr isotope ratios, and it is apparent that the mantle source has a more complex chemical structure and history. These variations result not only from slab-derived metasomatism, but also from mixing with isotopically old lithosphere and, in the northern Lau Basin, a component probably from the Samoan plume. Isotope data alone cannot fully constrain the nature of the components involved in back-arc basalt genesis. It is necessary to also consider major, minor and trace element data to further constrain the mantle components.

10.4.5 *Minor and trace elements*

In this section it will be demonstrated that many back-arc basalts have minor and trace element abundances intermediate between mid-ocean ridge and arc basalts. In effect, a MORB source signature is being overprinted with a subduction zone component.

Our argument will be primarily based on the abundances of six trace elements, Ba, Th, Ta, Nb, Ce and Zr. Data for back-arc basin basalts for these and other elements are given in Table 10.2, and plotted in a series of diagrams (Figures 10.12–10.16). All six elements are incompatible in basaltic systems, and we have listed them in decreasing order of incompatibility, so Ba is the most incompatible, Zr the least incompatible. Ba, Th and, to a lesser degree Ce, are LIL elements, a group which also includes K, Rb and Cs, and which tends to have low charge/radius ratio ionic states (Th^{4+} tends to form large complexes in basaltic melts). Most LIL elements are highly mobile during secondary alteration, which makes the interpretation of whole-rock data difficult. Th, Ba and Ce are less mobile than Rb, K, and Cs, and their primary igneous abundances may be preserved in slightly altered basalts. The elements Nb, Ta and Zr are HFS elements, whose ions have a high charge/radius ratio. They are considered as immobile in even strongly altered basalts, and therefore can be assumed to represent magmatic values with reasonable confidence.

Barium abundance data are plotted in bar chart form in Figure 10.12. In this and subsequent figures, normal or N-type MORB is distinguished from enriched or E-type MORB. The former are the depleted MORB found along most of the mid-ocean ridge system; E-type MORB resemble oceanic island basalts and are often, although not always, associated with mantle plumes (see Chapter 9). In Figure 10.12, N-type MORB clearly have the lowest contents of Ba; arc basalts tend to have higher concentrations, and back-arc basalts have intermediate values. Note, however, the wide range of Ba values in back-arc basalts.

In Figure 10.13, the Ba data are plotted against an index of fractionation, Zr. The purpose of this diagram is simply to demonstrate that the variation shown in Figure 10.12 is not due solely to fractionation during magma genesis; clearly, the arc and marginal basin basalts have higher Ba/Zr ratios than N-type MORB. Fractionation involving mafic phases would produce a range of Ba and Zr values; Ba/Zr ratios would also vary, but not to the extent seen in Figure 10.13. Clearly, some process is changing the Ba/Zr ratios of the basalts, and it is related to their tectonic setting. Alteration could increase Ba, but the samples plotted are unaltered or only slightly altered, and, in any case, why should back-arc basalts be more altered than MORB? Incorporation of Ba-rich arc-derived sediments during ponding of magma in sub-ridge chambers may also be possible, but is unlikely to be the sole explanation; as with the data on volatile elements, how do MORB escape this process (abyssal sediments contain even more Ba than arc detritus). The oxygen isotope data mentioned earlier also suggest that this model is not generally applicable. A different style of fractionation, perhaps involving a Zr retaining mineral, may be operative during back-arc basalt formation. This is also improbable,

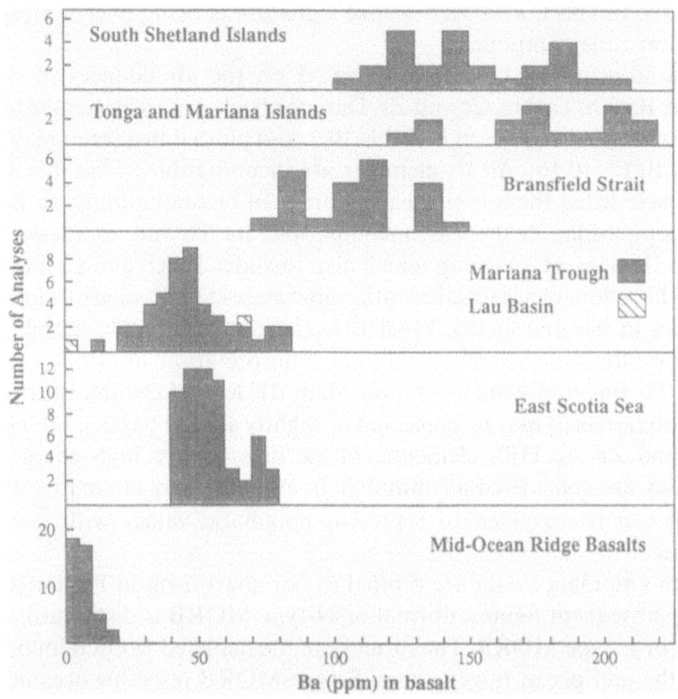

Figure 10.12 Distribution of Ba in basalts from a variety of tectonic settings. Data sources: N-type MORB: Saunders (1983 and unpublished data), Sun (1980), Sun and McDonough (1989); East Scotia Sea, Saunders and Tarney (1979); Mariana Trough and Lau Basin, Hawkins (1976), Volpe *et al.* (1987); Sinton and Fryer (1987); Bransfield Strait, Weaver *et al.* (1979); Tonga and Mariana Islands (oceanic island arcs) Ewart *et al.* (1973); Ewart and Hawkesworth (1987); Hole *et al.* (1984); Wood *et al.* (1980a, b; 1981); South Shetland Islands (continent-based island arc) Saunders and Tarney (unpublished data).

because in terms of major elements and Zr content, MORB and back-arc basalts show many similarities, and appear to be derived from a four-phase mantle source (olivine, clinopyroxene, orthopyroxene and spinel); there are no known chemical or mineralogical arguments that could support the concept of a Zr-rich residual phase in the mantle source of back-arc basalts. That leaves the conclusion that the high Ba/Zr ratios of arc and back-arc basalts are likely to be a characteristic of their source regions.

The argument can be taken further with other diagrams and other element pairs. Figure 10.14 is a set of multi-element diagrams, normalized to N-type MORB, for basalts from various back-arc basins and their associated island arcs. The enrichment in LIL elements in some back-arc basalts is clearly seen. We have previously argued that it is possible to relate the degree of LIL element enrichment in back-arc basalts to the width of the basin and to the maturity of the adjacent subduction zone; narrow basins associated with old subduction systems tend to show the greatest LIL enrichment

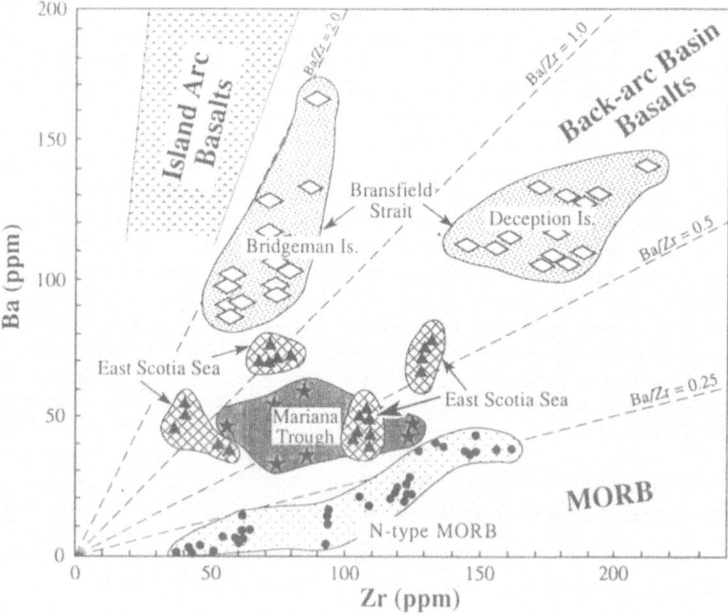

Figure 10.13 Ba versus Zr in basalts from various tectonic settings. Data sources as for Figure 10.12 (note that Deception and Bridgeman Islands are found in Bransfield Strait). An important observation from this diagram is the progressively higher Ba/Zr ratio in back-arc and arc basalts, compared with mid-ocean ridge basalts.

(Saunders and Tarney, 1984). However, it is important to stress that more data are required from back-arc basins to confirm this. A further observation from Figure 10.14a–d is the absence of large negative Nb anomalies in most back-arc basalts, unlike the associated arc lavas, but again more high precision, low concentration data are required to confirm this observation.

Figures 10.15 and 10.16 summarize the distribution of available trace element data in MORB, back-arc basalts and island arc basalts. Where the data are available, we have plotted arcs and back-arcs from the same systems. On both of these plots, the arc (= subduction?) component is clearly identified. Figure 10.15 is a plot of Ba/Zr versus Ce/Zr ratio. Note the wide range of Ba/Zr ratios, as shown by Figure 10.12, and the intermediate Ba/Zr ratios of back-arc basalts. As argued above, because Ba, Ce and Zr are incompatible elements in basalts, large variations in the ratios of these elements are unlikely to be produced during magmatic differentiation, so the fields on Figure 10.15 could represent the ratios of their parental magmas, and possibly of their mantle sources. The intermediate position of the back-arc basalts may be accounted for by mixing between N-type MORB and island-arc components. This is not a unique interpetation; alternatively, the back-arc basalts could, feasibly, be generated from end-members with N-type MORB and OIB compositions. The variations in Ce/Ba ratio between MORB, OIB and

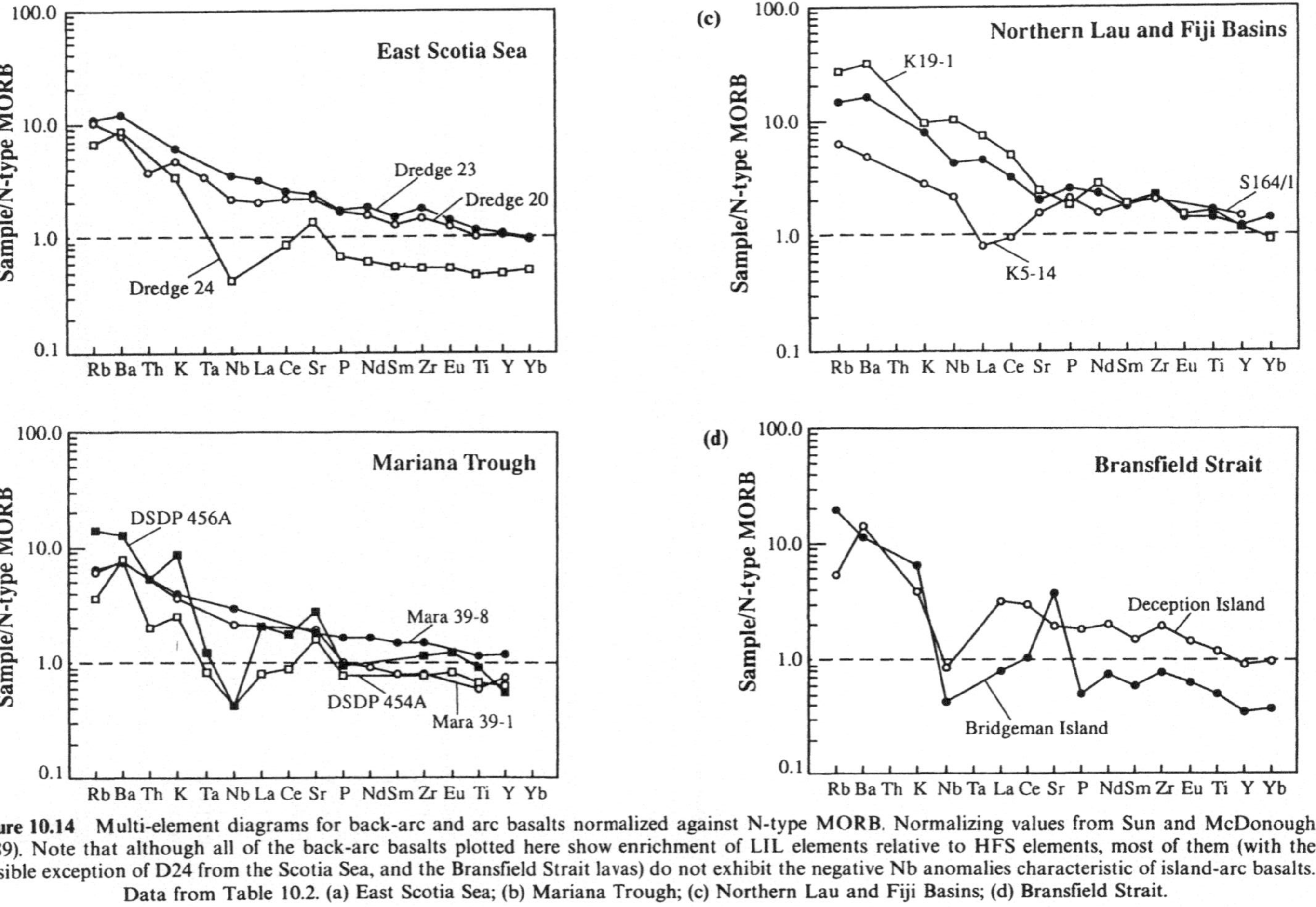

Figure 10.14 Multi-element diagrams for back-arc and arc basalts normalized against N-type MORB. Normalizing values from Sun and McDonough (1989). Note that although all of the back-arc basalts plotted here show enrichment of LIL elements relative to HFS elements, most of them (with the possible exception of D24 from the Scotia Sea, and the Bransfield Strait lavas) do not exhibit the negative Nb anomalies characteristic of island-arc basalts. Data from Table 10.2. (a) East Scotia Sea; (b) Mariana Trough; (c) Northern Lau and Fiji Basins; (d) Bransfield Strait.

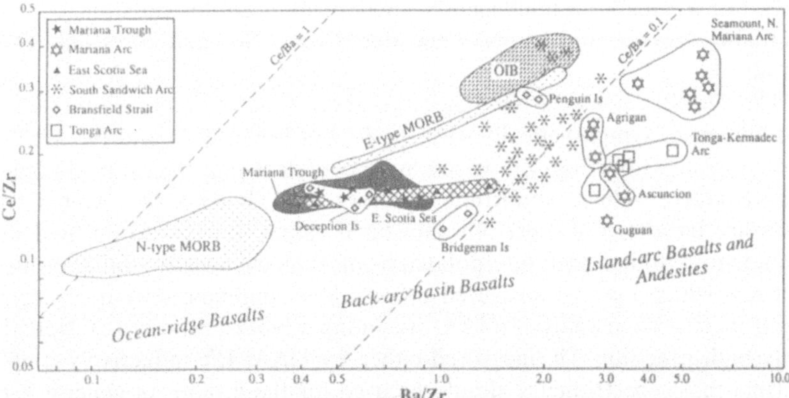

Figure 10.15 Ba/Zr versus Ce/Zr in various basaltic rocks. This diagram, like Figure 10.13, graphically illustrates the wide range of Ba/Zr ratios in terrestrial magmas, and superficially indicates that back-arc basalts are derived from mixtures of N-type MORB and arc-like mantle sources. This, however, is not a unique interpretation; note that they could equally represent a mixture between a subduction component and E-type MORB or OIB mantle component. Data sources as for Figure 10.12; plus oceanic island basalts (Palacz and Saunders, 1986); E-type MORB (Wood *et al.*, 1979). South Sandwich Islands from unpublished data by Baker (University of Leeds).

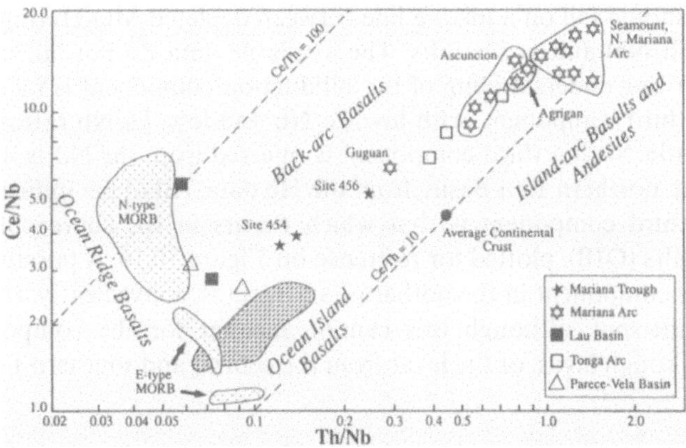

Figure 10.16 Ce/Nb versus Th/Nb. The topology of this diagram is similar to that of Figure 10.15, except that it illustrates the strong compositional variation (especially Ce/Nb ratio) in MORB. With the exception of one group of high Ce/Nb ratio samples from the Lau Basin (interpolated from the data and diagrams in Volpe *et al.*, 1988), back-arc basalts fall on a line between MORB with Ce/Nb about 2.5, and arc basalts. This suggests that back-arc basalts do not come from a simple two end-member mantle. Data sources as for Figure 10.12 and 10.15, plus Parece-Vela Basin from Wood *et al.* (1980b).

back-arc basalts is not sufficiently great to exclude this possibility for the Mariana Trough, and some of the East Scotia Sea and Bransfield Strait basalts.

The Th/Nb versus Ce/Nb diagram is more discriminating (Figure 10.16). Th/Nb varies sympathetically with Ba/Zr, but because Th and Nb are both more incompatible than Zr, we are more confident that the ratios do indeed reflect source values. Unfortunately, very few Th data are available for back-arc basalts, and most Nb data have values too low to be used with confidence (Nb has been interpolated from Ta, a chemically similar element, but one which may be measured to very low abundances by instrumental neutron activation analysis (NAA), assuming a Nb/Ta ratio of 16). Note that only high precision Th, measured either by NAA, ID, inductively-coupled plasma mass spectrometry should be used for these plots; in general X-ray fluorescence Th data are too imprecise at these levels).

An interesting feature of Figure 10.16 is that only one group of back-arc basalts (from the Lau Basin, low-volatile, 'depleted' group of Volpe *et al.*, 1988) has the high Ce/Nb and low Th/Nb ratios of N-type MORB (>2.5 and <0.07, respectively). Most back-arc basalts (admittedly only five data points!) fall on a line between arc basalts and MORB with Ce/Nb ratios of between 2.0 and 2.5. This result is not unexpected, bearing in mind earlier comments about the general absence of a strong negative niobium anomaly in back-arc basalts, although the small number of data points makes rigorous interpretation impossible.

Simplistically, if all back-arc basin basalts were derived from a source with two components (depleted MORB mantle and a subduction zone component), then they should fall on a mixing line between depleted MORB (high Ce/Nb and low Th/Nb) and arc basalts. The available data do not fall on such a line. Either our understanding of the subduction component is incorrect, or there is a third component, with low Ce/Nb and low Th/Nb ratios, present in the mantle. Such a third component is inferred from the Nd isotope data and, for the northern Lau Basin, from the He data. All of the indications are that this third component is that which occurs in the source of oceanic island basalts (OIB), plotted for reference on Figure 10.16. A possible source of the OIB component in the nothern Lau Basin is, conveniently, the nearby Samoan hot-spot, although this cannot account for the composition of Mariana Trough lavas, or the lavas from the central and southern Lau Basin.

10.5 Processes

The preceding review may be summarized as follows:

- The major, trace and isotopic data consistently indicate that many back-arc basin basalts show compositions transitional between MORB and island arc basalts. Relative to MORB, enrichment is seen in Ba, Rb, K, light REE, H_2O, Cl and $^{87}Sr/^{86}Sr$. Carbon and hydrogen isotopes are also different from MORB. Not all back-arc basalts show these characteristics. Some basalts recovered from the Lau Basin (and, within the available, limited data sets, the Parace-Vela Basin) are indistinguishable from MORB.

- The 'island-arc chemical signature' appears to be a feature of the basalts' mantle source. The observed trace element and isotope enrichments are consistent with derivation of the signature from the subducting oceanic lithosphere. It is this signature, in terms of LIL/HFS ratios, or even volatile contents, that allow the deduction of the tectonic setting of ophiolite terranes using discriminant analysis.
- Not all of the trace element and isotopic characteristics of back-arc basalts and island-arc basalts can be explained by a slab-derived component. The data indicate that a third component, with lower $^{143}Nd/^{144}Nd$ and Ce/Nb ratios than normal MORB, may be required. This component may be that sampled by oceanic island basalts and enriched MORB.
- The major element data indicate that the higher contents of volatiles may affect the fractionation history of back-arc basin magmas. However, the extent to which major element compositions are controlled by the thermal conditions of melting (a smaller extent of melting at lower temperatures and pressures) in the back-arc mantle, is as yet poorly constrained.

In this section the processes which may account for these observations are addressed. It is necessary to broaden the account to include arc magma genesis, because arc lavas should possess the strongest, and clearest, subduction signature. The broad model used here involves the production of arc and back-arc magmas from a mantle wedge peridotite 'host' variably contaminated or metasomatized by fluids from the subduction zone (Ringwood, 1974; Saunders *et al.*, 1980). The proportion of subduction-related contaminant is much lower in the back-arc source. A critical question is: what is the composition of the mantle 'host' into which the slab-derived fluids have been injected?

10.5.1 *Origin and nature of the slab-derived component*

Slab-derived material that may potentially be involved in arc and back-arc magma genesis includes the subducting basalt, a melt or fluid component from the basaltic crust, pelagic sediments (organic, carbonate, or clay-rich), and clastic sediments derived either from the adjacent arc or the continental landmass. Selective extraction of a fluid phase will lead to extensive fractionation of the chemical constituents, so the final slab-derived component will be a complex mixture from possibly all subducted materials. Most of these materials have concentrations of many elements several orders of magnitude greater than putative mantle concentrations, so only the smallest amounts of material need to be introduced into the mantle to perturb its trace element and isotopic composition; for most elements, this means less than 1–2% by mass. Bearing in mind that oceanic lithosphere is itself heterogeneous on a local and regional scale, it is surprising that arc and back-arc magma compositions are not more varied and complex.

The irregular morphology of the subducting seafloor, with its seamounts and horst and graben structures, provides an excellent vehicle for transporting sediments into the deeper mantle. Large volumes of material appear to be entering trench systems, and not all of it is accreted on the overriding plate. This is particularly true of 'decoupled' plate margins associated with back-arc basins, such as the Mariana Trench. Here, the Pacific plate, with a net convergence rate of 90 km per Ma, is carrying approximately 20 km^3 of pelagic sediment into each km of trench every 1 Ma (Karig and Kay, 1981; Hole *et al.*, 1984). This does not include the clastic material, and other trench-fill, derived from the arc itself. If these figures are extrapolated over the age of the subduction system, the incorporation is between 500 and 1000 km^3 of sediment into every kilometre of trench over a 40 Ma period. However, there is no large sedimentary accretionary prism in the Mariana forearc region; the sediment appears to be entirely subsumed beneath the forearc region (Hussong *et al.*, 1981).

The ultimate fate of this sediment is unclear; it may be subcreted beneath the forearc region, entrained into the subduction system proper, and thus ultimately cycled into the arc magma source, or even taken into the deep mantle. Whatever its fate, it is important to note that the volume of sediment entering the Mariana Trough far exceeds the volume required to account for those trace element and isotopic characteristics of the arc magmas which may be related to sediment subduction (Hole *et al.*, 1984). This suggests that the bulk of the subducted material is in some way isolated from the source of arc magmas, perhaps by entrainment of the sediment with in envelopes of mantle viscously coupled to the subducting slab (cf. Figure 10.2d).

Can sediment involvement be detected in arc and back-arc magma genesis using chemical tracers? A ubiquitous feature of arc basalts and andesites is the high content of LIL elements in relation to HFS elements (Figure 10.13). Many arc basalts and andesites have low contents of Nb (and Ta), leading to very high Th/Nb, Ba/Nb ratios (Saunders *et al.*, 1980; Gill, 1981; Pearce, 1983). This LIL element enrichment is not unique to the arc environment, but is common within it. There are several ways in which such enrichment may be accomplished. Fluids from dehydrating oceanic crust may be LIL element enriched. Submarine, low temperature alteration of oceanic basalt leads to the enrichment of most of the LIL elements, with the possible exception of the Th. The preferential extraction of these loosely bound species could occur during dehydration.

Pelagic sediments, with the possible exception of pure cherts and carbonates, will also contain LIL elements and have high LIL/HFS element ratios. Hole *et al.* (1984) and Karig and Kay (1981) have shown that the addition of small masses of sediment and slab-derived fluid can readily account for the LIL element enrichment in the Mariana Arc lavas. It is, however, difficult to evaluate the relative roles of basalt-derived fluid and sediment in arc and back-arc basalt genesis, on the basis of LIL elements alone. In general, pelagic sediment alone contains insufficient K and Rb to satisfy the arc requirement, but additional contributions of K- and Rb-rich fluids from the slab help to resolve this problem.

It is more useful to consider diagnostic chemical tracers to detect sediment subduction. This has been carried out successfully with Pb isotopes for several arc systems (Sun, 1980; Barreiro, 1983; White and Dupre, 1986; Woodhead and Fraser, 1985; Davidson, 1987; Woodhead *et al.*, 1987); Be isotopes (Tera *et al.*, 1986); and REE (Ce anomalies; Hole *et al.*, 1984). These studies are consistent with the incorporation of a few percent ($<2\%$) of sediment in the source of arc lavas (although, interestingly, lavas from the Mariana Islands do not carry the ^{10}Be excess found in other arcs). Unfortunately, none of these diagnostic tracers has been applied to back-arc basalts so it is not possible to state unequivocally state that either sediments, or sediment-derived fluids, were incorporated in their source. The data which lend the strongest support are the measurements of carbon isotopes in the Mariana Trough and Scotia Sea basalt glasses, indicating the incorporation of organic carbon in their mantle source (Mattey *et al.*, 1984).

10.5.2 *Slab-melting or dehydration?*

Critical to any model of arc-related magmatism is the way in which the subducted oceanic crust changes as it sinks into progressively hotter mantle. Most thermal models or experimental measurements indicate that the slab undergoes progressive dehydration. In addition to the cooling effects of the endothermic reactions associated with this dehydration, the removal of water causes the melting point of the slab material to rise. Consequently, under 'normal' circumstances, basaltic oceanic crust may be unable to melt (Anderson *et al.*, 1978, 1980). Water and other chemicals driven from the slab will react with and metasomatize the overlying mantle. Conditions may differ during subduction of an active spreading centre.

Tatsumi *et al.* (1986), in a modification of earlier models, suggest that the dehydration and metasomatism of the slab occur high in the subduction zone, possibly beneath the fore-arc-region. This is certainly consistent with the large volumes of water and thermogenic hydrocarbons which migrate along the decollements in the fore-arc and pre-arcs of some arcs. If metasomatism occurs in this high level region the model suggests that the mantle material may then be dragged down with the slab by viscous coupling, in a manner similar to that proposed by McKenzie (1969) and Jurdy and Stefanick (1983) for slab-generated mantle flow in back-arc regions. This secondary mantle flow may take the metasomatized mantle into the region of arc magma genesis, circulate it into the source of back-arc magmas, or remove it from the arc–back-arc region altogether. Mantle flow could also draw in 'new' mantle from regions hitherto far removed from the effects of the subducting slab; for example, the asthenosphere or remobilized lithosphere from beneath the overiding plate.

The available data (e.g. Figure 10.10 and 10.16), and trace element–isotope mixing relationships (Volpe *et al.*, 1988) preclude the generation of all back-arc basalt types solely from mixing between a subduction component and a depleted, N-type MORB mantle component. The previous section has shown that this binary association could account for some back-arc characteristics, particularly the high LIL/HFS ratios, but not all. Nor does the discussion consider when the metasomatism by slab-derived fluids actually occurred. It is usually tacitly assumed that the slab-derived component is introduced by the contemporaneous subduction zone, and, certainly, this is the simplest explanation, but proving this by definitive chemical or isotopic tracers (e.g. the short-lived isotope ^{10}Be, or by Th–U disequilibrium studies) has not been carried out (if indeed, it is possible). An alternative explanation is that the slab-derived component originates from a much older source, perhaps from the overhanging sub-continental lithosphere.

10.5.3 *Mantle wedge and magma formation in back-arc regions*

To account for the non-subduction related features of back-arc basalts, it is necessary to consider other mantle components, and to broaden the discussion to include other oceanic basalts. The low ^{143}Nd/^{144}Nd, low Ce/Nb and Th/Nb character of some back-arc basalts suggests the incorporation of a component seen in oceanic island basalts and E-type MORB. This component not only makes a substantial contribution to plume-related basalts but is indeed also detected in most asthenosphere-derived basalts. Even N-type MORB, with the possible exception of the most depleted varieties, probably contains a small 'OIB' component (this spectrum of MORB types can be seen in Figure 10.15). Is this ubiquitous component contributing to the back-arc mantle regions of back-arc basalts, just as it contributes to the mantle of MORB?

The origin of this 'OIB' component is much disputed, and a rigorous assessment is beyond the scope of this chapter, but studies of oceanic basalts worldwide show that it probably consists of a mixture of three main components (Zindler and Hart, 1986). Each of these components could be present in the mantle wedge of back-arcs and islands arcs:

- The depleted MORB mantle (DMM) component could form the bulk of the wedge, but the very low concentrations of incompatible elements ensure that its chemical signature is masked by other, volumetrically minor, components. The low $^{87}Sr/^{86}Sr$, $^{207}Pb/^{204}Pb$, $^{206}Pb/^{204}Pb$, $^{3}He/^{4}He$ and high $^{143}Nd/^{144}Nd$ ratios of some back-arc basalts do, however, support the idea that this component is present in the mantle wedge.

- Enriched mantle (EMI and EMII) components have been invoked to explain the enriched chemistry of certain plume-related oceanic islands (e.g. Samoa, Kerguelen, Gough, Tristan da Cunha). The origin of these components is obscure but necessitates the long-term isolation of the U–Pb system to generate distinctive Pb isotope ratios. The consensus is that the enriched mantle components represent recycled continental lithosphere, and/or recycled oceanic crust and sediment. Either model therefore invokes the injection of continental crustal material into the mantle (directly, or via the lithospheric mantle above a subduction zone), and the chemical consequences of both models could be similar to those produced by present-day slab-borne fluids. The isotopic consequences should, however, be different, because of the long time gap between the creation and tapping of the EM components.
- The HIMU component. Several oceanic islands are erupting basalts with characteristic high $^{206}Pb/^{204}Pb$ ratios and low $^{87}Sr/^{86}Sr$ ratios; this component has been termed HIMU after the requirement of a long-term high U/Pb ratio (μ) to generate the high $^{206}Pb/^{204}Pb$ ratios. To our knowledge, no basalts with this characteristic have been recovered from arcs or back-arc basins, but that alone does not rule out the presence of such a mantle component (it could be dominated by other components).

The majority of back-arc basalts show no evidence of having tapped either EMI, EMII or HIMU mantle components in significant amounts. In other words, the 'enriched' isotopic and trace element characteristics of oceanic island basalts (e.g. Hawaii, Samoa, Gough, St Helena) are not, in general, observed in back-arc basins, or in most island arcs. However, there are some exceptions. Kaersutite-bearing, alkaline, high-Nb basalts were recovered during Leg 58 of the DSDP from the Oki-Daito Basin in the north-west Shikoku Basin, an inactive back-arc basin.

The melt zone tapped during the formation of back-arc and arc basalts may have a very large volume. If melt separation from the residue begins at very small degrees of melting (McKenzie, 1985), then separation may commence at depths as great as 200 km beneath the spreading axis; melt separation may be delayed in back-arc basins because of the lower thermal gradient discussed earlier, but the source volumes scavenged could still be very large. The migration of small degree melts mobilizes the incompatible elements, although the major element and compatible element signature is controlled by lower pressure, shallower and more extensive melting beneath the ridge axis or arc. The important point is that the incompatible trace elements are extracted from a large volume of mantle, and any small-scale source heterogeneities (i.e. within the size of the melt extraction zone), for example, the subduction and wedge components, should be homogenized during the melting event. This may explain why, despite all of the variables present in the arc/back-arc region, the erupted magmas fall within a fairly restricted range of trace element and isotopic compositions.

Nevertheless, it will be apparent that the necessary corollary implied by the large scavenge zone, is that the melt zone does have the opportunity to sample a wide range of mantle types (depleted peridotite, slab-derived fluids, sub-lithospheric mantle drawn in by secondary flow), combine the melts, and produce an integrated melt whose original components cannot be readily unravelled. The best that can be said is that back-arc basalts represent transitional magma types, tapping a source similar to that beneath the mid-ocean ridge, but a source often contaminated by material from a subduction zone. In most instances, it is likely that the adjacent subduction zone is responsible for this contamination, but this is not necessarily true for all systems; material from ancient subduction systems, preserved in the sub-lithospheric mantle, may also be responsible. Recognizing the relative roles of these components is a major task for future research.

10.6 Concluding statements

1. Back-arc basins are a characteristic feature of many oceanic island arc systems, and they are also associated with several continent-based arcs. These basins form by lithospheric extension and separation at convergent plate boundaries. Such extension appears to be a function of relative plate motions, with hinge roll-back being an important process, particularly in the western Pacific. The majority of back-arc basins are floored by basaltic crust and are associated with high heat flow, suggesting that the mechanisms responsible for crustal generation are not that dissimilar to those occurring at mid-ocean ridges. However, the greater average depth and thinner crust of back-arc basins compared with oceanic basins indicates lower mantle temperatures beneath the back-arc region. This is consistent with the presence of the cool, subducting slab. Several sets of major element data, in particular those from the Mariana Trough, are consistent with, but do not prove, the extraction of melts rising along cool adiabatic gradients.
2. Back-arc basin basalts show a wide range of volatile, trace element and isotopic compositions. Some are indistinguishable from MORB, but others exhibit features transitional to island arc basalts (e.g. high LIL element and volatile contents). Displacement to higher $^{87}Sr/^{86}Sr$ ratios in some back-arc basalts is also consistent with the addition of a subduction-related component to their mantle source.
3. Some back-arc basalts contain another type of enriched signature,

indicating an oceanic island basalt component in their source. This component is indicated by the combination of low ε_{Nd} and high $^{87}Sr/^{86}Sr$ ratios, and low Th/Nb and La/Nb ratios.

4. The available trace element, isotopic and volatile data strongly support the idea that a slab-derived component is present in the source of many back-arc basalts. The exact nature of this component is poorly constrained by the available data; more work is required to evaluate the relative roles of subducted sediment and fluids from the oceanic crust. It is not clear why some basins should contain a higher proportion of arc-like basalts than other basins. This may be a function of the evolutionary state of the basin, the age (or maturity) of the adjacent subduction zone, the proximity of the back-arc spreading ridge to the island arc, or the availability of 'enriched' mantle from beneath the overriding plate.

11 Pacific ocean crust

RODEY BATIZA

11.1 Introduction

The Pacific Ocean basin comprises about one third of the earth's surface, or 49.8% by area of our planet's oceans and seas. It includes the earth's largest and fastest moving plate, the Pacific plate, in addition to the smaller Nazca, Cocos, Rivera, Juan de Fuca, Gorda and Explorer plates, three so-called microplates (Galapagos, Easter and Juan Fernandez), plus portions of neighbouring plates in the south and south-west Pacific (Figure 11.1). Despite its huge area, active volcanism and intrusion is currently occurring on only a tiny fraction of the Pacific basin; along the actively spreading ridges, at volcanic seamounts near these spreading axes and at active hot-spots. Elsewhere, the Pacific Ocean is floored by sediments that have accumulated on older oceanic crust and on several types of abundant volcanic constructions (Figure 11.1).

The history of the Pacific basin dates from at least 174 Ma (Renkin and Sclater, 1988) and is characterized by the appearance, disappearance and reorganization of oceanic plates and spreading centres (e.g. Tamaki and Larson, 1988; Mammerickx and Sharman, 1988; Watts *et al.*, 1988). Menard (1978) showed that much of the Tertiary tectonic history of the eastern Pacific was dominated by the break-up of the large, subducting Farrallon plate into small fragments. Indeed, reorganization of spreading geometry by rift propagation (Caress *et al.*, 1988; Hey, 1977a), microplate formation (Mammerickx *et al.*, 1988; Lonsdale, 1988a) and other mechanisms has dominated much of the geological history of the Pacific basin and continues to this day (Hey *et al.*, 1986). This has very important implications for the petrology of Pacific Ocean crust because the types of volcanic rocks found erupting today differ chemically according to their tectonic setting on and off the active ridges. This means that the older portions of Pacific basin crust cannot necessarily be thought of as vast regions of mid-ocean ridge basalt (MORB) that is homogeneous in chemistry. Instead, the crust of the Pacific basin is undoubtedly heterogeneous. The chemistry of the volcanic and plutonic rocks in each portion of the crust reflects a combination of the particular tectonic setting in which the crust formed, alteration processes and, commonly, off-axis intrusion and volcanism.

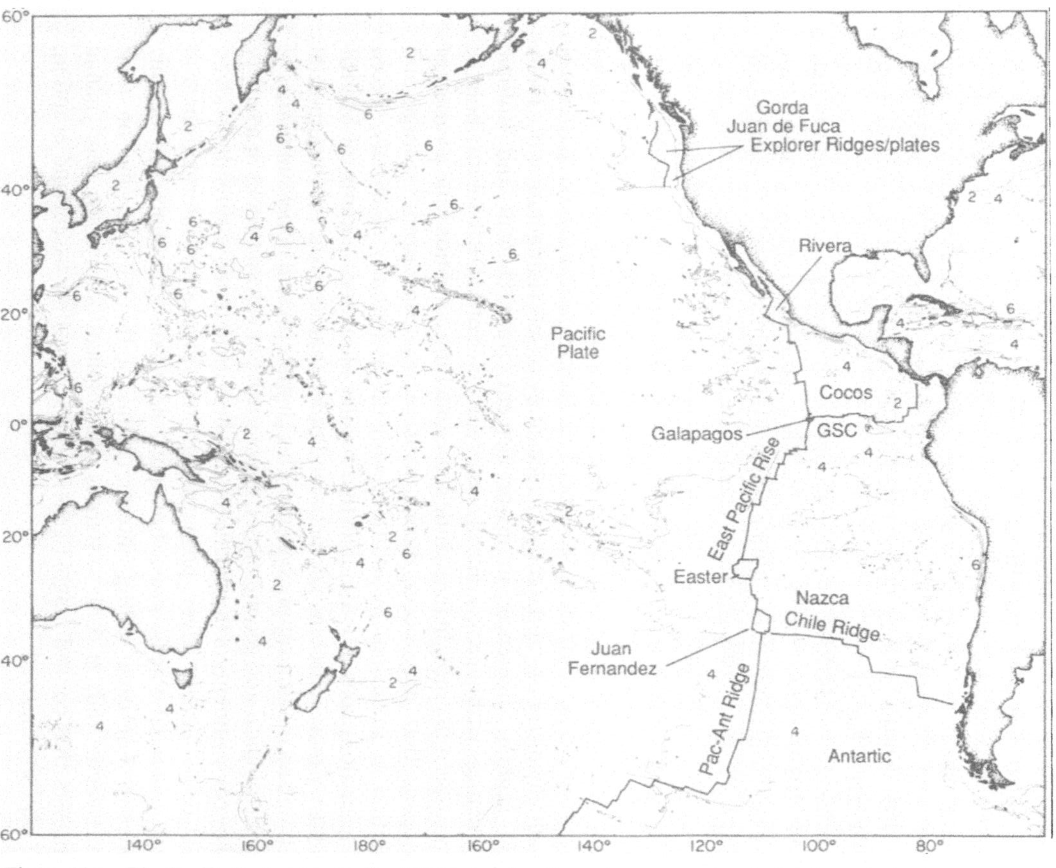

Figure 11.1 The Pacific Basin after Walker (1989) showing the main constructional plate boundaries from Lonsdale (1989b), Macdonald and Fox (1990) and Karsten (1988).

The oceanic crust of the Pacific was built predominantly at ridges spreading at moderate to fast rates (>40–50 mm y^{-1}). Consequently, it is thought to consist mostly of normal MORB (N-MORB) which is more evolved (lower magnesium number = Mg/Mg + Fe^{2+}) than MORB produced at slow spreading rates (Morel and Hekinian, 1980; Natland, 1980). A recent comparison of 1992 Pacific samples with 827 Atlantic samples showed that Atlantic samples have a magnesium number of 62.6 ± 9.7, whereas Pacific samples are only 58 ± 10.7). In contrast with ridge-produced crust, most of the large, high standing oceanic plateau and hot-spot volcanoes are built of basalts which are generally less depleted in incompatible trace elements (Mahoney, 1987), with the former being referred to as enriched MORB (E-MORB) and the latter as plume MORB (P-MORB) or oceanic island basalt (OIB) (see Chapter 9).

Although N-MORB from Pacific ridges is, on average, more evolved than N-MORB erupted at the Mid-Atlantic Ridge, considerable chemical variation may be found along-axis within a single 50–100 km segment of ridge in both oceans. In some instances these chemical variations are regular and define clear patterns; in other cases, the variation is highly irregular. All active ridge crests are segmented by tectonic offsets such as transform faults. In the Pacific, these offsets may serve as petrological boundaries, helping to define patterns of along-axis chemical variations. However, tectonic offsets may also exhibit only localized petrological 'edge-effects', and sometimes along-axis patterns of basalt chemistry are apparently unaffected by offsets (Langmuir *et al.*, 1986).

In general, differences in major element, trace element and isotopic abundances within geographically coherent suites of MORB are attributable to possible differences in the mantle source, melting conditions and a variety of processes that may occur during melt segregation, ascent, storage and eruption or intrusion. In addition, with the availability of closely spaced samples along-axis, it is thought that regular along-axis patterns of MORB chemistry must reflect the geometry and other characteristics of the sub-axial melting and magma supply systems. This chapter presents brief summaries of the petrological and geochemical characteristics of volcanic rocks in a variety of key environments of the Pacific basin. These include active spreading ridges, inactive (or failed) ridges, propagating rifts, ridge–offset intersections, old ridge-generated Pacific crust, hot-spots, near-axis seamounts and oceanic plateaux. For each setting, interpretations of the data bearing on their petrogenesis and, in some instances, the implications for the geodynamic processes that control magma generation, segregation, ascent and eruption, are also summarized.

11.2 Active ridges

The Pacific basin includes several distinct ridge systems (Figure. 11.1) including the East Pacific Rise (EPR), the Juan de Fuca–Gorda–Explorer

Ridges, the Galapagos Spreading Centre, the Chile Ridge and the Pacific–Antarctic Ridge. The latter two are very poorly sampled, so this discussion focuses primarily on the others. Table 11.1 gives some representative chemical analyses for lavas from Pacific spreading centres. In general, the ridges erupt N-MORB with marked depletions of incompatible trace elements, but small volumes of E-MORB and fractionated lavas (andesites and dacites) also occur.

Spreading ridges in the Pacific, on average, erupt more fractionated N-MORB than the Mid-Atlantic Ridge. Some of the petrographic characteristics of Pacific MORB are described in Chapter 5. In general, most Pacific MORB are sparsely phyric, although significant exceptions are found. Phenocryst assemblages are variable and may contain one or more of the phases olivine, plagioclase, clinopyroxene and spinel (Batiza, 1989a, 1989b; Perfit and

Table 11.1 Representative analyses from active Pacific ridge axes. Major oxides in wt%, trace elements in ppm

	Sample[a]							
Analysis	1 CH57-1[b]	2 CH60-3[b]	3 CH17-4[b]	4 E5-1	5 E9-52	6 E30-1	7 1120-2	8 995-3B
SiO_2	49.74	50.15	50.32	48.88	50.09	50.53	50.90	50.45
TiO_2	1.94	1.45	1.30	1.23	1.57	1.85	1.04	3.73
Al_2O_3	14.05	15.26	16.06	17.78	14.69	14.54	14.91	11.51
FeO*	11.29	9.96	8.82	7.65	9.19	9.87	8.93	18.02
MnO	0.20	0.19	0.15	0.15	0.18	0.18	0.19	0.18
MgO	7.38	8.37	8.65	7.62	7.34	6.65	8.16	3.90
CaO	11.40	12.36	11.90	12.26	12.01	11.08	12.60	8.70
Na_2O	2.71	2.62	2.68	2.56	2.98	3.10	2.23	2.87
K_2O	0.13	0.07	0.07	0.28	0.36	0.48	0.04	0.19
P_2O_5	0.19	0.11	0.05	0.18	0.24	0.26	0.08	0.46
Total	99.05	100.54	100.01	99.45	99.68	99.65	99.08	100.00
Mg number	58	64	68	64	59	55	64	30
La	4.63	2.76	3.01	—	—	—	2.24	11.4
Sm	4.38	3.30	2.99	—	—	—	2.85	12.5
Yb	3.86	3.18	2.84	—	—	—	2.81	12.7
Cr	170	320	340	228	210	307	—	5
Sr	65	90	160	192	194	231	73	66
Hf	3.64	2.51	2.28	—	—	—	1.69	7.62
Zr	160	90	63	94	124	93	62	300

[a](1) N-MORB, EPR, 6°48.1′N from JOI East Pacific Rise Synthesis (Tighe *et al.*, 1988);
(2) N-MORB, EPR, 8°12.2′N from JOI East Pacific Rise Synthesis (Tighe *et al.*, 1988);
(3) N-MORB, EPR, 12°51.5′N from JOI East Pacific Rise Synthesis (Tighe *et al.*, 1988);
(4) N-MORB, Endeavour Segment of Juan de Fuca, 47°41.6′N (Karsten, 1988);
(5) N-MORB, Endeavour Segment, 47°50.8′N (Karsten, 1988);
(6) N-MORB, Endeavour Segment, 48°2.5′N (Karsten, 1988);
(7) N-MORB, Equador Rift at 85°10′W (Perfit *et al.*, 1983);
(8) Fe-Ti basalt from Galapagos Rift-Inca transform intersection (Perfit *et al.*, 1983)
[b]Major elements by EMPA on glass chips at Lamont-Doherty Geological Observatory; trace elements by INAA at Washington University, St. Louis

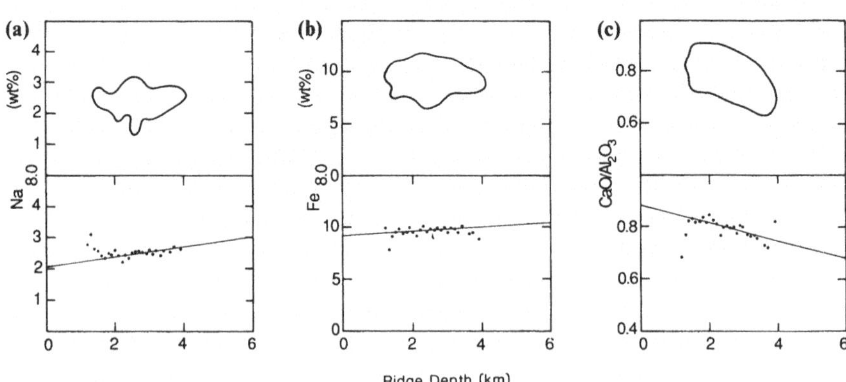

Figure 11.2 Chemistry of Pacific zero-age axial samples ($n = 1731$), after Brodholt and Batiza (1989). Fields enclose the range of observed (a) $Na_{8.0}$ Na_2O wt% corrected for fractionation according to the method of Klein and Langmuir, 1987), (b) $Fe_{8.0}$ and (c) CaO/Al_2O_3 versus ridge axis depth. Bottom panels show averages of data in 100 m depth intervals and the line is the best-fit linear regression through the data in the top panel. Pacific data exhibit the global correlations of chemistry and axial depth found by Klein and Langmuir (1987).

Fornari, 1983; Hekinian *et al.*, 1989; Bryan, 1983). Most Pacific MORB with MgO less than 8.0 wt% are multiply saturated with olivine, plagioclase and clinopyroxene (Nielsen, 1988; Klein and Langmuir, 1987); however, in contrast with the Atlantic basalts, clinopyroxene only rarely forms euhedral phenocrysts. Instead, clinopyroxene phenocrysts are usually subhedral to anhedral and complexly zoned.

Isotopically, Pacific MORB has the most radiogenic Pb but the least radiogenic Sr (White *et al.*, 1987) of any ocean basin, probably reflecting large-scale differences in mantle composition, history and convective flow among the main ocean basins. On a global basis, Klein and Langmuir (1987) have shown that MORB chemistry is related to the depth of the axes at which the basalts erupt; deep ridges produce basalts generated at relatively low pressure and low extents of melting whereas shallow ridges erupt basalt produced by deeper and larger extents of melting. Their spatially-averaged data, as well as unaveraged data (Brodholt and Batiza, 1989), indicate that this global trend is exhibited by Pacific ridges (Figure 11.2), even though they show a much narrower range of depth variation than Atlantic and Indian ocean ridges.

Characteristically, Pacific ridge axes exhibit along-axis undulations in depth (Figure 11.3) on wavelengths up to several hundred kilometres (Lonsdale, 1977; Macdonald *et al.*, 1984; Macdonald *et al.*, 1988a). Topographic lows commonly correspond to offsets in the axis, including transform faults, overlapping spreading centres and a variety of smaller offsets (Langmuir *et al.*, 1986; Batiza and Margolis, 1986). At several localities along the EPR, axial depth variations correlate with variations of the magnesium number of the axial lavas. For example, between 9°03′N and the Clipperton Transform,

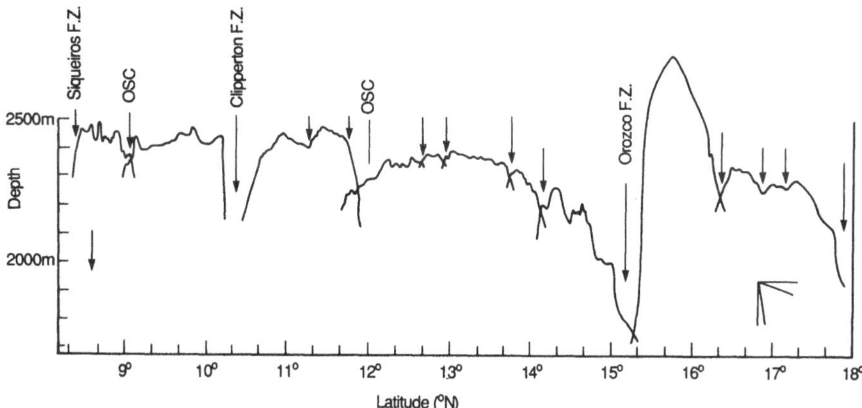

Figure 11.3 Depth of the East Pacific Rise axis from the Siqueiros transform (about 8°N) to 18°N (after Macdonald *et al.*, 1988a). Note that long wavelength undulations in depths are bounded by transforms and overlapping spreading centres (OSC). Undulations of smaller wavelength are bounded by small OSCs and other small offsets (arrows).

the lavas are all related by simple fractional crystallization (Figure 11.4), and high magnesium number lavas are found at the topographic high. South of the axial high, the magnesium number decreases regularly as the axis deepens. These changes also correlate with depth to the axial magma chamber mapped below the EPR axis with seismic techniques (Detrick *et al.*, 1987). As the axial magma chamber is continuous along this part of the EPR, these observations are consistent with either a laterally zoned chamber, cooling (and fractionation) in dykes, or a diapir-like central supply at the topographic high with slow along-axis lateral injection of magma or diminished supply as suggested by the models of Crane (1985), Schouten *et al.* (1985), Macdonald *et al.* (1988a, b) and Macdonald and Fox (1988) (Figure 11.5). Petrographic and modelling studies in progress favour a model with a magma chamber that is chemically zoned along-axis.

Whereas variation patterns similar to those above are observed elsewhere along the EPR (Thompson *et al.*, 1985), in many instances the patterns are much less regular (Langmuir *et al.*, 1986; Karsten *et al.*, 1990) as shown in Figure 11.6. In some instances, there is little or no correlation between depth and the chemistry of axial lavas, large variations in chemistry occur over short distances, and along-axis chemical variation is not affected by crossing offsets. This complexity has led to the suggestion that some axial segments are fed by a more complex system of magma supply, perhaps involving multiple vertical conduits and multiple, unconnected magma chambers (Perfit *et al.*, 1983). Alternatively, the geometry of the magma supply system could vary periodically or episodically over short time-scales, leading to closely-spaced eruptions of lava with widely different petrogenetic histories. So far, these

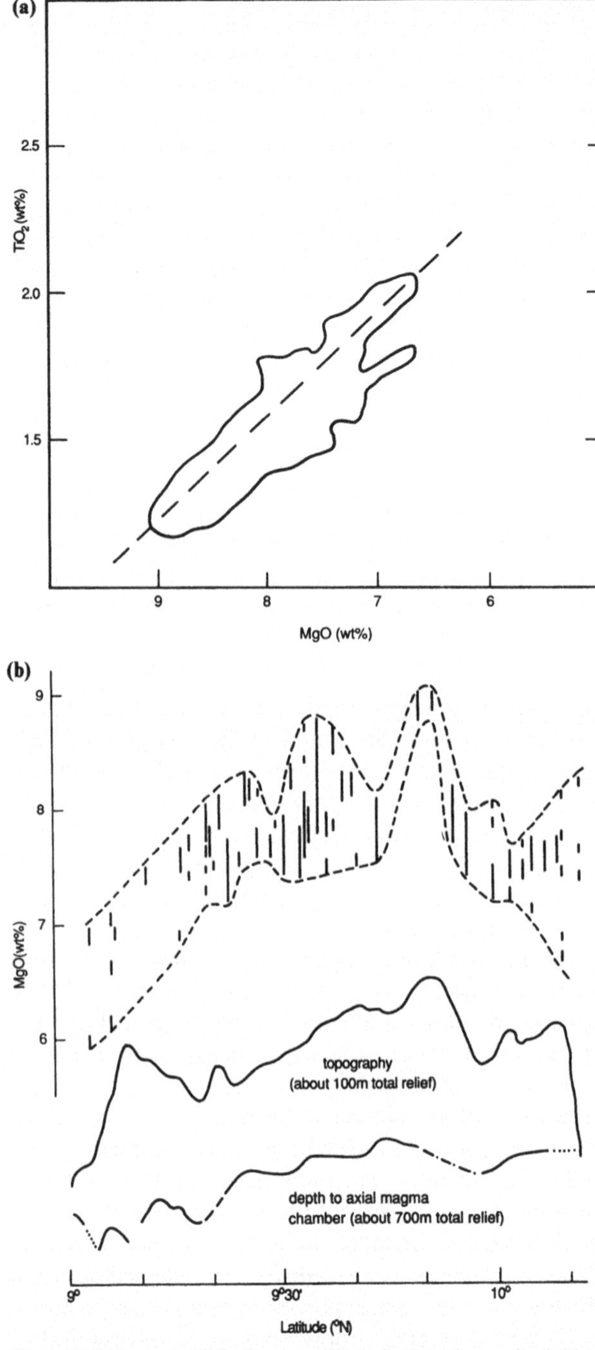

Figure 11.4 Plot of MgO versus TiO_2 for several hundred analyses of axial lavas for the EPR segment between 9°03′ and the Clipperton transform. The line through the data represents the results of least-squares fractionation models, which coincides with the predictions of the liquid line of descent models of Nielsen (1988). (b) Regular variation of MgO in EPR axial lavas from 9°–10°20′N. Note the rough correlation with topography (about 100 m relief) and depth to the roof of the axial magma chamber. The topography is taken from Macdonald and Fox (1990) and the seismic results from Detrick *et al.* (1987). Line segments show chemical analyses of several hundred basalt glasses from over 40 sampling sites.

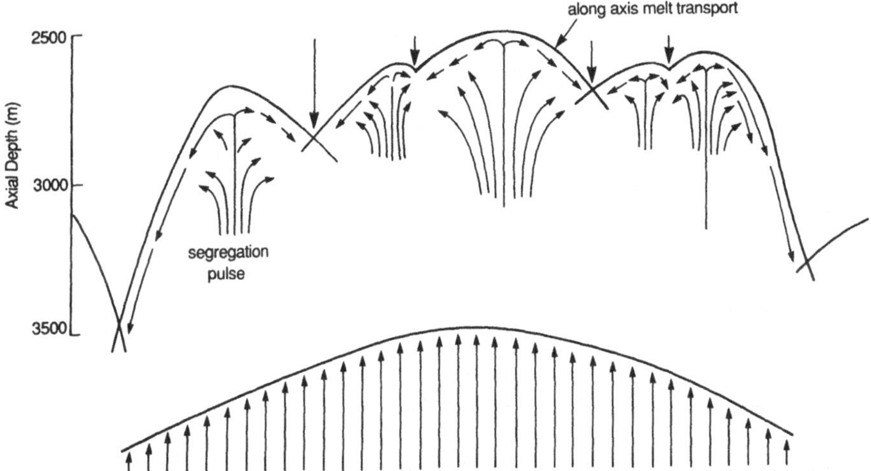

Figure 11.5 Magma supply model proposed by Macdonald *et al.* (1988a,b) to explain variations in axial depth, the behaviour of OSCs and some aspects of the regular along-axis chemical variation patterns. Broad regions of mantle upwelling supply the axis and help define broad wavelength undulations in depth. Shorter wavelength undulations are explained by smaller, diapiric melt segregation pulses that supply melt directly to topographic highs of small wavelength undulations. In this model, the distribution of melt to the deeper portions of the axis between axial highs is by lateral, along-axis transport as shown by the arrows.

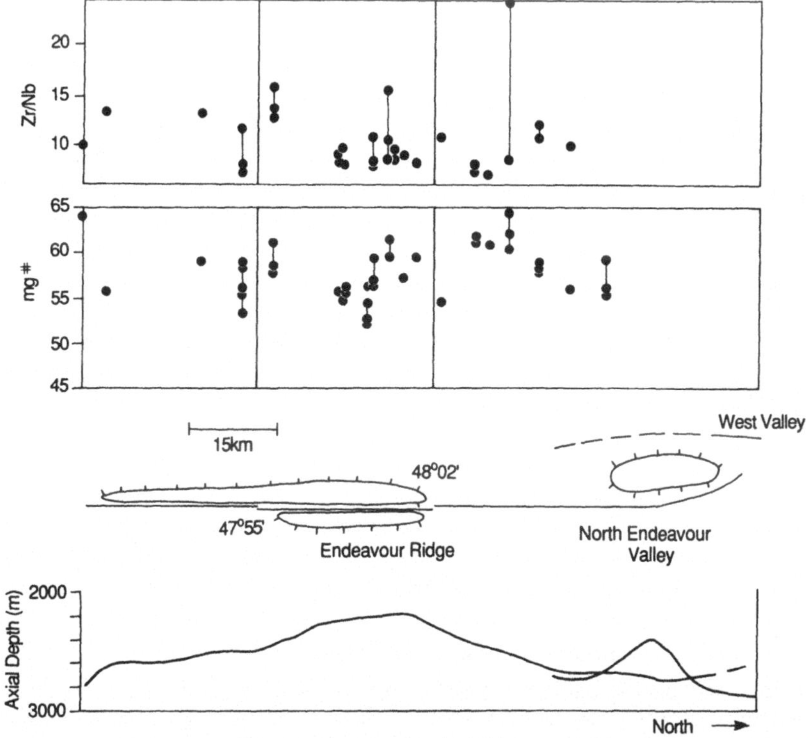

Figure 11.6 Axial depth and variation of magnesium number and Zr/Nb for basalt glasses of the Endeavour segment of the Juan de Fuca Ridge (after Karsten *et al.*, 1990). Note that the pattern of MgO variation with depth is less regular than in Figure 11.4(b).

various hypotheses have not been adequately tested. However, it is expected that future work in this area will shed light on both deep and shallow magma supply processes and geometry, thermo-mechanical conditions beneath active ridge crests and magma chamber processes.

11.3 Inactive or failed ridge crests

The complex tectonic evolution of the Pacific basin has resulted in the formation of numerous failed or inactive ridge crests, the abandonment of which is directly linked to the reorganization of spreading geometry. Most commonly, active rifts are abandoned as a direct consequence of rift propagation (Hey *et al.*, 1986), decapitation processes at overlapping rifts (Macdonald *et al.*, 1988b) and reorganization of microplates (Lonsdale, 1988a; 1990; Mammerickx *et al.*, 1988). Some failed rifts may result from spreading

Figure 11.7 Map showing some of the failed rifts that have been identified in the Pacific (after Mammerickx *et al.*, 1988). For clarity, the sizes of some failed rifts are exaggerated. Not shown are the failed rifts in the north-west Pacific (Lonsdale and Smith, 1986) and off the west coast of North America (Lonsdale, 1990).

centre 'jumps' (Mammerickx and Sanwell, 1986; Batiza, 1989a) that are unrelated to the aforementioned tectonic processes or else simple cessation of spreading caused by a change in plate-driving forces.

Failed rifts (Figure 11.7) are of great petrological interest because they may preserve evidence (in their youngest volcanic rocks) of the processes that attend rift failure. Understanding this significant departure from the steady state could enhance our understanding of both incipient oceanic rifts and active ridges. In the Pacific, only three failed rifts have been studied petrologically: the Mathematicians Ridge and the Guadalupe Trough (Batiza and Vanko, 1985), both of which are medium to large (> 100 km) failed rift systems, plus the small (7–20 km) failed rifts associated with the propagating rift at 95°30′W along the Galapagos Spreading Centre (Kleinrock *et al.*, 1989; Yonover, 1989; Yonover *et al.*, in press). The Mathematicians failed rift contains mostly very primitive N-MORB lavas and younger alkalic lavas erupted along the failed transforms (Table 11.2). The alkalic lavas are petrographically

Table 11.2 Representative analyses of Pacific failed rifts. Major oxides in wt%, trace elements in ppm

	Sample						
Analysis	7-3[a]	8-9[a]	12-3[b]	1-1[c]	2-1A[c]	6-1[c]	155-2[d]
SiO_2	49.87	48.02	49.79	47.73	49.66	49.47	49.36
TiO_2	0.71	1.27	1.66	1.52	1.67	1.83	1.07
Al_2O_3	16.68	16.45	15.60	18.23	17.80	16.90	16.36
FeO*	8.14	9.59	10.07	7.67	7.55	8.61	9.10
MnO	0.14	0.15	0.70	0.13	0.14	0.15	0.25
MgO	10.25	7.90	7.40	7.71	7.02	7.67	9.22
CaO	12.42	9.30	11.30	9.40	9.31	9.78	12.14
Na_2O	1.81	3.43	2.70	3.59	3.43	2.69	2.30
K_2O	0.09	0.30	0.26	1.34	1.68	1.20	0.06
P_2O_5	0.02	0.10	0.19	0.41	0.50	0.41	0.07
Total	100.13	96.51	99.67	97.73	98.76	98.98	99.93
Mg number	73	64	62	68	66	66	64
La	1.04	2.72	5.89	31.5	31.9	23.3	—
Sm	1.69	3.26	4.18	4.95	5.58	5.15	—
Yb	1.95	3.14	3.73	2.58	2.69	2.80	—
Cr	560	268	300	238	241	254	493
Sr	70	166	123	490	370	330	107
Hf	1.23	2.54	3.30	3.50	4.73	3.82	—
Zr		100	105	170	180	155	94

[a]Mathematician failed rift (Batiza and Vanko, 1985), Sample 8-9 is altered, with LOI = 2.54 wt%
[b]Guadalupe Trough (Batiza and Vanko, 1985). MnO is contaminated by very fine veins of Mn-rich material
[c]Alkali basalts from the failed transforms of the Mathematician failed rift (Batiza and Vanko, 1985)
[d]Failing rift of the 95.5°W propagating rift on the Galapagos Spreading Centre from Yonover (1989)

similar to some OIB (see Chapter 9). Interestingly, even though it was fast spreading prior to abandonment, it now has a deep rift valley like slow spreading ridges. This suggests that the spreading rate may decrease gradually before spreading ceases completely. The very primitive N-MORB lavas are also similar to the high MgO lavas erupted at the Mid-Atlantic Ridge. There is no petrological evidence for cooling magma chambers that might be stranded by the cessation of spreading. Volcanic reactivation of the failed transforms indicates that failed rifts may remain as zones of lithospheric weakness, serving as eruptive conduits for alkali basalt lavas (and their differentiates) produced by deeper and smaller extents of melting than MORB, but from an isotopically similar source (Graham *et al.*, 1988).

11.4 Propagating rifts

Propagating rifts are abundant along Pacific spreading centres. Many are currently active and abundant evidence indicates that they were also common

Table 11.3 Representative analyses of Pacific propagating rifts. Major oxides in wt%, trace elements in ppm

	Sample							
Analysis	D08 (D5)[a]	998 (D6)[a]	D64 (D6)[a]	997 (D6)[a]	17-1[b]	4-13[c]	29-6[c]	27-6[c]
SiO_2	50.62	52.35	57.09	70.71	49.68	51.02	50.49	49.14
TiO_2	1.04	2.49	1.76	0.61	0.86	1.50	1.87	2.00
Al_2O_3	16.49	12.43	13.48	12.30	16.38	14.13	13.44	17.25
FeO*	8.79	15.97	12.12	5.30	9.34	11.99	13.85	9.65
MnO	—	—	—	—	—	—	—	—
MgO	8.53	4.37	2.75	0.43	9.34	6.68	6.31	6.44
CaO	12.34	9.16	6.87	2.92	12.15	11.28	10.76	9.63
Na_2O	2.24	2.76	3.31	4.14	2.20	2.44	2.39	3.45
K_2O	0.04	0.28	0.58	1.30	0.04	0.18	0.08	1.07
P_2O_5	0.06	0.25	0.17	0.05	0.08	0.15	0.16	0.33
Total	100.15	99.90	98.13	97.82	100.07	99.37	99.35	98.96
Mg number	67	37	33	15	68	54	49	59
La	2.4	7.8	15.1	26.5	—	—	—	—
Sm	2.3	5.71	8.24	12.4	—	—	—	—
Yb	2.4	5.8	9.3	13.7	—	—	—	—
Cr	645	480	445	28	—	—	—	—
Sr	92	88	70	38	—	—	—	—
Hf	1.9	4.9	8.4	12.9	—	—	—	—
Zr	70	240	330	345	—	—	—	—

[a] From the 95.5°W propagating rift along the Galapagos Spreading Centre (GSC) (Clague *et al.*, 1981)

[b] Normal rift segment of the GSC from Christie and Sinton (1981)

[c] From the GSC (Christie and Sinton, 1981); 27.6 is an alkali-rich sample from the northern fault of the 95.5°W propagating rift

features in the geological past. The best studied propagating rifts are those at 95°30′W on the Galapagos Spreading Centre (Hey *et al.*, 1986), the Cobb offset of the Juan de Fuca system (Johnson *et al.*, 1983), the 'dueling' propagating ridges on the EPR near 20°40′S (Macdonald *et al.*, 1988b), and those associated with migrating overlapping spreading centres (OSCs) elsewhere along the EPR (Lonsdale, 1989a, b; Macdonald *et al.*, 1988a). Documentation of ancient propagating rifts is provided by the studies of Caress *et al.* (1988), Hey and Wilson (1982), Hey *et al.* (1985), Anderson-Fontana *et al.* (1986) and others.

The petrology of propagating rifts has been studied by Sinton *et al.* (1983), Christie and Sinton (1981), Byerly *et al.* (1976), Clague *et al.* (1981) and Christie and Sinton (1986). Table 11.3 gives representative chemical analyses and Figure 11.8 shows the regular spatial chemical variation patterns that are typically found near propagating rifts (including migrating OSCs of the EPR; see Langmuir *et al.*, 1986). Interpretations of the petrological data suggest that propagating rifts tap a relatively shallow mantle source and magmas at the tip of the propagator rise without a great deal of fractionation. A few kilometres behind the tip, however, are Fe-rich basalts and SiO_2-rich differentiated rocks, which are replaced, in a regular along-axis variation pattern, by less fractionated N-MORB well back from the propagating rift tip. This pattern is explained as a consequence of the balance between thermal maturation of the

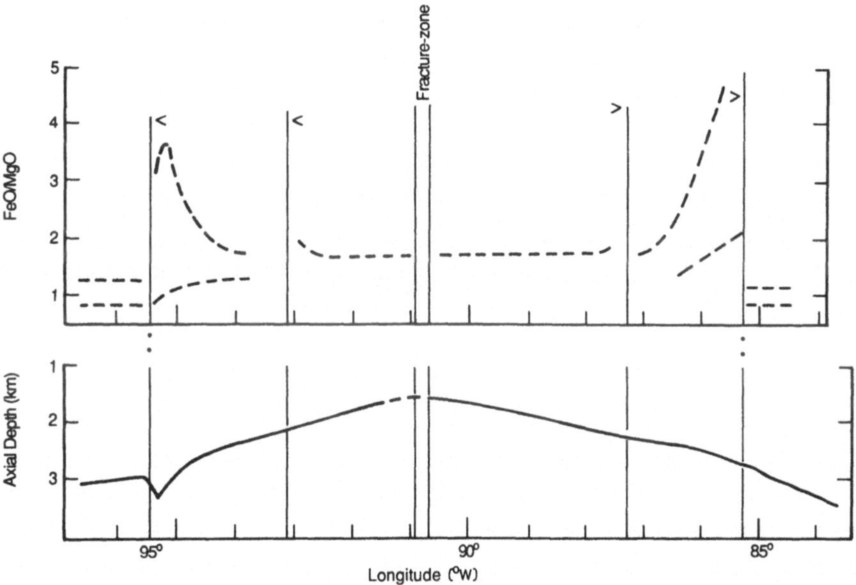

Figure 11.8 Depth of the Galapagos spreading centre (bottom panel) and range of FeO/MgO chemical variation along axis, after Christie and Sinton (1981). Note that chemical diversity increases near propagating rifts (arrows).

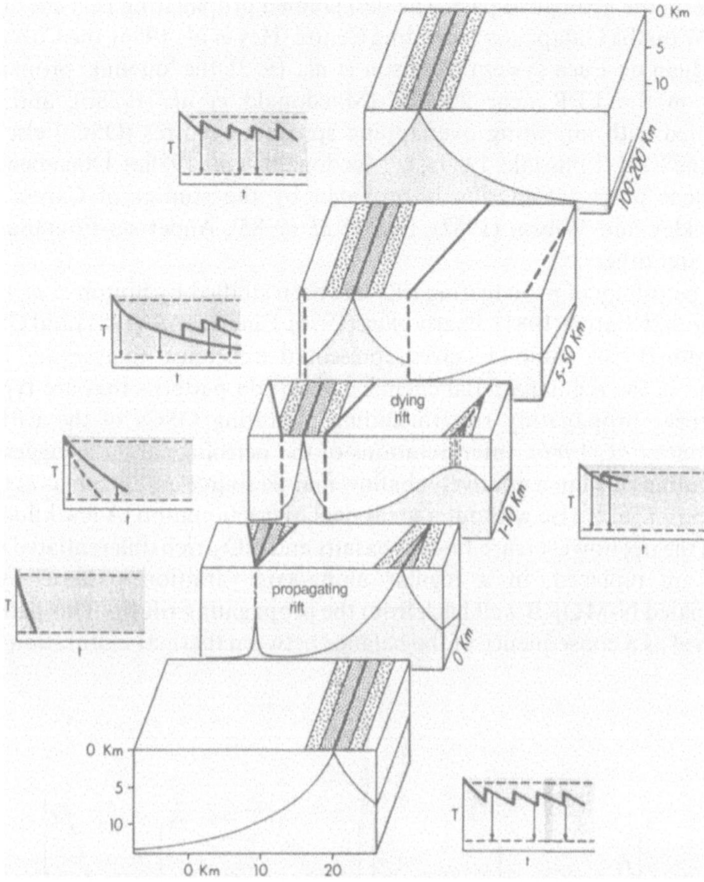

Figure 11.9 Schematic block diagram (from Hey *et al.*, 1989) showing the evolution of asthenospheric upwelling zones (stipple) beneath the 95.5°W propagator. Idealized cooling curves (after Christie and Sinton, 1981) show temperature (*T*) versus time (*t*) for different regions. Vertical arrows denote possible eruption times. Stipple on cooling curves shows regions where basalt magma can be partly liquid.

rift which is migrating into relatively cold lithosphere, and the advective heat from vertically upwelling hot mantle and magma which feeds the new spreading centre (Figure 11.9).

11.5 Edge effects at ridge offsets

As a result of the pronounced edge effects at propagating rift tips, it is perhaps not surprising that other types of ridge offsets, including transforms, OSCs and smaller offsets, also commonly show petrological and geochemical effects.

Table 11.4 Representative analyses of Pacific ridge-offset intersections. Major oxides in wt%, trace elements in ppm

	Sample						
Analysis	994-3A[a]	996-1B[a]	994-5B[a]	972-1[b]	975-1[b]	D11-6[b]	D12-5[b]
SiO_2	56.28	50.34	51.16	48.8	49.1	49.2	49.7
TiO_2	2.46	3.47	1.42	1.43	2.12	1.40	1.28
Al_2O_3	11.70	11.69	14.17	16.1	14.3	15.1	15.5
FeO*	15.26	17.93	11.55	9.55	11.3	9.48	9.43
MnO	0.26	0.21	0.23	0.11	0.21	0.14	0.15
MgO	2.68	4.37	7.77	8.89	7.16	8.37	8.53
CaO	7.43	8.93	11.49	12.6	11.4	12.0	12.3
Na_2O	3.36	2.83	2.15	2.50	2.86	2.25	2.42
K_2O	0.32	0.24	0.06	0.03	0.08	0.05	0.03
P_2O_5	0.40	0.37	—	0.06	0.03	0.05	0.06
Total	100.15	100.38	100.00	100.1	98.6	98.0	99.4
Mg number	26	33	57	62	53	61	62
La	13.9	10.2	2.91	—	—	—	—
Sm	13.1	8.56	3.60	3.43	4.98	3.24	2.85
Yb	13.6	8.64	4.09	3.28	1.06	3.35	3.08
Cr	8	8	165	386	231	393	400
Sr	120	64	64	112	156	96	78
Hf	11.1	5.62	2.00	—	—	—	—
Zr	425	261	78	103	171	95	76

[a]From the intersection of the Galapagos Rift and the Inca Transform (Perfit *et al.*, 1983)
[b]From the Tamayo Transform–EPR intersection (Bender *et al.*, 1984); 972-1 and 975-1 are from near the intersection; D11-6 and D12-5 are far from the intersection

These so-called 'edge effects' in axial basalt chemistry in the vicinity of offsets are variable in their characteristics and, although common, are not always present. There is apparently no relationship between offset length at transforms and the magnitude of edge effects, so a simple thermal cold-edge effect is an inadequate explanation in many instances. At the Galapagos Spreading Centre–Inca Transform boundary, the presence of abundant Fe-rich MORB and SiO_2-rich differentiates (Table 11.4) points to a cold-edge effect promoting fractional crystallization. In addition, MORB petrogenesis at the Inca Transform may also be affected by attempts of the Galapagos Rift to propagate across the transform. Perfit *et al.* (1983) invoke fractional crystallization as the dominant process to explain the observed chemical variation in the volcanic rocks there, but minor effects from differences in partial melting, magma mixing and other processes are also discernible. A similar kind of cold-edge effect may be present at the EPR just north of the Clipperton Transform, but is less pronounced on the south side of Clipperton (Langmuir *et al.*, 1986; Thompson *et al.*, 1985).

A distinctly different type of edge effect occurs at the Tamayo Transform (Bender *et al.*, 1984; Langmuir and Bender, 1984) (Figure 11.10). The chemistry

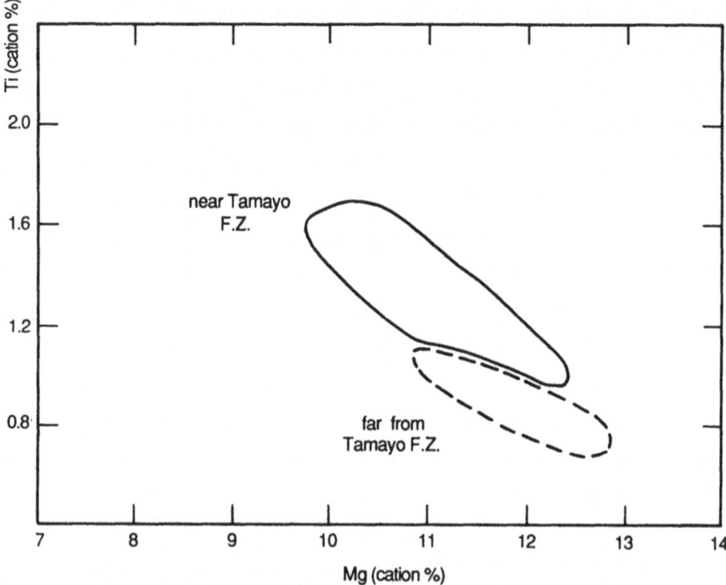

Figure 11.10 Plot of Ti versus Mg for EPR axial samples near (< 16 km) and far from the Tamayo transform after Bender *et al.* (1984). On the basis of these and additional data and arguments, the EPR samples near Tamayo are thought to be produced by smaller extents of partial melting at lower pressure than samples far (> 16 km) from the EPR–Tamayo intersection.

of samples near the transform intersection (Table 11.4) is interpreted to arise from smaller extents of partial melting, possibly at shallower depths, than EPR samples far from the transform. The edge effects at OSCs are variable. At migrating OSCs, the propagating limb may show effects similar to propagating rifts elsewhere, whereas the effects at the retreating limb are variable (Natland *et al.*, 1986; Langmuir *et al.*, 1986). Smaller offsets may show no chemical differences on opposing limbs, however, in some instances they act as petrological boundaries or appear to be preferred sites for the eruption of enriched MORB (Langmuir *et al.*, 1986; Sinton and Mahoney, personal communication).

11.6 Older ridge-generated Pacific crust

Volcanic rocks of older sedimented crust of the Pacific ocean can only be sampled by drilling. Thirty-three legs of the Deep Sea Drilling Project (DSDP) and the Ocean Drilling Programme (ODP) have been devoted to drilling the main Pacific basin (Figure 11.11). A total of 88 holes reached igneous basement. Many of these holes were drilled in the western Pacific and encountered younger alkalic volcanic rocks instead of true oceanic basement (Table 11.5).

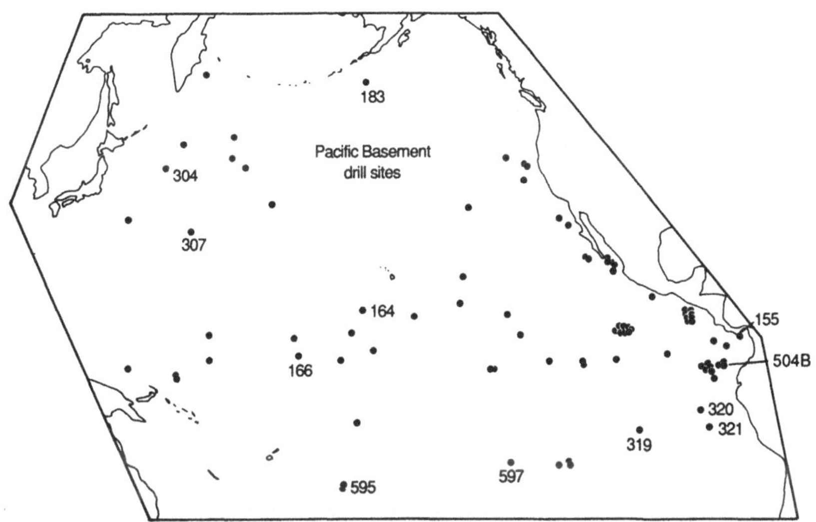

Figure 11.11 Map showing the location of Pacific drill sites which recovered basement or igneous rock. Sites with labels correspond to analyses in Table 11.5.

Others were on large volcanic constructions such as the Campbell Plateau (Leg 29), Shatsky and Hess Rises (Leg 32) and seamounts of the Hawaiian Chain (Legs 32, 33 and 55). Thus relatively few holes have penetrated deeply into ridge-generated volcanic rock; available drill samples are almost entirely N-MORB of variable magnesium number and state of alteration (Table 11.5).

An outstanding exception to the mostly shallow basement holes in normal Pacific crust is hole 504B, located 200 km south of the Costa Rica Rift on 5.9 Ma crust. Drilling during Legs 69, 70, 83 and 111 has resulted in a total penetration of 1350 or 1075 m into igneous oceanic crust. Drilling has progressed well into the sheeted dyke layer (Figure 11.12). The volcanic rocks and sills of 504B consist of very depleted N-MORB (Natland *et al.*, 1983), which is extremely homogeneous. This chemical homogeneity is ascribed to derivation of the crust from a well mixed, homogeneous magma chamber.

11.7 Hot-spot volcanoes

The Pacific has numerous active, hot-spot volcanoes, and island and seamount chains of hot-spot origin, most of which have OIB characters (see Chapter 9). Some of the active hot-spots are close to spreading ridges, such as the Galapagos hot-spot (Hey, 1977a; Verma *et al.*, 1983; Castillo *et al.*, 1988) and another whose chemical effects can be seen along the east rift zone of the Easter microplate (Hey *et al.*, 1985; Schilling *et al.*, 1985; Hanan and Schilling, 1989). In contrast, the active Hawaiian, Society, Samoa, Caroline, Austral, and

Table 11.5 Selected analyses of volcanic rock drilled on the Pacific basin. Major oxides in wt%, trace elements in ppm

Analysis	Sample															
	Leg 69 (Ave)[a]	Leg 70 (Ave)[a]	Leg 83 (Ave)[a]	Leg 111 (Ave)[a]	Leg 17 Site 164[b]	Leg 16 Site 155[b]	Leg 19 Site 183[b]	Leg 32 Site 307[b]	Leg 17 Site 166[b]	Leg 32 Site 304[b]	Leg 91 Site 595[b]	Leg 92 Site 597[b]	Leg 34 Site 319[b]	Leg 34 Site 320[b]	Leg 34 Site 321[b]	Leg 54 Site 420[b]
SiO_2	49.79	49.61	49.01	49.78	46.61	48.88	45.9	48.96	49.24	49.56	49.3	49.85	49.52	51.47	49.41	50.67
TiO_2	0.95	0.97	0.89	0.96	1.31	1.90	1.9	1.90	2.23	2.07	2.86	0.85	1.79	1.85	2.46	1.99
Al_2O_3	15.78	15.28	15.58	15.20	15.38	16.78	15.6	17.34	13.59	13.05	14.5	16.36	14.29	14.83	12.57	14.28
FeO*	8.73	9.18	8.90	9.34	9.22	8.61	11.43	9.79	12.00	12.68	11.67	8.03	10.64	8.75	12.37	9.9
MnO	0.17	0.17	0.19	0.18	0.16	—	0.20	0.02	0.23	0.22	0.20	0.14	0.17	0.18	—	0.19
MgO	8.12	8.35	8.62	8.42	3.99	7.85	7.7	2.38	5.73	6.57	6.33	9.17	7.09	7.37	5.94	6.30
CaO	12.83	12.82	12.55	12.89	12.50	5.13	8.6	0.59	10.08	10.60	8.86	13.78	11.15	11.64	9.45	11.20
Na_2O	2.25	2.24	1.95	2.00	3.24	3.02	2.9	1.29	3.45	2.78	3.35	2.31	2.71	2.85	2.37	2.80
K_2O	0.17	0.04	0.02	0.02	0.52	1.85	1.2	3.98	0.77	0.08	0.57	0.03	0.19	0.23	0.14	0.43
P_2O_5	0.07	0.07	0.07	0.06	0.12	—	0.46	0.24	0.20	0.22	0.20	0.08	—	0.19	—	0.16
Total	98.86	98.73	97.78	98.85	93.05	94.02	95.89	86.49	97.52	97.83	97.84	100.60	97.55	99.36	94.71	97.92
Mg number	67	66	67	66	48	66	59	34	51	53	54	71	59	64	51	58
Zr	55	50	59	48	—	—	—	—	—	—	—	—	—	—	—	—
Sr	88	70	62	59	—	—	—	—	—	—	—	—	—	—	—	—
Cr	356	294	326	289	—	—	—	—	—	—	—	—	—	—	—	—

[a]From Becker, Sakai *et al.* (1988)
[b]From the ODP data bank (via K.A. Lighty, ODP, Texas A&M University)

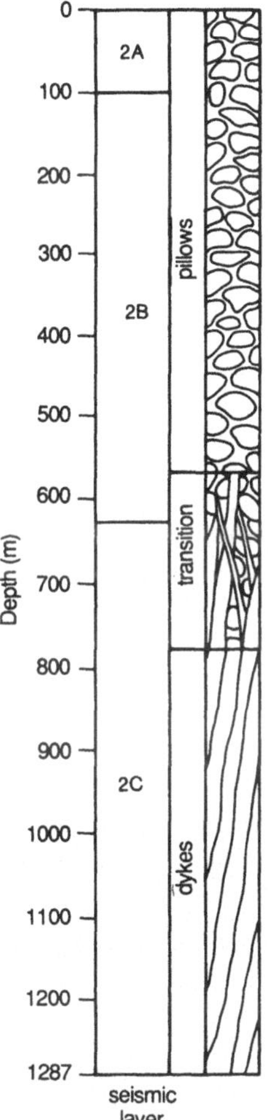

Figure 11.12 Lithology of volcanic and plutonic rocks recovered at site 504 B, after Becker *et al.* (1988). Site 504 B is the deepest penetration into the igneous crust of the Pacific Ocean.

possibly Louiseville hot-spots (Lonsdale, 1988b; Watts *et al.*, 1988) are located far from active spreading centres. Pacific hot-spots, in general, produce basalts that are more enriched in incompatible elements than MORB (Table 11.6). In addition, their isotopic abundances are distinct from MORB, and for near-axis hot-spot mixing between the hot-spot (plume?) source and the MORB source can be documented (Schilling, 1985).

The petrology and geochemistry of numerous individual Pacific basin hot-spot volcanoes and island or seamount chains are well documented (Duncan and Clague, 1985; Clague and Dalrymple, 1987; Zindler and Hart, 1986; Okal and Batiza, 1987). Hawaiian volcanoes, which are very well studied, apparently

Table 11.6 Selected analyses of Pacific hot-spot volcanics. Major oxides in wt%, trace elements in ppm

	Sample							
Analysis	1-12[a]	1-13[a]	5-4[a]	9-8[a]	k[b]	65HU-10[c]	C-129[c]	C-189[c]
SiO_2	53.0	42.5	45.8	46.2	48.9	46.62	42.53	39.91
TiO_2	1.21	2.72	2.64	3.57	2.34	2.37	2.92	2.76
Al_2O_3	17.3	17.9	15.2	13.3	12.1	14.72	12.43	9.13
FeO*	8.77	10.57	12.91	14.91	11.4	12.51	13.64	12.92
MnO	0.25	0.59	0.13	0.20	0.17	0.19	0.16	0.18
MgO	1.53	2.93	3.32	4.22	12.7	9.10	12.17	15.88
CaO	3.94	10.3	7.73	8.70	9.7	10.04	11.80	11.97
Na_2O	5.42	3.62	3.68	2.71	1.99	2.82	2.35	3.16
K_2O	2.80	0.88	1.55	0.92	0.44	0.91	0.81	1.53
P_2O_5	0.40	2.14	1.26	0.61	0.22	0.27	0.55	0.81
Total	94.63 (100.3)[d]	94.15 (100.4)[d]	94.22 (99.8)[d]	95.34 (100.2)[d]	99.96	99.55 (99.88)[d]	99.36 (99.83)[d]	98.25 (99.95)[d]
Mg number	27	37	35	38	70	61	66	66
La	58	66	39	20	—	—	—	—
Sm	12.5	11.0	15.0	7.8	—	—	—	—
Yb	5.0	7.6	5.5	4.7	—	—	—	—
Cr	2.5	25	1	38	—	—	—	—
Sr	755	755	410	245	—	—	—	—
Hf	13.2	6.4	9.0	6.0	—	—	—	—
Zr	605	280	270	250	—	—	—	—

[a]Sum of complete analysis with measured Fe_2O_3, H_2O, CO_2
[b]Alkalic differentiates from the Pratt–Walker chain and related volcanoes of the North East Pacific Gulf of Alaska (Dalrymple *et al.*, 1987)
[c]Average Kilauea tholeiite from Clague and Dalrymple (1987)
[d]Alkalic lavas from Hawaiian hot-spot volcanoes from Clague (1987); for trace elements see references cited therein

evolve through four petrological stages (see Chapter 9 and Figure 9.5): a pre-shield alkalic stage, a tholeiitic shield stage, a post-caldera alkalic stage and a post-erosional alkalic stage. Interpretation of the voluminous petrographic major, trace element and isotopic data (Clague, 1987) indicates that Hawaiian volcanoes are fed from several distinct mantle sources which melt to various extents during the history of the volcano. The petrogenesis of Hawaiian volcanoes is very complex in detail.

The petrological evolution of other well studied Pacific hot-spot volcanoes is equally complex (Dalrymple *et al.*, 1987; Mattey, 1982; Duncan *et al.*, 1986; Hawkins *et al.*, 1987), commonly involving numerous eruptive cycles that are petrologically and isotopically distinct. Multiple mantle sources are needed to account for hot-spot magmas (Zindler and Hart, 1986), indicating that the simple model of a homogeneous mantle source of plumes is inadequate to explain the data (see Chapters 9 and 10).

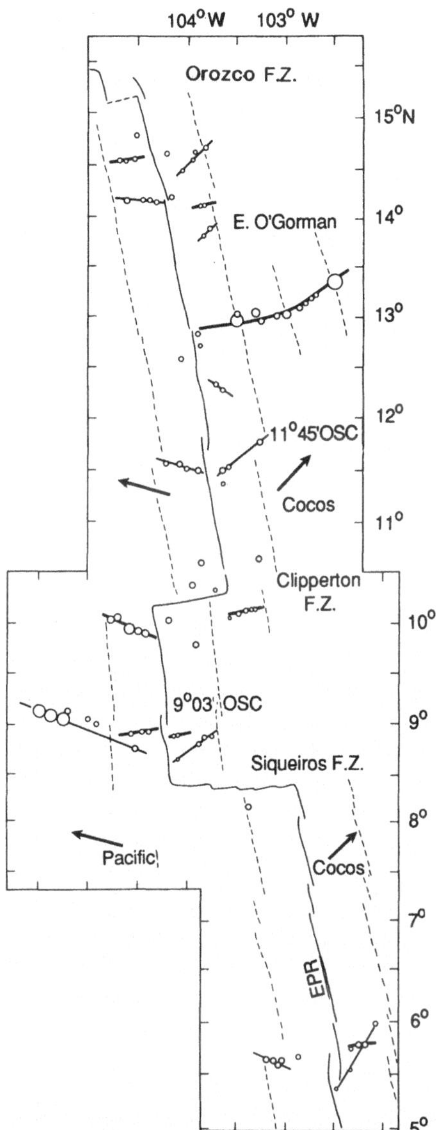

Figure 11.13 Map of the East Pacific Rise (from Batiza *et al.*, 1990 b), showing the locations of most near-axis seamounts of significant size (dots). The sizes of the dots roughly correspond to seamount sizes. Absolute plate motions are shown as large arrows. Note that most small linear volcanic chains are parallel to these arrows. The next most abundant (dashed lines) are chains whose orientations are parallel to relative plate motion.

11.8 Non-hot-spot seamounts

Not all of the central volcanoes of the Pacific basin form large linear chains parallel to absolute plate motion (Batiza, 1982). Indeed, most of the small (< 1 km high) and medium sized (1–3 km high) seamounts probably form near active ridge crests. Whereas near-ridge seamounts are also found near the Mid-Atlantic Ridge (Batiza *et al.*, 1989), those in the Pacific are much better known (Batiza and Vanko, 1984; Allan *et al.*, 1987, 1988, 1989; Fornari *et al.*, 1988b; Batiza *et al.*, 1990a, b; see also Chapter 9, section 9.8.4). Near the EPR (Figure 11.13), seamounts occur as isolated individuals and as small chains. The chains are mostly oriented parallel to absolute and relative plate motion and are much less commonly oblique to these orientations. They appear to be generated preferentially near ridge offsets and near along-axis topographic highs.

Petrologically, near-axis seamounts are mostly N-MORB, which differs from that generated at ridge axes in that seamount MORB is systematically more primitive than axial MORB, seamount MORB is more diverse

Table 11.7 Representative analyses of Pacific near-axis seamounts. Major oxides in wt%, trace elements in ppm

	Sample							
Analysis	1-1[a]	14-1[a]	20-1[a]	19-7[a]	1561-1622[b]	1387-1920[c]	1389-1810[c]	1389-2041[c]
SiO_2	50.42	51.56	49.90	49.39	48.55	48.26	49.84	51.77
TiO_2	0.97	2.01	1.34	2.33	0.94	1.86	2.35	3.00
Al_2O_3	16.09	14.61	16.65	17.69	17.84	17.38	17.79	16.97
FeO*	8.69	10.75	8.86	7.94	8.02	8.74	8.12	8.92
MnO	0.15	—	—	0.14	—	—	—	—
MgO	8.73	6.31	7.74	4.82	9.72	7.32	6.1	3.00
CaO	12.60	11.15	11.75	8.62	12.15	10.84	8.23	7.22
Na_2O	3.11	3.34	2.94	4.37	2.48	3.47	4.5	5.31
K_2O	0.06	0.21	0.28	2.01	0.04	0.47	1.74	2.74
P_2O_5	0.05	0.19	0.17	0.65	0.13	0.26	0.74	1.04
Total	100.87	100.13	99.63	97.96	99.87	98.6	99.41	99.97
Mg number	68	55	65	56	70	64	62	42
La	1.87	5.79	6.04	35.41	0.96	10.43	32.74	36.64
Sm	2.30	4.79	3.35	7.09	2.11	4.74	6.68	6.97
Yb	2.55	3.61	2.71	3.20	2.00	3.39	2.8	3.04
Cr	384	138	304	97	310	199	162	90
Sr	65	163	151	460	99.9	261	533	469
Hf	1.63	3.87	2.72	7.18	1.64	4.12	6.22	6.80
Zr	47	116	—	254	—	181	249	258

[a] Typical near-EPR seamount lavas showing a range of enrichments (Batiza and Vanko, 1984)
[b] Very depleted N-MORB from the Lamont seamount chain (Fornari *et al.*, 1988)
[c] Transitional and alkalic basalts from seamount 6 from Batiza *Et al.*, (1990a)

chemically and isotopically and, in some instances, seamount MORB is more depleted than nearby axial magmas (Table 11.7). These differences indicate that seamount lavas are fed independently of the axis, even though they commonly form only kilometres away from it. Their isotopic heterogeneity documents the heterogeneity of the MORB source (Zindler *et al.*, 1984; Graham *et al.*, 1988; Hekinian *et al.*, 1989) and also shows that seamounts are fed by small magma batches that rise with only minor chemical modification. Petrographically, seamount MORB is identical to axial MORB (Batiza and Vanko, 1984; Allan *et al.*, 1987, 1988, 1989), although they can be distinguished primarily on the basis of plagioclase (Batiza and Vanko, 1984) and spinel composition (Allan *et al.*, 1988). Some seamounts, especially those of relative motion-parallel chains, apparently evolved from a depleted N-MORB early stage to a late alkalic stage similar to Hawaiian volcanoes (Batiza *et al.*, 1990a).

The extreme depletion of some seamount magmas, especially those of chains parallel to absolute motion is perplexing. Whereas the orientation of such chains suggests that hot-spots may be involved (Karsten *et al.*, 1989; Fornari *et al.*, 1988), their small size, the absence of a topographic swell and their chemistry argue against large plumes. Melting of chemical heterogeneities has been invoked (Davis and Karsten, 1986; Fornari *et al.*, 1988); however, this hypothesis predicts more enriched basalts rather than more depleted rocks. Enriched basalts occur on near-axis seamounts, but usually they are from seamounts that belong to chains parallel to the relative motion (Batiza and Vanko, 1984). They could arise from more melting of the rising mantle material which provides MORB melt to the axis. Alternatively, they may be linked to small-scale upper mantle convection near active ridges (Haxby and Wiessel, 1986; Buck and Parmentier, 1986).

11.9 Oceanic plateaux

The Pacific basin contains numerous large, high standing plateau including the Shatsky Rise, Hess Rise, Ontong–Java Plateau, Magellan Rise and the Manihiki Plateau (Figure 11.14). They represent areas of thickened crust although their origin and composition has been debated for several decades (Nur and Ben-Avraham, 1982; Floyd, 1989). Their surfaces are mostly covered with sediment, therefore their petrological characteristics are known almost exclusively from drilled samples via DSDP and ODP operations.

Volcanic samples from these plateaux and correlative sill sheet flow complexes (Nauru basin; Castillo *et al.*, 1986; Floyd, 1986, 1989) are mostly basalts chemically similar to MORB although some alkalic differentiates have been recovered from Hess Rise (Vallier *et al.*, 1983) (Table 11.8). Isotopically, however, these volcanic rocks have affinities with ocean island basalts rather than MORB (Mahoney, 1987; Castillo *et al.*, 1990). Castillo *et al.* (1986, 1990)

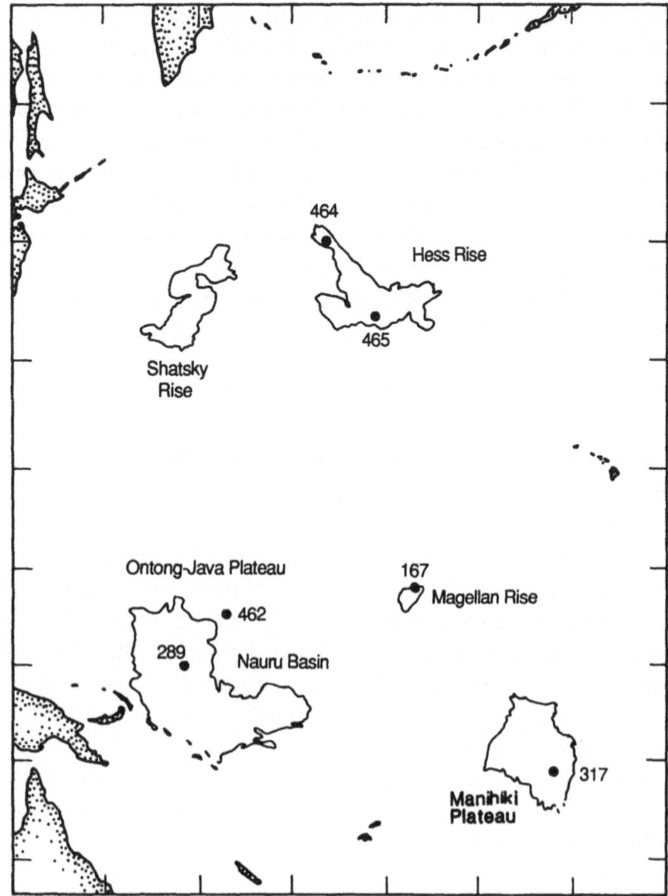

Figure 11.14 Map of the western Pacific showing the location of oceanic plateaux, after Mahoney (1987).

have shown that the Nauru Basin sill-sheet flow complex, which may be part of the Ontong–Java plateau, is stratigraphically zoned, with lavas derived from high extents of partial melting sandwiched between lava series derived from lower extents of melting. These data suggest that the large plateaux were probably built by near-axis hot-spots (Mahoney, 1989), possibly near ridge triple junctions or in the vicinity of active microplates.

11.10 Concluding statements

1. In the Pacific Ocean, the petrological and geochemical characteristics of the crust are a strong function of the tectonic environment in which the

Table 11.8 Selected analysis of Pacific hot-spot volcanics. Major oxides in wt%, trace elements in ppm

	Sample							
Analysis	34-4 65-67[a]	33-3 94-96[a]	94-3 78-84[b]	464 Basalt[c]	465 Trachyte[c]	132-3[a]	51-4 17-22[e]	102-5 19-21[f]
SiO_2	49.92	47.88	47.42	46.17	59.50	48.3	48.9	48.4
TiO_2	1.06	0.89	1.79	1.74	1.03	1.5	0.99	1.21
Al_2O_3	15.07	14.62	15.64	15.92	18.73	14.7	13.91	13.9
FeO*	9.67	9.54	9.49	12.55	2.57	12.07	11.23	12.23
MnO	0.18	0.13	0.16	0.27	0.04	0.19	—	0.23
MgO	7.29	8.41	6.92	7.86	1.00	6.9	7.81	7.08
CaO	11.47	10.85	9.43	4.32	2.44	11.6	11.72	11.8
Na_2O	2.19	2.13	3.84	3.23	5.10	2.2	2.06	2.18
K_2O	0.10	0.12	0.46	1.25	4.88	0.45	0.12	0.06
P_2O_5	0.13	0.12	0.20	0.09	0.36	0.12	—	0.11
Total	97.08 (100.22)[g]	94.69 (100.14)[g]	95.35 (101.50)[g]	93.40 (101.63)[g]	95.65 (99.82)[g]	98.03 (100.13)[g]	96.74	97.20 (98.70)[g]
Mg number	62	65	61	57	45	55	60	55
La	—	—	—	4.99	82.4	—	2.54	3.23
Sm	2.68[h]	2.01[h]	3.28[h]	3.22	10.60	3.27[h]	1.92	2.6
Yb	2.1	2.0	—	2.60	2.97	3	2.09	2.77
Cr	350	377	—	—	—	300	404	151
Sr	115.9[h]	99.3[h]	174[h]	167	306	131.7[h]	91.7[h]	—
Hf	1.6	1.1	—	3.16	16.90	—	1.50	2.00
Zr	83	66	109	111	718	70	—	—

[a] Manihiki Plateau, Site 317A (DSDP Leg 33) from Jackson *et al.* (1976)
[b] Magellan Rise, Site 167 (DSDP Leg 17) from Bass *et al.* (1973)
[c] Average analyses from Vallier *et al.* (1983), Hess Rise
[d] Ontong–Java Plateau, Site 289 (DSDP Leg 30) from Stoeser (1975)
[e] Nauru Basin Complex, Site 462A (DSDP Leg 61) from Batiza (1981)
[f] Nauru Basin Complex, Site 462A (DSDP Leg 89) from Castillo *et al.* (1986)
[g] Sum of original analysis
[h] Data for same flow unit from Mahoney (1987)

crust was initially created. In general, Pacific crust differs in significant ways from crust produced at slower spreading rates in the Atlantic and Indian Oceans. This is probably due fundamentally to the very large size and rapid spreading origin of the present Pacific plate and the ancestral oceanic plates of the Pacific Ocean. In many parts of the Pacific, additional modifications by off-axis intrusion and volcanism are also important. At actively spreading ridges such as the EPR, both relatively depleted and enriched MORB are erupted. In many instances, regular patterns of along-axis chemical variation are observed. Various kinds of stationary and migrating offsets along Pacific ridges (for example, propagating rifts) are commonly the loci of particular petrological and geochemical effects. These effects include apparent differences in the depth and extent of partial melting, crystal fractionation and possibly mantle source complexities in the vicinity of offsets.

2. Pacific MORB is, on average, more evolved than Atlantic MORB; the mean magnesium number of Pacific MORB is about 58, whereas Atlantic MORB is about 62. This small, but significant, difference is probably due to the greater mean cooling of Pacific MORB prior to eruption. In turn, this cooling could result from longer mean residence times in the crust, probably in the magma chambers. Despite the more primitive nature of Atlantic MORB, these tend to exhibit better development of well formed euhedral clinopyroxene crystals than Pacific MORB. Pacific and Atlantic MORB also differ isotopically, indicating the importance of regional and global mantle reservoirs and perhaps slight differences in the melt generation processes operating at slow- and fast-spreading rates.
3. Small seamounts are common near Pacific spreading centres. Their petrographic and chemical characteristics are similar to axial MORB, although they are systematically more primitive and more diverse. Near-axis seamounts, especially members of chains parallel to relative motion, may also contain small volumes of alkalic basalt similar to OIB.
4. Large island and seamount chains of hot-spot origin abound in the Pacific, as do large oceanic plateaux. Oceanic plateau lavas are typically depleted and MORB-like petrographically and geochemically, although isotopically they are more radiogenic than normal, non-plume, MORB. Large seamounts and volcanic islands may decorate plateaux or form chains and clusters. Their lavas range from tholeiitic basalts, typically with radiogenic isotope signatures, to more typical alkalic OIB and strongly undersaturated nephelinitic lavas and their differentiates.

12 Indian ocean crust

JAMES NATLAND

12.1 Introduction

In this chapter studies on Indian Ocean abyssal basalts are used as a vehicle to discuss magmatic lineages and melting processes, largely derived from detailed petrographic observations and mineral chemistry. The mineralogical aspects and the magmatic processes envisaged can be considered as adjunct topics to Chapters 5 and 7, respectively, in Part II. The first section considers the consequences of the existence of magmatic lineages, which are particularly well developed in abyssal basalts and related gabbroic assemblages from the Indian Ocean crust.

Magmatic lineages may be defined as sequences of basalts and successive differentiates following chemically and mineralogically distinctive fractionation pathways. The differentiates inherit or even enhance the distinctive attributes of parental basalts. The concept was developed from the study of alkaline lava suites (Coombs and Wilkinson, 1969) and is basically a refinement of the idea of petrographic provinces as discussed, for example, by Bowen (1928).

Abyssal tholeiites have often been considered to represent a single petrographic province in which one parental lava type predominates, or in which there are at best only a few, similar, parental types produced over a small range of physical conditions in the mantle. The differentiation of such a limited range of parental liquids in crustal magma chambers produces essentially the same liquid line of descent everywhere, amounting to a single magmatic lineage in the ocean basins. This approach owes much to the original impression of the chemical monotony of these basalts and has been eloquently argued from the experimental perspective (Presnall *et al.*, 1979; Presnall and Hoover, 1984, 1986, 1987; Fujii and Scarfe, 1985; Fujii, 1989). These authors suggest that the majority of abyssal tholeiites have non-picritic parents produced at fairly shallow depths, corresponding to 7–12 kbar astride the transition zone between the plagioclase lherzolite and spinel lherzolite facies in the mantle (see also Chapter 6).

Others, however, argue that the diversity of the so-called 'MORB array' (Figure 12.1) requires a variety of parental magmas produced over a wider

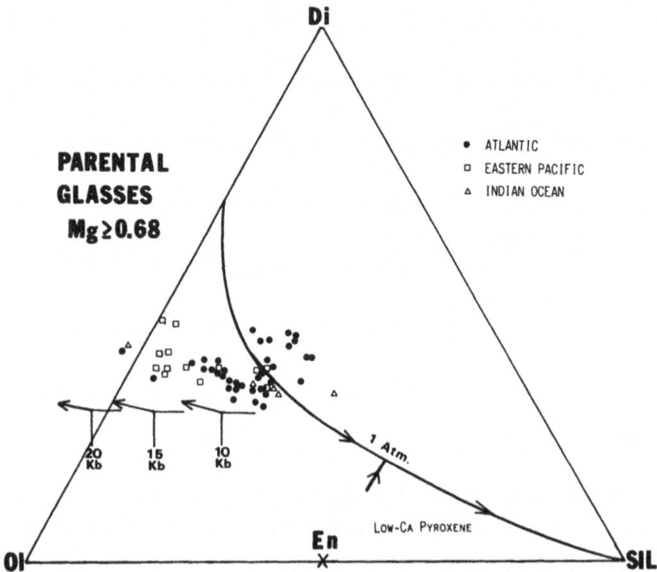

Figure 12.1 Proportions of normative olivine (Ol), diopside (Di) and SiO_2 (SIL) for abyssal-tholeiite glasses with magnesium number >0.68 taken from the literature, calculated using the procedure of Walker *et al.* (1979). Their 1 atm experimental cotectic is shown. High-pressure pseudo-invariant points are from Stolper (1980).

range of conditions, particularly depths, in the mantle. Most experimentalists agree that, for any abyssal basalt to be treated as primary in composition, it must once have been in equilibrium with both olivine and orthopyroxene in the mantle at some depth. Stolper (1980) documented a steady shift of the 1 atm pseudo-invariant point involving olivine, clinopyroxene and low Ca-pyroxene towards the olivine apex with higher pressure (Figure 12.1), in experiments equilibrating peridotite and a magnesian abyssal tholeiite. The shift of the pseudo-invariant point in the experiments outlines a range of melt compositions in equilibrium with appropriate mantle mineral assemblages at different pressures. These could potentially produce the range of parents for the MORB array, but for most of the compositions a substantial amount of olivine-dominated polybaric fractionation is required to reach the 1 atm cotectic. On this basis, Stolper (1980) argued that many primary abyssal tholeiites are picritic in composition (cf. O'Hara, 1968), even if few picrites actually reach the seafloor. Some experimental petrologists are strongly persuaded that the majority of abyssal tholeiites derive from high pressure (>20 kbar) parental picritic basalt (Elthon, 1989). The view that primary abyssal tholeiites are produced over a range of conditions in the mantle suggests that there may be more than one magmatic lineage along spreading ridges in the ocean basins (Chapter 6).

This viewpoint is clouded by uncertainties in methods of projection such as that used for Figure 12.1. The most serious of these, according to Klein and Langmuir (1987), is that the proportion of silica (SIL: the parameter most sensitive to pressure) is strongly dependent on Na_2O content, which is highly weighted in the projection algorithm (Walker *et al.*, 1979). The Na_2O content can vary significantly in parental liquids because of differences in the degree of partial melting (Fujii and Scarfe, 1985), which appears to vary regionally in the ocean basins (Klein and Langmuir, 1987; McKenzie and Bickle, 1988). This regional variability in itself suggests that magmatic lineages should indeed exist within the abyssal tholeiite suite, such that the sequence from the Galapagos Rift, for example, is only one such lineage (Chapter 5). However, relationships to depths of melting based on normative criteria, or other aspects of the physical conditions of melting, are still uncertain.

Dredge stations and drill sites in several ocean basins provide the evidence for the chemically and mineralogically distinctive magnesian abyssal tholeiites, with compositions little modified by fractionation from those arriving from the mantle, which erupt from spreading ridges in different places. Three general magmatic lineages are described using examples from Indian Ocean spreading ridges. It must be emphasized that these represent a continuum of chemically gradational basalt types, and the predominance of one type of basalt and its differentiates at some spreading ridge does not preclude the presence of other types at the ridge, or their participation in mixing processes in the immediately subjacent lower crust and upper mantle. Consideration of phenocryst assemblages leaves no doubt that there are consistently complex processes of magma coalescence in the mantle, and magma chamber mixing and shallow differentiation in the crystallization history of almost every porphyritic abyssal tholeiite.

12.2 Magmatic lineages of abyssal tholeiites in the Indian Ocean

The widest diversity of magnesian, near-parental, abyssal tholeiites exists in the Indian Ocean. Table 12.1 lists basalts glass compositions from three geographical provinces in the Indian Ocean (Figure 12.2). Each group includes the most magnesian glass in each province and one or more moderately fractionated glasses. Highly fractionated ferrobasalts, andesites and rhyodacites such as those found in the eastern Pacific (Chapter 5) are not found along these slowly spreading ridges in the Indian Ocean, probably because of the infrequency of eruptive events and the lack of persistent, recurrently replenished magma chambers at such ridges (Nisbet and Fowler, 1978; Natland, 1980).

Three types of basalts are distinguished, based on Na_2O and TiO_2 contents. Type 1 samples, from Eocene–Cretaceous Deep Sea Drilling Project (DSDP) sites, have very low Na_2O and TiO_2 contents at any given value of magnesium number (defined in Table 12.1). This is the least sodic group of samples found so far in all the major ocean basins. Three of the samples occur in the

Table 12.1 Selected glass analyses from Indian Ocean ridges. From Natland *et al.* (in press). Special calibrations for high precision MnO, K_2O and P_2O_5

	Sample[a]														
	Type 1				Type 2						Type 3				
Analysis	DSDP 212 (1)	DSDP 236 (1)	DSDP 221 (1)	DSDP 220 (1)	CIRCE 109-1 (3)	ANTP 128-2 (3)	CIRCE 116-1 (1)	ANTP 97-1 (6)	ANTP 89-2 (3)	ANTP 92-3 (2)	AII93-5 6-1 (19)	CIRCE 87-1 (1)	ANTP 111-1 (1)	CIRCE 110-4 (1)	ANTP 114-1 (1)
SiO_2	51.53	53.36	52.25	52.41	51.11	51.64	51.87	51.11	51.95	52.19	50.80	50.69	51.38	50.69	51.71
TiO_2	0.61	0.61	1.12	1.25	1.07	1.11	1.18	1.52	1.44	12.24	1.23	1.28	1.72	1.84	1.87
Al_2O_3	15.60	14.43	13.85	13.80	15.96	15.88	14.86	15.16	14.83	14.66	17.27	17.32	17.04	16.24	15.70
FeO*	7.86	9.93	12.22	12.91	8.26	8.55	8.96	9.90	10.52	10.98	7.83	7.53	8.40	9.07	9.64
MnO	0.153	0.192	0.244	0.225	—	0.176	0.151	0.150	0.194	—	0.153	—	0.184	—	—
MgO	8.97	7.80	6.66	6.07	8.11	8.13	7.64	7.20	6.95	5.21	8.53	8.25	7.27	7.08	6.28
CaO	13.48	12.38	11.11	10.48	12.37	11.86	12.11	11.12	10.92	8.95	10.15	10.25	9.62	9.64	10.02
Na_2O	1.42	1.65	2.17	2.19	2.44	2.55	2.48	2.63	2.65	3.43	3.64	3.71	4.34	3.95	3.78
K_2O	0.044	0.052	0.034	0.263	0.11	0.092	0.042	0.215	0.160	0.58	0.208	0.19	0.205	0.24	0.28
P_2O_5	0.056	0.054	0.143	0.152	0.12	0.148	0.087	0.150	0.161	0.27	0.153	0.15	0.197	0.20	0.24
Total	99.52	100.48	99.85	99.75	99.55	100.14	99.38	99.16	99.78	98.51	99.99	99.37	100.37	98.95	99.52
Magnesium number[b]	0.703	0.619	0.530	0.493	10.671	0.663	0.638	0.601	0.578	0.496	0.693	0.695	0.642	0.618	0.574
Latitude	19°11′S	1°45′S	7°58′S	6°31′N	24°53′S	13°01′S	23°31′S	18°43′S	17°38′S	17°10′S	27°36′S	9°03′S	28°51′S	26°37′S	29°18′S
Longitude	99°18′E	57°39′E	68°25′E	70°59′E	65°59′E	66°19′E	65°43′E	65°30′E	65°49′E	66°38′E	65°50′E	67°21′E	61°56′E	67°32′E	60°37′E
Depth (m)	6240	4489	4650	4036	4650	1885	3990	3330	5740	5245	4950	6060	5185	5100	4695
Location[c]	NE IO	NW IO	NW IO	NW IO	CIR	CIR	CIR	CIR	MCFZ CIR	MCFZ CIR	SWIR	VFZ CIR	SWIR	SWIR	MFZ SWIR
Crustal age (Ma)	~ 110	~ 60	~ 46	~ 51	—	—	—	—	—	—	—	—	—	—	—

[a]Number of spots or glass chips analysed in parentheses
[b]Calculated assuming $Fe^{2+}/(Fe^{2+}+Fe^{3+}) = 0.86$ (Presnall *et al.* (1979))
[c](IO) Indian Ocean; (CIR) Central Indian Ridge; (MCFZ) Marie Celeste Fracture Zone; (VFZ) Vema Fracture Zone; (SWIR) South-west Indian Ridge; (MFZ) Melville Fracture Zone.

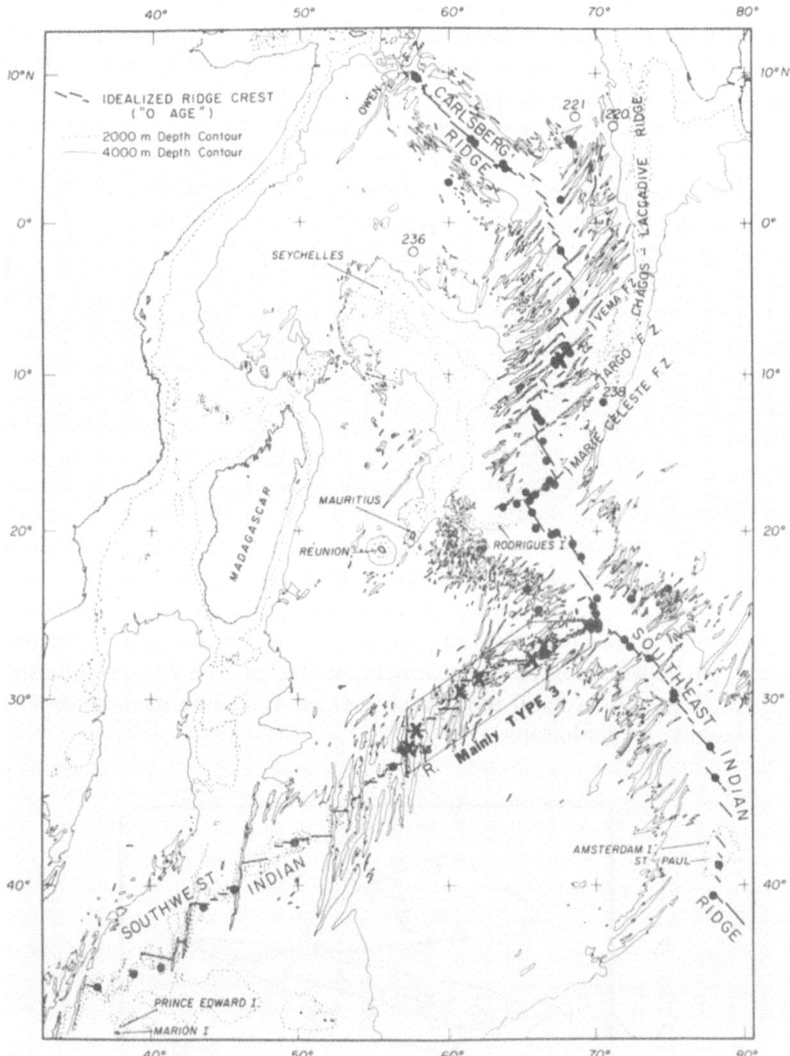

Figure 12.2 Distribution of basalt Types 1 (open circles), 2 (filled circles) and 3 (Xs) along and near spreading ridges and fracture zones in the western Indian Ocean. The base map is from Mahoney *et al.* (1989). The contour interval is 1000 m. Data are from Natland *et al.* (in press) and other sources.

north-west Indian Ocean, and the fourth, from much older crust in the eastern Indian Ocean, is grouped with them because of its similar distinctive composition.

Type 2 samples, from the Central and South-east Indian Ridges, are more typical of abyssal tholeiites worldwide in terms of Na_2O and TiO_2 contents;

those in Table 12.1 are representative of several dozen basalt glass types analysed from this part of the Indian Ocean Ridge system (Natland *et al.*, in press). Magnesian Type 2 basalts most resemble the primitive abyssal tholeiites studied experimentally.

Type 3 samples, from the very slowly spreading South-west Indian ridge near its triple junction intersection with the Central and South-east Indian Ridges, are the most sodic and titaniferous glasses sampled to date from spreading ridges. They strongly resemble sodic suites from tectonically isolated or geographically restricted locations such as the Cayman Trough in the Caribbean (Thompson *et al.*, 1980) and the Australian–Antarctic discordance (Klein *et al.*, 1988). Along the South-west Indian Ridge, the distribution of sodic basalts is abruptly terminated in the east by the triple junction shown in Figure 12.2. To the west, Type 3 basalts give way gradationally to Type 2 basalts at about the Atlantis II Fracture Zone, about 200 km from the triple junction (Natland *et al.*, in press).

All three basalt suites are abyssal tholeiites in that they have low abundances of K, Rb, Ba and Zr, in addition to depleted to flat rare earth element (REE) patterns (Frey *et al.*, 1980; Price *et al.*, 1986; Mahoney *et al.*, 1989). Although some of the basalts are mildly enriched isotopically, and all are distinct on the basis of Pb isotopes from basalts in the north Atlantic and eastern Pacific Oceans (Mahoney *et al.*, 1989), there is no systematic relationship between degree of enrichment and basalt type. Most significantly the Type 3 basalts are still moderately depleted, despite having Na_2O as high as many alkalic basalts.

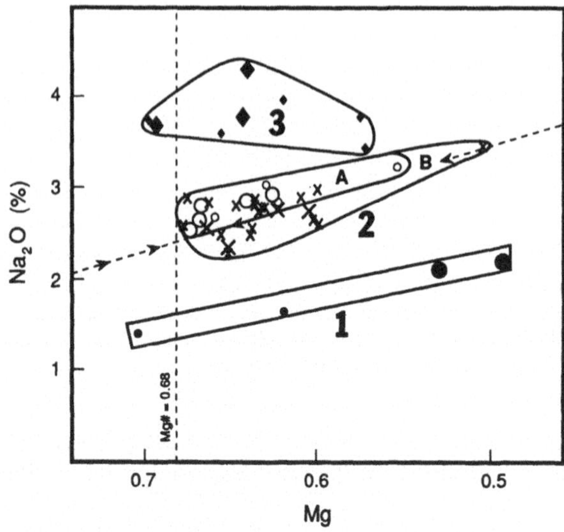

Figure 12.3 Na_2O versus magnesium number (defined in Table 12.1) for Indian Ocean Type 1, Type 2, and Type 3 basalt glasses. Data are from Natland *et al.* (in press). Bold symbols are samples analysed for Sr and Nd isotopes (Mahoney *et al.*, 1989).

The link between high-magnesium number glasses and more evolved compositions in the different Indian Ocean provinces is shown in terms of Na_2O contents in Figure 12.3. The differences in basalt compositions are clearly gradational, however characteristic the three suites in particular portions of the ridges may be. Moreover, a few basalts on the Central and South-west Indian Ridges are individually anomalous, being either more or less sodic than the other samples from the same ridges, and even within the same dredge hauls.

The glasses also clearly illustrate the complicating influence of Na_2O contents on normative projections (Figure 12.4). The highly sodic Type 3 glasses fall near the Ol-Di sideline, whereas the least sodic Type 1 glasses have high proportions of SIL throughout the same range in magnesium number. Type 2 basalt glasses fall in between. Within each group fractionation produces residual compositions plotting along trends which generally parallel the 1 atm cotectic, although some Type 2 basalts and most Type 3 basalts plot towards the Ol apex.

The three chemically distinct Indian Ocean basalt suites (Types 1, 2 and 3) differ mineralogically and thus qualify as separate magmatic lineages. The mineralogical contrast most clearly related to different Na_2O contents is that of the An content of plagioclase. At a given magnesium number, which can

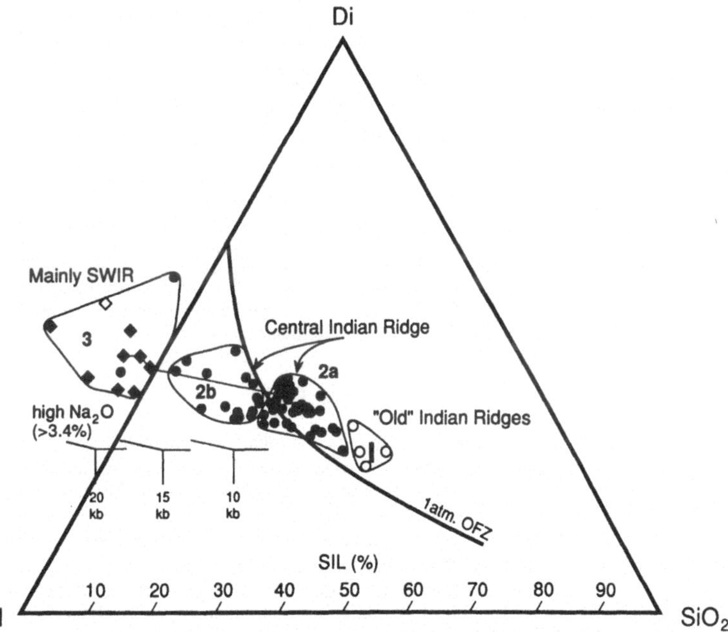

Figure 12.4 Proportions of normative Ol, Di, and SiO_2 for Indian Ocean Type 1, Type 2, and Type 3 basalt glasses. The 1 atm cotectic and high-pressure pseudo-invariant points are as in Figure 12.1.

be considered as an index of differentiation, small tabular plagioclases in glasses are systematically more albitic in the sequence of suites Type 1 to Type 2 to Type 3. As olivine compositions are simply related to liquid magnesium number (Roeder and Emslie, 1970) this means that plagioclases intergrown with olivines of a given Fo content are systematically more sodic

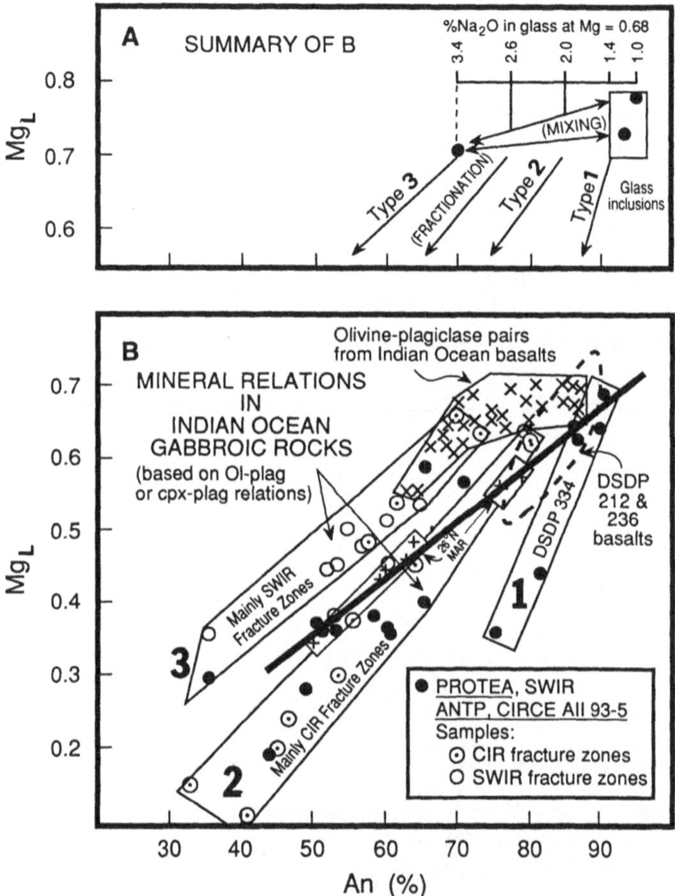

Figure 12.5 Calculated liquid magnesium numbers of liquids based on olivine and clinopyroxene compositions versus An contents of plagioclases coexisting in the same glomerocrysts in basalts (x) and intergrown minerals in gabbros (Bloomer *et al.*, 1989 and unpublished). Magnesium number calculations are based on Roeder and Emslie (1970) for olivines, and a modification of an equation by Duke (1976) for clinopyroxenes (see Chapter 5 for explanation). The top panel (A) shows general trends anticipated for Type 1, Type 2, and Type 3 basalts based on their parental Na_2O contents (Figure 12.3) and trends for plutonic rocks projected by Longhi (1982). The lower panel (B) also includes points (+) based on mineral compositions from Mid-Atlantic Ridge gabbros at 37°N (DSDP Site 334; Hodges and Papike, 1976) and 26°N (Tiezzi and Scott, 1980). Site 334 gabbros are inferred to have crystallized from Type 1 liquids (see text for further discussion).

in the same order (Figure 12.5A). Types 2 and 3 parental characteristics are also inherited by the highly evolved gabbroic rocks of the Central and South-west Indian Ridge, respectively (Figure 12.5B). There are no gabbros representing Type 1 parents sampled from the Indian Ocean, but gabbros drilled at DSDP Site 334 in the north Atlantic Ocean have the appropriate highly calcic plagioclases (Figure 12.5B; Hodges and Papike, 1976) and were obtained in the same DSDP drill hole as basalts only slightly more titaniferous and sodic than Indian Ocean Type 1 basalts (Aumento *et al.*, 1976). These rocks provide the best available plutonic equivalent for mineralogical comparisons to Type 1 basalts.

The three low, intermediate and high soda trends in the gabbro suites follow pathways consistent with the shallow fractionation of variably sodic parents (Longhi, 1982). The suggestion that Soret diffusion in thermally stratified magma bodies influences evolved abyssal gabbro compositions (Walker and Delong, 1982) was erroneously based on the assumption that north Atlantic gabbros with Type 1 and Type 2 characteristics belong to a single lineage (bold line in Figure 12.5b).

The three basalt suites in the Indian Ocean also differ in the occurrence and compositions of ferromagnesian silicates. The more magnesian Type 3 sodic basalts have olivine and Cr-spinel only, joined by fairly sodic plagioclase (An_{65}) on the liquidus at about magnesium number 0.65. The intermediate Type 2 basalts of the Central and South-east Indian Ridge invariably have olivine and plagioclase on the liquidus (An_{75-80}) regardless of magnesium number, together with Cr-spinel in the more magnesian samples. The most magnesian Type 1 low Na_2O basalt (from DSDP Site 212; Table 12.1) has olivine, extremely calcic plagioclase (An_{89-90}), Cr-spinel, and magnesian clinopyroxene occurring both as individual crystal in the glass, and in large glomerocrystic aggregates. The less magnesian Type 1 sample from DSDP Site 236 has phenocrysts and glomerocrysts with more iron-rich olivine and clinopyroxene, no spinel, more sodic (but still calcic) plagioclase (An_{85}) and magnesian orthopyroxene, specifically bronzite (Bloomer *et al.*, 1989). This is one of the very few abyssal tholeiites reported with two magnesian (high temperature) pyroxenes, the others also being from low Na_2O provinces in the north Atlantic (Sigurdsson and Brown, 1970; Sigurdsson, 1981) and the eastern Pacific (Perfit and Fornari, 1983).

The DSDP Site 334 gabbros also have crystallization sequences and ferromagnesian mineral compositions consistent with derivation from such basalts. In fact, the mineral compositions in these cumulates are virtually identical to those in the glomerocryst assemblages of the Type 1 basalts from the Indian Ocean. In addition to the similarities in olivine and plagioclase already mentioned, the gabbros contain low Ti Cr-spinel, magnesian clinopyroxene and intercumulus bronzite (Hodges and Papike, 1976). In Type 1 basalts, such minerals clearly crystallized at low pressure and therefore it is at least plausible to consider that magnesian pyroxenes in any abyssal gabbro crystallized at low pressure from similar liquids (Bloomer *et al.*, 1989), rather than at several kilobars pressure from Type 2 or Type 3 liquids, as argued by Elthon and Casey (1985) and Elthon (1984, 1989). In the actual gabbro suites in the Indian Ocean from Type 2 and Type 3 provinces, highly magnesian pyroxenes are not present; low Ca pyroxenes only crystallize at later stages of differentiation and they have a fairly iron-rich composition. They are usually pigeonite or hypersthene recrystallized from pigeonite, and are crystallized from liquids with magnesium numbers from 0.55 to 0.35 (Bloomer *et al.*, 1989; Natland *et al.*, in press).

12.3 Depths of partial melting

Experiments on magnesian abyssal tholeiites over ranges of pressure usually have been designed to determine the pressure at which the crystallization interval for silicate phases is at a minimum (Kushiro and Thompson, 1972; Bender *et al.*, 1978; Fujii and Kushiro, 1977; Fujii and Bougault, 1983). Typically this pressure is around 10 kbar, which is about the maximum pressure of plagioclase stability in the same liquids. The 10 kbar pressure is often interpreted to represent the last depth of liquid equilibration with mantle minerals and hence gives the depth of segregation of parental magmas. Presnall *et al.* (1979) cited these results in support of their hypothesis that parental magmas most commonly derive from depths corresponding to the transition between plagioclase lherzolite and spinel lherzolite in the mantle. Later, Presnall and Hoover (1984) determined that the addition of Na_2O to the system studied would spread out the transition interval between the mantle facies to a range of 7–12 kbar.

This hypothesis obviously cannot apply to magnesian Type 1 basalts with all silicate phases (and spinel) on the liquidus at 1 atm and high temperature (> 1220°C), and with bronzite joining the liquids at only a slightly lower temperature (1200°C). Such basalts have not been studied experimentally, probably because olivine-rich basalts with average Na_2O contents have seemed more appropriate parental compositions for most abyssal tholeiites. However, the magnesian Type 1 Indian Ocean basalts clearly exhibit an extremely low pressure convergence of silicate phase boundaries, and, by analogy, derive from parents which originated at shallow depths in the mantle (Fisk, 1982). The early crystallization of bronzite in particular indicates that the basalts at the seafloor are not far removed from a condition of equilibrium with olivine and orthopyroxene in the mantle (Figure 12.1).

The unusually high SiO_2 contents of the Type 1 basalts also suggests shallow depths of origin (2–5 kbar) based on the experimental results of Jaques and Green (1980). Elthon (1989) also considered > 50% SiO_2 in primitive basalts as an indication of shallow depths of partial melting, but suggested that such rocks are a minority among abyssal tholeiites, at least in the north

Atlantic and eastern Pacific, which have been comparatively well studied. However, almost all Indian Ocean basalt glasses, whether of Type 1, Type 2, or Type 3 characteristics, have SiO_2 contents in the range 50–52%, including the most magnesian glasses (Table 12.1). On this basis, Indian Ocean ridges appear regionally to reflect shallower depths of partial melting than the Mid-Atlantic Ridge and East Pacific Rise.

Within the Indian Ocean, and on a relative scale, average depths of partial melting of the parental magmas appear to increase in the sequence Type 1 to Type 2 to Type 3, as reflected in widening intervals of silicate mineral crystallization during differentiation. More specific depths of melting cannot be inferred from crystallization sequences because magnesian abyssal tholeiites from the Indian Ocean have not been studied experimentally, nor have magnesian basalts resembling either Type 1 or Type 3 parental compositions from any other ocean. The sequence is, possibly fortuitously, that predicted by analogy to the experimental data of Stolper (1980) (Figure 12.1), but the range of pressures suggested is probably considerably exaggerated because of the effect of Na_2O.

Based on the contrasting near-liquidus mineral assemblages and compositions in the three Indian Ocean basalt groups, approximate sequences of cumulates can be predicted for the ranges of compositions encompassed by the glasses. Type 3 (sodic) parental liquids, with the widest low pressure interval between the onset of olivine and plagioclase crystallization, would produce dunites, possibly chromitites, and troctolites as cumulates at high temperatures, then olivine gabbros and gabbro-norites. With Type 2 basalts, which have olivine, spinel and plagioclase on the liquidus at high magnesium numbers, dunites and chromitites would be rare, possibly not present at all, and high temperature cumulates would be dominated by troctolites and olivine gabbros. Type 1 parental liquids, which are multiply saturated in all silicate phases at high temperatures, would not produce even troctolites. Olivine gabbros and magnesian gabbro-norites would be the principal high temperature cumulates at low pressure. None of these parental liquids would produce wehrlite, bronzitite, or pyroxenite cumulates at low pressure because plagioclase precedes pyroxenes in the crystallization sequences of all three types. This is borne out by the absence of all these rocks among the abundant ultramafic rocks in dredge hauls from Indian Ocean fracture zones (Fisher *et al.*, 1986; Dick, 1989).

12.4 The mantle melting column

A general concept which has emerged in the past few years is that partial melting beneath spreading ridges occurs over a range of depths following the intersection of the lherzolite solidus by peridotite diapirs ascending adiabatically from the deeper mantle (Oxburgh, 1980; Chapter 7). In principle, melts should continue to aggregate in buoyant bodies of mantle rock as they ascend, but as high magnesium number melts are less dense and more buoyant than peridotites they will tend to concentrate in some places and abandon others as partial melting proceeds. The concentration of buoyancy forces ultimately initiates the propagation of fractures through the overlying mantle into magma chambers at the base of the crust. At this point mantle melts can stream along the fractures and into the crust (Nicolas, 1986).

One consequence of this process is that rigorous statements about depths of melting probably cannot be made based on the comparison of basalt liquid compositions with equilibrium experiments at elevated pressures which simulate batch partial melting. Such a process appears to be impossible based on evidence for U–Th disequilibrium during the melting of parental abyssal tholeiites (McKenzie, 1985), and the trace element compositions of residual clinopyroxenes in abyssal peridotites (Johnson *et al.*, 1990). Incremental partial melting, with the near instantaneous escape of the melt increments to some region of melt coalescence in the upper mantle or crust, is the only process compatible with these results, and is consistent with estimates of the fluid dynamic properties of peridotite undergoing partial melting (McKenzie, 1984).

Klein and Langmuir (1987) described partial melting as occurring in columns of mantle material, perhaps individually ascending beneath separate segments of spreading ridges defined by structural discontinuities (Whitehead *et al.*, 1984). Assuming a homogeneous mantle, the calculations of Klein and Langmuir (1987) show that high temperature gradients, such as those which exist at or near a hot-spot, tend to produce low soda magmas to the crust, exemplifying high degrees of partial melting. The mantle solidus in this situation is intersected by rising bodies of mantle material at comparatively great depth, and the average depth of partial melting, integrated over the length of the total melt column, is relatively deep. Under low geothermal gradients, the solidus is intersected at fairly shallow depths, the integrated degree of partial melting is not so extreme, melts are sodic, and the average depths of partial melting are less. On this basis, for the Indian Ocean, using parental soda compositions alone as the criterion, average depths and degrees of partial melting should increase in the sequence Type 3 to Type 2 to Type 1 basalts. This is the opposite to that implied by the crystallization sequences in the basalts.

McKenzie and Bickle (1988) recently endorsed the concept of the mantle melting column and argued for the necessity of shallow regions of mantle (within the plagioclase lherzolite stability field) contributing melt fractions to aggregating magmas, even if these arise from near the base of the melting column within the garnet stability mantle domain. The concept of depth of partial melting, as applied to any parental basalt composition by analogy to experimental petrology must therefore be construed from compositions that are in effect weighted averages, aggregated from melt fractions produced over a considerable range of depth. For individual basalts these might be strongly skewed to the compositions of predominant melt strains aggregated at specific storage levels in the mantle. For this reason, regional averages based on several basalt compositions are more indicative of the nature of the underlying melt column than an individual analysis.

An important question is how this type of process can actually give rise to basalts which are unquestionably consistent and uniform along particular segments of spreading ridges. The problem is more extreme for the minor and trace element abundances than for bulk compositions, as incremental partial melting is a very efficient mechanism for removing incompatible elements from source rocks at a very early stage. Somehow, melt strains must come back together to produce whatever level of consistency exists for incompatible element abundances and major oxide compositions along spreading ridges.

The compositions of phenocrysts and their glass inclusions in primitive porphyritic abyssal tholeiites provide evidence for the required processes of coalescence and homogenization in the mantle and in crustal magma chambers, which are discussed in the following section.

12.5 Mixing of parental magma stems

Many porphyritic abyssal tholeiites contain megacrysts of very calcic plagioclase, associated olivines, and usually either Cr-spinels or magnesian clinopyroxenes. These have been interpreted to represent (1) the early crystallization products of primitive basalts in crustal magma chambers (Muir and Tilley, 1964; Bryan and Moore, 1977), (2) xenocrystic fragments of the upper mantle (Donaldson and Brown, 1977; Wilkinson, 1982), and (3) crystallization of primitive magmas over a range of pressures (Donaldson and Brown, 1977; Sigurdsson and Schilling 1976; Natland *et al.*, 1983; Natland, 1989). The author considers the second viewpoint the least likely, although the arguments of Wilkinson (1982) on its behalf acutally come closest to stating the exotic nature of these minerals. He simply pointed out that the mineral compositions are inappropriate for them to have crystallized from common magnesian abyssal tholeiites, and noted the correspondence of the compositions to minerals in some dredged abyssal peridotites. The plagioclases, for example, are far too calcic, and the bright emerald green clinopyroxenes are too magnesian, with too little Na and Ti, to have crystallized from a plausible parental basalt (magnesium number 0.7) with about 1% TiO_2 and 2% Na_2O (i.e. a Type 2 parental liquid).

Petrographic evidence, however, indicates that glomerocrysts of these minerals, involving two or more phases, crystallized from melts (Natland *et al.*, 1983). The common occurrence of basaltic melt inclusions, especially in the calcic plagioclases, is incontrovertable proof that the minerals are magmatic (Natland, 1989). However, these melt inclusions do not have parental compositions related by crystal fractionation to the host glasses. Instead, the glass inclusions and the minerals containing them are evidence for the ubiquitous presence of extremely refractory (low TiO_2, low Na_2O) basaltic liquids in the make-up of almost every porphyritic abyssal tholeiite. This was first suggested on the basis of the study of basalts from DSDP Site 504, Costa Rica Rift (Autio and Rhodes, 1983), but the extremely refractory character of the liquids that produce many megacrysts is only now beginning to be understood. Integrating mineral data with the compositions of glass inclusions, it is found that the typical basaltic melts from which such megacrysts crystallize have magnesium numbers 0.74–0.63, TiO_2 0.2–0.7%, Na_2O 0.9–1.5% and CaO/Na_2O 7–10 (Fisk, 1984; Price *et al.*, 1986; Elthon and Casey, 1985; Natland, 1989; Meyer and Shibata, 1989). Comparison with data in Table 12.1 shows that these are the characteristics of the more magnesian Indian Ocean Type 1 basalts. Most reported individual megacryst compositions match those of phenocrysts in Type 1 basalts.

Figure 12.6 illustrates coalescence histories of Indian Ocean Type 2 basalts, first for one basalt studied in detail (Figure 12.6A), then for a compilation of several basalts (Figure 12.6B). Olivine–plagioclase pairs in glomerocrysts clearly demonstrate the pattern of magma mixing in the sample studied in detail, with normal zoning defining a range of refractory minerals in a high temperature mixing end-member and reverse zoning in a range of evolved minerals in a low temperature end-member. The host glass composition has about the magnesium number of the hybrid defined by the zoning relationships. The group of glomerocrysts containing calcic plagioclases defines a flat trend in Figure 12.6A, in which feldspars ranging from about An_{80}–An_{85} are intergrown with very similar olivines. This cannot represent a low pressure fractionation trend (Longhi, 1982), although the group of minerals in the low temperature mixing component obviously can, beginning at about the more sodic ($\{An_{80}\}$) end of the high temperature group. The range of mineral pairs in each group actually indicates that mixing was between a spectrum of magnesian magmas, and another spectrum of evolved magmas, rather than two simple end members. The high temperature spectrum was a range of similarly magnesian parental magma types with variable soda contents, as indicated by the differing plagioclase compositions. The glomerocrysts with the most calcic plagioclases crystallized from magma stems approaching Type 1 basalt compositions, whereas the low temperature mineral group nearly matches Type 3 basalt mineralogy. Figure 12.6B shows that these mixing relationships commonly occur in Indian Ocean Type 2 basalts, although most have only a few olivine–plagioclase glomerocrysts. One Type 2 basalt sample has glomerocrysts with very calcic plagioclase (An_{85-89}), olivines and magnesian clinopyroxenes virtually identical to phenocrysts in Type 1 basalts.

Similar deductions can be made using the Ti contents of spinels, for which compositions of isolated crystals in glass correlate strongly with bulk TiO_2

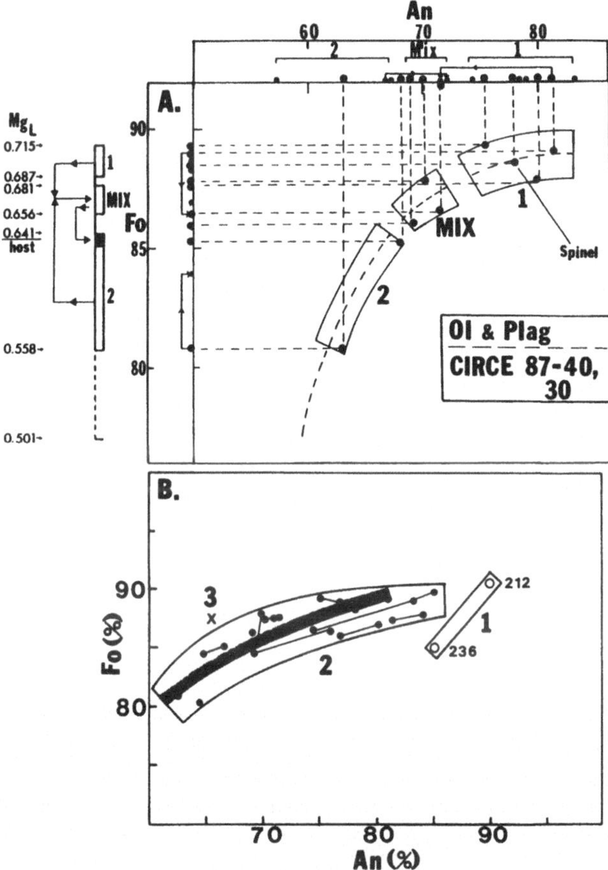

Figure 12.6 Compositions of co-precipitating (intergrown) plagioclase (An mol %) and olivines (Fo mol %) in glomerocrysts in Indian Ocean basalts. (A) Within a single Type 2 basalt (two thin sections, CIRCE 87-30 and 87-40) from Vema Fracture Zone, Central Indian Ridge. Individual plagioclases and olivines (not intergrown) are plotted at the top and left, respectively; coexisting pairs in glomerocrysts are plotted in the centre of the diagram. Zoning relations are indicated by arrows at the top and left, where estimated liquid magnesium numbers (Roeder and Emslie, 1970) are also indicated. (B) Within several other Indian Ocean basalts, indicated by Type. Mineral pairs in individual samples are linked by lines. Shaded area gives trend for CIRCE 87-30 and 40, from A (note difference in vertical scale). The nearly flat trend for Type 2 basalts is proposed to be related to mixing between parental Type 1, Type 2, and Type 3 magmas. See text for further discussion.

contents (Figure 12.7). Host glasses carrying spinel are generally fairly magnesian, with magnesium numbers greater than or equal to 0.6, and are not greatly removed from parental compositions by fractionation. Therefore it is possible to estimate near-parental melt TiO_2 contents for any spinel (i.e. different portions of zoned spinels or spinels enclosed in olivines and feldspars). This is achieved by the histograms illustrated in Figure 12.8. As

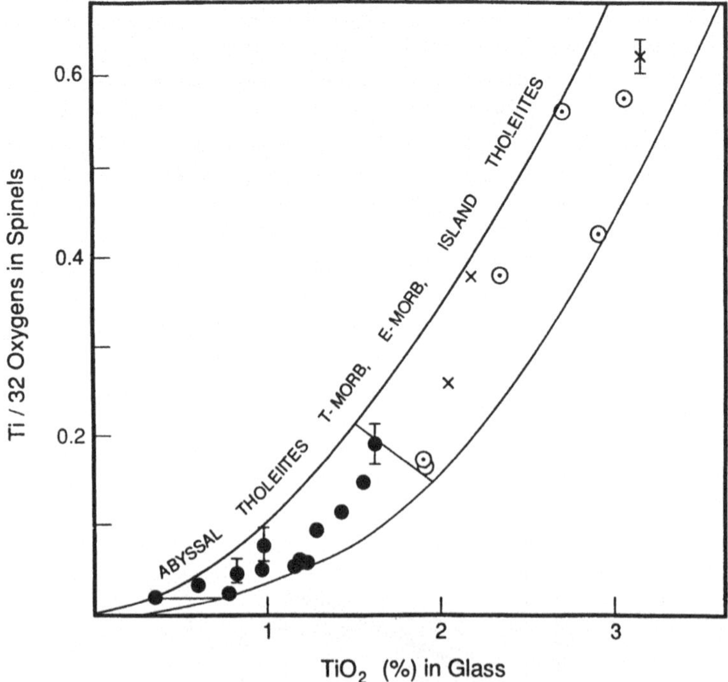

Figure 12.7 Ti/32 oxygens, computed from structural formulae, in magnesiochromites in glasses of abyssal-tholeiite, alkali basalts (E-MORB), and island-tholeiite compositions, versus TiO_2 contents of those glasses. Data are from Indian Ocean Type 1, Type 2, and Type 3 basalts, a dredged Lau Basin basalt (Hawkins and Melchior, 1985), dredged Hawaiian and Marquesan tholeiites and basalts from the eastern Pacific (Natland *et al.*, 1983; Natland, 1989; Natland *et al.*, in press and unpublished data). An upper limit of about 0.2 Ti/32 oxygens for abyssal tholeiites corresponds to Indian Ocean Type 3 basalts with about 1.5% TiO_2 contents. No abyssal tholeiite has spinels with less than 0.02 Ti/32 oxygens, corresponding to estimated glass compositions having about 0.3% TiO_2 contents.

overall spinel compositions are consistent with the TiO_2 contents of the host glasses, spinels in Type 1, Type 2 and Type 3 basalts are successively more titaniferous (on average) in that order although there is some overlap. Within individual basalts, the Ti contents of spinels are often variable, a consequence of the mixing history of the basalt. There are almost always spinels which crystallized from considerably less titaniferous liquids than the host glasses. Among Type 2 basalts, there are spinels which crystallized from melts with estimated 0.3–0.6% TiO_2 contents, overlapping the range for the one Type 1 basalt containing spinels (DSDP 212). At the other extreme, some spinels in Type 2 basalts crystallized from melts approaching Type 3 abundances of TiO_2. Spinels in one single Type 2 basalt span most of this range, as indicated in Figure 12.8.

Similarly, spinels in basalts from DSDP drill hole 504B (Costa Rica Rift,

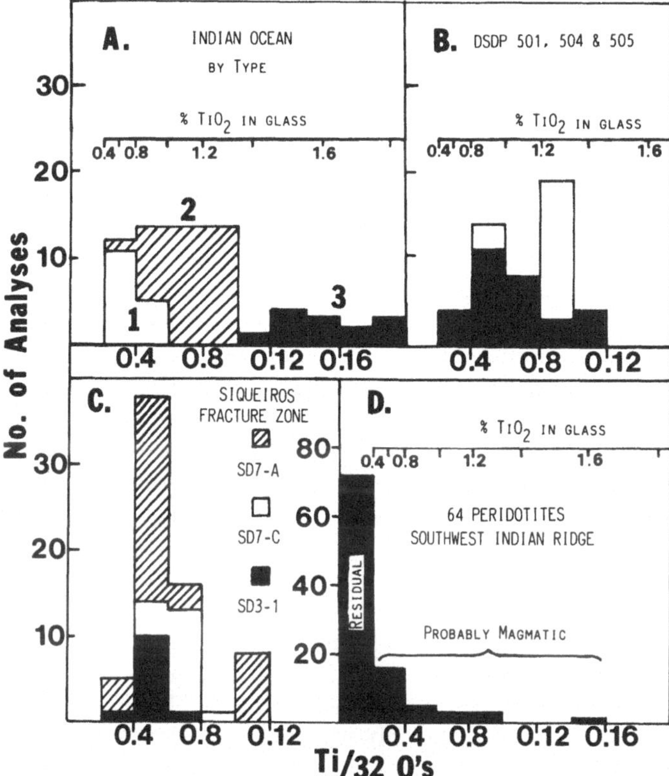

Figure 12.8 Histograms of Ti/32 oxygens for spinels, calibrated to TiO_2 contents of glasses, using the median curve of Figure 12.7. (A) In Indian Ocean Type, 1, Type 2, and Type 3 basalts, separately indicated. Individual spinels in sample ANTP 131-1, are shown by small triangles; the small arrow gives TiO_2 in ANTP 131-1 glass. (B) In basalts of DSDP Holes 501 + 504B (shaded), and 505 (not shaded) from Natland *et al.* (1983), Furuta and Tokuyma (1983), and Kempton *et al.* (1985). (C) In three magnesian abyssal tholeiites from the Siqueiros Fracture Zone (Natland, 1989, and unpublished); D. In abyssal peridotites of the Southwest Indian Ridge and America–Antarctic Ridge (Dick, 1989). The basalt spinels in A–C include phenocrysts and minerals enclosed in plagioclases and olivines, in addition to quench spinels in glass used to define the trend in Figure 12.7.

east Pacific) evidently crystallized from a range of magnesian liquid compositions with estimated TiO_2 contents ranging from about 0.3 to $>1.0\%$ (Figure 12.8B). The upper end of this range is about the TiO_2 content of the host glasses. Many of the low-Ti spinels occur within very refractory plagioclases (An_{88-91}) or associated olivines in glomerocrysts, and the feldspars often contain magnesian basaltic glass inclusions (Natland *et al.*, 1983). Many basalts of hole 504B also contain glomerocrysts of intergrown calcic plagioclase and clinopyroxene (chromian endipside), many rounded by partial resorption (Natland *et al.*, 1983). This is precisely what should happen to Type 1 phenocryst assemblages if they crystallized from refractory magmas and were later mixed into magnesian Type 2 or Type 3 magmas lacking those minerals on the liquidus. The eruptive liquidus assemblages (represented by small, euhedral olivines and tabular plagioclases in glass) are those of the more sodic hybrids, not of the refractory Type 1 component in the hybrids. The changes in crystallization sequences may also have been caused by a decrease in pressure (by about 2–4 kbar) during the ascent of magmas from the mantle (Natland *et al.*, 1983; Elthon, 1984, Tormey *et al.*, 1987).

Finally, several porphyritic magnesian basalts from the Siqueiros Fracture Zone each contain numerous spinels which also evidently crystallized from a range of liquid compositions, some having TiO_2 contents as low as 0.2% (Figure 12.8C). However, other spinels are as titaniferous as those in Indian Ocean Type 3 basalts. Siqueiros basalt SD–7a, for example, has strongly zoned spinels, containing refractory (low Ti, high Al) cores with very titaniferous and more Cr-rich skeletal rims (Natland, 1989). The host glass itself is fairly titaniferous and sodic (see Chapter 5, Table 5.1), approaching Type 3 compositions.

Although some of these spinels are extremely refractory, with very low Ti contents, they are not as refractory as most spinels in abyssal peridotites (Figure 12.8D) and are certainly not xenocryst fragments of the upper mantle. Indeed, the small percentage of spinels in abyssal peridotites that overlap the compositions of spinel phenocrysts are probably also magmatic, and crystallized from trapped, generally refractory (i.e. Type 1) melts in residual ultramafic rock.

In summary, a variety of mineralogical evidence demonstrates the pervasive involvement of extremely refractory basaltic melt strains in the development of parental abyssal tholeiites. Broadly speaking, the minerals are crystallized from melts similar to Indian Ocean Type 1 magnesian basalts, but which are, in some instances, even more refractory (with lower TiO_2 and Na_2O). These melt strains partially crystallized, then were mixed with other less refractory magmas, in some instances approaching Type 3 compositions, to produce the basalts which were eventually erupted. The isolated occurrence of some Type 3 basalts along the Central Indian Ridge, where Type 2 basalts predominate, and basalts with nearly Type 3 compositions in the Siqueiros Fracture Zone along the East Pacific Rise, shows that regions of mantle can produce diverse parental magmas. The basalt mineralogy shows that these diverse strains usually mix to produce the more typical basalts in a given region. The balance of the magnesian end-members (Type 1, Type 2 or Type 3) in the hybrids define abyssal tholeiite magmatic lineages. In all of this, Type 1 melt strains appear to be persistent components in many primitive abyssal tholeiites, derived from somewhere within mantle melting columns beneath many spreading ridge segments.

12.6 Mantle lithological heterogeneity and the melting column

Crystallization histories of porphyritic abyssal tholeiites such as those from the Indian Ocean impose three important conditions on models of partial melting.

(1) There is no evidence for crystallization of any phenocryst, megacryst, or glomerocrystic mineral aggregate from liquids with magnesium number > 0.75, whether or not those liquids had low TiO_2 and Na_2O. This is to say that picritic antecedent liquids cannot be inferred from the crystallization histories of any known abyssal tholeiite, despite the potential for such liquids to exist, as inferred from experimental petrology. The severity of this restriction is indicated by the fact that the majority of refractory magnesian liquids thus far found—those with the lowest TiO_2 and Na_2O contents—have been melt inclusions in plagioclases rather than olivines, and melt inclusions in olivines are no more magnesian than those in plagioclases.

(2) The principal mixing relations that can be inferred from the cores of phenocrysts, megacrysts, and glomerocrysts involve parental and near-parental magmas only, with a restricted range in magnesium number, from about 0.75 to 0.65 and with liquidus temperatures from about 1200–1230°C.

(3) At individual locations, liquids with this restricted range of temperatures and magnesium numbers precipitated minerals prior to and during the mixing process itself with a very wide range in their contents of TiO_2 (from 0.2–1.4%) and Na_2O (from $< 1\%$ to $> 3\%$). At these locations, and embodied in the crystallization histories of single specimens, the known global range in parental abyssal–tholeiite quenched glass compositions can be, and regularly is, exceeded. In most cases, the least titanian and sodic melt strains had the highest temperatures and magnesium numbers as inferred from the compositions of minerals and glass inclusions.

A consequence of these conditions is that, at the stage when melts are tapped from the mantle, temperatures are not greatly variable over the range of depths through which melts coalesce but melt compositions are. Experimental data show that during batch melting of homogeneous peridotite at given pressure, in order to produce such extents of variation in TiO_2 and Na_2O contents, there must be considerable variation in both temperature and magnesium number of liquids (e.g., Fujii and Scarfe, 1985). Such conditions are evidently not matched by natural circumstance. Other conditions prevail which buffer magnesium number and restrict the range of temperatures of coalescing melt strains while allowing extensive variation in TiO_2 and Na_2O contents over comparatively small distances in and near melt domains in the mantle.

Klein and Langmuir's (1987) melt-column model was developed to explain regional differences in average parental basalt compositions, most especially their Na_2O contents. But whereas decompression partial melting in ascending peridotite buffers magnesium number, it does not offer a simple explanation for diversity of melt compositions on a local scale. Assuming a homogeneous mantle to begin with, parental melt diversity on the scale of a single domain or cell of partial melting in the mantle requires tapping of melt packets which ascended over a great range of depths, with those having least TiO_2 and Na_2O having come from the greatest depths. However, the most common refractory megacryst assemblages crystallized from Type 1 melt packets which were multiply saturated in olivine, plagioclase, and clinopyroxene ($\pm$ spinel). Such assemblages imply that mantle sources of these low-TiO_2, low-Na_2O liquids did not begin to melt until they had ascended to quite shallow depths, despite the fact that these basalts are most prevalent near hot spots, where geothermal gradients are high. This is the opposite relationship of composition to depths of melting and geotherms that follows from the melt-column model.

An explanation may be that the mantle is not a homogeneous peridotite facies. Based on crystallization histories of olivine tholeiites from the Siqueiros Fracture Zone (Figure 12.8c), I suggested that the mantle is lithologically heterogeneous on the scale of a single melt domain, or diapir, in the mantle (Natland, 1989). That is, over small distances (metres to hundreds of metres), melt packets are derived from variably fertile (lherzolitic) to refractory (near-harzburgitic) peridotite lithofacies, with intrinsically different TiO_2 and Na_2O contents. Experimental results show that identical conditions of pressure and temperature can produce basaltic melts from fertile and refractory peridotite with very similar magnesium numbers and liquidus temperatures, but with very different abundances of TiO_2 and Na_2O (e.g. Jaques and Green, 1980). Crystallization and mixing of these melt strains during their ascent through overlying mantle can explain the principal features of refractory megacryst assemblages in all abyssal tholeiites, with differences in depths of melting of refractory peridotite components determining whether the assemblages are olivine-rich, as at Siqueiros Fracture Zone, or plagioclase- and clinopyroxene-phyric, as along portions of the Indian Ocean and the Costa Rica Rift at Hole 504B.

There are two corollary hypotheses. The first states that melt domains, or mantle diapirs, are probably zoned, with interiors consisting of more refractory peridotite than peripheral regions (Natland, 1989). Parental basalts in the Indian Ocean and elsewhere coalesced from high-temperature Type 1 melt strains carrying refractory phenocrysts, and cooler, less magnesian, more sodic and titanian melt strains which may have lacked minerals, but which now in any case are now evident only in the bulk compositions of hybrid parental basalts. Zonation of mantle melt domains in this manner may be a consequence of the buoyancy of low-iron, refractory peridotite within individual diapirs during convective processes (Jordan, 1979; Natland, 1989).

The second corollary states that for phase relations in natural basalts to contradict regional inferences based on melt-column models and geothermal gradients, lithological heterogeneity of the mantle must exist on a regional scale as well. In particular, refractory peridotite lithofacies appear to predominate in hot spot regions, based on the prevalence of fairly refractory and somewhat enriched average basalt compositions with Type-1 or partial Type-1 megacryst assemblages at places such as the FAMOUS area near the Azores in the North Atlantic and those portions of the Central Indian Ridge nearest Rodrigues and Reunion Islands. The only known Type 1 gabbros are also from the FAMOUS area. The low TiO_2 and Na_2O of the basalts is not a simple consequence of enhanced partial melting resulting from a higher geothermal gradient. The melt-column model still can be applied, but not on a regionally homogeneous mantle. The underlying explanation for geographically distinctive magmatic lineages among abyssal tholeiites is lithological heterogeneity of the mantle which is closely linked to the distribution of hot spots near spreading ridges.

12.7 Concluding statements

1. Abyssal tholeiites in the Indian Ocean do not represent a range of basalts produced by the differentiation of a common parental magma. Variations in depth of melting, degree of melting and local and regional lithological heterogeneity of the mantle determine the compositions of melts supplied to crustal magma chambers from the mantle, as well as the occurrence, proportions and compositions of phenocrysts.
2. Based on detailed petrographic and chemical evaluation of Indian Ocean abyssal tholeiites, it can be seen that three general types of parental basalts predominate in different regions of the ocean crust, each with a characteristic phenocryst assemblage, and each producing a distinctive magmatic lineage by crystallization differentiation. At comparably high magnesium numbers (0.68–0.72), the three parental types have low, intermediate and high abundances of Na_2O and TiO_2, respectively, with sequentially lower CaO/Al_2O_3. Only the high soda type is spinel and olivine-phyric at these magnesium numbers. The intermediate type also has calcic plagioclase on the liquidus and the low soda type has plagioclase plus magnesian clinopyroxene, together with olivine and Cr-spinel. Low Ca-pyroxene is the earliest to join low pressure crystallization sequences (at the highest temperature and magnesium number) in the low soda lineage, and latest in the high soda lineage. At a given magnesium number and Fo content of olivine, compositions of coexisting plagioclases are systematically more albitic in differentiated members of

the successively more sodic lava suites. The mineralogical contrasts are evident both in basalts and in the crystallization sequences observed in gabbros obtained from Indian Ocean fracture zones.

3. Despite the regional distinctiveness of the distribution of these basalt types, mixing histories deduced from phenocrysts consistently demonstrate the influence of low, intermediate and high soda primitive magmas in the composition of many individual porphyritic abyssal tholeiites. This is because magmas leaving the mantle coalesce from diverse magma strains produced over a range of thermal conditions within melting domains in the mantle which include diverse (refractory to fertile) peridotite lithofacies.
4. Regional variations in average or predominant parental basalt types extracted from multiple melting domains are a consequence of the large-scale thermal structure of the mantle (i.e. hot-spots versus normal ridge segments) and correlative contrasts in the large-scale lithological composition of the mantle.

13 North Atlantic ocean crust and Iceland

CHERRY WALKER

13.1 Introduction

This chapter summarizes geological, geophysical and geochemical observations from the North Atlantic basin, particularly 23°N (MARK area), 37°N (FAMOUS area), 40°N (the Azores), 45°N, 57–63°N (the Reykjanes Ridge), all relative to Iceland. Emphasis will be placed on the geochemistry of basalts from these areas, especially the nature of the source region from which the basalts originated and, to a lesser extent, what they can tell us about dynamic processes in the crust and mantle involved in their genesis. The geochemical data will also be considered in the light of other lines of evidence concerning the nature of the mantle under this ocean basin. Coverage of the South Atlantic Ocean is not within the scope of this chapter, but details can be found in Schilling *et al.* (1984), Le Roex *et al.* (1985a, b), Chaffey *et al.* (1989), Sun and McDonough (1989), ODP Leg 108 and references cited therein.

There is a general consensus about the presence of mantle plumes of various sizes under hot elevated areas in the Atlantic Ocean, and the fact that these are largely responsible for the variations in basalt geochemistry, as well as physiographic and geophysical parameters. The exact nature of the material constituting a plume mantle source, the origin of that material, and from which part of the earth's interior it may have risen, are still strongly debated such that each ocean basin may have to be treated as a separate dynamic mantle model. However, the source of any plume is geochemically unique and distinct from the globally extensive source region that feeds the mid-ocean ridge system. In the area under discussion, Iceland and the Azores are two such hot-spots fed by mantle plumes. Much research has been undertaken in this area, particularly by the Deep Sea Drilling Project (DSDP) and the subsequent Ocean Drilling Program (ODP), and this has produced vast amounts of data on numerous aspects of the Atlantic Ocean. This chapter reviews a large part of this, concentrating on the nature of the mantle domain, and it is proposed that a mixing model (Morgan, 1971; Vogt, 1971; Schilling, 1973a, b) is applicable to these hot-spots and their surrounding regions. This differs from the previous models by the complexity of the geochemistry, due to differing 'mixtures' of the various components available in each hot-spot source region and other local heterogeneities that may exist in the MORB source. Chemical and structural similarities between the Icelandic and Atlantic Ocean crust will be emphasized.

The opening of the Atlantic may be related to the development of the Icelandic and other plumes that underlie the ridge and flanks of the Mid-Atlantic Ridge (MAR) (Bott, 1988; Silver *et al.*, 1988; White and McKenzie, 1989a, b; White, 1990). However, plumes are not necessarily crucial to continental rifting, although the 1–2 km dynamic thermal uplift they provide helps gravity sliding (White and McKenzie, 1989a). In the North Atlantic, the Icelandic plume was active shortly before the opening of the ocean and small-scale rifting had been occurring for tens of millions of years before the Icelandic plume was initiated at approximately 62 Ma. After a 5 Ma period of plume activity, the rifting between Greenland and the Rockall Plateau became well established (White, 1988; White and McKenzie 1989a). Plate driving forces such as slab pull, ridge push and suction forces at well established subduction zones are thought to be collectively responsible for the opening of the Atlantic (White, 1990), rather than whole mantle convection associated with plumes and hot-spots as was earlier suggested by Morgan (1971).

13.2 Morphology and structure of the Mid-Atlantic Ridge

The Atlantic basin is deeper, older and spreading faster in the south than in the north. For example, half-spreading rates are 15 mm y^{-1} at the Kane Fracture Zone, 11–12 mm y^{-1} at the Azores and 10 mm y^{-1} on the Reykjanes Ridge and Iceland. The mid-ocean ridge (MOR) is intermittently segmented and offset by transform regions of various magnitudes. Topographic anomalies occur in the vicinity of hot-spots and may be represented by subaerial groups of oceanic islands developed at various distances from the ridge.

High resolution bathymetric studies (Luyendyk *et al.*, 1977) and submersible dives (first executed in the FAMOUS area, 37°N; Bellaiche *et al.*, 1974; Needman and Francheteau, 1974) have enabled earth scientists to study the nature of the seafloor topographic fabrics in great detail and to appraise and discuss their significance in relation to crustal and mantle structures.

The morphology and topography of the ocean basin can be described in terms of sets of parallel ridges and troughs at various orientations to the main spreading direction, which have varying widths and lengths (Figure 13.1a). In a typical slow spreading ridge environment, such as the North Atlantic, the intensity of the fabric is greater than at a fast spreading ridge. The arrangement, intensity and relative proportions of these lineaments are determined by such features as a median valley at the ridge crest (average 30 km wide), an axial neovolcanic zone (usually < 10 km) and ridge axis discontinuities such as transform faults. The MAR is segmented by large transform faults and by smaller non-transform discontinuities (Ramberg *et al.*, 1977; Schouten and

a)

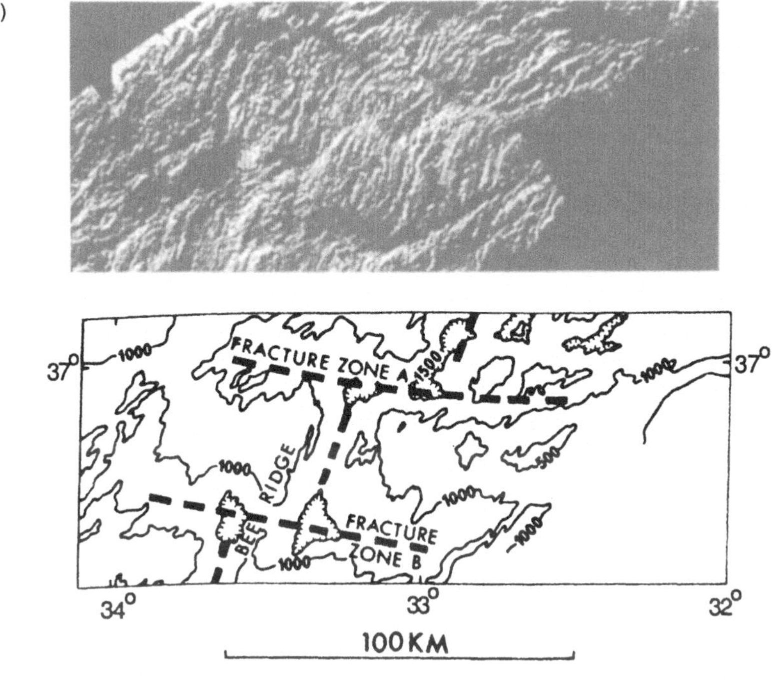

b)

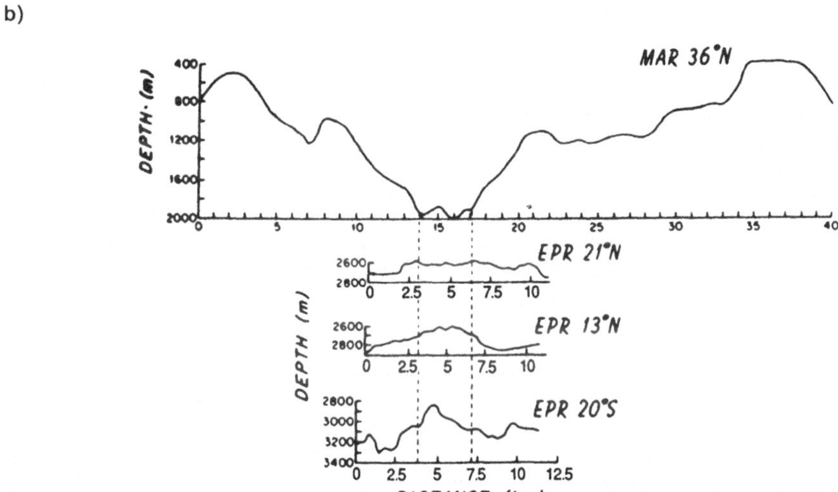

Figure 13.1 (a) Example of sonograph mosaic bathymetry and its interpretation at 32–37°N, FAMOUS area. Contour intervals at 500 fm (914.4 m) Broken lines indicate the spreading axis (Searle, 1979). (b) Topographic profiles across the central axis showing the elevation of the median valley and a comparison of the differences between the EPR and the MAR. The width of the FAMOUS Rift Valley inner floor is projected (dashed lines) onto the other profiles for scale. Vertical exaggeration of 1 × 6.45. (Francheteau and Ballard 1983).

White, 1980; Macdonald and Fox, 1983; Lonsdale, 1983; Macdonald *et al.*, 1984; Sempere and Macdonald, 1986) developed, on average, every 55 km (Macdonald, 1986). Recent work (Lin *et al.*, 1990; Sempere *et al.*, 1990) confirms that the median valley floor of a segment of ridge is itself elevated towards the centre of that segment (Figure 13.2). It has been demonstrated on the East Pacific Rise (EPR) that these segments behave as structurally and geochemically independent units (Thompson *et al.*, 1985; Langmuir *et al.*, 1986).

The plate boundary as represented by the axial rift zone may gradually migrate relative to a globally fixed reference frame, such as a hot-spot, by the normal spreading process. This should be distinguished from the processes that involve the relocation of the plate boundary, both laterally and axially, by ridge jump and rift propagation, respectively. The former is not uncommon in the Atlantic and has been reported from 36°N (DSDP Leg 37) and 23°N (ODP Leg 106/109). To the north of Iceland, the Kolbeinsey ridge is believed to have undergone several episodes of ridge jump (Johnson *et al.*, 1971) before the last jump occurred at 36 Ma (Oskarsson *et al.*, 1985).

The detail of the axial neovolcanic zone in a symmetrically spreading ridge is usually defined as an undulating central topographic ridge of young volcanic products erupted from fissures. It has been suggested that the slower spreading ridges with lower magma supply rate typically have more than one locus of volcanic activity (i.e. volcano) per segment, relative to the fast spreading ridges that typically have one volcano per segment (Searle, personal communication). The volcanic ridges become laterally removed from the axis by spreading, where they are faulted to form steep rift mountains up to 1500 m or more high (Bougault and Hekinian, 1974; Sempere *et al.*, 1990) and which mark the margins of the median valley typical of a slow spreading ridge. The topography on the immediate flanks of the median valley is thus very pronounced with successive parallel ridges and scarps, with several hundred metres of relief, which becomes less pronounced as a function of distance from the ridge axis, due to sediment burial and subsidence as the lithosphere cools (Parsons and Sclater, 1977), until the slope levels off at the foot of the ridge rise (Figure 13.1b).

These characteristic geomorphological features of the ridge are disturbed in the vicinity of a hot-spot. The depth to the ridge shallows, and the median valley and segmentation become less pronounced. This is demonstrated south of Iceland, on the Reykjanes Ridge (Talwani *et al.*, 1971), which is adjacent to the Iceland hot-spot, the largest plume in the North Atlantic. Vogt (1971) observed a topographic fabric oblique to the axial valley and the magnetic stripes (time-transgressive ridges) south of Iceland and near the Azores, but not near Jan Mayen (Vogt, 1974). A hot-spot often leaves a trail in the form of an aseismic ridge or a chain of islands (Chapter 9), depending on whether it coincides with a mid-ocean ridge segment. The aseismic ridge between the Faeroes and Greenland records the history of the Iceland hot-spot from

60 Ma. Assessment of the topography reveals a subdued rate of eruption from the plume between 20 and 30 Ma (Vogt and Avery, 1974).

Seismic refraction profiles through the oceanic crust have shown that there are principally three seismically distinct layers of material (White *et al.*, 1990). The crust is defined here to be from the surface down to the petrological Moho, represented by the base of the cumulates. The layers are distinguished by their different seismic P-wave velocities, which are determined by a combination of pressure, temperature, porosity, density and mineral composition and alignment. From this work the average crustal thickness of the oceanic lithosphere, away from interference of the ridge-axes, hot-spots, seamounts, or transform faults, is 6 km (White 1989). The thickness at the ridge-axis is variable. The crustal thickness along a single ridge segment is less in the vicinity of the ridge axis discontinuities, and greater at the centre of each segment. For example, in the MAR, 24 to 30°N, the crust is thinned

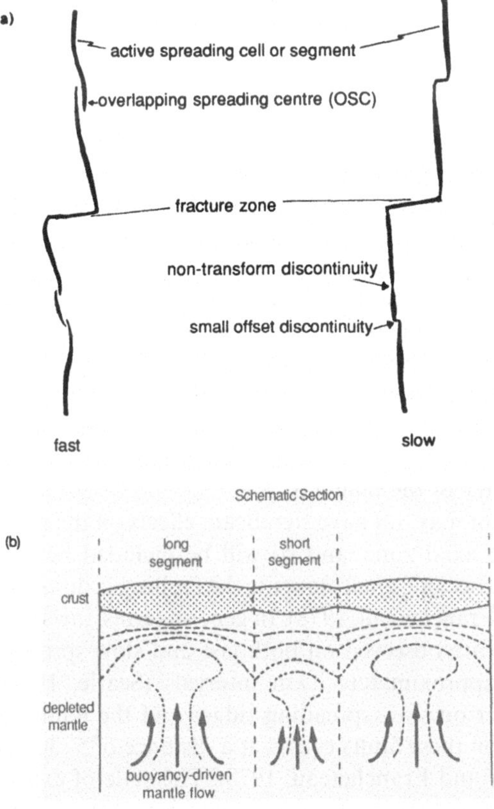

Figure 13.2 (a) Plan view of different styles of ridge segmentation between fast and slow spreading ridges. (b) Schematic cross-section of ridge segmentation (adapted from Lin *et al.*, 1990).

by as much as 50% at the segment boundary and it has been suggested that the longer the segment, the thicker the crust (Lin *et al.*, 1990) (Figure 13.2).

Earthquake epicentres have been reported at depths of 6 km in the Atlantic region (Lillwall *et al.*, 1978). However, fault spacing often suggests that the crust is 2 km thick and so not completely formed, but that it subsequently thickens rapidly within a short distance (about 15 km) from the ridge axis. This could indicate that either the crust is of variable thickness at the ridge axis, being thicker at slower spreading centres relative to fast ridges, or that the sub-ridge asthenosphere is capable of brittle fracture, suggested by the presence of basaltic dykes observed in ophiolite sections (Browning, 1984).

The Atlantic oceanic crust increases in thickness towards Iceland along the Reykjanes Ridge, and in the vicinity of other plateaux, such as the Azores Islands (Searle, 1976a), Cape Verdes Islands (Courtney and White, 1986) and other seamounts in the North Atlantic. This increase in crustal thickness is representative of an increased rate of magma production, which correlates with mantle potential temperature (White and McKenzie, 1986b). There is confusion over the relationship between the magma flux and the spreading rate. White (1989) states that due to the constant average thickness of oceanic crust (6–7 km), the magma flux is independent of the spreading rate and that asthenospheric processes exert a dominant control over the volume and distribution of the magma in the ocean basins. However, other workers (Dick, 1989) frequently describe low magma fluxes on slow spreading ridges. This parameter must be variable in some instances with episodic volcanic activity and due to the changes observed at plumes and within a single segment. On a typical spreading segment, in either the EPR or the MAR, there is a comparable average crustal thickness of 6–7 km, although local deviations occur. This average thickness will have taken longer to produce in the MAR relative to the EPR, and so it follows that the Atlantic must have an overall lower magma supply flux for a given period of time. Nevertheless, for any given unit volume of mantle, regardless of the spreading rate and time, the actual percentage of melt produced from that volume will be comparable at about 30%. It follows that the spreading rate should not significantly affect the bulk chemistry of the magmas.

Faulting may or may not have significant effects on the magmatic processes operating at the axial zone, and so will be included here, very briefly, for completeness. In oceanic environments the faulting is dominantly extensional (Taponnier and Francheteau, 1978). In general terms, inward dipping bundles of normal faults are observed on both fast and slow spreading ridges. They are formed at approximately 2 km intervals (Searle, 1984), although the spacing is greater on slow spreading ridges and the fault scarps are longer. The movement on these faults ceases at a distance of 5–30 km from the axial zone (Taponnier and Francheteau, 1978). The style of extension varies as a function of spreading rate and local magma flux. Extensional faulting in the Atlantic, and on other slow spreading ridges, at a time when the magma flux is very low, is believed to mimic that on the continents (Sempere *et al.*, 1990). Detachment faults similar to those seen in the Basin and Range Province (western USA) are observed in the MARK area of the Atlantic (23°N) (Brown and Karson, 1988). White *et al.* (1990) have documented many different styles of faulting, of which three major types have been observed to penetrate to the Moho (Chapter 3; Figure 3.6). Serpentinization occurs at the base of transform faults which penetrate the whole crustal sections, and this lowers the seismic Moho (layer 4 boundary) by as much as 2–3 km (White *et al.*, 1990). If any of these faults were activated in the vicinity of the active axial zone, then they could be a means of serpentinizing the lower crust and so producing apparently less variation in the seismic crustal thickness away from the axis. This is interesting because the crustal thickness at the axial zone is variable, yet at some distance from the axis it is reported to be more consistent and raises the question as to whether the seismic Moho is largely representative of a serpentinized upper mantle (Hess, 1962; Clague and Straley, 1977; Francis, 1981).

13.3 Morphology and structure of Iceland

The regional topography and morphology of Iceland is dominated by a central high (2000 m above sea level) in the vicinity of the Vatnajokull region (Figure 13.3). The Mid-Atlantic plate boundary is exposed on land, where it presents an excellent opportunity to study the oceanic crust. Before describing the details of the ridge crest in Iceland, it is necessary to introduce the basic geological characteristics of the island as these will be referred to throughout the chapter and are vital to the geochemical interpretation.

13.3.1 *Present plate boundary configuration*

At 64°N, the Mid-Atlantic plate boundary is expressed on land as a complicated arrangement of active volcanic and seismic zones (Figure 13.3a). These will be briefly described below, but for more detailed descriptions, see Ward (1971), Palmason and Saemundsson (1974), Saemundsson (1974, 1978, 1979) and Einarsson (1990). Throughout the literature the various volcanic zones have often changed their names (for example, Brooks *et al.*, 1974; McGarvie *et al.*, 1990); this chapter will follow the most widely used version, summarized in Figure 13.3a.

The main actively spreading volcanic zones are divided into the Western, Eastern and Northern neovolcanic zones. The Western Volcanic Zone (WVZ) is the on-land continuation of the Reykjanes Ridge. It extends inland, becoming progressively wider and trending dominantly north-east for half the distance across Iceland. It then bends due east at Langjokull to join the Eastern and Northern Volcanic Zone junction under Vatnajokull, (Figure 13.3). The

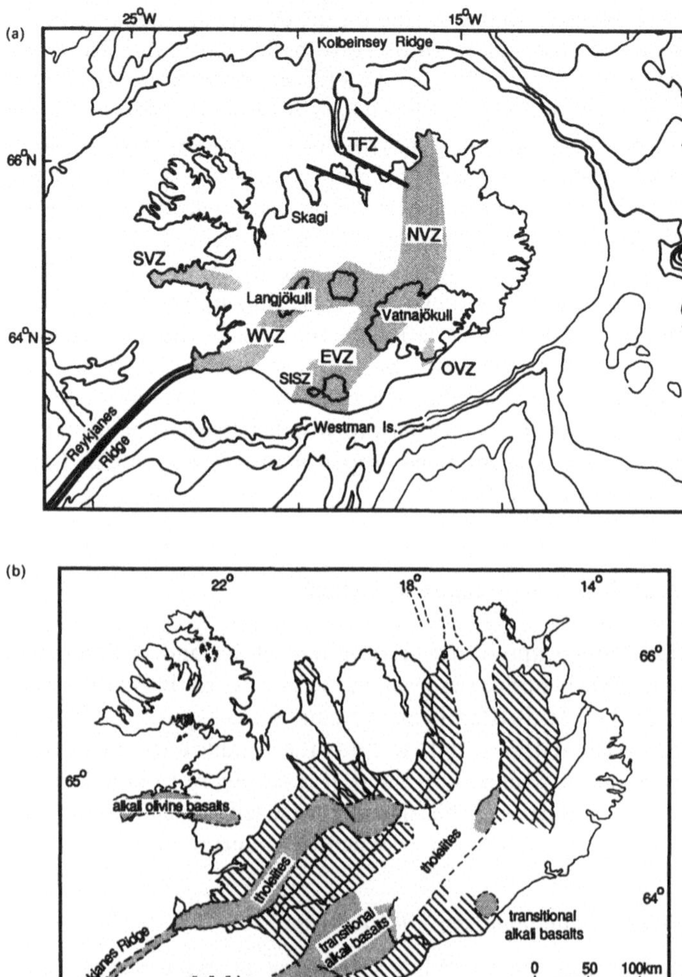

Figure 13.3 (a) Location map of Iceland, showing the distribution of the volcanic and fracture zones, and locations mentioned in the text. Bathymetry contours are drawn every 400 m (adapted from Einarsson 1990). (b) Map of Iceland showing the distribution of the different magma series (after Jakobsson 1979 b).

Northern Volcanic Zone (NVZ) has a more northerly trend, and extends north of the Vatnajokull glacier to the north coast of Iceland. The Eastern Volcanic Zone (EVZ) is the widest (72 km) and trends south-west parallel to the WVZ from Vatnajokull, as far as the island of Surtsey. A comparatively small volcanic zone protrudes from the south-east corner of Vatnajokull, and has been referred to as the Oraefajokull Volcanic Zone (OVZ) (Steinthorsson *et al.*, 1985). There is a possible fifth active volcanic zone, trending from east to west, on the Snaefellsnes Peninsula, west Iceland (SVZ). It has a seaward expression, in the form of a topographic ridge (Jokulbanki Ridge), that swings round from an east–west orientation to run parallel to the Reykjanes Ridge for at least 150 km (Brooks *et al.*, 1974). In much of the recent literature, the EVZ, OVZ and SVZ are referred to as lateral or flank zones (terminology of Saemundsson, 1979), as many volcano-tectonic properties differ from the rest of the neovolcanic zone, which is more representative of the typical mid-ocean ridge environment.

There are two zones of high seismicity which represent transform zones (Einarsson, 1991). In the north-east of Iceland the neovolcanic zone is sinistrally offset by the 75 km wide, 100 km long belt, known as the Tjornes Fracture Zone (TFZ) which connects the NVZ with the Kolbeinsey (or Iceland–Jan Mayen) ridge. In detail, there are at least three parallel north-west–south-east trending seismic belts roughly 30–40 km apart (Saemundsson, 1974; Einarsson, 1991). The dominant structural orientation within each belt is north–south and is displayed by a series of horsts and graben. In the south of Iceland there is an east–west trending, sinistral zone of high seismicity, known as the South Iceland Seismic Zone (SISZ), that is 10–15 km wide and 100 km long (Einarsson *et al.*, 1981). It joins the WVZ and EVZ without laterally offsetting either. The majority of the region is covered by an alluvial plain which hinders ground level examination of the region. Nevertheless, in the east, the topography is deflected from the normal north-east trend to an east–west orientation, and in the west, from the normal north-east trend to a north–south direction (Saemundsson, 1967; Walker, to be published).

Segmentation in Iceland, on a smaller scale, occurs in the form of volcanic systems (see later) (Saemundsson 1978). These are frequently arranged in an *en echelon* fashion in response to non-orthogonal spreading. It has been suggested that these behave as structurally and geochemically independent units, similar to segments on the EPR (Whitehead *et al.*, 1984) and possibly the MAR (Sempere *et al.*, 1990), although there is evidence to show that this is not always the case in the flank zones (McGarvie 1984; Blake 1984; McGarvie *et al.*, 1990).

13.3.2 *The neovolcanic zone*

The axis of the ridge crest, often referred to as the neovolcanic or axial zone in Iceland, has been described in detail by many workers (Saemundsson, 1978, 1979; Jakobsson *et al.*, 1978; Jakobsson, 1979a, b). The topography of the neovolcanic zone can be described as a slightly, but variably, elevated region, relative to the distal parts of Iceland. Geomorphologically, it is made up of a number of individual units or volcanic systems (Walker, 1963, 1974; Saemundsson, 1978) (Figures 13.4a and 13.5). Each system is composed of a set of parallel eruptive ridges or fissures, constituting a fissure swarm that

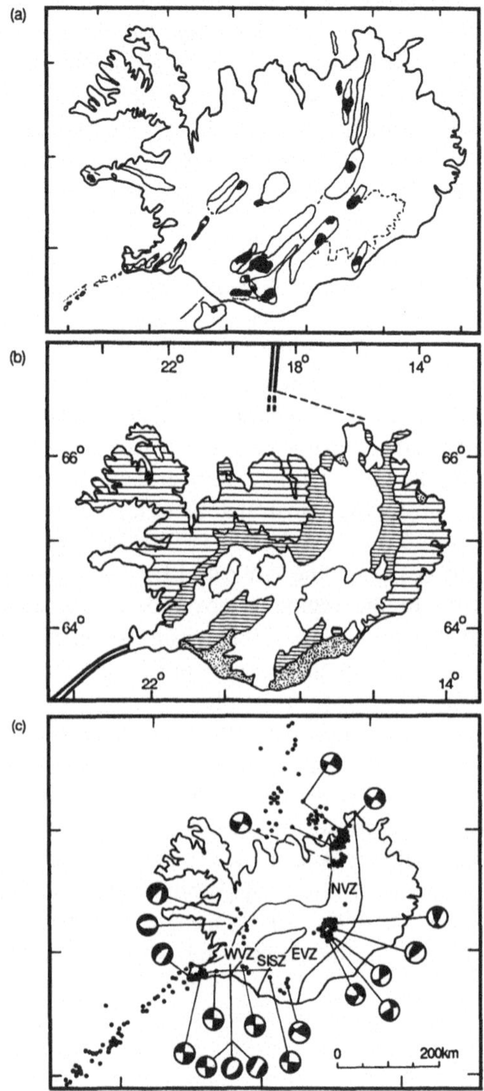

Figure 13.4 (a) Map of Iceland showing the distribution of the 29 volcanic systems (stippled) (after Jakobsson 1979b). (b) Map of Iceland showing the relative ages of the rocks. Wide horizontal = Tertiary (> 3.1 Ma); close horizontal = Plio–Pleistocene (0.7–3.1 Ma); clear = Upper Pleistocene and post-glacial (< 0.7 Ma); stippled = solid drift (Sæmundsson 1979). (c) Distribution of recent seismicity. Epicentres are taken from earthquake lists (PDE) of the US Geological Survey for the period of 1963–1987. Only epicentres determined with 10 or more stations are included. Focal mechanisms are shown schematically on the lower hemisphere stereographic projection of the focal sphere. Compressive quadrants are black (after Einarsson 1990).

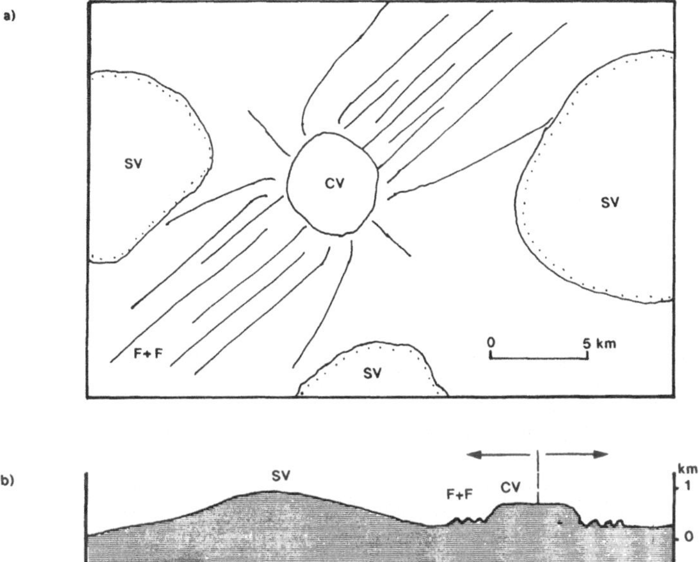

Figure 13.5 (a) Plan view of the volcanic system, as defined by Jakobsson (1978), showing the spatial relationship between the central volcano (CV), faults and fissures (F + F) and the off-axis lava-shield or shield volcano (SV). (b) Schematic cross section of the topography across a volcanic system (as observed in SW Iceland, e.g. CV = Hengill and SV = Skjaldbreiur or Ingolfsfjall). The spreading axis is indicated by arrows.

varies in length from 17 to 105 km, in width from 5 to 30 km (Jakobsson, 1979a), and which has a life span between 300 and 500 ka. These swarms may, with time, develop a central volcano (Saemundsson, 1978). The latter is characterized by being topographically elevated relative to the fissure swarm due to an increase in the magma extrusion at one site (Figure 13.5b). The system may also contain earlier slightly off-axis (5–10 km) shield volcanoes or lava shields of various sizes. These can be recognized by the unusually large accumulation of compositionally monotonous compound lavas and very low angles of slope on the flanks, typical of Hawaiian shield volcanoes or Skjaldbreidur (WVZ). It is important to appreciate that the volcanic morphology is greatly dependent on whether the eruption occurs under ice or subaerially (Thorarinsson, 1974).

It seems that there is often a correlation in Iceland between the composition of the lavas and the morphology of the volcano from which they were erupted (Jakobsson *et al.*, 1978, 1979a, b). Some of the variation between the different volcanic systems in Iceland can be attributed to their different stages of development (Jakobsson 1979a), and/or as a function of crustal thickness in that region. Away from the neovolcanic zone, in the Tertiary Icelandic crust, Walker (1963) identified similar volcanic systems, but the off-axis shield volcanoes appear to be less abundant (see Section 13.5.1).

The thickness of the crust in Iceland (Palmason, 1971; RRISP Working Group, 1980) varies from 8 to 16 km (compared with the range of 3–8 km for normal oceanic crust). The boundary between layers 3 and 4 is interpreted as the base of the crust (Figure 13.6). It is shallower beneath the central volcanoes of the neovolcanic zone and the extinct central volcanoes in the Tertiary and Skagi zones, suggesting that crustal and mantle lithospheric underplating occurs later under Iceland (Figure 13.6b). No crustal thinning

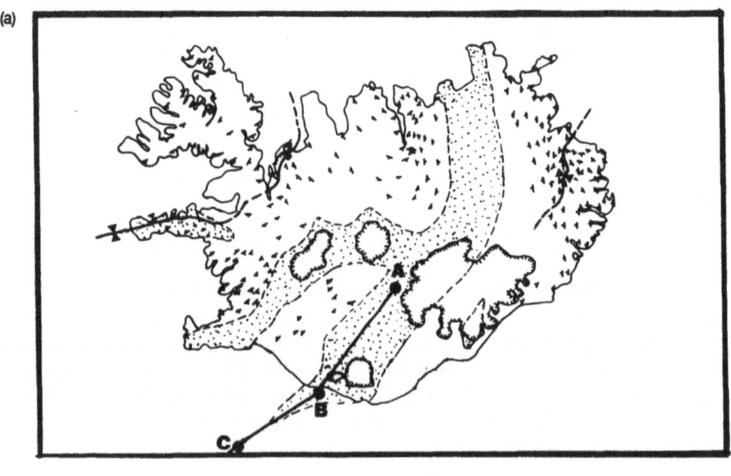

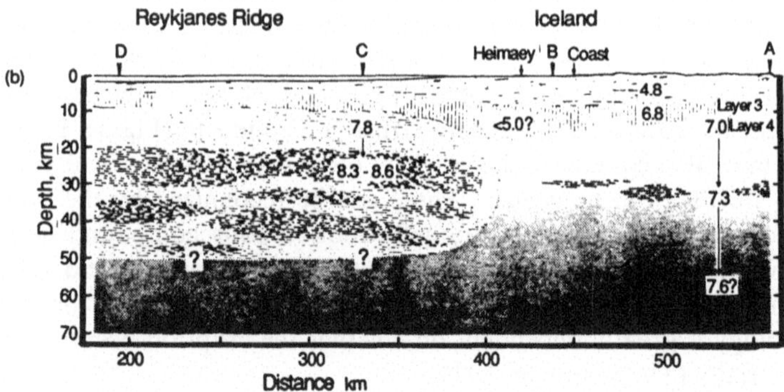

Figure 13.6 (a) Map of Iceland showing the neovolcanic zone (stippled) and the dips of the older lava successions (adapted from Pálmason and Sæmundsson, 1974). Note that the dip is dominantly towards the neovolcanic zone, but also two synforms are present and maybe interpreted as relict sites of the plate boundary. The traverse shown in 13.6b is indicated. (b) Seismic section of the south end of the EVZ and the east flank of the Reykjanes Ridge, showing a generalized crustal and upper mantle section. Letters indicate positions of large shots. Numbers of P-wave velocities are indicated in km/s. Crustal layers are continuous across the transition from the Reykjanes Ridge to Iceland, whereas there is a drastic change in the upper mantle structure close to the shelf slope, where a well developed oceanic lithosphere abuts on anomalous upper-mantle material (RRISP working group, 1980).

is observed in the vicinity of the transforms or the southward propagating tip of the EVZ and there is no observable change in thickness in the area of the hot-spot crest (RRISP Working Group, 1980). Areas where there is a noticeable increase in the thickness include northern and north-western Iceland, and the south-eastern corner under the proposed propagating EVZ (Palmason, 1971; Flovenz, 1980; RRISP Working Group, 1980).

Faulting in Iceland in the neovolcanic zone is dominated by normal extensional faulting, and is characterized by a horizontal minimum compressive stress in the north-west direction. The maximum stress varies between volcanic zones and between the latter and transform regions (Einarsson, 1990). The faulting varies in style and intensity over the neovolcanic zones, and seems to correlate with the morphological type of volcano. The volcanic systems of the WVZ and the NVZ show parallel fissure swarms (both eruptive and non-eruptive), well defined extensional normal faulting and shallow axial grabens (2 km wide). Superimposed on this regional stress in the more evolved volcanic systems, is a radial component induced by the presence of a central volcano (Ode, 1957). The off-axis shield volcanoes lack extensive normal faulting. The other main volcanic zones, or flank zones (EVZ and SVZ), lack the pronounced structural features described above, especially in the more distal regions of each zone (south and west, respectively) (Jakobsson 1979b). The structure of the Tertiary volcanic systems in eastern Iceland has been interpreted in terms of volcanic systems similar to those of the WVZ and the NVZ of the neovolcanic zone (Walker, 1963; Helgason, 1984, 1989). The lava piles on Iceland have been mapped and found to dip towards the present neovolcanic zones. There are two flexural monoclines in the Skagi region and in north-east Iceland (Figure 13.6a).

13.3.3 *Evolution of the Icelandic plate boundary*

The Iceland plateau began its evolution around 36 Ma (Oskarsson *et al.*, 1985) when the magma plume flux rate relative to the spreading rate increased (Vogt and Avery, 1974). From this period to the present, the location of the neovolcanic zone has never reached tectonic stability, as the relative position of the plume and the Mid-Atlantic plate boundary changed by a series of ridge jumps and ridge propagation episodes (Oskarsson *et al.*, 1985). The movement was initiated by the westward drift of the plate boundary relative to the stationary Icelandic plume. Although there are differing interpretations of the evolution of the plate boundary (Ward, 1971; Palmason, 1973; Saemundsson, 1974; Sigurdsson *et al.*, 1978; Johannesson, 1980; Vink, 1984; Oskarsson *et al.*, 1985; Jancin *et al.*, 1985), the following summary provides an outline of the origin of the currently active and recently extinct zones (Figure 13.7).

At 36 Ma, the plume was situated to the west of the plate boundary, and the Kolbeinsey ridge propagated northwards from the plume, causing

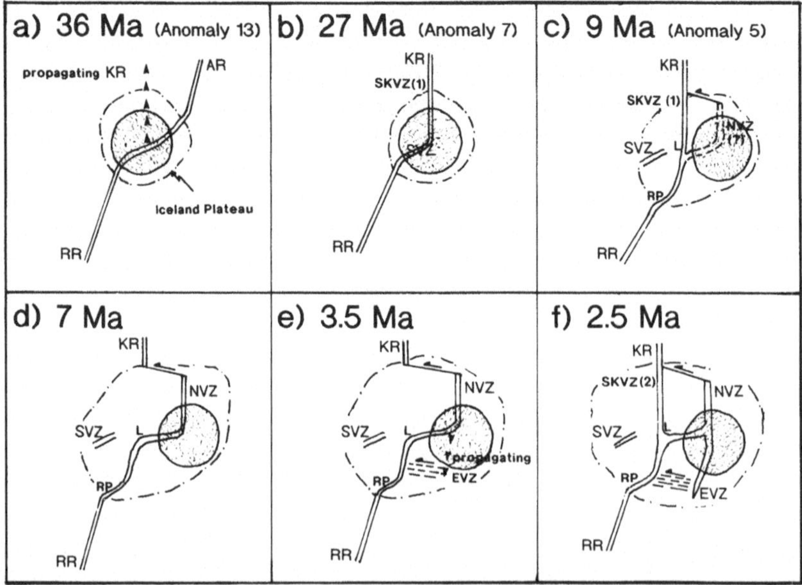

Figure 13.7 (a–f) Evolution of the plate boundary through time (36 Ma to present) (information extracted from Sæmundsson, 1974; Oskarsson *et al.*, 1985; Jancin *et al.*, 1985): KR = Kolbeinsey Ridge; RR = Reykjanes Ridge; AR = Aegir Ridge; SKVZ = Skaggi Volcanic Zone; SVZ = Snæfellsnes Volcanic Zone; NVZ = Northern Volcanic Zone; EVZ = Eastern Volcanic Zone; RP = Reykjanes Peninsula; L = Langjökull; Patterned area indicates the location of the centre of the head of the Icelandic plume.

extinction of a more easterly ridge segment (Figure 13.7a) (Vink, 1984). The northern end of the Reykjanes Ridge was represented by the Jokulbanki Ridge that bent into the Snaefellsnes peninsula. By 27 Ma this was connected to the Kolbeinsey Ridge via the Skagi Volcanic Zone (SKVZ) (Figure 13.7b). This configuration is thought to have remained stable for 17 Ma, during which time the plate boundary moved westwards relative to the plume. At 9 Ma, instability caused the Jokulbanki–Snaefellsnes Ridge to jump eastwards, producing the WVZ (Figure 13.7c). The NVZ and the TFZ may have been activated prior to this (Oskarsson *et al.*, 1985), or not until 7 Ma when the SKVZ became extinct (Jancin *et al.*, 1985) (Figure 13.7d). At 3.5 Ma the NVZ began to propagate southwards (Jakobsson, 1979b; Einarsson and Eriksson, 1982) at a rate of 3.5–5 cm y^{-1} producing the EVZ and the transient SISZ (Figure 13.7e) (Einarsson *et al.*, 1981). At 2.5 Ma (Figure 13.7f) the extinct SKVZ was reactivated for a short period (1.8 Ma) (Everts *et al.*, 1972), before the present day configuration was established. Since 0.7 Ma, the neovolcanic zones have changed very little, and have remained approximately in their current situation, with only minor oscillations of the order of 10 km.

The ridge jump has not always been from the west to the east, and temporally small oscillations have occurred, such as in the SKVZ (Sigurdsson *et al.*, 1978), and in north-east Iceland (Helgason, 1984, 1985, 1989), as indicated by the dips of the lavas (Figure 13.6a).

The Snaefellsnes peninsula (SVZ) has been interpreted as a transform region (Sigurdsson, 1970; Schafer, 1972; Saemundsson, 1974b; Sigvaldason, 1974; Sigurdsson *et al.*, 1978), based on east–west orientated volcanic systems (Piper, 1973) and off-axis seismicity between it and the Langjokull region, central Iceland, as reported by Einarsson (1990) (Figure 13.7c). The interpretation given here of the evolution of the neovolcanic zone suggests that it is analogous to the present day Reykjanes Peninsula, but is almost extinct. Unequal spreading rates have been suggested as a cause for the initiation of the SVZ 'fracture zone' as well as the SKVZ (Sigurdsson, 1970; Walker, 1975; Burke *et al.*, 1973). This has been challenged (Searle, 1976b) in that it is believed that it is a true volcanic zone undergoing extinction. It is not always possible to find direct modern oceanographic analogues for the plate configuration on Iceland, and this has led many workers to suggest that they may represent radial overflow of the plume (Einarsson *et al.*, 1977).

13.4 Mantle structure under the Atlantic and Iceland

13.4.1 *Asthenospheric mantle flow*

All thermal convection systems are driven by density differences which result from lateral temperature variations, and it has been proposed that the earth's mantle is convecting vigorously (McKenzie *et al.*, 1990). One of the foremost questions about the large-scale structure of the mantle is the exact nature of this vigorous convection: on what scale is the convection operating, and how many convective systems exist within the mantle?

The 'actual temperature' of the mantle increases with increasing depth. The 'potential temperature' is the actual temperature corrected to accommodate the effects of change in pressure and subsequent adiabatic decompression as one unit volume moves within the mantle via convection. As a block of mantle rises from (for example) the 1400°C horizontal isotherm, the particles expand and its real temperature decreases. Melting will occur when this temperature intersects the pressure and temperature sensitive solidus (McKenzie and Bickle, 1988, and references cited therein). This passive upwelling process is believed to occur beneath the mid-ocean ridges as the plates are moved apart by extension (McKenzie, 1984). More recent investigations reveal that the upwelling is not continuous along the ridge axis, and that it follows the pattern of ridge segmentation as envisaged by recent gravity work (Lin *et al.*, 1990).

Ballard *et al.* (1981), Whitehead *et al.* (1984) and Crane (1985) suggested that the central bathymetric high present in each ridge segment is due to thermal uplift and a local increase in the magma supply rate from an underlying magma chamber. This could also be achieved by localizing the magma production area in the mantle (Figure 13.2b). Mantle upwelling regions vary in size, and may control the size of the crustal segment above. Vogt and Johnson (1973) did not observe any segmentation of this sort on the Reykjanes Ridge, although it does occur in the vicinity of the Azores plume at 37°N (Searle, 1979) (Figure 13.1a). Segmentation on Iceland occurs on the scale of volcanic zones (Figure 13.3a) and/or on a smaller scale, represented by the volcanic systems (Figure 13.4a).

In concordance with mantle segmentation controlling crustal segmentation as discussed above, the distribution of the plate boundary should be initiated by asthenospheric mantle processes. Similarly, Vogt and Johnson (1973) postulated that the TFZ could dam up the plume asthenopheric flow but, in light of new evidence (Lin *et al.*, 1990) suggesting that transforms are caused by flow patterns in the asthenosphere, this seems unlikely. Large transforms (e.g. the Charlie Gibbs Fracture Zone) may be inherited from old continental structures at the time of continental break up, whereas small offsets and discontinuities may be controlled by mantle processes.

13.4.2 *Lithospheric thickness*

The average seismic P-wave velocity of the asthenosphere is 8.0–8.2 km s^{-1} under normal ocean basins and extends down to approximately 250 km. Bott (1965) postulated that there is anomalously low velocity mantle (7.2–7.4 km s^{-1}) under Iceland down to considerably greater depths, relative to normal mid-ocean ridges. Seismic tomography studies reveal that at 400 km depth (under the Azores and Iceland) this anomalous mantle of the plume heads is detected but below this the plume necks are not detected (Woodhouse and Dziewonski, 1984; Creager and Jordan, 1984, 1986; Grand, 1987). The normal asthenosphere may mix with that of the plume. Francis (1973) noted a decrease in the seismicity from 59.5°N on the Reykjanes Ridge towards Iceland, which was later interpreted in terms of the low velocity mantle from the plume extending this far south (Francis, 1975).

The lithosphere is the outer brittle layer that overlies the more plastic asthenosphere, and strictly includes all the solid material from the top of the crust down through the upper mantle to the top of the plastically deforming asthenosphere, and represents a solid rigid plate that may move independently. This boundary cannot be detected convincingly by refraction profiles alone, which only record changes in the seismic P-wave velocities, such that gravity surveys must be employed. At oceanic spreading centres the lithosphere thickens away from the axial zone, from 2 km at the ridge-axis (Searle, 1984) to 130 km thick at ages $>$ 100 Ma (Parsons and Sclater, 1977).

In Iceland the lithosphere–asthenosphere boundary has been interpreted as the boundary between seismic layers 3 and 4 (Palmason, 1971). More recent work confirms the observation that low velocity values occur below the Icelandic crust (7.4 km s^{-1} relative to 8.3 km s^{-1} on the Reykjanes Ridge) (RRISP Working Group, 1980). It appears that the upper mantle constituent of the lithosphere beneath Iceland, that would be expected to have P-wave seismic velocities of 8.3 km s^{-1}, is either mostly absent (Figure 13.6b), or there are some other peculiarities reducing the velocities to 7.4 km s^{-1}. In view of the cumulative evidence that suggests that this low velocity mantle extends down to depths greater than 250 km (Bott, 1965; Francis, 1969; Long and Mitchell, 1970; Woodhouse and Dziewonski, 1984; Creager and Jordan, 1984, 1986; Grand, 1987), it seems likely that the first interpretation is more accurate, as it satisfies the majority of the evidence. Underplating, which adds the cooling upper mantle to the base of the crustal lithosphere at ocean spreading centres, does not seem to occur under Iceland (Eysteinsson and Hermance, 1985).

Asthenospheric flow is frequently envisaged as radial from a plume (Morgan, 1971) and as lateral rolls from the MOR (Lin *et al.*, 1990). The radial flow of plume mantle may be represented by the distribution of the active volcanic and spreading zones and, in addition, physiographical and geochemical observations (Morgan, 1971; Vogt, 1971; Schilling, 1973a; Dewey and Burke, 1974; Einarsson *et al.*, 1977; Wyss, 1980), features which are exhibited by Iceland and the surrounding region.

Vogt (1971) and Schilling (1973a) suggested that variations in the topography along the Reykjanes Ridge and other regions near hot-spots were indicative of fluctuations in the rate of the asthenospheric mantle flow longitudinally underneath the ridge axis. Schilling (1973b) commented on the fact that the flow rate of the plume must be substantially greater than the flow beneath MORs, and suggests that this is why these hot-spot regions have anomalous crustal thicknesses. The degree of overflow varies from plume to plume, and temporally about a single plume. Vogt (1972) assessed the topography of the aseismic ridge between the Faeroes and Greenland and showed that between 50 and 60 Ma the discharge rate was very high and that it subsequently decreased to a minimum in the mid-Tertiary, before increasing again in the Late Tertiary.

Recent views of mantle processes, however, suggest that the upwelling plume mantle is 150–200°C hotter than that upwelling beneath the MOR, and an increase of 100°C in the melting region will produce double the volume of magma (White and McKenzie, 1989b). The consequence of this is that the variations in the topography produced by crustal thickness variation have to be a direct response of the mantle potential temperature, and not fluctuations in the discharge rate by the plume. This accentuates the importance of the morphology in interpreting mantle dynamics that are directly relevant to geochemical models.

13.4.3 *Existence of axial magma chambers*

Considerable research has been directed towards discovering the existence and nature of magma chambers at ocean ridges (Iyer, 1984; Browning, 1984; Orcutt *et al.*, 1984; Detrick *et al.*, 1987; Macdonald, 1989; Foulger and Toomey, 1989; Kent *et al.*, 1990; and for a recent concise review, Langmuir, 1990), although it is not within the scope of this chapter to review such work in any detail. In summary, however, very small crustal magma chambers have been indentified at the fast spreading EPR (<2 km depth), whereas none have been found at the MAR, although they have been imaged in Iceland (at about 3 km depth). If small chambers (<200 m thick) are present in the Atlantic, then they would be extremely difficult to image seismically. Contrasting views exist over the number of chambers per crustal segment in the EPR (either one, as suggested by Ballard *et al.*, 1981; Whitehead *et al.*, 1984; Crane, 1985; White, 1989; or more than one, Thompson *et al.*, 1989), and the existence of ponded magma at the base of the crust as well as or instead of a crustal chamber (Cann, 1970; White, 1989; Dick, 1989). The most recent efforts in the MAR (MARK area) have failed to image a crustal magma chamber (Detrick *et al.*, 1990). In Iceland, magnetotelluric investigations have suggested there is a large ponded magma reservoir under the neovolcanic zones (15% partial melt in this zone of the mantle), that increases in size towards the centre of the island (Eysteinsson and Hermance, 1985). It seems then that the crustal thickness, spreading rate, and the requirement of sustained mantle temperatures may control the existence of a crustal chamber; the latter may be the most important.

The presence of a magma reservoir is significant to basalt petrogenesis because of the differentiation and mixing processes that may occur within it and which can mask the original source geochemical signature. However, compatible elements can be used to identify differentiation processes and the extent to which they have operated on the magmas. In contrast, the incompatible elements are relatively unaffected by these processes and, together with radiogenic isotopes, can be used to characterize the source and partial melting processes occurring in the mantle.

13.5 Petrographic series

On the basis of modal mineralogy, CIPW norms (Yoder and Tilley, 1962), total alkali contents (Macdonald and Katsura, 1964) and TiO_2 abundances (Chayes, 1965), basin oceanic basalts can be divided into different magmatic series. In the Atlantic and Iceland these are the tholeiitic, transitional alkali and alkali basalt series (Jakobsson, 1979a). Their characteristic geochemical features are summarized in Figure 13.8. The tholeiitic series is by far the most

(a)

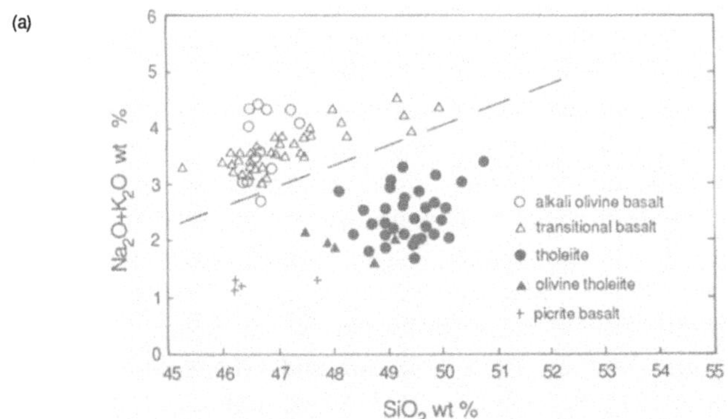

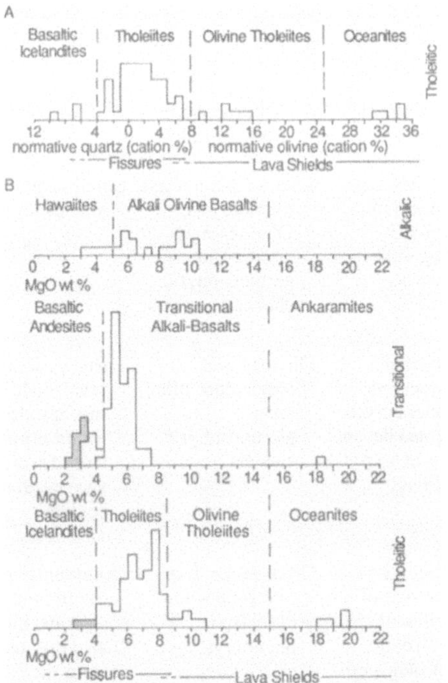

Figure 13.8 (a) Alkali:silica diagram for Postglacial basalts from the Western Volcanic Zone and the Eastern Volcanic Zone discriminating between the different series (Jakobsson 1979a). (b) A: frequency distribution of postglacial rocks from the Western Volcanic Zone with respect to the normative content (cation %) of either olivine or quartz tholeiites. B: frequency distribution of postglacial rocks from the Eastern Volcanic Zone (alkali and tholeiite series) and the Western Volcanic Zone (tholeiite series) with respect to the MgO content (Jakobsson 1979b).

abundant in the ocean basins, followed by the transitional alkali and the alkali series, with basaltic rocks being the most voluminous.

Although magmatic series contain more evolved members than the dominant basalt, there is a distinct absence of acidic rocks in the North Atlantic. Oceanic andesites have been reported from aseismic ridges (Faeroes–Greenland), and acid and intermediate rocks form 9% of the erupted product on Iceland associated with the central volcanoes (Saemundsson, 1979). The bulk of the evolved rocks in Iceland are produced in the transitional and alkalic series. On the Reykjanes Ridge, evolved basalts and andesites occur in DSDP hole 409 in 2 Ma crust (Wood *et al.*, 1979a). There is controversy over the origin of these more evolved rocks, especially the acidic members (Moorbath and Walker, 1965; O'Nions and Gronvold, 1973; Sigvaldason, 1974; O'Hara, 1975; Sigurdsson, 1977; Macdonald *et al.*, 1987; Storey *et al.*, 1989; Thompson *et al.*, 1989), although assimilation fractional crystallization (AFC) or fractional crystallization are the two main processes invoked

Table 13.1 Characteristic petrography within the tholeiitic series. Compiled from various authors (see text)

Basalt groups	Mineralogical assemblages[a]	Characteristic petrography	Volcanic form on Iceland	Comment
Picrite	Abundant Mg-rich ol phenocrysts in a Mg-rich ol + plag + cpx groundmass	Lack of plagioclase phenocrysts. Olivines may often be in glomerophyric clusters	Small lava shields	There is often some dispute over the origin of these magmas, and many believe that they are of a cumulate nature
Olivine tholeiite	Plag + Mg-rich ol phenocrysts, microphenocrysts of plag needles in a plag + ol + cpx groundmass	Olivines and plagioclase often form glomerophyric clusters	Large to small lava shields, but are also found erupted from fissure swarms and central volcanoes	In the hand specimen these are very distinctive, and are the lava type of the large shield volcanoes in Iceland
Tholeiite	Plag phenocrysts ± cpx. Mg-rich ol rarely present >3%. Groundmass of plag + cpx	Often more than one generation of feldspar phenocryst	Dominantly on the fissure swarms, but also erupted from the central volcanoes	The mineralogy and texture may be extremely variable
Quartz tholeiite	Plag phenocrysts ± cpx. Mg-rich ol never present. Groundmass of plag + cpx	Phenocrysts very abundant and up to three generations may be present	Central volcanoes and mature fissure swarms	These are probably over-represented in Iceland

[a](ol) olivine; (plag) plagioclase; (cpx) clinopyroxene

(Chapter 5). In general, the evolved compositions are more characteristic of the Atlantic oceanic islands (e.g. the Azores, Canaries, Tristan da Cunha) than the oceanic ridge (Chapter 9).

13.5.1 *Tholeiitic basalt series*

The dominant basalts erupted in the North Atlantic and Iceland fall into two petrological groups (Table 13.1 and Figure 13.8)—the olivine tholeiites and the tholeiites—both of which are found intercalated in a core section from the FAMOUS area (37°N, MAR, DSDP Leg 37) (Bougault and Hekinian, 1974; Blanchard *et al.*, 1976), and have also been reported from 45°N (Aumento, 1967), 53°N (Hekinian and Aumento, 1973) and 25°N in Cretaceous crust (Rice *et al.*, 1980).

In Iceland, these two groups are identified as the dominant eruptive products. The olivine tholeiites are erupted mainly in the off-axis shield volcanoes, but are also observed in the axial region, whereas the tholeiites are confined to the latter within the central volcanoes and fissure swarms (Figure 13.5). Within the present day rift zone, the olivine tholeiites are confined to the WVZ and the NVZ, and may become more abundant towards the centre of Iceland in the Langjokull region (Sigurdsson *et al.*, 1978; Schilling *et al.*, 1978; Meyer *et al.*, 1985). Within a single volcanic system, Jakobsson *et al.* (1978) suggested a temporal cyclic relationship between the two magma types. The large off-axis shield volcanoes are much rarer in the Tertiary lava pile, and it has been suggested that these large voluminous eruptions are due to the decompression associated with isostatic rebound towards the end of and after deglaciation.

Which of these two petrological groups is the dominant erupted product in both Iceland and the MAR is unclear. On Iceland, the occurrence of the olivine tholeiites in the axial region, as well as off-axis, suggests that they could be dominant overall in late glacial and post-glacial times. There are contrasting reports of the dominant basalt type in the MAR, that could reflect a sampling bias, and the lack of refined correlation between bathymetry, location and relative ages of the samples. If such shield volcanoes occur in the Atlantic as off-axis volcanoes, then the same may apply. Off-axis volcanism has been reported by Leg 49, on the Reykjanes Ridge (Luyendyk *et al.*, 1979), but this may be too close to Iceland to be typical. Another occurrence is at 37°N within the median valley, at comparable distances observed in Iceland of 15 km (Figure 13.1) (Needham and Francheteau, 1974), but this region may also be atypical.

Many workers consider these two types of basalt to be derived from each other by simple fractional crystallization processes, but there is frequently a chemical hiatus between them (Bougault and Hekinian, 1974; Blanchard *et al.*, 1976; Jakobsson *et al.*, 1978), suggesting that they are not from the same batch of magma and have different fractional crystallization histories.

The parental magmas for the tholeiites are very rarely, if at all, sampled. The geochemical and field evidence on Iceland suggests a tectonic control for the distribution differences in these two magmas, with the olivine tholeiites being derived directly from the mantle, with very little modification en route. Either way these two tholeiitic types are petrographically representative of typical mid-ocean ridge basalts.

13.5.2 *Alkali basalt series*

The mineralogy and petrology of the alkali and transitional alkali series are much more varied. The series characteristically contain undersaturated feldspathoid minerals and can also be recognized by their alkali pyroxene compositions. Nevertheless, the North Atlantic basalts rarely possess feldspathoid minerals with the normative nepheline 'hidden' in the pyroxene. Bass (1971) and Ridley *et al.* (1973) pointed out that slow spreading ridges have a greater abundance of alkali rocks. In the Atlantic Ocean they have been reported from: (1) Reykjanes Ridge, in the same drill hole as tholeiitic lavas (Wood *et al.*, 1979a); (2) 45°N (Aumento, 1968; Wood *et al.*, 1979b; Tarney *et al.*, 1979); (3) the FAMOUS region (37°N) (Aumento, 1967) and (4) from various oceanic islands, such as the Azores (40°N) (Assuncao *et al.*, 1970; Schminke, 1973; White *et al.*, 1979; Davies *et al.*, 1989; Storey *et al.*, 1989) and the Cape Verdes (14.6–17°N) (Davies *et al.*, 1989) etc. In Iceland, the lateral zones (EVZ, SVZ, OVZ and the extinct SKVZ) are the only places that basalts with this alkali affinity have been erupted (Figure 13.3).

13.5.3 *Relationship and origin of the different series*

There has been considerable dispute over the origin of the alkali basalt series. To help resolve this problem the spatial and temporal relationships between the two series may provide some additional constraints. Iceland provides a perfect opportunity to examine such relationships. There are 29 volcanic systems currently active in Iceland; only four of these are alkaline, seven are transitional and the remainder are tholeiitic (Figure 13.8) (Jakobsson, 1979a). In view of the current interpretation of the plate boundary configuration (section 13.3.1; Saemundsson, 1974a; Einarsson, 1990), these alkali and transitional alkali rocks occur in the propagating tip of the EVZ, and in a zone that may represent either a 'leaky' transform zone or a diminishing volcanic zone (SVZ), (Sigurdsson, 1970). Alternatively, they can collectively be seen as an expression of the radial flow from the plume. The alkali volcanic systems are much less productive than the tholeiitic systems.

There is strong evidence to support mixing of the tholeiitic and alkalic series at crustal levels (McGarvie, 1984; Blake, 1984). Recently, McGarvie *et al.* (1990) have shown that the tholeiitic magmas of a volcanic system in the north of the EVZ travel up to 100 km to mix with the transitional alkali series (rhyolites and basalts) in the south and that this process of mixing commenced and became more frequent in post-glacial times (supporting the idea of a propagating rift). The alkali volcanic systems have noticeably fewer primitive equivalents to the tholeiitic picrites outcropping in the post-glacial eruptives of the WVZ and lack the off-axis shield volcanoes, as well as displaying a very high percentage of acid rocks (Jakobsson, 1979a).

The tholeiitic volcanic systems in Iceland display well pronounced tectonic fabrics akin to those in the Atlantic, but these are reduced in the alkali systems. It should be pointed out that the magmas erupting above the area of the hot-spot, at Vatnajokull, show no alkali characteristics. There is no consistent correlation between the erupted volumes of these two magma types in a single eruption, as has previously been suggested (Sigvaldason, 1974), but it varies regionally and temporally within and between single volcanic systems (Jakobsson, 1979a; Sigurdsson *et al.*, 1978).

Within the Tertiary volcanics (16–3.1 Ma) currently exposed, at least 44 volcanic systems have been identified, all of which are tholeiitic (Walker, 1963). This may be due to the obvious potential sampling bias of the limited exposure of the Tertiary rocks in Iceland. There can be a gradual change in the type of series being erupted within a single volcanic zone with time, which may be either towards alkalinity or the reverse. In the Plio-Pleistocene and Upper Pleistocene (3.1–0.7 Ma), which was the time of onset of glaciation in Iceland, the EVZ and the SKVZ became active, and the SVZ began to erupt more alkali rocks. Both the SVZ (Sigurdsson, 1970) and the SKVZ (Sigurdsson *et al.*, 1978) demonstrate a trend towards alkalinity during approximately a 2–3 Ma period (Figure 13.7e and f). The southern part of the EVZ has always erupted alkali and transitional alkali rocks (Jakobsson, 1979a) since the Pleistocene.

Sigvaldason (1974) and O'Nions *et al.* (1973) have shown that both alkali and tholeiitic magmas can be derived from one homogeneous source, as indicated by the Sr isotope data from the alkaline Westerman Islands and tholeiites from the rest of the neovolcanic zone. This suggests that other processes are responsible for the petrographic differences. However, other workers (e.g. Hart *et al.*, 1973) have argued that remelting a metasomatically-enriched mantle could produce alkali basalts, in a similar manner to the process proposed for Hawaii (Frey and Roden, 1986). This theory would require virtually all the lower oceanic lithosphere to be metasomatized if all the alkali basalts from the Atlantic oceanic islands are produced in this way. This seems to be drawing on a process that is far more complicated than necessary. The Cape Verdes hot-spot presents a problem to this theory as it has been penetrating the same oceanic lithospheric section for 125 Ma (Courtney and White, 1986), being situated near the African pole of rotation. This theory can satisfy neither the geophysical evidence for these areas nor the evidence in Iceland. There is no upper mantle part of the lithosphere under the EVZ in Iceland (Figure 13.3), suggesting that metasomatic enrichment is not the cause of the eruption of alkali basalts in this area. An alternative model whereby the alkali basalts in Iceland are produced by melting the thicker lower crustal layers of the flank zones, which contain alkali amphibole (Oskarsson *et al.*, 1985; Steinthorsson *et al.*, 1985), can be discounted on isotopic evidence as the observed ratios would not have had time to develop in the crust following remelting (Meyer *et al.*, 1985).

If it is considered that the alkali magmas are derived from the same mantle source as the tholeiites by a melting mechanism alone, then the percentage of partial melt must be very small and the area undergoing melting must be much larger than that which produces a tholeiitic magma batch of the same volume (Gast, 1968). In this instance MOR tholeiites are believed to be derived by 20–30% partial melting whereas alkali magmas represent only 4–6% melting. Similarly, irrespective of the depth of generation of primitive MORB (Chapter 6), alkali basalts, if from a similar homogeneous source, must have originated from even greater depths than the tholeiites, or equilibrated with the mantle at higher pressures.

Can inferences concerning source homogeneity and degree or depth of melting of alkali and tholeiitic melts be related to the distribution relationships observed in Iceland? The EVZ shows a gradual increase in alkalinity towards the south, which corresponds to an observed increase in crustal thickness in this direction (Palmason, 1971) and increased distance from the melting column of the plume that must be rising beneath Iceland. The reason for the presence of the alkali rocks in this region could, therefore, be attributed to smaller degrees of partial melting, at greater depths in the mantle, reaching this region relative to the rest of the neovolcanic zone in Iceland. This is in accordance with a propagating tip penetrating an older and thicker crust.

The SVZ erupted tholeiitic magmas throughout the Tertiary until 2.5 Ma. There has been a gradual change in the amount and chemistry of the magma produced in this region towards a smaller volume of more alkali magma (Sigurdsson, 1970). Sigurdsson *et al.* (1978) demonstrated that the SKVZ also erupted magmas that became progressively more alkalic with time within this geological period. Palmason (1971) reports that the depth to the asthenosphere in this region increases northwards from 8 km at Langajokull to 16 km under the SKVZ. The reduced proximity of the melting regions and heat source (centre of the plume) with time in these volcanic zones would be compatible with a ridge jump initiated by the plume as the Mid-Atlantic plate boundary drifted westwards. This gradual reduction in the heat flow to the melting source regions would cause a decrease in the percentage of partial melt produced, and the thickening crust clogging up the volcanic zone would possibly lead to extinction, as in the case of the SKVZ. The SVZ, although further away, is orientated in an east–west direction and could represent a weak radial flow line of the plume (Einarsson *et al.*, 1977). The OVZ may be a similar flow direction, running sub-parallel to the EVZ

(Figure 13.3). The most recent geophysical evidence from the Atlantic (Lin *et al.*, 1990) may support the idea of mantle flow being the cause of ridge segmentation, that is, the cause of the distribution of the volcanic activity. This would suggest that the different series in the EVZ and the SVZ may be derived from the same source region, although geochemical evidence indicates that this may not be the case for the SVZ (see later).

13.5.4 *Clinopyroxene-phyric basalts*

Clinopyroxene phenocrysts are anomalously abundant in basalts from the vicinity of 34–45°N and 60–68°N, peaking on the Azores and Iceland. This has been related in various ways to the presence of the plumes in these areas (Cann, 1970; Moore and Schilling, 1973; Bougault and Hekinian, 1974). There are exceptions to this general observation; for example, at 45°N on the MAR, which has many of the geochemical characteristics of hot-spots but no clinopyroxene anomaly (Schilling *et al.*, 1983). The relative abundance of clinopyroxene between hot-spot areas varies, as in the lavas from the Azores relative to those from Iceland. Schilling *et al.* (1983) interpret this in terms of these regions having distinct fractional crystallization histories. The pyroxene-bearing basalts from the FAMOUS area are more evolved than the tholeiitic suites from the same dredge haul, suggesting they formed by late fractionation in a magma chamber. However, Bougault and Hekinian (1974) postulated that 'pyroxene basalts' are derived from a different parental magma than the two usual tholeiitic groups.

A smaller percentage of partial melting of a uniform source would be expected to produce abundant clinopyroxene in the magmas. This is the case in the highly undersaturated alkali basalts, but the alkali olivine basalts and the olivine tholeiites are equally saturated with respect to clinopyroxene, but rarely exhibit it as a phenocrystic phase, that is, Mg-rich magmas usually do not possess clinopyroxene phenocrysts, whereas Mg-poor evolved magmas may. The occurrence of clinopyroxene as a phenocryst phase is rare in typical MORB, but it can occur (Rice *et al.*, 1980) and exhibits two forms, the equant type in evolved rocks with low MgO concentrations and the anhedral ('resorbed') type that has a disputed origin.

What is important to establish is the order of crystallization and not just the presence of clinopyroxene. It is assumed that the equant type phenocrysts have formed in the usual way by fractionation of olivine + plagioclase + clinopyroxene, possibly in a magma reservoir. The resorbed type phenocrysts are thought to crystallize before most of the olivine and all the plagioclase, and to be derived from 'somewhere else'. It is the latter that is a matter of dispute; the clinopyroxene could be derived by mixing of a new, relatively unevolved, magma batch with phenocrysts of a previously more evolved batch. This undoubtedly occurs in MORB genesis, as is displayed by plagioclase phenocryst generations (Blanchard *et al.*, 1976), but analysis of the compositions of some of the clinopyroxenes involved reveals that they have Cr- and Mg-rich undifferentiated compositions. Experimental evidence suggests that with an increase in pressure, the order of fractional crystallization will alter to clinopyroxene + olivine. The relict phenocrysts could have crystallized on the walls of dykes at depths of approximately 20 km before being carried up to the surface as xenocrysts. This raises the question of whether the so-called 'clinopyroxene anomaly' towards Iceland is due to increased crustal thickness, enabling differentiation to occur either in the magma chamber to produce the equant type phenocrysts (Luyendyk, Cann *et al.*, 1979), and/or at deeper levels in the crust for resorbed-type phenocrysts. With the information on the crustal structure that is already known, answers to the above questions can be derived. In the off-axis plumes such as the Azores and the Cape Verdes, the presence of the sub-oceanic lithosphere could provide a place for the high pressure xenocrystic clinopyroxene to crystallize.

13.6. Geochemical variation

13.6.1 *Basaltic chemical types and the plume model*

Morgan (1971, 1972) and Vogt (1971) introduced the idea of hot convective mantle plumes rising beneath oceanic islands (Chapter 9). Initially MORB tholeiites were thought to be remarkably uniform and, against this reference point, a progressive increase in light rare earth element (REE) enrichment, large ion lithophile elements (LIL), halogens and the radiogenic isotopes Pb and Sr, coupled with a decrease in the Nd isotopes, was observed towards the plume regions (in particular Schilling, 1973a, b, 1975; but also by Tatsumoto *et al.*, 1965; Gass, 1970; Peterman and Hedge, 1971; Hart *et al.*, 1973; Sun *et al.*, 1975; White 1976; White and Schilling, 1978; Schilling *et al.*, 1983). It was proposed that this chemical gradient was due to the presence of two distinct source regions, a depleted low velocity zone (DLVZ) source region that is globally present under the ocean ridge systems, and a primordial hot mantle plume (PHMP), which rises up from deep in the mantle. The two sources mix in varying proportions beneath the ridge axis, but progressively less mixing takes place with increasing distance from any one plume. The plumes in the Atlantic were thought to define mantle domains and attempts were made to map these using the major element composition of glasses (Melson and O'Hearn, 1979; Dmitriev *et al.*, 1979; Sigurdsson, 1981), incompatible elements and isotopes (Schilling *et al.*, 1983). The latter study confirmed Sigurdsson's (1981) observations, but noted that although the domain boundaries are not sharp, differentially enriched plumes and 'normal' segments along the MAR between the plumes could be distinguished. Fracture zones may act as natural barriers to plume activity, such as the Hayes Fracture Zone which cuts off the Azores plume domain (Bougault and Cande, 1985; White and Schilling, 1978; Bougault and Treuil, 1980).

A chemical terminology for basalts was established based on what was essentially the Atlantic model for slow spreading mid-ocean ridges. The latter can be described in terms of the degree of elevation of the median valley relative to the norm, which shows an approximate positive correlation with the shallowness of the ridge in the Atlantic. Elevated (E-), transitional (T-) and normal (N-) ridge segments are a physiographical terminology generally applied to the MAR (and other mid-ocean ridges). It was then discovered that a good correlation exists between the degree of elevation and shallowness of the segments on the one hand, and the chemistry of the basalts erupted in the vicinity of a plume on the other (Schilling, 1973a, b). 'Elevated' was loosely redefined as a genetic term as 'plume' (P-) (see Schilling, 1975, for detailed definitions; subsequently used by Schilling *et al.*, 1983; Ito *et al.*, 1987).

The relative enrichment of the MAR basalts was estimated in terms of the degree of enrichment of the light REE. This can be demonstrated by the La/Sm concentration ratio, or related functions, such as the ratio of the enrichment factor relative to a chondritic composition, that is, $[La/Sm]_{EF}$ (Schilling, 1973a). Unless otherwise stated in this chapter, the relative degree of elemental enrichment and depletion is related to chondritic values. Normal ridge segment basalts (N-MORB) were defined as possessing an $[La/Sm]_{EF}$ of less than one, transitional ridge segment basalts (T-MORB) of about 1, and plume or elevated ridge segments (P-MORB or E-MORB) greater than one. The terminology was applied to other ocean basins regardless of the spreading rate and linked composition with ridge physiography. Later Schilling *et al.* (1983), redefined N-MORB as $[La/Sm]_{EF} < 0.7$, the transitional segments as having a sharp gradient of enrichment in the $[La/Sm]_{EF}$, and the 'plume' elevated segment displaying a maximum in $[La/Sm]_{EF}$ with arbitrary cut-off values for individual plumes. The terminology is weakened, however, by the frequent fluctuation of the degree of enrichment of basalts from within a single transitional or enriched segment drill hole (Wood *et al.*, 1976b; Tarney *et al.*, 1979). Also, Le Roex *et al.* (1985b) observed transitional enriched basalts in the very slow spreading segments of the American–Antarctic ridge far from the Bouvet plume. This emphasizes the pitfalls of applying genetic significance to the terminology of the basalts.

As the preceding section has outlined, much of the variation within E-MORB and T-MORB areas of the MAR was originally accounted for by interaction between enriched plume sources and depleted MORB sources. However, when data from the Atlantic increased sufficiently in the 1970s to highlight the variability of enrichment (DSDP Legs 37 and 49), there was a fashion of opposition to the mantle plume hypothesis. Various alternatives were proposed including: (1) a veined source (Dick, 1977; Hanson, 1977; Wood,

Table 13.2 Geochemical data for different MORB types from the Mid-Atlantic Ridge (MAR) relative to examples of Icelandic basalts. Average N-MORB and E-MORB from Sun and McDonough (1989)

	Basalt type, locality and (sample number)								
	N-MORB	T-MORB	E-MORB	E-MORB	Icelandic basalts				
Analysis	MAR 37°N S. HFZ (82-556-2-1)[a]	MAR 63°N Reykj. Ridge (49-407-45-3)[b]	MAR 63°N Reykj. Ridge (49-409-24-2)[b]	MAR 63°N Reykj. Ridge (49-407-47-1)[b]	Tholeiite (jak 97a-2)[c]	Transitional alkali basalt (jak 97a-6)[c]	Hawaiite (jak 97a-10)[c]	Average N-MORB[d]	Average E-MORB[d]
Major oxides (wt%):									
SiO_2	50.56	48.69	49.86	—	49.3	47.08	48.46	—	—
TiO_2	1.04	2.15	1.78	—	2.28	4.63	3.00	—	—
Al_2O_3	15.96	15.23	14.32	—	13.51	12.71	16.40	—	—
Fe_2O_3*	7.83	11.92	12.75	—	14.57	15.47	13.37	—	—
MnO	0.15	0.13	0.18	—	0.24	0.22	0.26	—	—
MgO	7.89	8.10	7.70	—	6.34	5.06	3.82	—	—
CaO	12.74	10.60	11.39	—	10.71	9.91	7.71	—	—
Na_2O	2.44	2.65	2.44	—	2.46	3.08	5.11	—	—
K_2O	0.16	0.19	0.23	—	0.24	0.72	1.19	—	—
P_2O_5	0.11	0.23	0.19	—	0.21	0.57	0.62	—	—
Trace elements (ppm):									
Ba	30	—	70	75	—	—	—	6.3	57
Co	—	71	48	40	—	—	—	—	—
Cr	326	313	317	122	—	—	—	290	46
Hf	1.8	3.69	3.1	3.2	—	—	—	2.05	2.03
K	—	—	1494	3155	—	—	—	600	2100
Nb	1.6	—	11.5	17.7	—	—	—	2.33	8.3

Ni	140	146	106	48.4	—	—	—	138	32
Pb	—	—	—	261	—	—	—	0.3	0.6
Rb	2	—	4.4	6.3	—	—	—	0.56	5.04
Sc	—	45.6	45.3	17	—	—	—	44	39
Sr	112	—	103	173	—	—	—	90	155
Ta	0.09	0.90	0.68	0.87	—	—	—	0.132	0.47
Th	0.14	0.61	0.59	0.67	—	—	—	0.12	0.60
Ti	—	12769	10431	11750	—	—	—	7600	6000
V	267	383	377	357	—	—	—	—	—
Y	26	—	41.8	35.9	—	—	—	28	22
Zr	62	140	116	130	—	—	—	74	73
Rare earth elements (ppm):									
La	1.84	8.4	6.1	8.4	—	—	—	2.5	6.3
Ce	6.45	24	18.6	23.1	—	—	—	7.5	15
Nd	6.88	17	13.9	16.1	—	—	—	7.3	9.0
Sm	2.59	5.2	4.71	5.02	—	—	—	2.63	2.60
Eu	0.98	1.9	—	—	—	—	—	1.02	0.91
Gd	3.66	—	—	—	—	—	—	3.68	2.97
Tb	0.71	0.91	0.97	0.93	—	—	—	0.67	0.53
Tm	0.43	0.6	—	—	—	—	—	0.456	0.356
Yb	2.77	3.22	3.94	3.21	—	—	—	3.05	2.37
Lu	0.45	0.52	—	—	—	—	—	0.455	0.354

[a] Weaver *et al.* (1982)
[b] DSDP Leg 49
[c] Jakobssen (1979a)
[d] Sun and McDonough (1989)

1979; Wood *et al.*, 1979a, b; Tarney *et al.*, 1979, 1980; Dick *et al.*, 1984; Le Roex *et al.*, 1983); (2) a streaky source (Zindler *et al.*, 1982, 1984; Fitton and James, 1986); (3) a 'marble cake' mantle (Allegre and Turcotte, 1986), and (4) fluid or gaseous phases inducing metasomatism (Green, 1972; Frey and Green, 1974; Lloyd and Bailey, 1975; Frey *et al.*, 1978; Schilling *et al.*, 1980; Menzies, 1983; Schilling *et al.*, 1983; Dick *et al.*, 1984; Stolz and Davis, 1988).

The plume model (e.g. Schilling *et al.*, 1983), however, apparently satisfies the following features: (1) thermally-related geoid bulge (Morgan, 1971); (2) geophysical evidence suggesting that melting occurs down to 250–400 km depth beneath Iceland (Bott, 1965, 1988; Tryggvason, 1964; Francis, 1969; Long and Mitchel, 1970; Hermance and Grillot, 1970; Woodhouse and Dziewonski, 1984; (3) crustal thickness increase (Palmason, 1971); (4) flow fabrics and/or temperature variation (Vogt, 1971; McKenzie, 1984); (5) most of the isotopic geochemical variation, and (6) the temporal persistency of two source regions (for 60 Ma in Iceland). The idea of MOR magmatism existing as a passive response to extensional spreading, and large plumes (like Iceland) having a different origin of a more forceful nature were being widely, but theoretically, discussed in the early 1970s (Morgan, 1971; Vogt, 1971; Schilling, 1973a, b; Schilling and Noe-Nygaard, 1974). These ideas formed the basis to similar and currently accepted views of plume activity (e.g. White and McKenzie, 1989a, b).

13.6.2 *Normal ridge segment: N-MORB tholeiites*

It is useful to use N-type MORB as a reference frame, as this is globally the most abundant basalt type within oceans. The chemical characteristics of the different MORB types are listed in Table 13.2.

N-MORB typically displays a depleted nature, relative to chondritic values, as demonstrated by the low abundances of incompatible trace elements relative to the compatible elements. The $[La/Sm]_{EF}$ is low (< 1). The radiogenic isotope ratios, $^{87}Sr/^{86}Sr$ (0.70234–0.70245) and $^{206}Pb/^{204}Pb$ (18.53–18.74) are also depleted, whereas $^{143}Nd/^{144}Nd$ ratios are enriched (0.51318–0.51322) (Park and Staudgel, 1990). As demonstrated globally there is usually a good negative correlation between $^{143}Nd/^{144}Nd$ and the $^{87}Sr/^{86}Sr$ ratios for MAR basalts, for example, from 11°N and 6°S (Hart, 1976; O'Nions *et al.*, 1977). By definition, N-MORB should only show minor deviations in composition spatially or temporally, as is generally observed, suggesting that the source region for these basalts is remarkably uniform on the scale sampled by the erupted magmas.

The MARK area in the Atlantic (23°N) is one of the most thoroughly studied areas in terms of overall geology (Purdey *et al.*, 1979; Detrick *et al.*, 1984 for the bathymetry; Mayer *et al.*, 1985 for a general account). From 22 to 24°N (Kane Fracture Zone) the ridge is divided into three segments separated by non-transform discontinuities. ODP Leg 106/109 sunk holes in two of these segments on zero-age crust (Detrick *et al.*, 1990; Donato *et al.*, 1990) and found only very minor variations in the degree of depletion between the three sites. DSDP Legs 51, 52, and 53 sampled Cretaceous crust (108 Ma) at 25°N, on the same flow line as Leg 106/109, and retrieved N-MORB, suggesting that the MORB source has remained uniform for this length of time to the present. However, Cretaceous crust from the Caribbean displays some anomalous features and has been compared to thickened segments of oceanic crust such as oceanic plateaux (Floyd, 1989).

13.6.3 *Transitional ridge segments: T-MORB tholeiites*

These basalts are enriched in LIL elements, $^{87}Sr/^{86}Sr$ and $^{206}Pb/^{204}Pb$ ratios and depleted in $^{143}Nd/^{144}Nd$ ratios relative to N-MORB, and the change should occur with a steep gradient such as seen along the Reykjanes Ridge (Figure 13.9). Fluctuations and variations in the same location from light REE

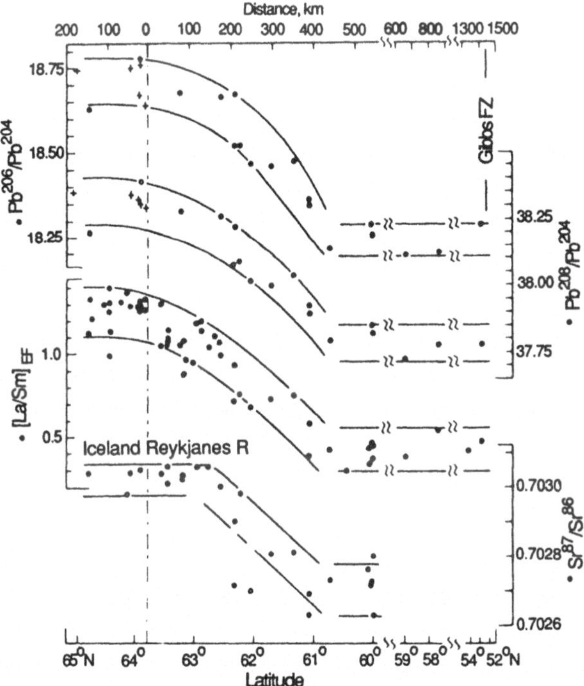

Figure 13.9 Chemical gradients observed along the Reykjanes Ridge, which formed the basis for Schilling's model. Ridge basalts were dredged by R.V. Trident during cruises TR41 and TR101 in 1967 and 1971. Note the broken scales south of 60°N. La and Sm enrichment factors relative to chondrite meteorite concentrations from Sun *et al.* (1979).

depleted to light REE enriched seems to be more typical of basalts in the transitional regions. Joron *et al.* (1980) demonstrate that there are comparable La/Ta ratios for the Reykjanes Ridge, 37°N and 45°N and classifies them all as T-MORB accordingly, although Schilling *et al.* (1983), considered 45°N as an E-MORB segment.

13.6.3.1 *Reykjanes Ridge* (*MAR*, 61°*N to* 63°*N*). One of the two best known examples of T-MORB occurrences is the Reykjanes Ridge (Schilling 1973a,b; Hart *et al.*, 1973; Hart and Schilling, 1973; Sun *et al.*, 1975) (Figure 13.9). This area shows a gradual change from N-MORB, south of 61°N, to E-MORB at 63°N, 400 km south of the Iceland coast (Figure 13.9). Segmentation occurs on the Reykjanes Ridge only in the form of *en echelon* fissure swarms, very similar to those seen on Iceland. The median valley becomes less pronounced and shallower towards Iceland.

The major element compositions of the tholeiites show only slight variations (Sigurdsson, 1981), whereas the incompatible and the highly incompatible elements become progressively more enriched towards Iceland (Schilling, 1973b). Chondrite-normalized REE patterns change from light REE depleted, flat, and then slightly light REE enriched towards and on Iceland. This gradient occurs over 400 km of the Reykjanes Ridge and correlates well with other elements. Sr isotope work carried out on the same batch of samples analysed by Schilling (1973b) shows that the truly transitional area is nearer 200 km long, and that the gradient is more of a scatter (Figure 13.9) (Hart *et al.*, 1973). Northwards from 200 km south of Iceland (62.5°N), the $^{87}Sr/^{86}Sr$ ratios are relatively uniform and high (0.70304), and from 400 km south of Iceland (61°N) the lower uniform ratios averaged at 0.70273 are equivalent to N-MORB (Hart, 1971). There may also be a positive correlation between $^{87}Sr/^{86}Sr$ ratios and depth (Hart *et al.*, 1973; Schilling and Noe-Nygaard, 1974; Flower *et al.*, 1975).

Lead isotope data on quartz and olivine normative tholeiites from the same area provide supporting evidence for this gradient and show a closer correlation with the LIL elements (Figure 13.9) (Sun *et al.*, 1975). The $^{206}Pb/^{204}Pb$ data give high ratios over Iceland (18.70) and low ratios south of 61°N (18.30) to the Charlie Gibbs Fracture Zone (53°N), with a progressive decrease in the transitional zone on the Reykjanes Ridge. The gradient mimics the $[La/Sm]_{EF}$ smooth convex upwards curve, but not the $^{87}Sr/^{86}Sr$ steep gradient. The data of O'Nions and Pankhurst (1974) for Sr and Pb from their own samples were concordant with the smooth progression reported by Schilling (1973b).

The $^{143}Nd/^{144}Nd$ isotope ratios (O'Nions *et al.*, 1977) in basalts north of 63°N are indistinguishable from Icelandic samples. The usual negative correlation between $^{143}Nd/^{144}Nd$ and $^{87}Sr/^{86}Sr$ ratios (the mantle array) breaks down (O'Nions *et al.*, 1977), and so is not consistently present in transitional segments. The consequence of this observation on the terminology applied by Schilling (1975) is that T-MORB starts further north with regard to the Sr data relative to the Nd and Pb data.

DSDP Leg 49 (Wood *et al.*, 1979b; Tarney *et al.*, 1979) tested the temporal relationship of the Reykjanes Ridge enrichment by drilling three holes (407, 408 and 409) perpendicular to the axis (Figure 13.10). Chondrite-normalized REE patterns are highly variable from light REE depleted to light REE

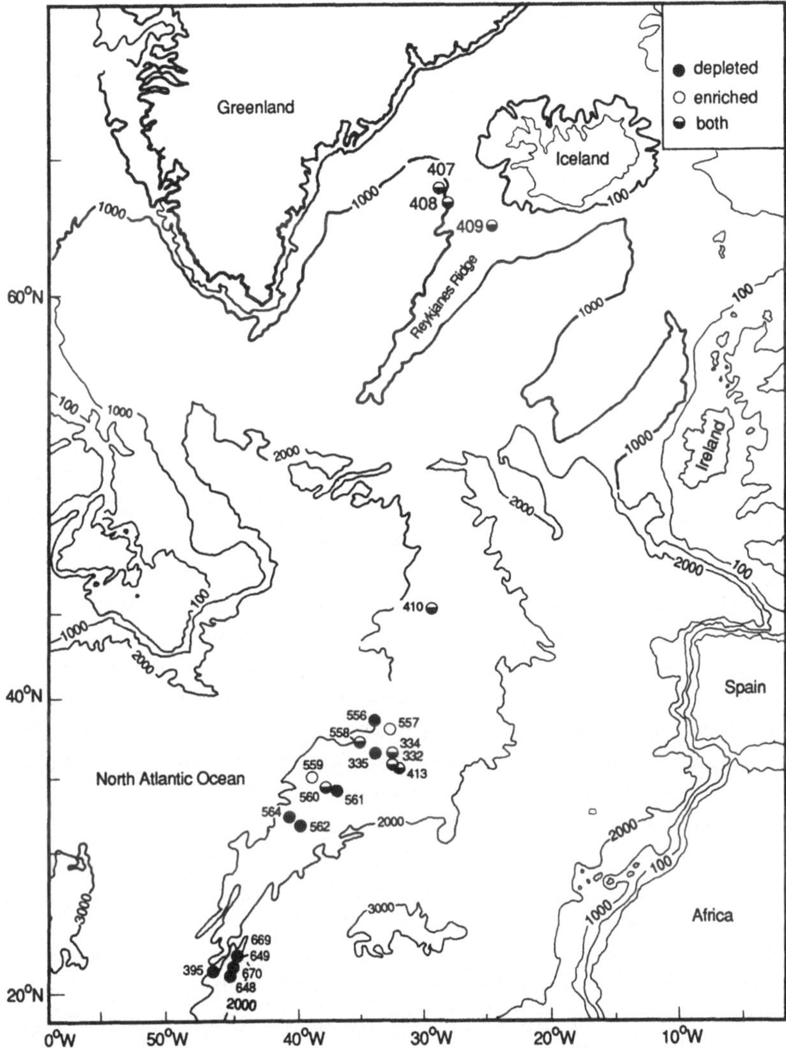

Figure 13.10 A map to demonstrate the relative locations of the DSDP and ODP drill holes (collected from DSDP and ODP volumes mentioned in the text), and to summarize the degree of depletion found in the basalts recovered from each drill hole, based on $[La/Sm]_{EF}$ or similar ratios.

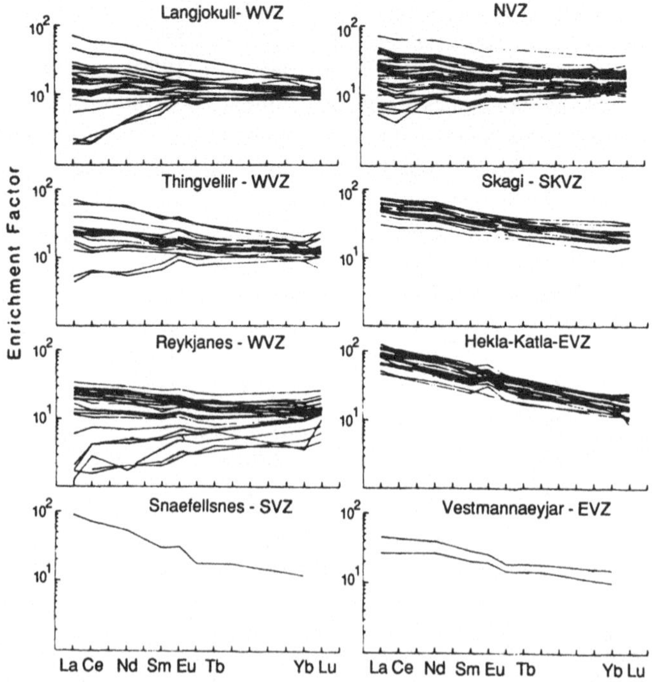

Figure 13.11 REE patterns for basalts from different volcanic regions in Iceland. Enrichment factor = (REE concentration in sample)/(REE concentration in chondrites) (Meyer *et al.*, 1985). Data for the Vestmannaeyjar are from O'Nions *et al.* (1973), and for the Snæfellsnes from O'Nions *et al.* (1977).

enriched (Wood *et al.*, 1979). The older basalts in all three holes (35–40 Ma) are more enriched in light REE relative to the younger samples near the ridge. They also report alkali basalts from the same hole as tholeiites. On closer inspection of the data, the distal hole (407) has the higher K_2O and TiO_2 abundances and the basalts with alkali affinity are the samples with the light REE enrichment. It seems then that if smaller degrees of partial melting produced these alkalic basalts, then it follows that they should have the more enriched light REE patterns, and this is in accordance with a period of low output of the plume (Vogt, 1971; Schilling, 1975). Pb isotopic data support the notion that the Icelandic plume geochemical anomaly was present at 20 Ma, and possibly even at 28 Ma (Mattinsson, 1979).

13.6.3.2 *FAMOUS area* (*MAR*, 37°*N*). The other well studied location is the FAMOUS area (37°N; DSDP Legs 37, 49 and 82) (Figure 13.10). The area is composed of 40–50 km long segments that produce both tholeiitic and alkali basalts. According to Schilling (1973a,b) the basalts are light REE enriched and this feature increases along the ridge axis towards the Azores in the north and temporally towards the present day axis. The REE patterns from 2 Ma crust (Leg 49, hole 411) are variable, with either light REE enriched concave upwards patterns, or depleted light REE contents (Tarney *et al.*, 1979) (Figure 13.11b).

DSDP Leg 37 was designed to test the chemical variability perpendicular to the ridge axis in close proximity to the Azores hot-spot. A distal hole drilled in 16.5 Ma crust (Leg 37, hole 335) has light REE depleted patterns ($[La/Sm]_{EF} < 1$) and depleted abundances of LIL elements (Puchelt *et al.*, 1977; Schilling *et al.*, 1977), suggesting N-MORB characteristics. Although O'Nions and Pankhurst (1977) found various degrees of light REE enrichment in hole 335 (16.5 Ma), N-MORB patterns are still dominant. More proximal holes (334 and 332) show dominantly light REE enrichment patterns (O'Nions and Pankhurst, 1976; Puchelt *et al.*, 1977; Schilling *et al.*, 1977). The $^{87}Sr/^{86}Sr$ ratios range from 0.70287 to 0.70316 (O'Nions and Pankhurst, 1976) with the higher values from the older 13 Ma crust, and this supports the LIL element enrichment observed by O'Nions and Pankhurst (1977), but is clearly in contrast to Schillings' predicted enrichment trend from old to young crust. The $^{143}Nd/^{144}Nd$ values for the same proximal (Hole 332) samples (O'Nions *et al.*, 1977) are low and range from 0.51309 to 0.51315, but appear to follow the systematic variation displayed with the Sr isotopes in the mantle array. Pb isotopic data are heterogeneous spatially and temporally (Mattinson, 1979).

DSDP/IPOD Leg 82 (Bougault and Cande, 1985) returned to this area (Figure 13.10 and 13.11) and drilled 35 Ma crust (Hole 556, anomaly 13). They recovered N-MORB with slightly enriched $^{206}Pb/^{204}Pb$ ratios. Closer to the Azores (hole 557, 18 Ma crust) enriched basalts were recovered. Further south, hole 558 (35 Ma old crust) was drilled on the same flow line as hole 335 of Leg 37 (16.5 Ma crust). The 35 Ma crust, which contains both depleted and enriched basalts within the same hole, is similar in this respect to Hole 413, Leg 49. Continuing further south on the same anomaly (35 Ma), the compositions are enriched, both isotopically and in the light REE and other LIL elements. On the same flow line, but closer to the axis (Figure 13.10), Hole 561 showed a variation in the degree of element enrichment. South of 33°N and the Hayes Fracture Zone, three more holes were sunk on the same anomaly (Holes 562 to 564), and consistently produced N-MORB. This suggests that the Hayes Fracture Zone acts as a natural barrier to the chemical effect of the Azores plume (Bougault and Cande, 1985; White and Schilling, 1978; Bougault and Treuil, 1980).

13.6.4 *Enriched ridge segments: E-MORB tholeiites*

13.6.4.1 45°*N region, MAR.* This area has a well developed rift valley and flank morphology akin to typical MOR (Figure 13.1b). Schilling *et al.* (1983) reported high $[La/Sm]_{EF}$ ratios of 2.33 and so defined it as E-MORB. Site 410, DSDP Leg 49, was drilled in 10 Ma crust and revealed considerable variations in the degree of enrichment (Wood *et al.*, 1979b) (Figure 13.10). The consistently high trace element abundances are interpreted as a result of either high degrees of fractional crystallization or low degrees of partial melting (Tarney *et al.*, 1979), the latter being compatible with the presence of alkali basalts. The 45°N region can be distinguished from the Reykjanes Ridge on the incompatible element characteristics, but to a lesser extent from the FAMOUS area. This region of the MAR has high $^{87}Sr/^{86}Sr$ ratios (White and Schilling, 1978), and the Pb isotopic data from Hole 410 (Mattinson, 1979) exhibit very high $^{206}Pb/^{204}Pb$ ratios that are within the range of Pb ratios from oceanic islands in the South Atlantic.

There has been some dispute about the relationship of this segment of ridge relative to the Azores hot-spot, as it does not display any of the morphological features suggestive of a hot plume influence, although much of the geochemistry suggests that it is E-MORB. There is evidence for a plate boundary relocation in this region based on the structure of the area and nearby seamounts, which may have masked the original features or prevented their development (Searle and Laughton, 1977; Searle and Whitmarsh, 1978).

13.6.4.2 *Iceland.* Icelandic basalts should by definition (Schilling, 1975) demonstrate a consistent enrichment in LIL elements relative to MAR segments. The mainland Iceland olivine tholeiites and tholeiites are characterized by high Fe and Ti, low K/Rb ratios and LIL element abundances, and light REE enrichment (Shimokawa and Masuda, 1972; Brooks *et al.*, 1974). There is some scatter in the concentrations of incompatible elements such as K, Ti, Rb and Zr over Iceland relative to the MAR (Sigvaldason, 1974), such that the enrichments are seen more as a general increase in concentrations. The absolute variation in incompatible elements is probably representative of the effects of fractional crystallization.

Earlier research (Shimokawa and Masuda, 1972; Schilling, 1973a; O'Nions and Gronvold, 1973; Hart *et al.*, 1973) found that a uniformity existed in the degree of light REE enrichment and isotopic ratios which supported the classification of E-MORB proper. However, as the data set increased from both the neovolcanic zone and older rocks on Iceland, the true variability of this enrichment came to light (Figure 13.11) (Sun and Jahn, 1975; O'Nions *et al.*, 1976). Tholeiitic basalts with N-MORB characters have been sampled from the NVZ post-glacial basalts and both LIL depleted and LIL enriched characteristics have been reported from within the products of a single fissure swarm (Sigvaldason *et al.*, 1976). The same is true for $^{87}Sr/^{86}Sr$ ratios, which were thought to have a uniform average of 0.70315 for glacial and post-glacial basalts (O'Nions *et al.*, 1973) (except for the alkali basalts of the SVZ).

Subsequent studies revealed a range from 0.70291 to 0.70341 (O'Nions *et al.*, 1976), excluding the SVZ, which is similar to the range observed on the Reykjanes Ridge. In general there is a positive correlation between the $^{87}Sr/^{86}Sr$ ratios and the degree of LIL element enrichment in these basalts

(Figure 13.9b). In the Tertiary, the LIL element enrichment and the $^{87}Sr/^{86}Sr$ ratios are distinctly higher and a progressive decrease in the latter from 0.7036 (Tertiary, 15 Ma) to 0.70315 (Pleistocene, 2 Ma) is observed (O'Nions and Pankhurst, 1973). The degree of decrease in the gradient increases sharply at 4 Ma (Figure 13.12a). The $^{143}Nd/^{144}Nd$ isotope ratios for Tertiary basalts are significantly lower than those for the post-glacial basalts (Figure 13.12b) (O'Nions *et al.*, 1977), and therefore the correlation with the Sr isotope ratios exists as it does for the post-glacial basalts. However, the Tertiary basalts are less radiogenic in Pb (Sun and Jahn, 1975; Welke *et al.*, 1968). The usual correlation between the Sr, Nd and Pb systems is not apparent in the Tertiary samples, with the Pb data deviating from the usual trend.

The geochemistry shows a sharp change either side of the Tjornes Fracture Zone (Jakobsson 1979b; Schilling *et al.*, 1983). O'Nions and Pankhurst (1974)

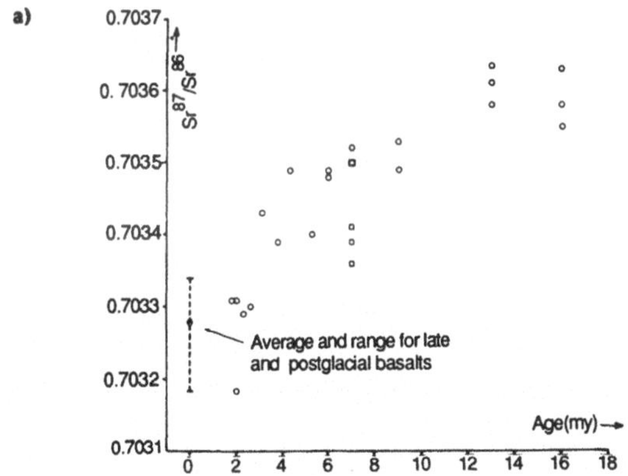

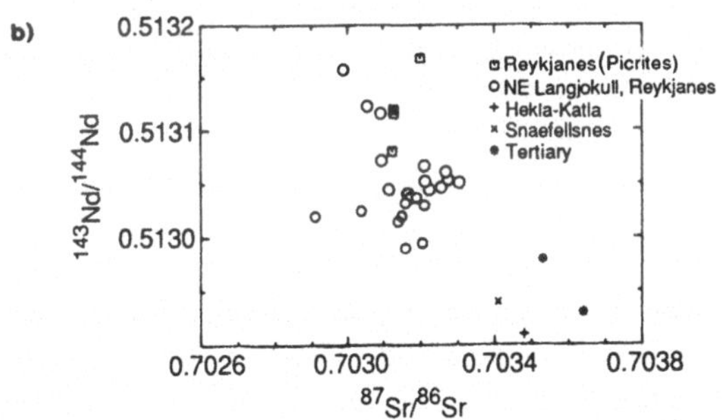

Figure 13.12 (a) Sr ratios versus age. Circles are basic rocks, squares are acid rocks (O'Nions and Pankhurst 1973). (b) Sr and Nd isotopic ratios, showing that the transitional alkali basalts of the EVZ, the Tertiary tholeiitic basalts and the Snæfellsness alkali basalts have the higher Sr and lower Nd ratios. From Meyer *et al.* (1985).

report low $^{87}Sr/^{86}Sr$ ratios similar to N-MORB (based on one sample from the Kolbeinsey ridge) and relative to the variation just described within the neovolcanic zone on Iceland. However, to suggest that the geochemical anomaly does not exist on this ridge segment is a little premature. A similar discontinuity was reported by Sigurdsson *et al.*, (1978) between the Skagi and the Langjokull volcanic zones (65°10′N) on the basis of only La/Sm ratios. It may be that these are merely sampling discontinuities or, in the latter case, a direct response to differing degrees of partial melting.

13.6.5 *Alkali basalts from Iceland and other Atlantic oceanic islands*

The alkali basalts exposed on Iceland are all younger than 3.5 Ma. O'Nions *et al.* (1973) observed a higher degree of light REE enrichment and higher total abundances of REE in the alkali olivine basalts of the Westmann Islands relative to mainland Iceland. They interpret this in terms of smaller degrees of partial melting at the propagating tip of the EVZ compared to the tholeiites of the rest of the neovolcanic zone. The $^{87}Sr/^{86}Sr$ ratios from the same region are similar to the rest of the tholeiites on Iceland (Hart *et al.*, 1973), suggesting that the $^{87}Sr/^{86}Sr$ ratios are representative of the source, whereas light REE enrichment is not (Galer and O'Nions, 1986).

Further research suggests that Sr isotopic values and the REE enrichment are much more varied than supposed by O'Nions *et al.* (1976) for the rest of Iceland. For example, the $^{87}Sr/^{86}Sr$ ratios (average 0.70341) and the degree of LIL element enrichment for the Snaefellsnes Peninsula alkali basalts is higher than anywhere else within the neovolcanic zones on Iceland, and are equivalent to those found on Jan Mayen (north of Iceland) (O'Nions and Pankhurst, 1974; O'Nions *et al.*, 1976). The degree of LIL element enrichment is higher than that in the Tertiary, whereas the $^{87}Sr/^{86}Sr$ values are comparable, and it has been suggested that these alkali basalts have a similar source to the Tertiary lavas. Comparable values have been reported from Torfajokull (EVZ), where assimilation fractional crystallization is believed to have occurred (O'Nions and Gronvold, 1973), and this will affect the Sr and Nd ratios (Hermond *et al.*, 1988).

The $^{143}Nd/^{144}Nd$ isotope ratios for two post-glacial alkali basalts from the Snaefellsnes Peninsula are comparable to the contemporary tholeiites (O'Nions *et al.*, 1977; O'Nions and Pankhurst, 1973). This indicates that the isotope composition of the source region in Iceland may have changed on a temporal scale and may even be spatially heterogeneous, if crustal contamination has not occurred. The lack of correlation between the Pb and the Sr and Nd isotope ratios was used to argue against contamination of the Tertiary lavas. However, Pb may behave as a mobile element during hydrothermal alteration (Dickin, 1981), which may explain the lack of correlation and support crustal influences on the isotopic ratios of the Icelandic basalts.

Oceanic island basalts and Icelandic alkali basalts yield a much wider range and generally higher values of Sr and Nd isotope ratios than MORB (Chapter 9), and although the ratios within a single island group may show little variation (O'Nions *et al.*, 1977), large differences occur between different hot-spots. St. Helena and Ascension have low $^{87}Sr/^{86}Sr$ ratios, similar to that of N-MORB (Tatsumoto *et al.*, 1965). A $^{207}Pb/^{204}Pb$–$^{206}Pb/^{204}Pb$ plot demonstrates that the ratios for many oceanic islands near a MOR segment are extremely variable, but they all seem to show linear trends towards N-MORB values, suggesting that mixing is a regular phenomenon (Sun *et al.*, 1975). Almost every oceanic island source requires a unique combination of Rb, Sr, U and Pb in the source to account for the observed isotopic ratios and abundances.

The variation between the Azores and Iceland hot-spots is discussed by Schilling *et al.* (1983). These regions both possess elevated morphology, unusual tectonic characteristics and geochemical signatures. The radial extent to which basalts with N-MORB affinities are affected varies from plume to plume. The mantle beneath the islands is enriched in LIL elements, H_2O and halogens. This enrichment is more pronounced in the Azores than Iceland, yet the elevation is greater in Iceland. Enrichment in Fe, Mg and Ca, and depletion in Na and SiO_2 for a given Mg-value, is more characteristic of Iceland. The isotopic evidence shows that the source region of the Azores (Chapter 9) has had a distinct evolution and different periods of isolation relative to Iceland (Sun and McDonough, 1989). The combined evidence indicates that there are three distinct mantle sources present in the North Atlantic, the DLVZ, the Iceland plume and the Azores plume, and that others exist to supply plume-dominated basaltic magmas to oceanic islands in the South Atlantic (Weaver *et al.*, 1987; Chaffey *et al.*, 1989; Davies *et al.*, 1989; see Chapter 9).

13.7 Comparison of the North Atlantic and Iceland

Both regions display episodic activity, both tectonically and volcanically, within a dominantly extensional environment, with extensive fissures and normal faulting occurring even directly over the plume head in Iceland. The basic layered crustal structure is oceanic in nature and oceanic analogues can be called on to describe the present plate boundary configuration (Figure 13.3) (Fornari *et al.*, 1989). The *en echelon* arrangements of the volcanic systems in parts of Iceland have their counterparts on the Reykjanes Ridge. The EVZ can conveniently be interpreted in terms of a propagating tip of the volcanic zone. In the EPR, propagating rifts dominantly propagate away from plumes.

The other zones (SVZ, SKVZ and OVZ) are difficult to find modern oceanic counterparts for, and thus collectively they are more likely candidates for the manifestations of radial flow of the mantle. The transform zones in normal oceanic basins rarely have classic textbook form. Ridge jumps occur in a normal oceanic environment and are not just a response to the plume, although they are more likely to occur in such an unstable environment. MOR magmatism is of a passive nature and occurs in response to extension (Schilling, 1973b; McKenzie *et al.*, 1990). Mantle plumes are the result of self-maintained mantle convection and not a passive response to plate motion and so Iceland may not be the ideal place to study such passive processes.

When using the term 'volcano' in Iceland, Jakobsson (1979b) rightly suggests that this should apply to a volcanic system as a whole. Each eruption in that system can be thought of as a monogenetic crater eruption. This notion draws the scales of volcanoes in the MAR (Brown and Karson, 1988), the EPR and Iceland closer together, and fits the idea of a large volcano in each segment having distinctive morphological, volcanic–tectonic character as well as a distinctive chemistry. Lateral magma flow occurs on Iceland within a system (and possibly between systems) for distances of 70 km (Bjornsson, 1977; Sigurdsson and Sparks, 1978; Saemundsson, 1978; Steithorsson, 1978). In addition, it is suggested that mixing of magmas occurs in a basal crustal magma reservoir (Eysteinsson and Hermance, 1985). Lateral magma movement within single oceanic segments (Lin *et al.*, 1990) probably also occurs in MAR segments in addition to magma mixing.

It seems that Icelandic rocks are more differentiated than those of the rest of the Atlantic, but, as indicated earlier, these processes can be identified and their effects subtracted from the overall geochemical characteristics, bringing the Atlantic and Iceland closer together on a comparative scale. The existence and abundance of the central volcanoes and the acid and intermediate rocks in Iceland may be attributed to thicker crust, which either enables the magmas to differentiate further in crustal chambers or to assimilate crustal material to produce acid rocks. Alternatively, these features may be a result of the sustained high geothermal gradient provided by the plume. There are calderas reported from the summits of volcanoes on normal segments of the MAR (Detrick *et al.*, 1990) comparable to calderas such as Krafla (NVZ). The alkali rocks are due to the different melting regime at the plume and are not typical of the oceanic environment, although they are also found in other areas (45°N) of the MAR.

Towards Iceland, along the Reykjanes Ridge, there is an observable gradation towards higher mantle temperatures that has produced thicker crust, and may account for some of the changes in the chemistry of the erupted lavas. It seems that the effects caused by the hot-spot can be identified, and as long as the plume associated processes are acknowledged, it is acceptable to use Iceland as a place to study processes at MORs.

13.8 Concluding statements

1. The structure, composition and nature of convection in the mantle are absolutely crucial to the accuracy of any geochemical model. The sub-Atlantic mantle may be interpreted as having three scales of convection: (1) large-scale, whole mantle that emerges at the surface in plumes such as Iceland; (2) smaller scale that occurs passively at the mid-ocean ridge; and (3) minor internal 'mixing' that occurs in the mantle to account for the isotopic variability of Atlantic MORB. The small-scale convection occurring at the mid-ocean ridge does not completely obliterate radial asthenospheric flow from adjacent plumes or vice versa. The Atlantic Ocean provides a natural laboratory within which to observe the interaction between deep mantle plumes and the mid-ocean ridge tectonics.
2. Atlantic Ocean ridge basalts fall into three petrological series covering the full range of tholeiites to alkali basalts. The alkali basalts are predominantly produced by a smaller degree of partial melting from a similarly depleted source as the associated tholeiites, which is distinct from the more enriched plume source of Atlantic oceanic islands. The Mid-Atlantic Ridge and Icelandic tholeiites show various degrees of isotopic, LIL and light REE depletion and enrichment, and are often chemically characterized as normal (N-) MORB, transitional (T-) MORB and enriched (E-) MORB or plume (P-) MORB. Although the original classification related to the consistent correlation with the degree of elevation of the median valley, or the shallowness of the ridge, this is no longer the case. It is stressed that such genetic terms as 'plume' are misleading and the descriptive term 'enriched' is preferable in relating to different MAR segments. It is also important to appreciate that the three MORB types naturally grade into each other and, in view of the problems in defining the chemical limits to each type, it might be more appropriate to use only N-MORB (depleted features) and E-MORB (variably enriched features).
3. The degree of LIL and light REE enrichment in Icelandic basalts and the Reykjanes Ridge varies considerably, both spatially and temporally. The isotopic data are more consistent and show a positive gradient towards Iceland from the south, and some uniformity within the Icelandic volcanic zones. Temporal variation exists in Iceland from the Tertiary lavas to the present and must reflect a progressive change in the composition of the source region, or represent a degree of efficiency of mixing the depleted MORB source with the plume source. Anomalous isotopic ratios occur in the SVZ and Tertiary crustal assimilation has been suggested as a means of producing these features. The volcanic zones and segments in Iceland may represent finger-like expressions of radial flow that are preferentially

channelled along lines of weakness. The correlation observed between the ridges and troughs on the Reykjanes Ridge with the high and low $^{87}Sr/^{86}Sr$ ratios, respectively, agrees with the idea of the plume asthenospheric source producing higher volumes of magma and thus providing the higher isotopic ratios in the magma mixture. This suggests that mixing occurs between the two mantle sources below Iceland.

4. The presence of plumes and their interaction with the active ridge are a feature of the Atlantic Ocean crust. The nature of components identified in the plume source are varied and each hot-spot should be investigated separately. There is a general consensus that subducted oceanic lithosphere, stored and isolated in the mantle over long periods of time, is involved and has been identified as a contributor to the source regions of the Cape Verdes and Azores hot-spots. Melting of enriched sub-oceanic lithosphere may also be a contributing factor, although this source does not apply to Iceland. Differences between the Azores and Iceland hot-spots, such as the anomaly in the residual gravity/residual elevation could be achieved by an increase in the temperature of the mantle beneath Iceland, and/or an increase in the percentage of partial melting present. The latter would be compatible with the presence of the different magma series of these two islands. The high LIL element enrichment in the Azores relative to Iceland supports the notion of sub-oceanic lithospheric enrichment under the Azores, and in a similar manner at 37°N and 45°N MAR.

PART IV SOURCES

14 Stable and noble gas isotopes

RICHARD EXLEY

14.1 Introduction

There are four main reasons why stable isotopes provide important information about the petrogenesis and source regions of oceanic basalts.

- Hydrogen, carbon, nitrogen, oxygen and sulphur are essential components of the volatiles which drive all magmatism. Direct mantle nodule samples provide sparse evidence on the host phases of these elements. The study of stable isotopes in oceanic basalts provides a window on these elements in the mantle.
- Large fractionations of stable isotope ratios exist between the crust, mantle, hydrosphere, atmosphere and the biosphere. Stable isotopes thus provide a tool for studying the dynamics of exchange between these reservoirs.
- By linking stable isotopes with, in particular, noble gases and radiogenic isotopes, the earth's degassing history may be studied.
- Very large stable isotopic effects are observed in meteorites. These provide models of unprocessed planetary material which indicate the possible isotopic heterogeneities in the deep earth. Again, the study of mantle stable isotopes via the vehicle of oceanic basalts may provide a window into deep earth history.

14.2 Stable isotopes

Natural variations in stable isotope ratios are usually caused by fractionation effects. These are either equilibrium fractionations, caused by the effect of different atomic masses on the energy levels of chemical bonds, or kinetic fractionations, caused by the effect of the different isotopic masses in rate-controlled processes such as diffusion. The theory of these effects was described in the classic paper by Urey (1947). Even for elements with large mass differences between their stable isotopes, the ratios vary only by a maximum of 1–2% (except for hydrogen/deuterium). Hydrogen, carbon, nitrogen, oxygen and sulphur have sufficiently large mass differences between their stable isotopes for both kinetic and equilibrium fractionation effects to cause variations in their isotopic compositions. The chemistry of these elements, which are ubiquitous in fluid and gaseous phases, ensures that they are involved in geochemical processes which have significant associated isotopic fractionations.

Carbon, nitrogen and sulphur are trace elements in oceanic basalts, whereas oxygen is the major element of the crust and mantle. Sulphur occurs as minor sulphides in basalts. In all submarine geochemical processes reactions with seawater are of overriding importance. Hydrogen and oxygen isotopes are thus subject to drastic effects resulting from water–rock interactions. Seawater contains large amounts of dissolved sulphate which can affect the isotopic composition of sulphur. Carbonate, dissolved nitrate and organic contaminants can similarly affect the carbon and nitrogen isotopic compositions.

As a result of the secondary effects observed and the difficulties of obtaining unaltered materials, the literature on carbon, nitrogen and hydrogen stable isotopes is limited. There are very few rocks for which comprehensive sets of trace element, radiogenic and stable isotope data are available, and data for stable isotope compositions in oceanic basalts are less extensive than for strontium, neodymium and lead isotopes. As tools for the investigation of mantle sources, stable isotopes therefore lag behind radiogenic isotopes. However, because the field is still in its infancy, rapid developments in our understanding of the stable isotopic compositions of basalt source regions are taking place.

The distribution of each stable isotope in the oceanic environment will be considered briefly. Appendix C describes the derivation of the per mil (‰) notation used in reporting stable isotope data.

14.2.1 *Sampling and speciation*

Fresh whole rocks and glasses with little or no evidence of seawater alteration provide suitable samples for the study of stable isotopes of oxygen. However, hydrogen, carbon, nitrogen and sulphur occur as species such as H_2O, CO_2, CH_4, SO_2 and H_2S, which form the so-called 'magmatic volatiles'. Equilibria between these are dependent on pressure, temperature and redox conditions (P, T, fO_2). These P–T, fO_2 equilibria produce large variations in the isotopic compositions which further complicate the use of stable isotope ratios as mantle tracers. At high pressures volatiles are held in solution in basaltic magmas, but as the magma ascends, the volatiles follow solubility curves resulting in either partial or complete exsolution from the silicate liquids (degassing). The liquid immiscibility of sulphide and/or carbonate liquids provides another means of fractionating these elements from silicate magmas. Silicates, except for some hydrous minerals, do not form compounds with these elements, and so magma suffers a nearly complete loss of these elements on final crystallization. The degassing and crystallization processes are accompanied by isotopic fractionation, generally following Rayleigh Laws, and pristine isotopic compositions for the magmatic volatiles should be rare. This is particularly so for oceanic island basalts erupted subaerially. Glass inclusions in phenocrysts are perhaps the only possible source of ancient magmatic volatiles in such rocks. Modern oceanic islands such as Hawaii (Chapter 9) and Iceland (Chapter 13) display active volcanism which allows the direct sampling of magmatic volatiles, but the sampling process is difficult, hazardous and prone to fractionation effects.

The best samples of magmatic volatiles are provided by basaltic glasses from the mid-ocean ridges, erupted at water depths greater than 2 km. Eruption under water has two effects: (1) the water pressure retains volatiles in solution in the magma (i.e. prevents volatile exsolution); and (2) the chilling effect quenches the magma, freezing in the vesicles and the dissolved gases. As the volatiles are trace components, relatively large samples are required to study both isotopic fractionation and problems of volatile speciation (e.g. dissolved and trapped gases). This simple problem of sample availability has again restricted the development of this branch of geochemistry, and has particularly hampered the comparison of mid-ocean ridge basalts (MORB) and oceanic island basalts (OIB), and of oceanic and continental mantle sources.

14.2.2 *Nitrogen*

Nitrogen is one of the rarest magmatic volatiles and also one of the most difficult to measure. The nitrogen contents of basaltic glasses vary up to about 2 ppm, although there has been controversy over possible higher concentrations. Early work by Becker and Clayton (1977) suggested that the nitrogen retained by MORB glasses is ^{15}N-enriched compared to the nitrogen of the atmosphere. Studies by Exley *et al.* (1987) and by Zhang and Clayton (1988) have suggested an average δ^{15}N value near $+7.00$‰ for MORB. The limited data for OIB samples suggest similar or possibly higher δ^{15}N. Zhang and Clayton (1988) have produced evidence from the east Pacific to suggest that recycled nitrogen occurs in subduction zone magmas (e.g. back-arc basin basalts), but that the return flux of nitrogen to the mantle is low. Comparison of δ^{15}N data for oceanic basalts with data for the sub-continental mantle, mainly for diamonds (Javoy *et al.*, 1984; Boyd *et al.*, 1987), suggests that a much greater complexity is preserved in the diamond sample. This presumably reflects the much more homogeneous nature of the better stirred sub-oceanic mantle. It is possible that in the chemical systems typical of the sub-continental mantle samples (diamond, mica-bearing rocks) nitrogen has a phase chemistry which results in isotopic effects not observed under the redox conditions and mineral assemblages characteristic of the sub-oceanic mantle.

Nitrogen is of considerable importance in that its geochemical behaviour as an unreactive molecular gas in oceanic magmatism should produce analogous behaviour to noble gases. As there are very few samples with combined data for nitrogen and for helium (and argon neon, xenon) isotopes, this role as a link with the noble gases has been limited. Possible relationships were examined by Exley *et al.* (1987), but no firm conclusions could be drawn. Numerous groups are working to accumulate sufficient data to enable the future elucidation of the role of nitrogen in mantle source regions.

14.2.3 *Carbon*

Throughout the 1980s there was disagreement over the speciation of carbon, its concentration, and the interpretation of the isotopic data (DesMarais, 1986). The relative importance of fractionation processes and their effect on isotopic ratios has been the main feature of this controversy, which has made it difficult to interpret carbon isotopes in terms of mantle source variations. The difficulties of obtaining suitable samples from many regions of the oceanic crust have hindered the development of carbon isotope geochemistry as a tool for the study of the mantle.

There are large isotopic fractionations between reduced and oxidized forms of carbon, reflected in differences in the $\delta^{13}C$ values between organic and inorganic materials. These variations are particularly extreme in the continental crust, with high values in limestones, and very low values in organic-rich materials. As a result, sedimentary components which might be recycled back to the upper mantle by subduction, have isotopic signatures which could be used as tracers of recycling processes.

The isotopic composition of 'primordial' carbon from the sub-oceanic mantle is of primary importance in understanding the chemical balance of the mantle–crust system. Whereas oceanic basalts are in general undersaturated with respect to H_2O at their depths of eruption, CO_2 saturation levels and solubility decrease rapidly as eruption depths approach. Basalts degas large amounts of CO_2 as they are depressurized. This dramatic loss is a possible cause of carbon isotopic fractionation which might obscure mantle source variations.

The $\delta^{13}C$ values of CO_2 in magmatic volatile samples in active volcanic regions are fairly heavy (–2 to –6‰). Magmatic carbon forming methane (typically 1% of the CO_2 in amount) has values of typically –20 to –30‰, compatible with the calculated isotopic fractionation factor for this system. Values of $\delta^{13}C$ for carbonatites, diamonds and samples of mantle carbon also fall in a range convergent on –6‰ (Deines, 1980).

Mass balance calculations using the isotopic compositions of crustal reservoirs lead to a value of about –7‰ for crustal carbon; if this has resulted from mantle degassing, oceanic basalts should be characterized by a $\delta^{13}C$ of about –7‰ if the crust–mantle system has remained in a state of equilibrium with respect to isotopes. The carbon isotopic composition of the mantle has been reviewed by Kyser (1986) and Mattey (1987). Early workers suggested low $\delta^{13}C$ values (Craig, 1953; Wickman 1956). More recent workers have adopted analytical approaches designed to take account of the speciation of carbon in basaltic glass samples. These have included acid treatments to remove carbonates, vacuum crushing to release trapped volatiles from inclusions or vesicles and fusion to release carbon dissolved in the glass. Various workers have adopted a stepped heating procedure to resolve components present on the surface of the samples from those resident in the glass. Combustion and pyrolysis have been used to examine the relationships between reduced and oxidized phases.

Data for N-MORB suggest an average $\delta^{13}C$ value for the oxidized carbon in basalt glass of around $-6.5 \pm 1.0‰$ (Pineau and Javoy, 1983; Mattey *et al.*, 1984; DesMarais and Moore, 1984; Sakai *et al.*, 1984; Exley *et al.*, 1986). Data for oceanic islands, with a sampling bias towards Hawaii and Iceland and including data for fresh glasses (Kilauea East Rift, Loihi Seamount), suggest a slightly higher $\delta^{13}C$ around -5.0 and up to $-3.0‰$ (Gerlach and Thomas, 1986; Exley *et al.*, 1986). These values characterize the CO_2 in vesicles and the majority of the dissolved carbon in the glass; many recent studies of carbon speciation in basalt glass by infrared absorption spectroscopy have shown that carbon is present as the carbonate anion (Fine and Stolper, 1986).

Data for subduction-related volcanics is sparse, but data for back-arc basin basalts from the Scotia Arc (Mattey *et al.*, 1984) and for the Lau and Fiji Basins (Exley *et al.*, 1986) support evidence from hydrogen isotopes (Poreda, 1985) for a slab-derived recycled component in these basalts.

14.2.4 *Hydrogen*

The two isotopes of hydrogen have a 50% mass difference, and because hydrogen is so abundant in seawater, contamination and degassing effects can dominate the systematics of hydrogen isotopes in oceanic basalts. As with all stable isotope variations of magmatic volatiles, there is debate as to whether the sources studied are 'undegassed', or rather 'relatively undegassed'. The consensus value for the mantle is $\delta D_{SMOW} = -80‰$ (Boettcher and O'Neil, 1980). Seawater contamination drives this towards SMOW, either by hydrothermal alteration or by magmatic addition. Loss of methane or hydrogen leads to a higher δD. Loss of H_2O and low temperature hydration decrease δD.

In general, most oceanic basalts have values of δD which can be reconstructed to give values near the $-80‰$ suggested for mantle hydrogen from hydrous minerals in alkali basalts and mantle nodules. Estimates of degassing and fractional crystallization are vital to an understanding of the δD of oceanic basalts (Kyser and O'Neil, 1984).

Studies of subduction-related magmas and oceanic island basalts have suggested that recycled volatiles from the subducted slab are present in the original magmas (Poreda, 1985; Kyser *et al.*, 1986).

14.2.5 *Sulphur*

Sulphur forms two major compounds in magmatic systems: sulphide minerals in the magma and SO_2 gas in the exsolved volatiles; hydrogen sulphide is a minor component. The effects of the competing processes of sulphide precipitation and SO_2 degassing on sulphur isotopic compositions are dependent on oxygen fugacity. Subaerially erupted lavas have a considerably reduced sulphur content as a result of SO_2 loss on eruption.

The isotopic composition of sulphur in meteorites has an extremely restricted range (average 0.1‰), which is shared by the isotopic data for direct mantle samples such as xenoliths. When the data for oceanic basalts are examined, a slightly larger, but still restricted, range is observed (Grinenko *et al.*, 1975; Sakai *et al.*, 1984). The MORB values are slightly less ^{34}S enriched than OIB (± 0.1 and $+0.4$‰ respectively). The sulphur data are, as for all oceanic basalts, biased by the concentration of work on Hawaii as the dominant representative of oceanic island basalts.

Data for subduction-related basalts and andesites show a wider range in $\delta^{34}S$. This has a variety of possible causes. One of these is the variation in the source of such magmas compared to the source for MORB and OIB. The variations include the effect of high H_2O on oxygen fugacity and hence on the sulphur isotopic fractionations, or the addition of sedimentary sulphur with high $\delta^{34}S$ values (Harmon and Hoefs, 1986).

14.2.6 *Oxygen*

The oxygen isotopic composition of the mantle has been reviewed by Kyser (1986). Oxygen is the major element of the crust and mantle, and so its behaviour in magma generation is not affected by trace element type processes in the ways that carbon, sulphur and nitrogen are. The amount of oxygen lost by degassing is trivial and cannot affect the measured $\delta^{18}O$ values in rocks. However, hydrothermal alteration in seawater-dominated systems, either by magmatic assimilation or by post-extrusive secondary processes, can have very important effects on oxygen isotopic compositions.

Temperatures in the mantle are sufficiently high that significant mineral–mineral (olivine–orthopyroxene–clinopyroxene) and mineral–melt ^{18}O fractionations are not expected. The alteration of oxygen isotopic values by exchange with metasomatic fluids, whether CO_2- or H_2O-rich is another possible mechanism for the alteration of mantle $\delta^{18}O$ values; however, mass balance considerations imply unrealistically high fluid to rock ratios to effect changes over large volumes of the mantle.

It therefore seems probable that any observed variations in $\delta^{18}O$ found in either direct mantle samples or in mantle-derived magmas reflect true mantle heterogeneity. MORB $\delta^{18}O$ values have a tight distribution averaging about

7.0‰. Whereas there is an overlap with MORB samples, basalts from oceanic islands (particularly tholeiites), show lower $\delta^{18}O$ values of about -5.5‰. There is an apparent trend in $\delta^{18}O$ in OIB, with alkali basalts showing values much closer to the MORB average than the OIB tholeiites. Direct comparison with mantle nodule $\delta^{18}O$ is difficult as a result of the problems of measuring oxygen isotopic compositions in minerals such as olivine in mantle peridotite. However, data for eclogite mantle xenoliths, many of which are increasingly regarded as representing subducted and recycled material, show an overlap with the lower values found for OIB. This agreement provides evidence for the presence in the mantle of recycled material which contributes to the source of some OIB (Chapter 9).

As a result of the extensive hydrothermal circulation systems that develop in zones of active magmatism, oceanic basalts have, in general, undergone extensive interaction with seawater. These complex hydrothermal processes lead to changes in oxygen isotopic compositions in the oceanic crust which have been extensively reviewed by Muehlenbachs (1986). These isotopic variations are related to the structure and age of the oceanic crust as it spreads from the active mid-ocean ridge. Studies of ophiolites have been particularly important in relating oxygen isotopic compositions to hydrothermal circulation patterns and the magmatic structure of the oceanic crust. The exchange of oxygen between seawater and the oceanic crust may have a pivotal role in controlling the oxygen isotope value of the oceans. The upper mantle, which is the earth's largest oxygen reservoir, acts as a buffer in these exchange reactions.

14.3 Noble gases

The noble or rare gases (helium, neon, argon, krypton and xenon) show a wide variation in their isotopic composition in geochemical systems. The importance of the noble gases in understanding the source regions of oceanic basalts lies in what they can tell us about the degassing of the mantle (e.g. Ozima, 1975; Staudacher and Allegre, 1982). Evidence from both argon and helium isotopes shows that the upper mantle continues to degas primordial volatiles. As a result of the same (or greater) sampling and analytical difficulties outlined for the stable isotope volatiles, data are even less extensive for the non-volatile isotopes. However, both argon and helium isotopes have been used to indicate mantle degassing history and structure. Basalt glasses erupted on the seafloor contain excess ^{40}Ar and display $^{40}Ar/^{36}Ar$ ratios as high as 25 000 (Sarda *et al.*, 1985), with typical values between 8000 and 10 000. These high ratios are indicative of an early, thorough degassing of the earth followed by the accumulation of ^{40}Ar from ^{40}K decay. Helium isotopic ratios in the atmosphere are very depleted in the primordial isotope ^{3}He and are dominated by radiogenic ^{4}He. Extensive studies of basalt glasses have shown that MORB glasses have about 100 times atmospheric $^{3}He/^{4}He$ (Lupton and Craig, 1976; Kurz *et al.*, 1982). Data from the Hawaiian hot-spot in particular have shown distinctive noble gas isotopic compositions, with $^{40}Ar/^{36}Ar$ about 350 and up to 30 times atmospheric $^{3}He/^{4}He$ ratios (Allegre *et al.*, 1983; Kurz *et al.*, 1983; Rison and Craig, 1983). Neon isotopes show excess ^{20}Ne in MORB samples (Craig and Lupton, 1976; Poreda and di Brozolo, 1984). Work by Sarda *et al.*, (1988) shows that Loihi Seamount neon is isotopically indistinguishable from atmospheric, whereas MORB samples show correlated ^{20}Ne and ^{21}Ne excesses, possibly indicative of a recycled sedimentary component in the MORB source (Sarda *et al.*, 1989).

These results have been interpreted as showing that the Hawaiian magma source was not degassed to the same extent as the bulk of the sub-oceanic mantle, and that hot-spot or 'plume' magmas are sampling a deep-seated source of primordial volatiles (Chapter 9). Fisher (1985) suggests that the data are consistent with a recycled component of possible subduction-related origin.

14.4 Mantle models

The integration of mantle models derived from trace element and radiogenic isotope data with information obtained from stable isotopes has several pitfalls. Leaving aside oxygen isotope variations, there is a general tendency to regard the behaviour of volatiles as being decoupled from the silicate component of magmatic systems. Conversely it has been fashionable to appeal to the migration of incompatible element-rich fluids and/or small degree partial melts to explain the apparent enrichment of the source region of many alkali basalts.

It has been suggested that because of the large fractionations that take place in stable isotopes within crustal systems, such isotopes may be used as tracers of recycling processes. However, an important distinction must be drawn between the use of stable isotopes and of radiogenic isotopes. Both the parent and daughter elements of radiogenic isotopic systems reside in mineral phases. Parent/daughter fractionations over a long period lead to isotopic heterogeneity within source domains. The recognition of such heterogeneity is dependent on processes such as magma extraction, operating on relatively local volumes over long time periods at very elevated temperatures, that bring basalt melts to the surface. In contrast, hydrogen, carbon, nitrogen and sulphur form trace phases and can have transport mechanisms potentially independent of magmatism. Thus the time constants for the remixing/homogenization of stable isotope compositions are likely to be much shorter than those for radiogenic isotopes.

The simple two-component model derived from radiogenic isotope data, of a depleted upper mantle MORB source and a less depleted lower mantle OIB source, has been refined as data for new isotopic systems and localities have become available. The data for noble gas isotopes has so far proved to be generally consistent with the data for neodymium, strontium, lead and hafnium isotopes, as evidence has emerged of a degassed upper mantle and a lower mantle characterized by higher $^3He/^4He$ and lower $^{40}Ar/^{36}Ar$ ratios. Noble gas isotopes provide a means of linking stable isotope magmatic volatile data with radiogenic isotopes. However, they are imperfect for this purpose because of their failure to form compounds which might be recycled to the mantle. A comparison of carbon and nitrogen isotopic data for MORB and OIB with that for the noble gases (Exley *et al.*, 1986a, b; 1987) indicates probable relationships with the noble gas isotopes which are consistent with a much lower degree of degassing for OIB hot-spot sources such as Hawaii.

Recent data for the sub-continental lithosphere (Mattey *et al.*, 1989a,b) suggest that relationships between stable isotopes and neodymium, strontium and helium isotopes can occur. These may well be related to the migration of H_2O- and CO_2-rich supercritical fluids which act as potent solvents for incompatible elements. The introduction of lower mantle volatiles is another potential mechanism.

In general, the evidence for carbon, nitrogen and hydrogen recycling in subduction zone magmas implies that the volatiles of the oceanic crust can re-enter the mantle, but conversely that the ability of these elements to recycle may be limited. The extent of these processes will be dependent on the development of stable metamorphic mineral hosts for the volatile components in the subducting slab. The geothermal conditions of the subduction zone and subduction rate will thus be determining factors.

As pointed out by Kyser (1986), oxygen is a major element in the mantle, and the fact that there are observable differences in $\delta^{18}O$ values (e.g. between MORB and Hawaiian OIB) implies important stable isotope heterogeneities in the mantle source regions of oceanic basalts. If the $\delta^{18}O$ values for OIB resulted from the recycling of an eclogitic component to the OIB (Hawaii) source, then the relatively undepleted noble gas and radiogenic isotope compositions imply a decoupling of major and trace components. Proposals for a three-component OIB source for Hawaii involving a deep undepleted mantle plume, a recycled component and a relatively depleted component, are likely. Comparison of the data for oceanic and sub-continental basalt sources and for mantle xenoliths (Mattey *et al.*, 1989b) suggest that carbon isotopic compositions and strontium–neodymium isotopic compositions may be coupled by fluid migration in the mantle, but are subsequently decoupled by the retention of the radiogenic components. Mantle fluid migration or the movement of low degree partial melts couples stable isotopes with parent/daughter fractionations in the radiogenic isotope systems. Helium isotopes are coupled with carbon isotopes because the migration of helium in the mantle is strongly related to the movement of $CO_2^-(H_2O \pm CH_4)$-rich fluid phases. Oxygen isotopes preserve relict recycled silicate component compositions because of the mass balance effects involved and the difficulty of mixing-out such major element heterogeneity.

14.5 Concluding statements

1. Studies of trace amounts of stable isotopes and noble gas isotopes in oceanic basalts are one of the most important sources of information on mantle volatiles, and can help to elucidate one of the chemical driving forces of magmatism.
2. The integration of stable isotope and noble gas data with trace element and radiogenic isotope data has added to our understanding of mantle structure. The database for many isotopic species is inadequate, and its interpretation is one of the more controversial fields of geochemistry. However, data for carbon, hydrogen, nitrogen, sulphur, helium and argon isotopes shows that the sub-oceanic mantle contains both regions relatively rich in primordial volatiles and more thoroughly degassed regions. The evidence for recycled components is increasing, especially when linked to data for radiogenic isotopes in subduction-related basalts.
3. Oxygen isotope variations provide evidence of the recycling of oceanic crust to the mantle via the subduction cycle, and are one of the main geochemical indices of the interaction of oceanic crust with seawater in hydrothermal systems.

15 Oceanic peridotites

MARTIN MENZIES

15.1 Introduction

'......(the) mineralogical constitution (of St. Paul's rocks) is not simple (being of) a felspathic nature, including thin veins of serpentine...' (Darwin, 1845)

In the 1800s Charles Darwin recognized the existence of serpentine on the island of St. Paul's rocks in the Atlantic Ocean and correctly concluded that the island was not of volcanic origin. Around the turn of the century peridotites broadly similar to those on St. Paul's rocks were found as basalt- and kimberlite-borne xenoliths in ocean basins and on continental crust. Indeed it was the correlation between the seismic properties of such ultramafic xenoliths and earthquake seismic data that led to the suggestion by Bowen (1928) that much of the upper mantle was similar in composition to peridotite. Bowen (1928) synthesized much of the available geological and geophysical data and concluded that the sub-crustal regions were peridotitic and that partial melting of garnet peridotite would lead to the production of a basaltic magma and a 'barren' or residual peridotite. This idea of a partial melting continuum [i.e. garnet peridotite (source)–dunite (residue)–basalt (partial melt)] gained wide acceptance and was fundamental in any consideration of upper mantle evolution.

Rock collections from the Darwin and *Challenger* cruises were investigated by Tilley (1947), who proposed that the serpentinized spinel and hornblende peridotite mylonites from St. Paul rocks were of 'plutonic' origin and had formed in a deep oceanic environment. Serpentinized peridotite was also found in a dredge haul from the Mid-Atlantic Ridge (Shand, 1949) and it was proposed that the elevation of the oceanic ridges above the surrounding seafloor was due to the volume increase associated with serpentinization (Hess, 1955). Although this was later shown to be incorrect, it did emphasize the importance of peridotites in sub-oceanic environments and for the first time suggested a form of convective overturn in the oceanic mantle. Ringwood (1962) believed that undepleted or undifferentiated mantle material upwelled beneath part of the oceanic crust (i.e. ridges) and that the extraction of basaltic magma from this undifferentiated mantle meant that elsewhere the oceanic crust was floored by depleted or differentiated peridotites. Dunites and harzburgites are examples of such depleted material which frequently contains basaltic schlieren or segregations produced by the polybaric crystallization of basaltic melts. Ringwood (1962) proposed that such hybrid materials graded downwards into undepleted garnet peridotite.

In an equally important contribution, Green and Ringwood (1967) outlined a model for the origin of basaltic magmas involving partial melting of undifferentiated garnet or spinel lherzolite and the production of a harzburgite residue, thus verifying the earlier suggestion of Bowen (1928). Continued investigation of the St. Paul peridotites (Tilley, 1966) substantiated the earlier assertion (Tilley, 1947) that they were indeed deep or plutonic in origin and that they comprised part of the oceanic mantle (Melson *et al.*, 1967). Ultramafic rocks encountered in some of the first drill cores from the ocean basins (e.g. Mayaguez, Puerto Rico) and from several oceanic fracture zones (e.g. Chain, Romanche and St. Paul) were interpreted as either (1) undepleted source peridotites, (2) depleted residual peridotites, (3) polybaric magmatic derivatives, or (4) crustal cumulate peridotites (Aumento, 1970; Aumento and Loubat, 1971; Shih, 1972). In many instances the interaction of oceanic peridotites with seawater modified their elemental and isotopic chemistry making it difficult to assign any genetic significance to their geochemistry (Bonatti *et al.*, 1970).

Detailed investigations of the Jurassic oceanic lithosphere (i.e. ophiolites; see Chapter 4) played a fundamentally important role in unravelling the complexities of modern oceanic lithologies (Coleman, 1977; Gass *et al.*,1984). In particular, the relationship between cumulus and non-cumulus peridotites demonstrated the nature of the oceanic crust–mantle boundary. Davies (1968) noted that within the Papuan ophiolite, cumulus peridotites graded upwards into cumulus and non-cumulus gabbros, in turn overlain by dolerite dykes and submarine volcanic rocks, and non-cumulus peridotites provided an apparently inert 'floor' for the plutonic sequence. The non-cumulus rocks had apparently experienced a metamorphic event prior to the deposition of the cumulus rocks above them. Davies (1971) interpreted the non-cumulus (metamorphic or tectonite) peridotites as pre-existing oceanic mantle, deformed by metamorphic processes during convective overturn. In contrast, the cumulus peridotites were interpreted as the lower part of the oceanic crust formed by fractional crystallization of basaltic magmas extracted from the upper mantle and ponded in magma chambers beneath the oceanic crust. This distinction between cumulate (igneous) and metamorphic (tectonite) peridotites was of fundamental importance in helping to unravel the complexities of dredged and xenolithic peridotite suites as both are essentially random collections of material where the relationships between different rock types are frequently unknown.

15.2 Oceanic peridotites

The earth's upper mantle is believed to be primarily peridotitic (olivine and pyroxene) and perhaps eclogitic (pyroxene and garnet) at depth. Most of the oceanic peridotites discussed in this chapter represent, at most, the uppermost 150 km of the upper mantle. The change in peridotite facies with depth allows the approximate assessment of the depth of crystallization of a particular peridotite on the basis of its mineralogy: (1) Plagioclase peridotites—abyssal, dredged and basalt-borne xenoliths from < 30 km (< 10 kb); (2) spinel peridotites—abyssal, dredged and basalt-borne xenoliths from < 50 km (< 17 kb); and (3) garnet peridotites—basalt-borne xenoliths from > 50 km (> 17 kb).

As partial melting models play a significant role in determining the petrology and chemistry of upper mantle peridotites it is important to realise that a plethora of terms are used to describe the chemical state of a peridotite. Terms such as depleted, refractory, residual, or differentiated invariably apply to peridotites that have experienced partial melting events, whereas terms such as undepleted, primitive, parental, or undifferentiated usually refer to peridotites that have experienced little or no partial melting and as such are closer in composition to estimates of primary mantle composition (Table 15.1). For simplicity, the terms depleted and undepleted are used in this chapter. In contrast, terms such as enrichment or metasomatism both refer to changes in the chemistry of peridotites but only in the case of metasomatism is there an identifiable mineralogical change.

The petrology and geochemistry of oceanic peridotites will be discussed under four headings which can be thought of as tectonic settings.

- Rift to passive margin peridotites. Peridotites from rift to passive margin localities are found as exposures on islands in the Red Sea, as basalt-borne xenoliths on islands off the coast of Alaska, as possible xenoliths off the south-west coast of Australia and as abyssal peridotites from the eastern margins of the Atlantic Ocean. These peridotites can be thought of as transitional between continental and oceanic mantle.
- Mid-ocean ridge peridotites. The bulk of mid-ocean ridge peridotites occur as abyssal peridotites near ridge axes in the Atlantic and Indian Oceans and to a lesser extent as basalt-borne xenoliths. Mid-ocean ridge peridotites are important because they represent those mantle rocks most closely associated, in time and space, with mid-ocean ridge volcanism.
- Intra-plate peridotites. Intra-plate peridotites occur primarily as basalt-borne xenoliths on oceanic islands such as Hawaii, Tahiti and Malaita. They provide a vital database with which to unravel the evolution of the oceanic lithosphere since formation at a mid-ocean ridge.
- Active margin peridotites. Active margin peridotites are found as dredged materials and as obducted fragments of the oceanic lithosphere. These

Table 15.1 Major element composition (wt%) of oceanic and ophiolitic peridotites according to tectonic setting compared to estimates of primary mantle compositions

Analysis	Samples[a]																			
	Pre-oceanic			Mid-ocean ridge				Intra-plate			Active margin			Primary mantle						
	1	2	3	4	5	6	7	8	9	10	11	12	13	14	15	16	17	18	19	20
SiO_2	44.65	43.6	44.4	43.63	43.6	44.64	43.9	44.41	44.35	44.0	39.82	43.47	44.62	45.16	45.0	44.71	45.6	45.1	46.2	45.8
TiO_2	0.15	—	0.13	0.11	0.02	0.03	0.04	0.08	0.08	0.1	0.01	0.01	0.03	0.71	0.09	0.16	0.2	0.22	0.23	0.1
Al_2O_3	3.85	1.16	2.38	1.85	1.18	2.2	1.4	3.72	3.41	4.0	0.87	0.47	0.64	3.54	3.5	2.46	3.3	4.14	4.75	3.3
Cr_2O_3	—	—	—	0.42	0.22	0.49	0.58	0.54	0.53	0.2	0.46	0.39	0.44	0.43	0.41	0.42	0.4	0.45	0.43	0.4
MgO	38.0	46.19	42.06	42.58	45.2	42.38	45.9	39.47	38.88	38.0	48.6	45.72	44.96	37.47	39.0	41.0	38.5	38.0	35.5	38.3
CaO	3.18	0.66	1.34	2.15	1.13	1.85	0.95	3.26	2.77	3.5	0.37	0.77	0.67	3.08	3.25	2.42	3.1	3.54	4.36	2.8
MnO	0.14	0.13	0.17	0.16	0.14	—	0.11	0.08	0.15	—	0.10	0.15	0.17	0.14	0.11	0.18	0.15	0.13	0.13	0.16
FeO	8.63	7.80	8.31	8.66	8.20	8.27	7.5	8.09	8.26	9.0	8.86	8.68	8.16	8.04	8.0	8.15	8.1	7.82	7.7	8.0
Na_2O	0.34	0.07	0.27	0.14	0.02	0.05	—	0.39	0.17	0.6	0.37	0.006	0.01	0.57	0.28	0.29	0.4	0.36	0.40	0.2
K_2O	—	—	—	0.02	—	—	—	—	0.05	<0.06	—	0.002	0.00	—	—	—	—	—	—	—
NiO	—	—	—	0.28	0.22	—	0.32	—	0.25	—	—	0.27	0.75	0.20	0.25	0.26	—	0.27	0.23	0.2

[a](1) Bonatti *et al.* (1986) average spinel lherzolite, Zabargad, Red Sea;
(2) Hutchison and Gass (1971) spinel lherzolite, Kod Ali, Red Sea;
(3) Maaloe and Aoki (1977) average basalt-borne spinel lherzolite (oceanic);
(4) Menzies and Allen (1974) average spinel lherzolite, Othris ophiolite, Greece;
(5) Dick and Fisher (1984) computed abyssal spinel lherzolite, Atlantic Ocean;
(6) Michael and Bonatti (1985) computed less refractory abyssal spinel peridotite, Atlantic Ocean;
(7) Green *et al.* (1979) calculated primitive MORB residue;
(8) Reid and Woods (1978) spinel lherzolite (Group A) Salt Lake Crater, Hawaii
(9) Melson *et al.* (1967) spinel peridotite mylonite, St. Paul's rocks, Atlantic Ocean;
(10) Sen (1987) estimated least depleted source of Hawaiian lavas (partial fusion model);
(11) Hess and Otalara (1964) Mayaguez peridotite, Puerto Rico;
(12) Menzies and Allen (1974) harzburgite, Troodos ophiolite, Cyprus;
(13) Quick (1981) harzburgite, Trinity ophiolite, California;
(14) Ringwood (1962, 1979) hypothetical mixture of basalt (high Ti) and peridotite;
(15) Hutchison (1974) basalt-borne spinel lherzolites;
(16) Maaloe and Aoki (1977) basalt-borne and kimberlite-borne spinel and garnet lherzolites;
(17) Jagoutz *et al.* (1979) basalt-borne spinel lherzolite xenoliths;
(18) Wanke (1981) basalt-borne spinel lherzolites;
(19) Palme and Nickel (1985) basalt-borne spinel lherzolites;
(20) Boyd hypothetical mixture of kimberlite-borne low temperature peridotite and Yilgarn komatiite (1989)

samples allow the study of processes active in the lithospheric wedge above subduction zones and the comparison of such processes with intra-plate and mid-ocean ridge processes.

15.2.1 *Rift to passive margin transition*

Oceanic peridotites from passive margins (e.g. the eastern Atlantic Ocean) or pre-oceanic basins (e.g. the Red Sea) are compositionally different from the bulk of abyssal peridotites from ridge mountains and fracture zones and, in many instances, bear a remarkable resemblance to basalt-borne xenoliths entrained from beneath Phanerozoic or Proterozoic continental crust.

Zabargad Island in the Red Sea is composed of coarse-grained spinel lherzolites (Figure 15.1) (65% ol, 16% opx, 16% cpx, 3% sp), amphibole peridotites (> 3% magnesio-hornblende), plagioclase peridotites (> 3% calcic plagioclase) and minor amounts of dunite and wehrlite. The ultramafic rocks exposed on the island are a small fraction of a much larger body that extends to at least 8 km (Styles and Gerdes, 1983). On the basis of geochemistry, Bonatti *et al.* (1986) proposed that the spinel peridotites are essentially undepleted mantle material (Figures 15.2, 15.3), the plagioclase peridotites are hybrid rocks containing a melt component, and the amphibole peridotites are the

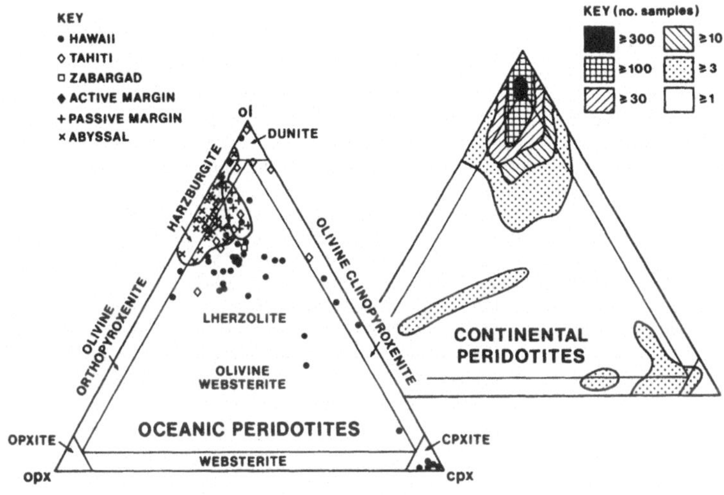

Figure 15.1 (a) Modal composition of oceanic and basalt-borne peridotite xenoliths. Oceanic peridotites (i.e. active margin, passive margin and abyssal) are samples of the shallowest (depleted) oceanic mantle and are predominantly lherzolites and harzburgites with few, if any, dunites (Michael and Bonatti, 1985 and references cited therein; Bonatti *et al.*, 1986). Basalt-borne xenoliths (i.e. Hawaii and Tahiti) are from the deeper (less depleted) oceanic mantle and are composed of pyroxenites and lherzolites with relatively few harzburgites (Sen, 1987; Tracy, 1980). (b) Modal composition of basalt-borne peridotite xenoliths from the western USA (Wilshire *et al.*, 1985). Note the predominance of dunites, lherzolites and the lesser amounts of pyroxenites and harzburgites.

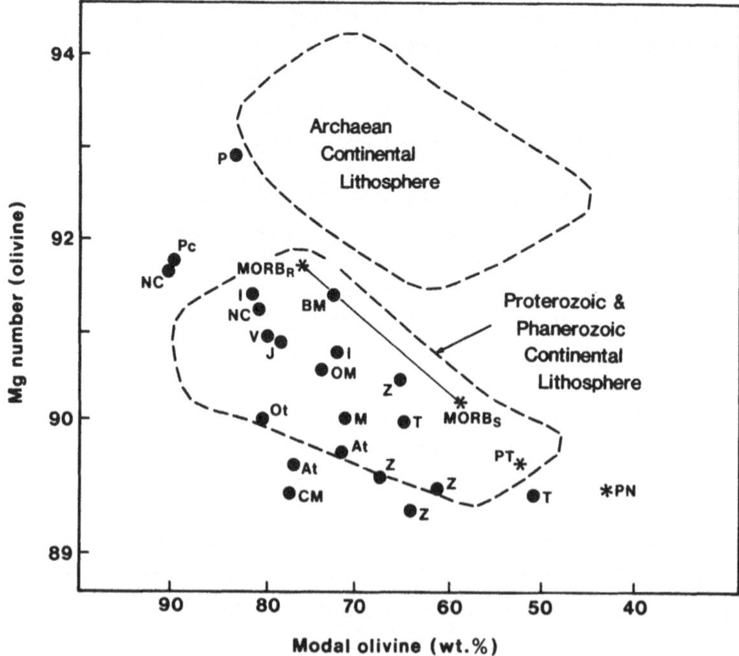

Figure 15.2 Magnesium number in olivine versus modal percentage of olivine (after Boyd, 1989) for peridotites from Phanerozoic oceanic mantle compared to peridotites from beneath Phanerozoic, Proterozoic and Archaean continental localities. Note the uniqueness of Archaean lithosphere and the similarity between the Proterozoic and Phanerozoic continental mantle and oceanic mantle. The oceanic data show a range in composition from relatively undepleted lherzolites in pre-oceanic settings (i.e. rift to passive margin) on the bottom right of the diagram to depleted harzburgites in active margins (i.e. supra-subduction) on the top left of the diagram. Data sources are from the following. (a) Abyssal: the Pacific (Pc), Indian (I) and Atlantic (At) oceans (Dick and Fisher, 1984; Bonatti and Michael, 1989). (b) Basalt-borne xenoliths or exposures: Zabargad (Z), Tahiti (T) and Malaita (M) (Bonatti *et al.*, 1986; Tracy, 1980; Neal, 1988). (c) Ancient oceanic lithosphere: Canyon Mountain (CM), Papua (P), New Caledonia (NC), Josephine (J), Oman (Om), Burro Mountain (BM), Othris (Ot) and Vulcan Peak (V) (Dungan and Ave'Lallement, 1977; Jaques and Chappell, 1980; Guillon, 1975; Dick, 1977; Boudier and Coleman, 1981; Loney et al., 1971; Menzies and Allen, 1974; Himmelberg and Loney, 1973). (d) Hypothetical mantle source compositions: source (s) and residue (r) for MORBs, Pyrolite (Py) and Palme and Nickel's (PN) estimate of undifferentiated mantle (Green *et al.*, 1979; Ringwood, 1979; Palme and Nickel, 1985).

metasomatic by-product of the interaction of hydrous fluids and mantle peridotite. Isotopic analyses of the Zabargad peridotites (Figure 15.4) indicate a considerable range in Sr and Nd isotopic ratios similar to mid-ocean ridge and oceanic island basalts, basalt-borne peridotites from adjacent continental locations (Menzies and Murthy, 1980), and orogenic peridotite massifs (e.g. Beni Bouchera). To explain the isotopic ratios, processes such as partial melting must be considered. Mantle melting can remove incompatible elements (e.g.

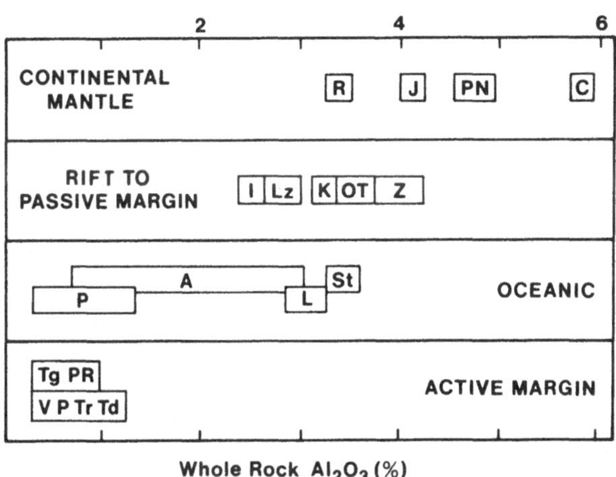

Figure 15.3 Aluminium content of mantle peridotites from different tectonic environments. The gradual change from aluminium-rich peridotites (lherzolites) to aluminium-poor peridotites (harzburgites) occurs in line with a change of tectonic setting from rift to passive margin to mid-oceanic to active margin. Continental mantle is defined on the basis of several estimates of undepleted mantle composition [R = Ringwood (1962); J = Jagoutz *et al.* (1979); PN = Palme and Nickel (1985); C = Carter (1970)]. Rift to passive margin mantle is defined on the basis of abyssal, xenolithic and ophiolitic peridotites [I = Iberia (Bonatti and Michael, 1989), Lz = Lanzo (Boudier, 1978), K = Kod Ali (Hutchison and Gass, 1971), Ot = Othris (Menzies and Allen, 1974), Z = Zabargad (Bonatti *et al.* 1986)]. Oceanic mantle is defined on the basis of abyssal and ophiolitic peridotites [P = Pacific (Serri *et al.*, 1985), A = Atlantic (Bonatti and Michael 1989), L = Ligurides (Beccaluva *et al.*, 1980), St = St Paul's rocks (Melson *et al.*, 1972)]. Active margin mantle is defined on the basis of dredged and ophiolitic peridotites [Tg = Tonga (Bloomer and Fisher, 1987), PR = Puerto Rico (Bowin *et al.*, 1966), V = Vourinos (Beccaluva *et al.*, 1980), P = Papua (Davies, 1971), Tr = Trinity (Quick, 1981), Td = Troodos (Menzies and Allen, 1974)].

Rb, Nd) from an undepleted source peridotite and produce a depleted residue (low Rb/Sr and a high Sm/Nd ratio); with time this loss of Rb and Nd registers as a low $^{87}Sr/^{86}Sr$ and a high $^{143}Nd/^{144}Nd$ ratio. The Zabargad peridotites have clearly been partially melted at some time, thus accounting for their present day isotopic ratios. The incompatible element concentrations in the amphibole peridotites may, however, betray a more recent metasomatic event as the elemental enrichments have not resided in the peridotite long enough for their presence to be registered in the isotopic ratios.

To what extent the basalt-borne xenoliths erupted on the island of Nunivak off the south coast of Alaska can be considered to have formed in a passive margin and not an active margin setting is not known. It could be inferred from its position at some distance from the active Aleutian arc that active margin processes are perhaps not as important as passive margin processes and therefore it will be included in this section. Seventy-five per cent of some 5000 basalt-borne xenoliths studied by Francis (1976) are lherzolites, in many instances amphibole-bearing lherzolites. Isotopic data were used to demonstrate a similarity in the source region of the metasomatic fluid or

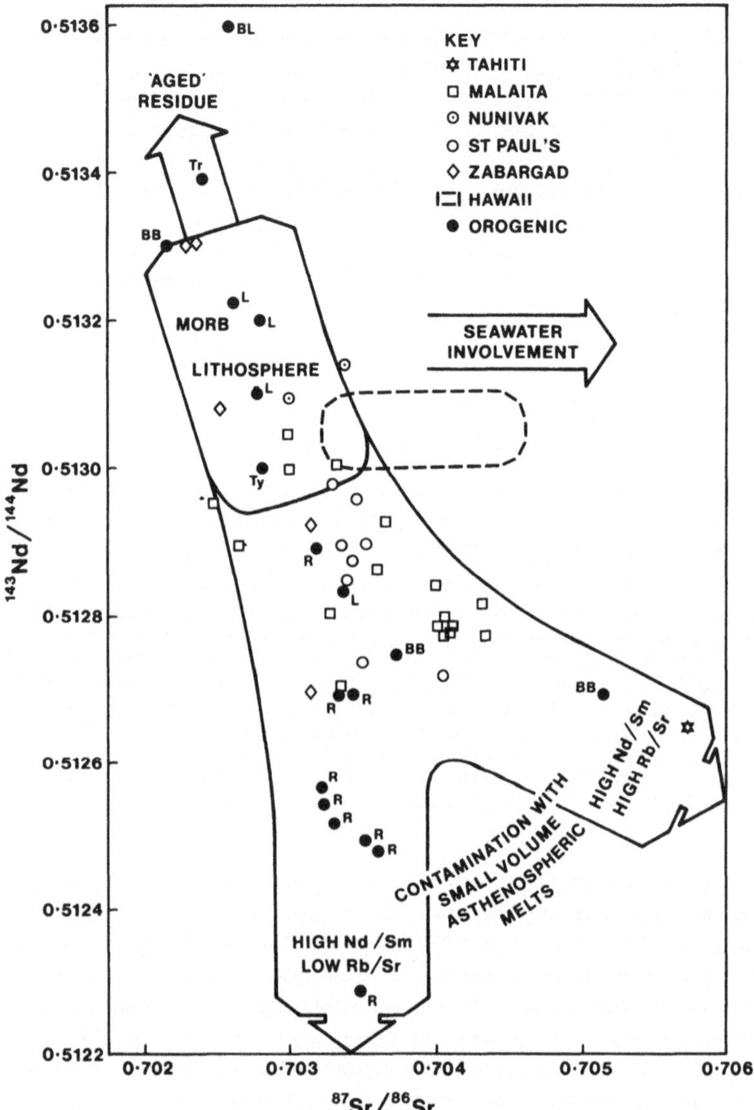

Figure 15.4 Sr and Nd isotopic composition of oceanic and ophiolitic peridotites. Exposures or basalt-borne xenoliths found in ocean basins are shown by open symbols [Tahiti (Menzies and Hawkesworth, 1987); Malaita (Neal, 1988); Nunivak (Roden *et al.*, 1985); St. Paul (Roden *et al.*, 1984); Zabargad (Brueckner *et al.*, 1988); Hawaii (Brouxel *et al.*, 1988)]. Closed symbols are orogenic peridotites (Bl = Baldiserro, Tr = Troodos, BB = Beni Bouchera, L = Lanzo, Ty = Trinity and R = Ronda) (Menzies 1984; Richard and Allegre, 1980; Reisberg and Zindler, 1986). The extreme range in isotopic composition is believed to be the result of modification of MORB lithosphere by upwelling small volume melts from the asthenosphere. In the case of Ronda the small volume melt was probably enriched in Nd and not Rb (e.g. carbonatite?) and in the case of Tahiti and Malaita in Rb and Nd (e.g. alnoite). The high $^{143}Nd/^{114}Nd$ and low $^{87}Sr/^{86}Sr$ encountered in the Baldissero, Beni Bouchera and Trinity peridotites indicates the existence of 'aged' residua within the ocean basins that are not related to the genesis of recent MORBs. This may indicate recycling of ancient residua as suggested by Polve and Allegre (1980).

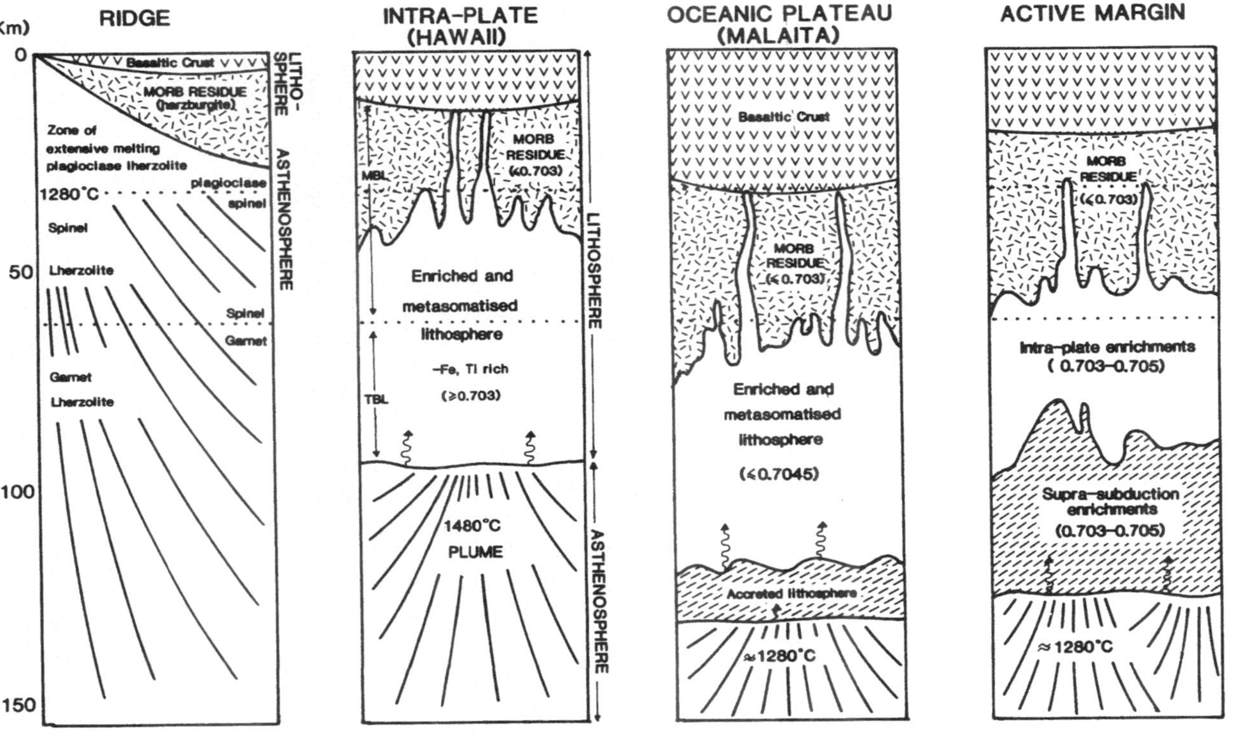

Figure 15.5 Transformation of the oceanic lithosphere as a result of intra-plate and suprasubduction processes. Note that the harzburgite protolith generated as a result of partial melting at a mid-ocean ridge may be transformed by interaction with upwelling asthenospheric melts, particularly in areas of plume activity (e.g. Hawaii), and with subduction related melts or fluids in active margin environments (e.g. Malaita, Japan). The end-product is chemically stratified oceanic lithosphere consisting of relict harzburgite protolith and chemically discrete horizons with intra-plate and supra-subduction trace element enrichments. It is unlikely, however, that sufficient time will have elapsed since chemical transformation occurred for the development of extreme isotopic heterogeneities.

melt responsible for the growth of the amphibole and the alkaline magmas that entrained the xenoliths (Menzies and Murthy, 1980; Roden *et al.*, 1984). It was suggested by these workers that the source for the metasomatic fluid or melt was sub-lithospheric and may be located within the asthenosphere. As the unmetasomatized spinel lherzolites from Nunivak Island have an isotopic composition similar to MORB, they may represent the depleted lithospheric protolith that existed prior to invasion of the lithosphere by asthenospheric magmas (Figure 15.5). In addition, the distinct difference between the Pb isotopic composition of the Nunivak host magmas and some of the spinel lherzolites (Zartman and Tera, 1973; Ben Othman *et al.*, 1990) supports a genetically distinct origin for the lithospheric peridotites and the upwelling alkaline magmatism. The alkaline magmatism may have a source within the MORB asthenosphere and the lithospheric protolith may be a by-product of an earlier depletion event related to crustal extraction in the Archaean–Proterozoic.

Xenoliths similar to those encountered on Nunivak Island have been reported from another passive margin, the Naturaliste Plateau off the south-west coast of Australia (Nicholls *et al.*, 1981). These spinel lherzolites and clinopyroxenites contain kaersutite and are compositionally unlike abyssal peridotites but similar to exposures of oceanic peridotite (e.g. St. Paul's rocks) and basalt-borne xenoliths from within the ocean basins (e.g. Hawaii, Nunivak, Mid-Atlantic Ridge) and the continents (e.g. the USA).

In a consideration of the petrology and chemistry of oceanic peridotites Bonatti and Michael (1989) demonstrated that peridotites from rift to passive margin localities (e.g. Iberia and Spitzbergen, Atlantic; Zabargad, Red Sea; and Naturaliste, Australia) tended to have higher modal clinopyroxene, higher abundances of whole rock Al_2O_3 and FeO (Figure 15.3), and lower abundances of MgO than their abyssal counterparts (Table 15.1). Moreover the composition of spinel, olivine and orthopyroxene in rift to passive margin peridotites compared favourably with undepleted peridotite xenoliths entrained by basaltic magmas erupted through continental crust. It can be speculated that the continental lithosphere occurring beneath Proterozoic or Phanerozoic continental crust may be sampled in a rift to-passive margin setting. Consequently the similarity between the Nunivak and Naturaliste oceanic xenoliths and continental xenolith suites is to be expected.

15.2.2 *Mid-ocean ridge processes*

Petrological and geochemical data for abyssal plagioclase and spinel harzburgites provide important information about the nature of the depleted peridotites generated by partial melting processes at mid-ocean ridges. In contrast, spinel lherzolites occurring as basalt-borne xenoliths on the Mid-Atlantic Ridge (Sinton, 1979) provide an invaluable insight into the nature of deeper undepleted portions of the sub-oceanic lithosphere and asthenosphere. Such xenoliths are compositionally distinct from the bulk of abyssal spinel lherzolites, which indicates a possible vertical variation in the petrology and chemistry of the mantle (Figure 15.5), from depleted mantle underlying the oceanic crust to undepleted mantle at depth (Bowen 1928; Ringwood, 1962). Basalt-borne xenoliths from ridge locations may thus constrain the chemistry of the source material responsible for the genesis of oceanic basalts and the residual spinel and plagioclase peridotites found at shallow depths.

Dredged samples of plagioclase and spinel peridotites have been recovered from the rift mountains and fracture zones of the Atlantic and Indian Oceans but not the Pacific Ocean. Dredged peridotites are commonly associated with basaltic and gabbroic rocks that presumably constitute disrupted parts of the overlying crust. Most olivine-bearing assemblages are, however, variably serpentinized. For example, serpentinization varies from 30 to 100% in peridotites from the Owen Fracture Zone, Indian Ocean (Hamlyn and Bonatti, 1980), resulting in the preservation of minerals in the order spinel ≫ clinopyroxene > orthopyroxene > plagioclase > olivine. Despite this extensive alteration, careful mineralogical studies have demonstrated that of over 300 abyssal peridotites, 60.5% were harzburgites and 39.5% were lherzolites (Dick and Fisher 1983). Variations in the modal mineralogy of North Atlantic abyssal peridotites have also been documented (60–86% olivine, 13–36% orthopyroxene, 0–12% clinopyroxene, 0.2–0.9% spinel) and a plagioclase-free spinel harzburgite (74.8% olivine, 20.61% orthopyroxene, 3.57% clinopyroxene, 0.51% spinel and 0.88% plagioclase) (Figure 15.1) was identified as the most common oceanic peridotite (Dick *et al.*, 1984, Michael and Bonatti, 1985). Although considerable variation in peridotites was apparent at individual dredge sites (i.e. harzburgite to lherzolite), in general spinel harzburgite and lherzolite constituted 70% of the abyssal peridotite population (Dick and Bullen, 1984). Peridotites associated with the Owen Fracture Zone in the Indian Ocean also consisted of spinel lherzolites, harzburgites and dunites. Plagioclase peridotites were found to be less abundant than spinel peridotites (about 30%) in the Romanche Fracture Zone and at St. Paul rocks and in most instances were considered to be hybrid peridotites consisting of a mixture of melt and residue (Figure 15.5).

No garnet peridotites have been reported from abyssal populations but some of the spinel lherzolites have textural features indicative of the breakdown of a precursor garnet lherzolite mineralogy (Hamlyn and Bonatti, 1980). Dunites are rare in abyssal peridotite suites (Dick and Bullen, 1984) (Figure 15.1) and much more common in basalt-borne xenolith suites. In detail, abyssal peridotites consist of harzburgite and lherzolite whereas basalt-borne oceanic xenolith suites are mainly dunites and lherzolites (e.g. Nunivak, Tahiti, and Hawaii). The intimate association of protogranular (coarse) and porphyroclastic (sheared) abyssal peridotites is not too unexpected given their location in fracture zones. Diapiric upwelling associated with emplacement could lead to such a textural association. Coarse and sheared assemblages within kimberlite and basalt-borne spinel and garnet peridotite suites similarly have been interpreted as a by-product of mantle diapiric upwelling where a sheared envelope of mantle peridotite surrounds a central core of granular peridotite.

Hamlyn and Bonatti (1980) noted that the Indian Ocean crust was underlain by lherzolite with subordinate amounts of harzburgite and dunite. This is consistent with the dominance of spinel and plagioclase lherzolites in ophiolites (see Chapter 4) believed to have formed at mid-ocean ridges (Beccaluva *et al.*, 1980; Serri *et al.*, 1985) and experimental studies on the origin of MORB and the production of a lherzolitic residue at 21–35 km (Bender *et al.*, 1978). Hamlyn and Bonatti (1984) proposed that the peridotitic assemblages found in the Owen Fracture Zone are not related by a source–melt–residue model. They also discounted an igneous origin for the peridotites because of the lack of cumulate textures and the need for a parental magma saturated in olivine, orthopyroxene and clinopyroxene to account for the mineralogy of the lherzolites. The preferred interpretation was one of mantle heterogeneity where the harzburgite and lherzolite were intermingled as a result of mechanical recycling processes. Mantle heterogeneity was also invoked by Dick and Fisher (1984) to explain the range in mineral chemistry found in abyssal peridotites, in particular the correlation between mineral and model compositions.

Dick and Bullen (1984) found that spinels from abyssal (and ophiolitic) peridotites could be used as petrogenetic markers and that the compositional similarities between abyssal peridotites and obducted peridotites indicated that several orogenic and ophiolitic peridotites may have initially formed at a mid-ocean ridge (i.e. Ronda, Beni Bouchera, Balmuccia, Baldissero, Lanzo, Liguria, Trinity and White Hills). The sub-crustal peridotites in Liguria (Beccaluva *et al.*, 1980) are dominantly spinel and plagioclase lherzolite, with associated dunite and harzburgite as in the case beneath modern ridge systems (Hamlyn and Bonatti, 1980). Certain of the ophiolitic peridotites (Ottonello *et al.*, 1979) were relatively undepleted in major and trace element abundances and their overall chemistry approximated to that of undepleted mantle (Table 15.1), a feature that is consistent with the presence of undepleted mantle at depth beneath ridge axes.

Peridotites produced at modern or ancient mid-ocean ridges tend to have much lower modal amounts of clinopyroxene than the rift to passive margin peridotites, resulting in lower abundances of Al_2O_3, FeO and higher abundances of MgO (Figure 15.3 and Table 15.1). Michael and Bonatti (1985) noted the existence of a more chromiferous spinel, more magnesian olivine and a less aluminous orthopyroxene. In general peridotites produced at mid-ocean ridges tend to contain greater amounts of magnesian olivine than their rift to passive margin counterparts (Figure 15.2). Certain geochemical parameters (Figure 15.2 and 15.3), however, indicate a continuum of compositions between rift to passive margin peridotites and mid-ocean ridge peridotites. The limited data from the Pacific Ocean indicate that abyssal peridotites are more depleted than Atlantic abyssal peridotites (Figure 15.3) and that some compositional overlap exists between mid-ocean ridge peridotites and basalt-borne xenoliths erupted through Phanerozoic or Proterozoic continental crust (Figure 15.2).

Bonatti *et al.* (1970) reported very high strontium isotopic ratios for peridotites from the equatorial Mid-Atlantic ridge. This, in conjunction with low Rb/Sr ratios and a depletion in incompatible elements, led them to suggest that oceanic peridotites constituted part of a unique layer similar to that found in alpine massifs and unrelated to basalt genesis. Subsequent work has shown that these high strontium isotopic ratios are the result of post-consolidation processes and are not necessarily indicative of mantle processes. In contrast, clinopyroxenes from orogenic and ophiolitic peridotites have retained information pertinent to mantle processes (Brueckner, 1974; Menzies and Murthy, 1976; Polve and Allegre, 1980; Richard and Allegre, 1980) (Figure 15.4). Very few isotopic studies of abyssal peridotites have been undertaken because of the serpentinized nature of the rocks, however, the available isotopic data for orogenic and ophiolitic peridotites indicate Sr and Nd isotopic heterogeneity which overlaps with that observed in rift to passive margin peridotites and basalt-borne continental peridotite xenoliths (Figure 15.4). This is consistent with the derivation of abyssal peridotites from a heterogeneous oceanic mantle source (asthenosphere) similar to that found in rift to passive margin environments.

On the basis of isotopic studies of orogenic peridotites, Polve and Allegre (1980) proposed that the oceanic mantle was progressively more depleted, mixed and recycled with time. Partial melting at ridges and subduction zones produced a highly depleted peridotite which was mechanically intermingled with less depleted material and remelted (Hamlyn and Bonatti, 1980). Such a recycling model is consistent with depletions in the high field strength elements (e.g. Ti and Zr) found in many peridotites from the ocean basins (Salters and Shimizu, 1988). As such elemental features are normally characteristic of active margin processes, it could be taken as evidence that all oceanic peridotites are not necessarily simply related to the genesis of MORB.

15.2.3 *Intraplate processes*

Oceanic peridotites found near sites of intraplate volcanism have been chemically modified by asthenosphere–lithosphere interaction and are consequently different from the depleted peridotites produced at mid-ocean ridges or undepleted peridotites sampled at rift to passive margins.

The mineralogy and chemistry of oceanic abyssal peridotites changes between the equator and 60°N (Dick *et al.*, 1984; Michael and Bonatti, 1985).

Around 34–45°N the peridotites have lower modal amounts of clinopyroxene (< 1%) and orthopyroxene (< 25%) and higher modal amounts of olivine (> 75%) than the abyssal peridotites to the north or south. The modal amount of clinopyroxene (0–6%) and orthopyroxene (17–35%) increases and the modal amount of olivine (65–80%) decreases away from this intraplate location in the Atlantic Ocean. Further, the peridotites at 34–45°N contain spinels richer in chrome, and orthopyroxenes poorer in aluminium and iron and richer in magnesium than abyssal peridotites to the north or south. Dick *et al.* (1984) elegantly demonstrated the change in spinel composition in relation to the geoid anomaly map of the North Atlantic. In particular, spinel compositions in abyssal peridotites increased systematically near the region associated with the Azores triple junction. The geographical coincidence of geoid anomalies, trace element enrichments in basalts and highly depleted peridotites led Dick *et al.* (1984) and Michael and Bonatti (1985) to propose greater extents of partial melting on the Mid-Atlantic Ridge near the Azores thermal anomaly. Indeed their data suggest that peridotite compositional variations are inextricably linked to variations in the thermal structure of the seafloor. Dick *et al.* (1984) also noted that significant heat transport must have occurred at sub-lithospheric depths against the mantle flow direction as determined by plate movements.

If high degrees of partial melting (and low abundances of incompatible elements) are associated with hot-spot activity, as implied by these studies, then the high incompatible element abundances commonly found in hot-spot magmatism must represent input from a discrete sub-lithospheric source. Indeed the passage of asthenospheric magmas through the overlying mechanical boundary layer or lithosphere could deplete the lithosphere due to the elevation of the isotherms, but also enrich it as a result of the influx of incompatible elements (Figure 15.5). St. Paul's rocks in the Atlantic may be a case in point. The variety of mylonitized rocks including amphibole (pargasite) peridotite, spinel peridotite and hornblendites (kaersutite) (Melson *et al.*, 1972) are isotopically similar to oceanic island basalts (Figure 15.4) and have been interpreted as a possible source for alkaline magmas. The existence of alkaline intrusives in the vicinity of the island and the presence of a fracture zone may indicate that sub-lithospheric melts could have reached the surface, allowing for high level transformation of the peridotites by the interaction of lithospheric MORB peridotites with conduits of upwelling alkaline magmatism (OIB) (Figure 15.5).

The interplay between oceanic lithosphere and intraplate asthenospheric volcanism adjacent to the Azores hot-spot is also apparent in basalt-borne xenolith suites associated with the Hawaiian hot-spot (Figure 15.5). Basalt-borne xenoliths from Hawaii (Jackson and Wright, 1970; Reid and Woods, 1978; Sen 1987) are predominantly spinel lherzolites with variable amounts of dunite, harzburgite, olivine websterite, olivine clinopyroxenite and pyroxenites (Figure 15.1). The petrology of the Hawaiian xenolith suite is similar to abyssal peridotite populations but tends to have less dunites than continental basalt-borne suites where dunite and lherzolite are the common rock types (Figure 15.1). The petrogenetic variety within individual xenolith suites is believed to be controlled by the development of magma storage reservoirs under the Hawaiian islands (Sen, 1987; Clague 1987). The alkalic basalts and basanites, erupted early in the history of the islands, contain basalt-borne xenoliths that are mainly metamorphic dunites and lherzolites with minor amounts (< 1%) of cumulate peridotites. The predominance of metamorphic (mantle?) xenoliths may be because the host magmas are small volume melts that have travelled relatively uninterrupted from their mantle source region to the surface. In contrast, the main shield-building event on these islands is characterized by the eruption of tholeiitic magmas that contain little or no metamorphic xenoliths but a preponderance of dyke, sill and vein fragments (75%) and cumulate xenoliths (25%). Such an association may be adequately explained by the accumulation and fractionation of tholeiitic magmas within high-level magma chambers and the subsequent disruption, during eruption, of crustal cumulates and magmatic feeder systems (Clague, 1987). Post-caldera alkalic basalts contain 35–57% cumulate xenoliths and 33–62% metamorphic dunites, indicating that these mantle-derived magmas encountered both mantle rocks and crustal magma chambers en route to the surface. Finally, by the time the post-erosional alkalic basalts are erupted, the magma chambers that fed the tholeiitic magmas have crystallized and therefore the post-erosional nephelinites come directly from the mantle with 99% metamorphic xenoliths and 1% cumulate xenoliths. Thus, the earliest and latest volcanism on Hawaii contains the deepest xenoliths and the intermediate phases of volcanism contain cumulate xenoliths that betray residence times for the tholeiitic magmas in sub-caldera or deeper magma chambers.

Sen (1987) proposed that the dunite xenoliths on Oahu were shallow crustal cumulates from magma chambers related to the accumulation and eruption of the Koolau tholeiities. The spinel lherzolites are believed to be fragments of sub-Hawaiian lithosphere modified by metasomatic processes contemporaneous with upwelling of small volume asthenospheric melts that produced the post-caldera or post-erosional alkaline basalts. With the use of Sr, Nd and Pb isotopic data, Brouxel *et al.* (1988) substantiated this assertion and demonstrated that spinel lherzolites from Kauai were fragments of the depleted MORB lithosphere (Figure 15.5) and, as such, were not genetically related to their host magmas. Whereas the spread in Sr isotopic composition shown in Figure 15.4 is believed to be due to seawater contamination, the range in Nd more accurately reflects mantle compositions as Nd isotopes are not affected by alteration. The sub-Hawaiian lithosphere has a Nd isotopic composition identical to spinel peridotites from Zabargad and Malaita and basalt-borne xenoliths from several continental regions. It is also important to note that the Sr and Nd isotopic composition of lithospheric spinel peridotites from Hawaii overlap with the isotopic data for post-erosional alkaline basalts. These alkaline melts have been interpreted as small volume melts of the sub-Hawaiian lithosphere (Frey and Roden, 1987, for review). In contrast, some lherzolites from the Hawaiian islands have Pb isotopic compositions similar to their host magma (Zartman and Tera, 1973), indicating that they are either derivatives of alkali basalts, the source (Jackson and Wright, 1970) or residue involved in the genesis of alkali basalts, or most likely, fragments of the lithosphere that have been metasomatized or enriched by the passage of asthenospheric melts (Figure 15.5).

Basalt-borne lherzolites and harzburgites from Ua Huka, Marqueses (Berger, 1981) have spinel compositions very similar to the spinel lherzolites and pyroxenites from Hawaii (Sen, 1987) and continental basalt-borne xenoliths. Moreover, the range in spinel composition covers the range shown by abyssal peridotites (Dick and Bullen, 1984). Harzburgite xenoliths from Tubuai are believed to be fragments of depleted (magnesian) oceanic lithosphere and lherzolites are believed to be part of the undepleted (iron-rich) asthenosphere (Berger, 1981). In contrast, a cumulate origin is proposed for the dunites from Tubuai in shallow magma chambers similar to the Hawaiian dunites. In some instances the Tubuai peridotites contain spinels with a composition that overlaps with the dunites from Koolau (Sen, 1987) and as these rocks have temperatures of 1160–1235°C, a magmatic origin seems to be the most appropriate (Berger, 1981). High level magmatic processes may also explain the occurrence of basalt-borne dunites and wehrlites in Tahiti (e.g. Papenoo and Faatua). These xenoliths are not believed to be related to the host magmatism (Tracy, 1980), but may be related to the previous accumulation of magma in high-level magma chambers.

Malaita in the Solomon islands is part of the uplifted edge of the Ontong Java Plateau. Such plateaux are found in the Pacific and Indian Oceans and are characterized by thick crust and low seismic velocities. It is in many instances difficult to define their origin as they may represent remnants of arc ridges, uplifted oceanic crustal sections consisting of oceanic island basalts, or detached continental fragments (see Chapter 11). In the case of Malaita, the plateau may have resulted from the coincidence of a spreading ridge and a plume. The alnoite intrusions on Malaita contain spinel and garnet-bearing lherzolite xenoliths (50–70% olivine, 10–35% orthopyroxene and 5–40% clinopyroxene) and a discrete megacryst suite of garnet, clinopyroxene, orthopyroxene, clinopyroxene–ilmenite intergrowths, ilmenite and zircon (Nixon and Boyd, 1979; Nixon and Coleman, 1978; Nixon *et al.*, 1980; Bielski-Zyskind *et al.*, 1984; Neal, 1988; Nixon and Neal, 1987). As most of the peridotites in the western Pacific (ophiolitic and abyssal) are extremely depleted (section 15.2.4), the elemental and isotopic enrichments in the Malaita xenoliths require the involvement of another component. It could be speculated that the Ontong Java Plateau was overthickenend as a response to plume–ridge overlap and that the xenoliths reflect a hybridization of

MORB lithosphere with upwelling small volume melts from the asthenosphere or a deep mantle plume (Figure 15.5) (Neal and Davidson, 1989). If the plume–ridge collision occurred some time ago in the history of the Pacific, the elemental heterogeneity observed at Malaita may reflect an accumulation of incompatible elements due to asthenosphere–lithosphere interaction. Sr and Nd isotopes for the Malaita xenoliths (Neal, 1988) show a range from MORB values for anhydrous peridotites (unmetasomatized lithosphere) to OIB values for hydrous peridotites (metasomatized lithosphere = lithosphere–asthenosphere hybrid).

A similar xenolith suite to that on Malaita has been found in New Zealand. The Kakanui Mineral Breccia, New Zealand (Dickey, 1968) comprises spinel lherzolites, garnet pyroxenites, eclogites and a megacryst suite (i.e. pyropic garnet, olivine, pyroxene, amphibole and spinel). Isotopic data for the Kakanui xenoliths (Gamble and Menzies, unpublished data) point to the presence of MORB lithosphere with a later influence of asthenospheric melts. Indeed several of the Kakanui kaersutite megacrysts have Pb isotopic compositions similar to other asthenospheric-derived kaersutites (Ben Othman *et al.*, 1990).

Oceanic lithosphere can become contaminated in several ways by intraplate processes and the constructional volcanism associated with oceanic islands (Figure 15.5): (1) depletion adjacent to hot-spots leading to conversion of the residual lherzolite within the oceanic plate to a more harzburgitic residue (e.g. Azores and Hawaii); (2) metasomatism of the MORB lithosphere by small degree melts from the asthenosphere or deeper mantle (e.g. St. Paul's, Malaita, Hawaii) producing lithosphere–asthenosphere hybrid peridotites (hydrous peridotites), which display considerably more heterogeneity in Sr, Nd and Pb isotopes than the anhydrous lithospheric protolith; and (3) development of magma chambers within the oceanic crust and on the crust–mantle interface producing cumulate peridotites and pyroxenites (e.g. Hawaii, Tubuai, Réunion and Marquesas).

15.2.4 *Active margin processes*

Oceanic peridotites from active margins are highly depleted and represent the compositional antithesis of oceanic peridotites from rift to passive margin environments. In addition, active margin peridotites are chemically distinct from the bulk of abyssal peridotites from ridge mountains and fracture zones and, in some instances, they bear some resemblance to kimberlite-borne xenoliths entrained from beneath Archaean continental crust.

Ultramafic rocks dredged from active margins are limited in number and in many instances badly serpentinized. Antigorite–talc and serpentine-bearing rocks from the Puerto Rico Trench were interpreted as oceanic peridotites by Bowin *et al.* (1966), and Eggler *et al.* (1973) reported antigorite–serpentinites from the Cayman Trough in the Caribbean Sea. As these rocks dominated the dredge hauls they concluded that they must be a dominant basement rock in this region. Ultramafic rocks have also been reported from the vicinity of the Tonga and Mariana Trenches (Bloomer, 1983; Bloomer and Fisher, 1987; Bloomer and Hawkins, 1983; Shcherbakov and Savelyeva, 1984). In the case of the Tonga Trench some of the peridotites are extremely fresh and consist of harzburgite with minor dunite and lherzolite. These peridotites have a low modal concentration of clinopyroxene (3.6%) and the coexisting mineralogy is highly magnesian compared to peridotites from the mid-oceanic ridges, e.g. the spinel compositions are outside the field defined by abyssal peridotites. Several of the Tonga Trench peridotites have a highly magnesian mineralogy (Fo_{91}) and are equivalent to peridotites from the Mariana Trench (Bloomer, 1983). These highly magnesian peridotites have been interpreted as the residues after the removal of boninitic melts.

Metamorphic peridotites from the west Pacific ophiolites are also highly depleted (e.g. New Caledonia and Papua) and contain high modal amounts of olivine with magnesium numbers in excess of 91. Similarly, ophiolites that are believed to have been emplaced in active margin environments (Dick and Bullen, 1984) [e.g. Troodos (Greenbaum, 1972) and Josephine (Dick, 1977)] are characterized by a predominance of highly magnesian harzburgites (Table 15.1) that, like the Papuan ophiolites, occasionally contain orthopyroxenite veins. These pyroxenite veins may be the final melt fraction produced within the oceanic lithosphere in an active margin environment (Dick, 1977). Some of the peridotites from the Vourinos, Papua, Troodos and Trinity ophiolites have mineralogical and chemical variations that overlap with the Tonga and Puerto Rico active margin peridotites (Figures 15.2 and 15.3).

Detailed petrographic and mineralogical studies of the Trinity ophiolite California (Quick, 1981) reveal the presence of plagioclase lherzolite, plagioclase harzburgite, lherzolite, harzburgite and dunite with less abundant pyroxenites, wehrlites and websterites. The plagioclase lherzolite is composed of 70–80% olivine,. 15–20% orthopyroxene, 2–10% clinopyroxene, 1–2% spinel and 2–10% plagioclase and the lherzolite 70–80% olivine, 15–20% orthopyroxene, 1–10% clinopyroxene and 1–2% spinel, which compares favourably with the typical abyssal peridotite (74.8% olivine, 20.61% orthopyroxene, 3.57% clinopyroxene, 0.51% spinel and 0.88% plagioclase). These rocks have equilibrated at < 15 km, within the plagioclase stability field, and plagioclase-rich dyke compositions coincide with the 5 kb cotectic partial melt coexisting with olivine, orthopyroxene and clinopyroxene (Quick, 1981). Structural and stratigraphic considerations indicate, however, that the Trinity ophiolite formed in an active margin setting (Quick, 1981), and consequently the undepleted nature of some of the Trinity lherzolites must reflect a relic of MORB processes. Other orogenic peridotites found in active margin environments contain significant amounts of undepleted peridotite (Arai and Takahashi, 1989) similar to the Trinity ophiolite. On the island of Hokkaido, orogenic peridotite massifs consist of plagioclase and spinel lherzolite and spinel harzburgite that have been metasomatized such that there is now extensive development of mica. This again may be due to asthenosphere–lithosphere interaction.

Garnet lherzolites, spinel lherzolites and amphibole-bearing peridotites and pyroxenites have been reported from Itinome-gata, Japan (Tanaka and Aoki, 1979) and micaceous dunites from the Philippines (Flower, personal communication). The material from Itinome-gata is believed to be a derivative of arc volcanism, but could equally well be thought of as fragments of sub-arc lithosphere modified by upwelling melts and fluids from the active margin (Figure 15.5). The pyroxenite–glimmerite xenoliths from the Philippines may have resulted from the influx of hydrous fluids into the wedge above the subduction zone.

Bonatti and Michael (1989) compiled the available data from active margins and noted that the peridotites are virtually clinopyroxene-free and tend to be harzburgites or dunites (Figure 15.1). Relative to abyssal and rift to passive margin peridotites, the peridotites from active margins contain the lowest abundance of aluminium and iron (Figure 15.3) and tend to be the most magnesian (Table 15.1). Of all the oceanic peridotites these have the lowest abundance of magmatophile elements and are thus the most depleted. In addition, the active margin peridotites contain the most chromiferous spinel and the most magnesian olivine, as do the metamorphic peridotites that floor active margin ophiolites (Figure 15.2). Little or no isotopic data is available for active margin dredged peridotites because of the poor state of preservation, and limited data are available for the metamorphic peridotites from active margin ophiolites (Figure 15.4). Clinopyroxenes from the Troodos and Trinity ophiolites are isotopically depleted, indicating a time-integrated depletion in Rb and Nd similar to the Zabargad spinel peridotites and the spinel peridotites from Hawaii.

Active margin processes further deplete the oceanic lithosphere such that the most refractory oceanic peridotites are found in association with arc environments (e.g. Tonga, Mariana, Papua, New Caledonia, Trinity, Troodos). Hydrous melting above subduction zones is believed to be the cause of this depletion. It may also account for the isotopically depleted character of tectonite peridotites from active margin ophiolites in that much of the intra-plate enrichment, which may have been added between the ridge and the trench, has been removed during melting above the subduction zone. Indeed this intra-plate component has probably been added to arc volcanic rocks.

15.3 Petrogenetic models

15.3.1 *Oceanic mantle under continents or* vice versa?

It was initially thought that continental regions were underlain by undepleted mantle and ocean basins by depleted mantle (Nicolas and Jackson, 1972). The petrological provinciality of several Mediterranean ophiolitic and orogenic peridotites was taken as evidence of a sub-continental mantle origin for the lherzolites from the western Mediterranean and a sub-oceanic mantle origin for harzburgites from the eastern Mediterranean. The general scheme proposed by Nicolas and Jackson (1972) required revision because of the presence of lherzolitic rocks beneath many of the world's ophiolites (Menzies and Allen, 1974; Ottonello *et al.*, 1979; Quick, 1981) and the abundance of spinel lherzolites in abyssal peridotite suites. Further, Reid and Woods (1978) reported the presence of 'oceanic mantle' beneath continental crust in the south-western USA indicating that some petrogenetic similarities existed between the evolution of the mantle underlying young continental regions (= 2500 Ma) and that beneath the ocean basins (= 200 Ma). Boudier and Nicolas (1985) revised the earlier model of Nicolas and Jackson (1972) and proposed a lherzolite subtype and a harzburgite subtype for modern and ancient oceanic lithosphere. They proposed that the harzburgite subtype was the result of fast spreading (Pacific Ocean) and the lherzolite subtype the result of slow spreading (Atlantic and Indian Oceans). This is in part substantiated by Pacific Ocean abyssal peridotites which are more depleted than the Atlantic and Indian Ocean abyssal peridotites (Figures 15.2 and 15.3).

It is now apparent that the asthenosphere or adiabatic interior beneath the lithospheric plates is geochemically homogeneous and that the major differences between oceanic and continental upper mantle exist within the lithospheric mantle (i.e. mechanical and thermal boundary layer). It has been demonstrated that sub-continental Archaean lithosphere is petrologically (Boyd, 1989) and geochemically (Menzies, 1990) unique relative to lithospheric mantle beneath Proterozoic and Phanerozoic continental crustal regions which is not that different from that which underlies the modern ocean basins (Boyd, 1989). It should be noted, however, that Archaean lithospheric mantle and active margin lithospheric mantle are in some instances chemically similar (Fig. 9.2). Perhaps the chemically depleted character of Archaean lithosphere has been further depleted by subduction processes active over several billion years.

15.3.2 *Heterogeneous oceanic mantle?*

Bonatti and Michael (1989) believe that the variation in the petrology and chemistry of abyssal peridotites is related to a variation in the thermal regime beneath different parts of the ocean basins. Peridotites from pre-oceanic, passive margin and mid-ocean ridge locations tend to be increasingly more depleted in magmatophile elements, such as, CaO, Al_2O_3 and FeO (seen as higher modal amounts of olivine and lower modal amounts of clinopyroxene and orthopyroxene). This constitutes a change from lherzolites to harzburgites and eventually to dunites (Figures 15.2 and 15.3). Bonatti and Michael (1989) used several parameters to distinguish peridotites from these different tectonic environments and, as the most depleted peridotites (Table 15.1) are found in supra-subduction zone environments, they concluded that hydrous melting must result in a final extraction of silicate melt from lithospheric peridotites.

Rift to passive margin or pre-oceanic peridotites from the Red Sea and elsewhere compare favourably with the Lanzo and Othris ophiolitic and orogenic peridotites (Table 15.1), which are believed to have formed in small ocean basins. The Lanzo and Othris peridotites are undepleted relative to oceanic lithosphere as represented by abyssal peridotites from the Atlantic and Pacific Oceans (MORB). Abyssal peridotites compare favourably with the ophiolitic peridotites from the Ligurides, Italy, which are interpreted as MORB ophiolites. Active margin peridotites from the Tonga, Mariana and Puerto Rico trenches are compositionally similar to ophiolitic harzburgites from Vourinos, Papua, Trinity and Troodos.

Note that this interpretation is only valid if it can be shown that active margins are underlain by a predominance of harzburgites. The available evidence is sparse and depends to a large extent on badly serpentinized material or ophiolites which are interpreted to have formed in an active margin setting. Peridotites from Hokkaido, Japan, have a large proportion of undepleted material which is not compatible with extensive depletion in an active margin environment. Perhaps this indicates that the refractory protolith which initially formed at the ridge axis was subsequently metasomatized and enriched as a result of intraplate and active margin processes (Figure 15.5). Examples can also be found of ophiolites (e.g. Trinity and Troodos) with relatively undepleted lherzolitic pockets (Table 15.1) in extremely depleted harzburgite. It is very difficult to assign any petrogenetic significance to the dredged peridotites from active margins until there is more information about the volumetric significance of harzburgite relative to lherzolite.

The inferred lateral changes in upper mantle petrology and chemistry may hold only for the uppermost portions of the oceanic lithosphere as basalt-borne peridotite xenoliths from the ocean basins indicate that the deeper mantle regions in ocean basins are relatively undepleted (Figure 15.5), regardless of tectonic setting (e.g. Mid-Atlantic Ridge, Hawaii, Japan). Many of the studied abyssal, ophiolitic, orogenic and basalt-borne peridotites have high $^{143}Nd/^{144}Nd$ and low $^{87}Sr/^{86}Sr$ ratios compatible with partial melting or depletion processes. The range in isotopic composition in oceanic peridotites indicates considerable heterogeneity in the oceanic lithosphere. What is the cause of this heterogeneity? The residual protolith produced at mid-ocean ridges constitutes the bulk of the oceanic lithosphere and has an isotopic composition equivalent to MORB (Figure 15.5). This is the case for the lithosphere beneath the Pacific Ocean (Malaita, Nunivak, Hawaii), the Atlantic Ocean (St. Paul), the Red Sea (Zabargad) and several ophiolites (Lanzo, Trinity). Those peridotites with lower $^{87}Sr/^{86}Sr$ ratios and higher

$^{143}Nd/^{144}Nd$ ratio than MORB cannot be simply related to the recent extraction of MORB and must therefore relate to older depletion events. The existence of such 'aged' residues may lend support to the idea that a considerable amount of recycling and intermingling of materials has occurred within the asthenosphere. In contrast to the 'aged' residua, a considerable number of oceanic peridotites plot outside the MORB field towards higher $^{87}Sr/^{86}Sr$ ratio and lower $^{143}Nd/^{144}Nd$ ratios. This is inconsistent with a partial melting origin involving MORB asthenosphere and requires a more complex model. Interestingly, in the case of Zabargad, Malaita and St. Paul's many of these peridotites contain hydrous minerals and/or have enhanced concentrations of incompatible elements. This may indicate the migration of small volume melts (rich in Rb, LREE and U) into the lithosphere from the underlying asthenosphere (Figure 15.5). It may be necessary to invoke the upward passage of compositionally different melt fractions to account for the different isotopic compositions. A melt rich in LREE and depleted in Rb, such as a carbonatite or a nephelinite, may explain the vertical array and a melt rich in the LREE and Rb, such as an alnoite or a kimberlite, may explain the isotopic features observed at Malaita.

The petrological, mineralogical and chemical heterogeneity observed within the oceanic lithosphere appears to be an artefact of the interaction of a carapace of cold, depleted lithosphere with underlying hot, undepleted asthenosphere.

15.4 Concluding statements

1. Oceanic peridotites provide vital information about the nature of the oceanic lithosphere in a variety of tectonic settings. The oceanic lithosphere (crust and mantle) produced at spreading centres is initially related to the origin of MORBs and at mid-ocean ridges the oceanic crust is underplated by a depleted harzburgite, the product of melt extraction from an undepleted spinel or plagioclase lherzolite. This accounts for the predominance of harzburgites in fracture zones on the ocean floor and the occurrence of relatively undepleted lherzolites as xenoliths in alkaline magmas erupted on ridge segments. In addition, plagioclase lherzolites constitute a significant proportion of dredged material and these peridotites provide evidence for the incomplete removal of melt from the peridotite matrix. As the oceanic plate moves away from the spreading centre the depleted mantle accreted at sub-Moho pressures and temperatures cools conductively by heat loss through the crust and the resultant density increase causes the crust to sink. Cooling of the depleted peridotite leads of the stabilization and thickening of the oceanic lithosphere ($<1280°C$), reaching thicknesses of approximately 120–150 km beneath the older parts of the ocean basins.

2. Throughout its lifetime the lithosphere (mechanical boundary layer) is continually exposed, at its base, to the asthenosphere (adiabatic interior), where interference from small volume melts leads to thermal and chemical transformation (thermal boundary layer). Seafloor spreading also transports this thickening mechanical boundary layer into the path of upwelling deep mantle plumes, resulting in further modification of the lithospheric protolith. In Hawaii, the petrology and geochemistry of plagioclase, spinel and garnet peridotites and pyroxenites allows the speculation that the crust is underlain by a residual MORB protolith formed at the ridge axis and that the upwelling of asthenospheric small volume melts has modified this residue and may have produced chemically stratified lithosphere. For example, the presence of high temperature pyroxenite xenoliths, which are believed to be derivatives of sub-lithospheric alkaline magmas, points to the upwelling of alkaline melt fractions into the lithosphere. The change in spinel composition with depth may be due to equilibration with Fe-rich melts from the asthenosphere. Similarly, in the case of Malaita, the MORB residual protolith is underlain by enriched garnet peridotites, the chemistry of which may reflect the influx of incompatible element enriched melts.
3. It appears that the greatest inventory of trace elements (and by inference the greatest potential for isotopic heterogeneity) occurs where the lithosphere is at its thickest (120 km) and oldest (< 200 Ma). Eventually the lithosphere is exposed to hydrous melting at active margins where it is very effectively stripped of any elemental heterogeneity inherited by the MORB protolith since formation at a ridge.

Appendix A

The approximate positions of the main localities mentioned in the text are shown in the four maps contained in this appendix. These maps correspond to the four parts of the book—Structure, Processes, Environments and Sources—and reference should be made to the appropriate map. It should be noted that for reasons of clarity more common localities, which can generally be found in a standard geographical atlas, may not be shown.

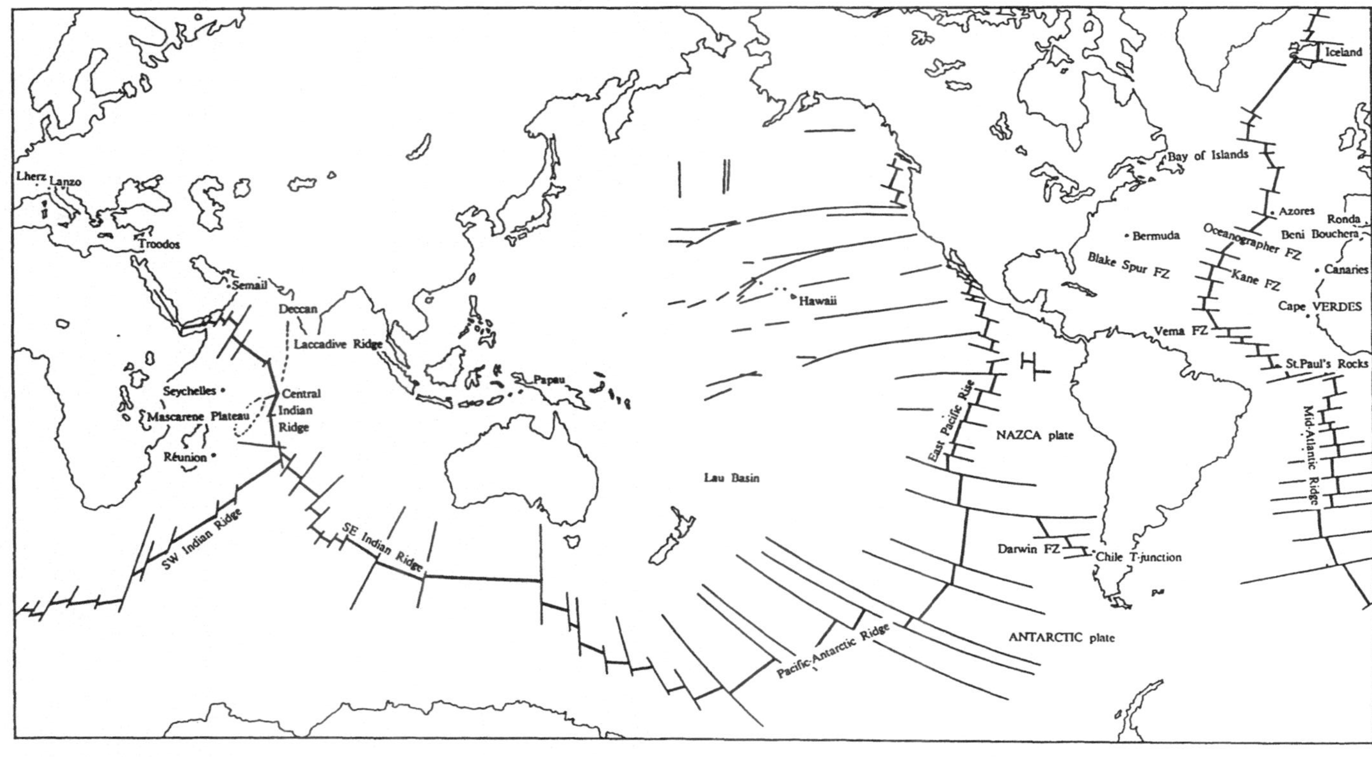

Figure A.1 World map with major oceanic spreading centres and fracture zones indicated and showing the approximate position of the main localities mentioned in Part I.

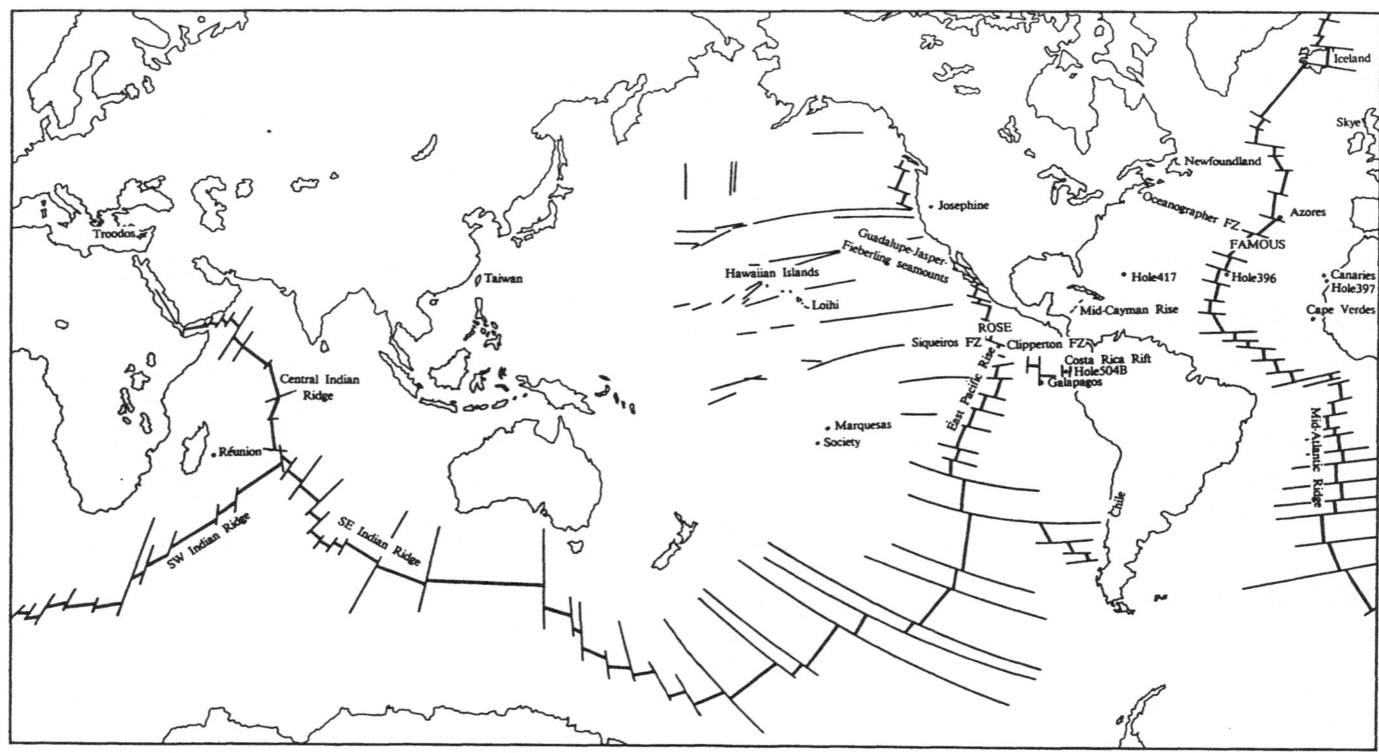

Figure A.2 World map with major oceanic spreading centres and fracture zones indicated and showing the approximate position of the main localities mentioned in Part II.

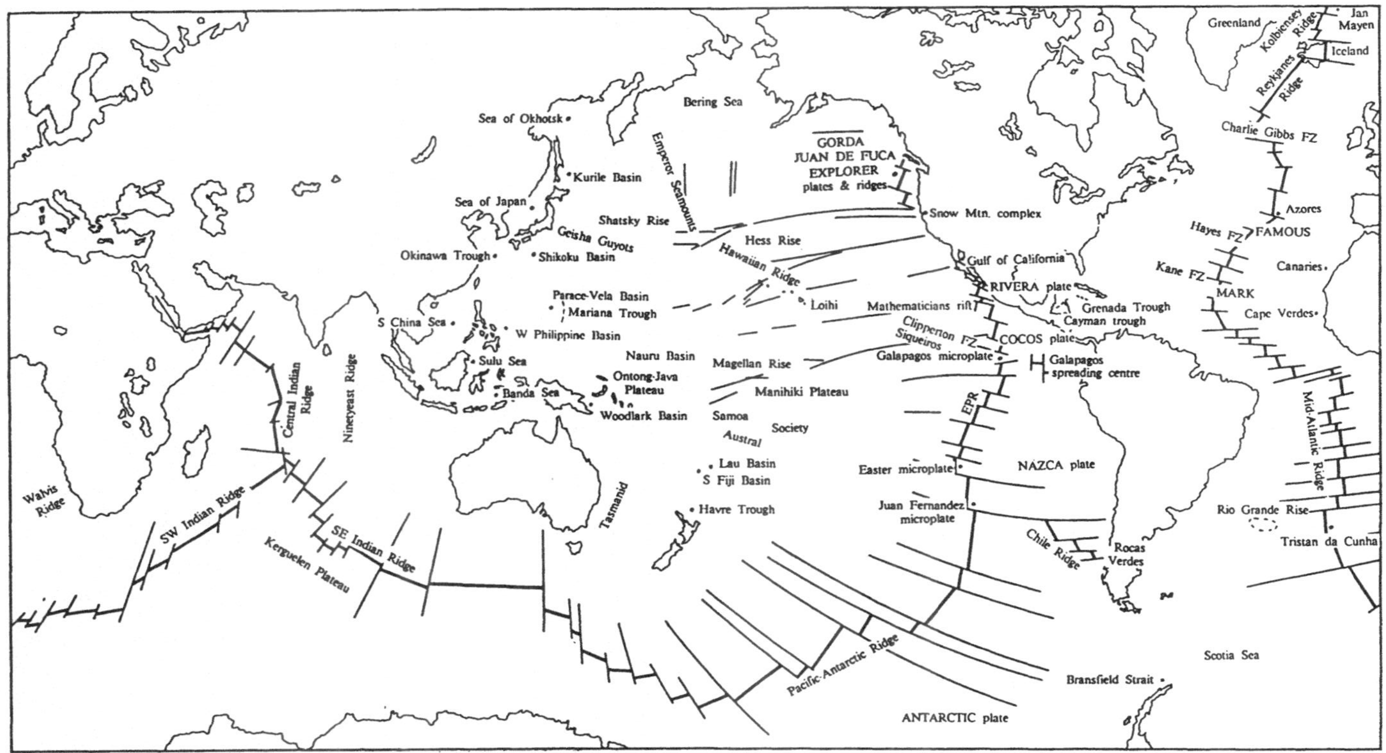

Figure A.3 World map with major oceanic spreading centres and fracture zones indicated and showing the approximate position of the main localities mentioned in Part III.

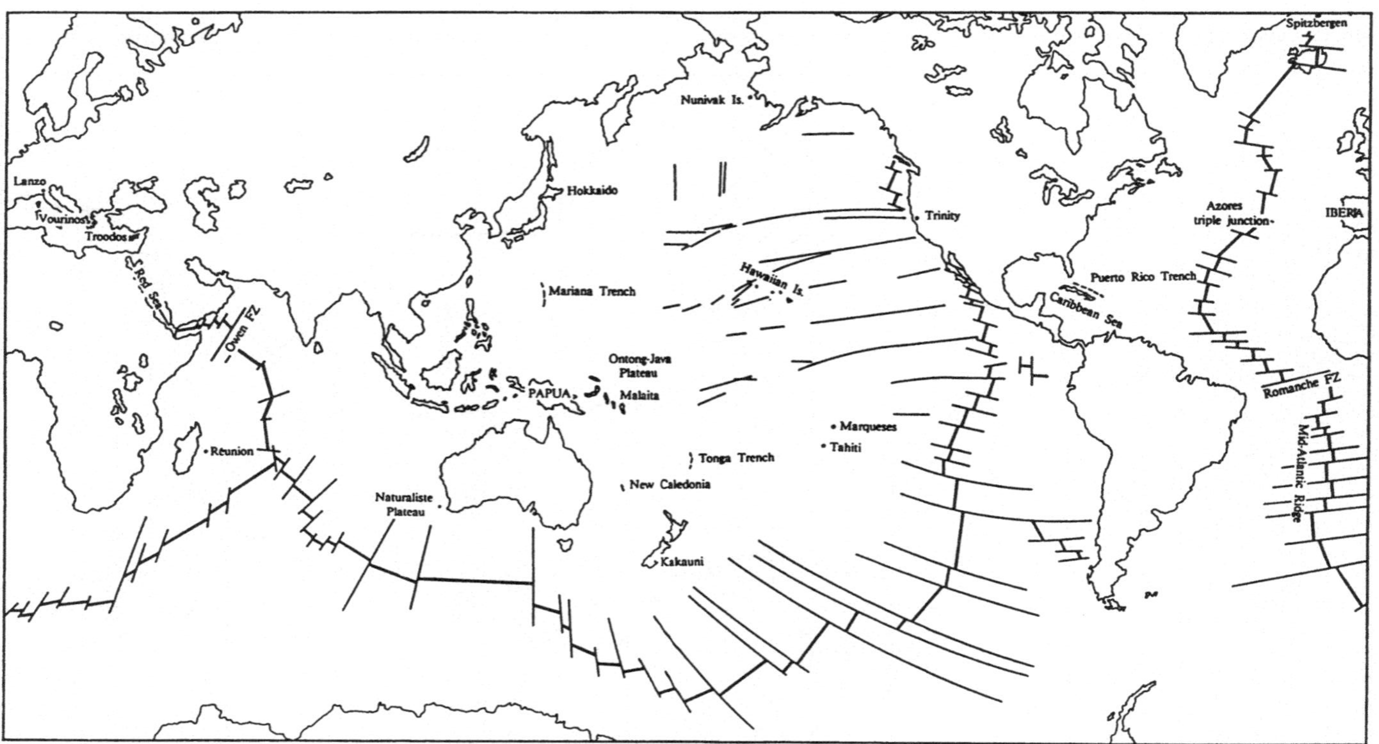

Figure A.4 World map with major oceanic spreading centres and fracture zones indicated and showing the approximate position of the main localities mentioned in Part IV.

Appendix B

Abundance of the elements in the earth (Ganpathy and Anders, 1974) and mean carbonaceous chondrite, C1 (Anders and Grevesse, 1989)

Atomic number	Element	Symbol	Abundance: Earth (ppm)	Abundance: C1 chondrite (ppm)
1	Hydrogen	H	78	20200[a]
3	Lithium	Li	2.7	1.5
4	Beryllium	Be	0.056	0.0249
5	Boron	B	0.47	0.87
6	Carbon	C	350	34500[a]
7	Nitrogen	N	9.1	3180[a]
8	Oxygen	O	285000	464000[a]
9	Fluorine	F	53	60.7
11	Sodium	Na	1580	5000
12	Magnesium	Mg	132100	98900
13	Aluminium	Al	17700	8680
14	Silicon	Si	143400	106400
15	Phosphorus	P	2150	1220
16	Sulphur	S	18400	62500
17	Chlorine	Cl	25	704
19	Potassium	K	170	558
20	Calcium	Ca	19300	9280
21	Scandium	Sc	12.1	5.82
22	Titanium	Ti	1030	436
23	Vanadium	V	103	56.5
24	Chromium	Cr	4780	2660
25	Manganese	Mn	590	1990
26	Iron	Fe	358700	190400
27	Cobalt	Co	940	502
28	Nickel	Ni	20400	11000
29	Copper	Cu	57	126
30	Zinc	Zn	93	312
31	Gallium	Ga	5.5	10
32	Germanium	Ge	13.8	32.7
33	Arsenic	As	3.6	1.86
34	Selenium	Se	6.1	18.6
35	Bromine	Br	0.134	3.57
37	Rubidium	Rb	0.58	2.30
38	Strontium	Sr	18.2	7.80
39	Yttrium	Y	3.29	1.56
40	Zirconium	Zr	19.7	3.94
41	Niobium	Nb	1.00	0.246
42	Molybdenum	Mo	2.96	0.928
43	Ruthenium	Ru	1.42	0.712

[a]Values from Orgueil carbonaceous chondrite

Appendix B (*continued*)

Atomic number	Element	Symbol	Abundance Earth (ppm)	C1 chondrite (ppm)
45	Rhodium	Rh	0.32	0.134
46	Palladium	Pd	1.00	0.560
47	Silver	Ag	0.080	0.199
48	Cadmium	Cd	0.021	0.686
49	Indium	In	0.0027	0.080
50	Tin	Sn	0.71	1.72
51	Antimony	Sb	0.064	0.142
52	Tellurium	Te	0.94	2.32
53	Iodine	I	0.017	0.433
55	Cesium	Cs	0.059	0.187
56	Barium	Ba	5.1	2.34
57	Lanthanum	La	0.48	0.2347
58	Cerium	Ce	1.28	0.6032
59	Praseodymium	Pr	0.162	0.0891
60	Neodymium	Nd	0.87	0.4524
62	Samarium	Sm	0.26	0.1471
63	Europium	Eu	0.10	0.0560
64	Gadolinium	Gd	0.37	0.1966
65	Terbium	Tb	0.067	0.0363
66	Dysprosium	Dy	0.45	0.2427
67	Holmium	Ho	0.101	0.0556
68	Erbium	Er	0.29	0.1589
69	Thulium	Tm	0.044	0.0242
70	Ytterbium	Yb	0.29	0.1625
71	Lutetium	Lu	0.049	0.0243
72	Hafnium	Hf	0.29	0.104
73	Tantalum	Ta	0.029	0.0142
74	Tungsten	W	0.250	0.0926
75	Rhenium	Re	0.076	0.0365
76	Osmium	Os	1.10	0.486
77	Iridium	Ir	1.06	0.481
78	Platinum	Pt	2.1	0.990
79	Gold	Au	0.29	0.140
80	Mercury	Hg	0.0099	0.258
81	Thallium	Tl	0.0049	0.142
82	Lead	Pb	0.13	2.470
83	Bismuth	Bi	0.0037	0.114
90	Thorium	Th	0.065	0.0294
92	Uranium	U	0.018	0.0081

Appendix C

Isotopic ratios commonly used in magmatic petrogenesis

Radiogenic isotopes

Parent nuclide	Relative isotopic abundance (%)	Decay mode	Daughter (stable) nuclide	Ratio used	Ratio values: Primordial (4.5 Ga)	Ratio values: bulk earth (today)	References
^{87}Rb	27.835	β	^{87}Sr	$^{87}Sr/^{86}Sr$	0.699	0.7037	Faure and Powell (1972)
^{147}Sm	15.0	α	^{143}Nd	$^{143}Nd/^{144}Nd$	0.505828	0.511847	DePaolo (1988)
^{176}Lu	2.59	β	^{176}Hf	$^{176}Hf/^{177}Hf$	0.27978	0.28286	Patchett (1981)
^{187}Re	62.60	β	^{187}Os	$^{187}Os/^{186}Os$	0.805	~1.03–1.04	Luck *et al.* (1980)
						—	Allegre and Luck (1980)
						—	
^{235}U	0.7200	Chain	^{207}Pb	$^{207}Pb/^{204}Pb$	10.218	—	Oversby (1970)
^{238}U	99.2745	Chain	^{206}Pb	$^{206}Pb/^{204}Pb$	9.346		Oversby (1970)
^{232}Th	100	Chain	^{208}Pb	$^{208}Pb/^{204}Pb$	28.96		Oversby (1970)

Rare gas isotopes

The rare gases helium, neon, argon, krypton and xenon have numerous isotopes, some of which may be radiogenic. However, anomalous ^{3}He, ^{20}Ne, ^{21}Ne, ^{36}Ar and ^{129}Xe enrichments in mantle-derived rocks are considered to reflect derivation from a primordial, undifferentiated (lower) mantle source, where isotopic ratios are distinct from corresponding atmospheric ratios. 'Primordial' values quoted below are generally based on those found in chondrites (e.g. Mazor *et al.*, 1970; Black, 1972; Manuel and Sabu, 1981; Wacker, 1986).

Isotopic ratio	Approximate primordial values	Atmospheric value
$^{3}He/^{4}He$	1.2×10^{-4}–4×10^{-4}	1.4×10^{-6}
$^{20}Ne/^{22}Ne$	13.5–15.5	9.81
$^{21}Ne/^{22}Ne$	0.03–0.04	0.029
$^{36}Ar/^{38}Ar$	5–7	5.35

Cosmogenic isotopes

Cosmic neutron interaction with atmospheric oxygen and nitrogen produces ^{10}Be (e.g. $^{14}N[n, p\alpha]^{10}Be$) that decays over a short period (half-life 1.5 Ma) to ^{10}B via β emission. Anomalously high $^{10}Be/^{9}Be$ in young island arc lavas are often considered to reflect a subducted sediment component in volcanic arc sources.

Sample	^{10}Be 10^6 atoms/g	$^{10}Be/^{9}Be \times 10^{-11}$	References
MORB, OIB, CFB	0.1–0.9	~1	Tera *et al.* 1986
Island arc volcanics	0.1–24.0	0.4–22.0	Morris and Tera, 1989
Pelagic sediments	5000		Brown, 1984

Stable isotopes

Nuclide	Relative isotopic abundance (%)	Ratio used	Primordial or mantle values (‰)	References
^{1}H	99.9966			
^{2}D	0.0034	$^{2}D/^{1}H$	$\delta D = -0.8$	Hoefs (1973)
^{12}C	98.90			
^{13}C	1.10	$^{13}C/^{12}C$	$\delta^{13}C = -3$ to -8	Hoefs (1973)
^{16}O	99.762			
^{17}O	0.038			
^{18}O	0.200	$^{18}O/^{16}O$	$\delta^{18}O = 5.5$–6.5	James (1981)
^{32}S	95.02			
^{33}S	0.75			
^{34}S	4.21			
^{36}S	0.02	$^{34}S/^{32}S$	$\delta^{34}S = 0.0$–0.6	Hoefs (1973)

Relative isotopic abundances from Anders and Grevesse (1989).

Notation for neodymium and strontium isotopes

Comparison of the initial isotopic ratio of a sample formed *t* years ago and that of a source, such as represented by the chondritic uniform reservoir (CHUR, equivalent to estimate for the bulk earth) at the same time, is given by the epsilon notation (ε):

$$\varepsilon_{Nd} = \left[\frac{^{143}Nd/^{144}Nd_{sample[t]}}{^{143}Nd/^{144}Nd_{bulk\ earth[t]}} - 1 \right] \times 10^4$$

$$\varepsilon_{Sr} = \left[\frac{^{87}Sr/^{86}Sr_{sample[t]}}{^{87}Sr/^{86}Sr_{bulk\ earth[t]}} - 1 \right] \times 10^4$$

Notation for hydrogen, carbon, oxygen and sulphur isotopes

Ratios of stable isotopes are usually reported as the per mil (‰) deviation of the isotopic ratio in the sample relative to that in a standard [standard mean ocean water (SMOW) for hydrogen and oxygen isotopes; PeeDee Belemnite carbonate for carbon isotopes; Canon Diablo meteoritic troilite for sulphur isotopes], and given the delta notation (δ). For example, with oxygen isotopes:

$$\delta^{18}O = \left[\frac{{}^{18}O/{}^{16}O_{[\text{sample}]}}{{}^{18}O/{}^{16}O_{[\text{standard}]}} - 1 \right] \times 10^3$$

and similarly for δD, $\delta^{13}C$ and $\delta^{34}S$.

Appendix D

Normalization factors for rare earth and incompatible elements used in the construction of multi-element (spidergram) diagrams

Rare earth elements

	Chondrite normalization factor (ppm)[a]				
REE	1	2	3	4	5
La	0.32	0.329	0.30	0.33	0.237
Ce	0.94	0.865	0.84	0.88	0.612
Pr	0.12	0.122	0.12	0.112	0.095
Nd	0.6	0.63	0.58	0.60	0.467
Sm	0.2	0.203	0.21	0.181	0.153
Eu	0.073	0.07	0.074	0.069	0.058
Gd	0.31	0.276	0.32	0.249	0.2055
Tb	0.05	—	0.058	0.047	0.0374
Dy	0.31	0.343	0.31	—	0.2540
Ho	0.073	0.076	0.073	0.07	0.0566
Er	0.21	0.225	0.21	0.20	0.1655
Tm	0.033	—	0.033	0.03	0.0255
Yb	0.19	0.22	0.17	0.20	0.170
Lu	0.031	0.0339	0.031	0.034	0.0254

[a](1) Herrmann, 1970; (2) Nakamura, 1974; (3) Graham and Nichols, 1969; (4) Frey *et al.*, 1968; (5) Sun and McDonough, 1989; (6) Anders and Grevesse, 1989; (7) Thompson, 1982; (8) Wood *et al.*, 1979; (9) Saunders and Tarney, 1984; (10) Pearce, 1983

Incompatible elements

	Normalization factor (ppm)[a]							
	Chondrite			Primitive mantle		N-MORB		
Element	5	6	7	5	8	5	9	10
Cs	0.188	0.187	—	0.032	0.019	0.007	—	—
Rb	2.32	2.3	0.35	0.635	0.86	0.56	1.0	2.0
Ba	2.41	2.34	6.9	6.989	7.56	6.30	12	20
Th	0.029	0.0294	0.042	0.085	0.096	0.12	0.2	0.2
U	0.008	0.0081	—	0.021	0.027	0.047	—	—
Nb	0.246	0.246	0.35	0.713	0.62	2.33	2.5	3.5
Ta	0.014	0.0142	0.02	0.041	0.043	0.132	0.17	0.18
K	545	558	120	250	252	600	830	1500
La	0.237	0.2347	0.328	0.687	0.71	2.50	3	—
Ce	0.612	0.6032	0.865	1.775	1.9	7.50	10	10
Pb	2.47	2.470	—	0.185	—	0.30	—	—
Pr	0.095	0.0891	—	0.276	—	1.32	—	—
Sr	7.26	7.80	11.8	21.1	23	90	136	120
P	1220	1220	46	95	90.4	510	570	1200
Nd	0.467	0.4524	0.63	1.354	1.29	7.30	8	—
Sm	0.153	0.1471	0.203	0.444	0.385	2.63	—	3.3
Zr	3.86	3.94	6.84	11.2	11	74	88	90
Hf	0.1066	0.104	0.2	0.309	0.35	2.05	2.5	2.4
Eu	0.058	0.0560	—	0.168	—	1.02	1.2	—
Ti	445	436	620	1300	1527	7600	8400	15000
Gd	0.2055	0.1966	—	0.596	—	3.68	—	—
Tb	0.0374	0.0363	0.052	0.108	0.099	0.67	—	—
Dy	0.2540	0.2427	—	0.737	—	4.55	—	—
Y	1.57	1.56	2	4.55	4.87	28	35	30
Er	0.1655	0.1589	—	0.480	—	2.97	—	—
Tm	0.0255	0.0242	0.034	0.074	—	0.456	—	—
Yb	0.170	0.1625	0.22	0.493	—	3.05	3.5	3.4
Lu	0.0254	0.0243	—	0.074	—	0.455	—	—

Elements listed downwards in order of decreasing incompatibility (after Sun and McDonough, 1989). Authors may have their own preference for listing elements in normalized diagrams that may slightly change the order to that given above. Somewhat different from the rest, Pearce (1983) groups together mobile elements (Sr, K, Rb, Ba) and immobile elements (Ta, Nb, Ce, P, Zr, Hf, Sm, Ti, Y, Yb) to the left and right of the diagram, respectively, such that the incompatibility of each group increases from the outside towards the centre

[a]See footnote to REE normalization table.

References

Abers, G.A., Parsons, B. and Weissel, J.K. (1988) Seamount abundances and distributions in the southeast Pacific. *Earth Planet. Sci. Lett.* **87**, 137–151.

Abrams, J.L., Detrick, R.S. and Fox P.J. (1988) Morphology and crustal structure of the Kane Fracture Zone Transverse Ridge. *J. Geophys. Res.* **93**, 3195–3210.

Agee, C.B. and Walker, D. (1988) Aluminium partitioning between olivine and ultrabasic silicate liquid to 6 GPa. *Trans. Am. Geophys. Union, EOS* **69**, 1511.

Aggrey, K.E., Muenow, D.W. and Batiza, R. (1988a) Volatile abundances in basaltic glasses from seamounts flanking the East Pacific Rise at 21° and 12-14°N *Geochim. Cosmochim. Acta* **52**, 2115–2120.

Aggrey, K.E., Muenow, D.W. and Sinton, J.M. (1988b) Volatile abundances in submarine glasses from the North Fiji and Lau back-arc basins. *Geochim. Cosmochim. Acta* **52**, 2501–2506.

Aherne, J.L. and Turcotte, D.L. (1979) Magma migration beneath an ocean ridge. *Earth Planet. Sci. Lett.* **45**, 115–122.

Ailin-Pyzik, Z.B. and Sommer, S.E. (1981) Microscale chemical effects of low temperature alteration of DSDP basaltic glasses. *J. Geophys. Res.* **86**, 9503–9510.

Alabaster, T. and Storey, B.C. (1990) A modified Gulf of California model for South Georgia, north Scotia Ridge, and implications for the Rocas Verdes back-arc basin, southern Andes. *Geology* **18**, 497–500.

Albarede, F. (1988) Further merits of the equilibrium melting model. *Chem. Geol.* **70**, 152–159.

Allan, J.F., Batiza, R. and Lonsdale, P. (1987) Petrology and chemistry of lavas from seamounts flanking the East Pacific Rise axis, 21°N: implications concerning the mantle source composition for both seamount and adjacent EPR lavas. In: Keating, B.H., Fryer, P., Batiza, R. and Boehlert, G.W. (eds) *Seamounts, Islands and Atolls, Am. Geophys. Union Geophys. Monogr.* **43**, pp. 255–282.

Allan, J.F., Batiza, R., Perfit, M.R., Fornari, D.J. and Sack, R.O. (1989) Petrology of lavas from the Lamont Seamount Chain and adjacent East Pacific Rise, 10°N. *J. Petrol.* **30**, 1245–1298.

Allan, J.F., Sack, R.O. and Batiza, R. (1988) Cr-rich spinels as petrogenetic indicators: MORB-type lavas from the Lamont Seamount Chain, eastern Pacific. *Am. Mineral.* **73**, 741–753.

Allegre, C.J. (1982) Chemical geodynamics. *Tectonophysics* **81**, 109–132.

Allegre, C.J. and Luck, J-M. (1980) Osmium isotopes as petrogenetic and geological traces. *Earth Planet. Sci. Lett.* **48**, pp. 148–154.

Allegre, C.J. and Turcotte, D.L. (1986) Implications of a two-component marble-cake mantle. *Nature, London* **323**, 123–127.

Allegre, C.J., Hamelin, B., Provost, A. and Dupre, B. (1987) Topology in isotopic multispace and origin of mantle chemical heterogeneities. *Earth Planet. Sci. Lett.* **81**, 319–337.

Allegre, C.J., Staudacher, T., Sarda, P. and Kurz, M. (1983) Constraints on evolution of Earth's mantle from rare gas systematics. *Nature, London* **303**, 762–766.

Alt, J.C. and Honnorez, J. (1984) Alteration of the upper oceanic crust, DSDP site 417: mineralogy and chemistry. *Contrib. Mineral. Petrol.* **87**, 149–169.

Alt, J.C., Honnorez, J., Laverne, C. and Emmerman, R. (1986a) The structure and evolution of a submarine hydrothermal system: DSDP Site 504B. *J. Geophys. Res.* **91**, 10309–10335.

Alt, J.C., Muehlenbachs, K. and Honnorez, J. (1986b) An oxygen profile through the upper kilometer of the oceanic crust, DSDP Hole 504B. *Earth Planet. Sci. Lett.* **80**, 217–229.

Anders, E. and Grevesse, N. (1989) Abundances of the elements: meteoritic and solar. *Geochim. Cosmochim. Acta* **53**, 197–214.

Anderson, A.T. and Greenland, L.P. (1970) Phosphorus fractionation diagram as a quantitative indicator of crystallization differentiation of basaltic liquids. *Geochim. Cosmochim. Acta* **33**, 493–505.

Anderson, R.N. and Hobart, M.A. (1976) The relationship between heat flow, sediment thickness and age in the eastern Pacific *J. Geophys. Res.* **81**, 2968–2989.

Anderson, R.N., Clague, D.A., Klitgord, K.D., Marshall, M. and Nishimori, R. (1975) Magmatic and petrologic variation along the Galapagos Spreading Centre and their relationship to the Galapagos melting anomaly. *Bull. Geol. Soc. Am.* **86**, 683–694.

Anderson, R.N., Delong, S.E. and Schwartz, W.M. (1978) Thermal model for subduction with dehydration in the downgoing slab. *J. Geol.* **86**, 731–739.

Anderson, R.N., Delong, S.E. and Schwartz, W.M. (1980) Dehydration, asthenospheric convection and seismicity in subduction zones. *J. Geol.* **88**, 445–451.

Anderson-Fontana, S., Engeln, J.F., Lundgren, P., Larson, R.L. and Stein, S. (1986) Tectonics and evolution of the Juan Fernandez microplate at the Pacific-Nazca-Antarctic triple junction. *J. Geophys. Res.* **91**, 2005–2018.

Andrews, A.J. (1977) Low temperature fluid alteration of oceanic layer 2 basalts, DSDP Leg 37. *Can. J. Earth Sci.* **14**, 911–926.

Andrews, A.J. (1978) Petrology and geochemistry of alteration in layer 2 basalts, DSDP Leg 37. Unpublished PhD thesis, University of Western Ontario, Canada.

Andrews, A.J. (1980) Saponite and celadonite in layer 2 basalts, DSDP Leg 37. *Contrib. Mineral. Petrol.* **73**, 323–340.

Anonymous (1972) Penrose field conference on ophiolites. *Geotimes* **17**, 24–25.

Arai, S. and Takahashi, N. (1989) Formation and compositional variation of phlogopites in the Horonian peridotite complex, Hokkaido, northern Japan: implications of origin and fractionation of metasomatic fluids in the upper mantle. *Contrib. Mineral. Petrol.* **10**, 165–175.

Arndt, N.T. (1977) Partitioning of nickel between olivine and ultrabasic and basic komatiite liquids. *Carnegie Inst. Washington Yearbook* **76**, 553–557.

Assunacao, C., Torre, D.E. and Canhilo, M.H. (1970) Notas sobre petrografia comparada das ihas Atlanticas. *Bull. Mus. Lab. Mineral. Geol. Fac. Sci., Lisbon Univ.* **11**, 305–342.

Atherton, M.P., Pitcher, W.S. and Warden, V. (1983) The Mesozoic marginal basin of central Peru. *Nature, London* **305**, 303–306.

Augevine, C.L., Turcotte, D.L. and Ockendon, J.R. (1984) Geometrical form of aseismic ridges, volcanoes and seamounts *J. Geophys. Res.* **89**, 11287–11292.

Aumento, F. (1967) Magmatic evolution on the Mid-Atlantic Ridge. *Earth Planet. Sci. Lett.* **2**, 225–230.

Aumento, F. (1968) The Mid-Atlantic Ridge near 45°N. II: Basalts from the area of Confederation Peak. *Can. J. Earth Sci.* **5**, 1–21.

Aumento, F. (1970) Serpentine mineralogy of ultrabasic intrusions in Canada and on the mid-Atlantic Ridge. *Geol. Surv. Can. Pap.* **69**.

Aumento, F. and Loncarevic, B. (1969) The Mid-Atlantic Ridge near 45°N, III. Bald Mountain. *Can. J. Earth Sci.* **6**, 11–23.

Aumento, F. and Loubat, H. (1971) The Mid-Atlantic Ridge near 45°N. XVI: Serpentinized ultramafic intrusions. *Can. J. Earth Sci.* **8**, 631–663.

Aumento, F., Loncarevic, B. and Ross, D.I. (1971) Hudson geotraverse: geology of the mid-Atlantic ridge at 45°N. *Phil. Trans. R. Soc., London* **A268**, 623–650.

Aumento, F., Melson, W.G. *et al.* (Eds) (1976) *Initial Reports of the Deep Drilling Project* **37**, 1008 pp.

Autio, L.K. and Rhodes, J.M. (1983) Costa Rica Rift Zone basalts: geochemical and experimental data from a possible example of multistage melting. In: Cann, J.R., Langseth, M., Honnorez, J., Von Herzen, R.P., White, S.M. *et al.* (eds) *Initial Reports of the Deep Sea Drilling Project* **69**, 729–745.

Ave Lallemant, H.G. (1976) Structure of the Canyon Mountain (Oregon) ophiolite and its implication for sea floor spreading. *Geol. Soc. Am. Spec. Pap.* **173**, 49.

Bailey, D.K. (1977) Lithosphere control of continental rift magmatism. *J. Geol. Soc., London* **133**, 103–106.

Bailey, D.K. (1987) Mantle matasomatism—perspective and prospect. In Fitton, J.G. and Upton, B.G.J. (eds) *Alkaline Igneous Rocks. Geol. Soc. London Spec. Publ.* **30**, pp. 1–13.

Baker, I. (1969) Petrology of the volcanic rocks of St. Helena Island, South Atlantic. *Bull. Geol. Soc. Am.* **80**, 1283–1310.

Baker, P.E. (1973) Islands of the South Atlantic. In: Nairn, A.E.M. and Stehli, F.G. (eds) *The Ocean Basins and Margins, Vol. 1: The South Atlantic*, Plenum Press, New York, pp. 493–553.

Baker, P.E., Gass, I.G., Harris, P.G. and Le Maitre, R.W. (1964) The volcanological report of the Royal Society Expedition to Tristan da Cunha, 1962. *Phil. Trans. R. Soc., London* **A256** 439–575.

Ballard, R.D., Francheteau, J., Juteau, T., Rangan, C. and Normark, W. (1981) East Pacific Rise at 21°N: the volcanic, tectonic and hydrothermal processes of the central axis. *Earth Planet. Sci. Lett.* **55**, 1–10.

Baragar, W.R.A., Lambert, M.B., Baglow, N. and Gibson, I. (1987) Sheeted dykes of the Troodos ophiolite, Cyprus. In: Hall, H.C. and Fahrig, W.F. (eds) *Mafic Dyke Swarms, Geol. Assoc. Can. Spec. Pap.* **34**, 257–272.

Bargar, K.E. and Jackson, E.D. (1974) Volumes of individual shield volcanoes along the Hawaiian-Emperor chain. *US Geol. Surv. J. Res.* **2**, 545–550.

Barker, P.F. (1972) A spreading centre in the east Scotia Sea. *Earth Planet. Sci. Lett.* **15**, 123–132.

Barker, P.F. and Burrell, J. (1977) The opening of the Drake Passage. *Mar. Geol.* **25**, 15–34.

Barker, P.F. and Griffiths, D.H. (1972) The evolution of the Scotia Ridge and the Scotia Sea. *Phil. Trans. R. Soc. London* **A271**, 151–183.

Barker, P.F. and Hill, I.A. (1981) Back-arc extension in the Scotia Sea. *Phil. Trans. R. Soc., London* **A300**, 249–262.

Barreiro, B.A. (1983) Lead isotopic compositions of South Sandwich Island volcanic rocks and their bearing on magmagenesis in intra-oceanic island arcs. *Geochim. cosmochim. Acta* **47**, 817–822.

Bartholomew, D.S. and Tarney, J. (1984a) Geochemical characteristics of magmatism in the southern Andes (45-46°S). In: Barreiro, B. and Harmon, R.S. (eds) *Andean Magmatism*, Shiva Publications, Nantwich, pp. 220–229.

Bartholomew, D.S. and Tarney, J. (1984b) Crustal extension in the southern Andes (45-46°S). In Kokelaar, B.P. and Howells, M.F. (eds) *Marginal Basin Geology, Geol. Soc. London, Spec. Publ.* **16**, 195–205.

Bartholomew, I.D. (1983) The primary structures and fabrics of the upper mantle and lower crust from ophiolite complexes. Unpublished PhD thesis, The Open University, Milton Keynes, 523 pp.

Basaltic Volcanism Study Project (BVSP) (1981) *Basaltic Volcanism on the Terrestrial Planets* Pergamon Press, New York, 1286 pp.

Bass, M.N. (1971) Variable abyssal basalt populations and their relationship to sea-floor spreading rates. *Earth Planet. Sci. Lett.* **11**, 18–22.

Bass, M.N. (1975) Secondary minerals in oceanic basalts. *Carnegie Inst. Washington Yearbook* **74**, 234–240.

Bass, M.N. (1976) Secondary minerals in oceanic basalt, with special reference to Leg 34, Deep Sea Drilling Project. In: Yeats, R.S., Hart, S.R. *et al.* (eds) *Initial Reports of the Deep Sea Drilling Project* **34**, 393–432.

Bass, M.N., Moberley, R., Rhodes, J.M., Shih, C.Y. and Church, S.E. (1973) Volcanic rocks cored in the central Pacific, Leg 17 Deep Sea Drilling Project. In: Winterer, E.L., Ewing, J.I. *et al.* (eds) *Initial Reports of the Deep Sea Drilling Project* **17**, 492–503.

Batiza, R. (1977) Petrology and chemistry of Guadalupe Island: an alkalic seamount on a fossil ridge crest. *Geology* **5**, 760–764.

Batiza, R. (1980) Origin and petrology of young oceanic central volcanoes: are most tholeiitic rather than alkalic? *Geology* **8**, 477–482.

Batiza, R. (1981) Trace element characteristics of Leg 61 basalts. In: Larson, R.L., Schlanger, S.O. *et al.* (eds) *Initial Reports of the Deep Sea Drilling Project* **61**, 689–696.

Batiza, R. (1982) Abundance, distribution and size of volcanoes in the Pacific Ocean and implications for the origin of non-hot spot volcanoes. *Earth Planet. Sci. Lett.* **60**, 195–206.

Batiza, R. (1989a) Failed rifts. In: Winterer, E.L., Hussong, D.M. and Decker, R.W. (eds) *Geology of North America, Volume N: The Eastern Pacific Ocean and Hawaii.* Geological Society of America, Boulder, Co. pp. 177–186.

Batiza, R. (1989b) Petrology and geochemistry of eastern Pacific spreading centers. In: Winterer, E.L., Hussong, D.M. and Decker, R.W. (eds) *Geology of North America, Volume N: The Eastern Pacific Ocean and Hawaii.* Geological Society of America, Boulder, Co, pp. 145–159.

Batiza, R. and Johnson, J. (1980) Trace element and isotopic evidence for magma mixing in transitional and alkalic basalts from the East Pacific Rise at 8°N. In Rosendahl, B.R., Hekinian, R. *et al.* (eds) *Initial Reports of the Deep Sea Drilling Project* **54**, 63–69.

Batiza, R. and Margolis, S.H. (1986) A model for the origin of small non-overlapping offsets (SNOO's) of the East Pacific Rise. *Nature, London* **320**, 439–441.

Batiza, R. and Vanko, D. (1983) Volcanic development of small oceanic central volcanoes on the flanks of the East Pacific Rise inferred from narrow-beam echo-sounder surveys. *Mar. Geol.* **54**, 53–90.

Batiza, R. and Vanko, D. (1984) Petrology of young Pacific seamounts *J. Geophys. Res.* **89**, 11235–11260.

Batiza, R. and Vanko, D.A. (1985) Petrologic evolution of large failed rifts in the Eastern Pacific: petrology of volcanic and plutonic rocks from Mathematician Ridge area and the Guadalupe trough. *J. Petrol.* **26**, 564–602.

Batiza, R. and Watts, A.B. (1986) Scientific seamount drilling report, *JOI-USSAC Sponsored Workshop*, 1–13, Lamont-Doherty Geological Observatory.

Batiza, R., Fornari, D.J., Vanko, D.A. and Lonsdale, P. (1984) Craters, calderas and hyaloclastites on young Pacific seamounts. *J. Geophys. Res.* **89**, 8371–8390.

Batiza, R., Fox, P.J., Vogt, P.R., Cande, S.C., Grindlay, N.R., Melson, W.G. and O'Hearn, T. (1989) Morphology, abundance and chemistry of near-ridge seamounts in the vicinity of the Mid-Atlantic Ridge ~26°S. *J. Geol.* **97**, 209–220.

Batiza, R., Niu, Y. and Zayac, W.C. (1990) Chemistry of seamounts near the East Pacific Rise: implications for the geometry of sub-axial mantle flow. *Geology* **18**, 1122–1125.

Batiza, R., Rosendahl, B.R. and Fisher, R.L. (1977) Evolution of oceanic crust 3: petrology and chemistry of basalts from the East Pacific Rise and Siqueiros transform fault. *J. Geophys. Res.* **82**, 265–276.

Batiza, R., Smith, T. and Niu, Y. (1989) Geologic and petrologic evolution of seamounts near the EPR based on submersible and camera study. *Mar. Geophys. Res.* **11**, 169–236.

Beccaluva, L., Piccardo, G.B. and Serri, G. (1980) Petrology of northern Apennine ophiolites and comparison with other Tethyan ophiolites. In: Panayiotou, A. (ed) *Ophiolites—Proceedings of the International Ophiolite Symposium, Cyprus, 1979*, pp. 314–331.

Becker, K., Sakai, H. *et al.* (1988) Site 504: Costa Rica Rift. In: Becker, K., Sakai, H. *et al.* (eds) *Proceedings of the Ocean Drilling Program, Initial Reports* (*Part A*) **111**, 35–251.

Becker, K., Sakai, H., Adamson, A.C., Alexandrovich, J., Alt, J.C., Anderson, R.N., Bideau, D., Gable, R. *et al.* (1989) Drilling deep into young oceanic crust. Hole 504B, Costa Rica rift. *Rev. Geophy.* **27**, 79–102.

Becker, R.H. and Clayton, R.N. (1977) Nitrogen isotopes in igneous rocks. *Trans. Am. Geophys. Union, EOS* **58**, 536.

Bellaiche, G., Cheminee, J.L. and Francheteau, J. (1974) Inner floor of the Rift Valley: first submersible study. *Nature, London* **250**, 558–560.

Bence, A.E., Bayliss, D.M., Bender, J.F and Grove, T.L. (1979) Controls on the major and minor element chemistry of mid-ocean ridge basalts and glasses. In: Talwani, M., Harrison, C.G., and Hayes, D.E. (eds) *Deep Drilling Results in the Atlantic Ocean: Ocean Crust, Am. Geophys. Union, Maurice Ewing Ser.* **2**, 331–341.

Bence, A.E., Taylor, S.R. and Fisk, M. (1980) Major and trace element geochemistry of basalts from Ojin, Nintokm and Suiko Seamounts of the Emperor Seamount chain, DSDP-IPOD Leg 55. In: Jackson, E.D., Koizumi, I. *et al.* (eds) *Initial Reports of the Deep Sea Drilling Project* **55**, 599–605.

Bender, J.F., Hodges, F.N. and Bence, A.E. (1978) Petrogenesis of basalts from the project FAMOUS area: experimental study from 0 to 15 kbars. *Earth Planet. Sci. Lett.* **41**, 277–302.

Bender, J.F., Langmuir, C.H. and Hanson, G.N (1984) Petrogenesis of basalt glasses from the Tamayo region, East Pacific Rise. *J. Petrol.* **25**, 213–254.

Ben Othman, D., Tilton, G. and Menzies, M.A. (1990) Pb, Nd and Sr isotopic investigations of kaersutite and clinopyroxene from ultramafic nodules and their host basalts: the nature of the sub-continental mantle. *Geochim. Cosmochim. Acta* **54**, 3449–3460.

Benson, W.N. (1926) The tectonic conditions accompanying the intrusion of basic and ultrabasic igneous rocks. *US Natl. Acad. Sci. Mem.* **1**, 1–90.

Berger, E.T. (1981) Enclaves ultramafiques, megacristaux et leurs basaltes-hotes en contexts oceanique (Pacific Sud) et continental (Massif Central Francais). *Thesis, Universite de Paris-Sud, Centre D'Orsay*, 469 pp.

Berndt, M. and Seyfried, W.E. (1986) B, Li and associated trace element chemistry of alteration minerals, holes 597B and 597C. In: Leinen, M., Rea, D.K. *et al.* (eds) *Initial Reports of the Deep Sea Drilling Project* **92**, 491–497.

Berndt, M., Seyfried, W.E. and Beck, J.W. (1988) Hydrothermal alteration processes at mid-ocean ridges: experimental and theoretical constraints from Ca and Sr exchange reactions and Sr isotopic ratios. *J. Geophys. Res.* **93**, 4573–4583.

Berndt, M., Seyfried, W.E. and Janecky, D.R. (1989) Plagioclase and epidote buffering of cation ratios in mid-ocean ridge hydrothermal fluids: experimental results in and near the supercritical region. *Geochim. Cosmochim. Acta* **53**, 2283–2300.

Bibee, L.D., Shor, G.G. and Lu, R.S. (1980) Interarc spreading in the Mariana Trough. *Mar. Geol.* **35**, 183–197.

Bielski-Zyskind, M., Wasserburg, G.J. and Nixon, P.H. (1984) Sm-Nd and Rb-Sr systematics in volcanics and ultramafic xenoliths from Malaita, Solomon Islands and the nature of the Ontong-Java Plateau. *J. Geophys. Res.* **89**, 2415–2424.

Bienvenu, P.R., Bougault, H., Joron, J.L., Treuil, M. and Dmitriev, L. (1990) MORB alteration: rare earth element/non-rare earth hygromagmaphile element fractionation. *Chem. Geol.* **82**, 1–14.

Bischoff, J.L. (1969) The Red Sea geothermal deposits: their mineralogy, chemistry and genesis. In: Degens, E.T. and Ross, D.A. (eds) *Hot Brines as Recent Heavy Metal Deposits of the Red Sea*, Springer-Verlag, New York, pp. 368–401.

Bischoff, J.L. and Dickson, F.W. (1975) Seawater–basalt interaction at 200°C and 500 bars: implications for the origin of seafloor heavy metal deposits and regulation of seawater chemistry. *Earth Planet. Sci. Lett.* **25**, 385–397.

Bjornsson, A., Saemundsson, K., Einarsson, P., Tryggvason, E. and Gronvold, K. (1977) Current rifting episode in northern Iceland. *Nature, London* **266**, 318–323.

Black, D.C. (1972) On the origins of trapped helium, neon and argon isotopic variations in meteorites. I: Gas-rich meteorites, lunar soil and breccia. *Geochim. Cosmochim. Acta* **36**, 347–357.

Blackinton, J.G., Hussong, D.M. and Kosalos, J. (1983) First results from a combination side-scan sonar and seafloor mapping system (SeaMarc II). *Offshore Technology Conference, OTC 4478*, pp. 307–311.

Blake, S. (1984) Magma mixing and hybridization processes at the alkalic, silicic, Torfajokull central volcano triggered by tholeiitic Veidivotn fissuring, South Iceland. *J. Volcanol. Geotherm. Res.* **22**, 1–31.

Blanchard, D.P., Rhodes, J.M., Dungan, M.A., Rodgers, K.V., Donaldson, C.H., Brannon, J.C., Jacobs, J.W. and Gibson, E.K. (1976) The chemistry and petrology of basalts from Leg 37 of the Deep-Sea Drilling Project. *J. Geophys. Res.* **81**, 4231–4246.

Bloomer, S.H. (1983) Distribution and origin of igneous rocks from the landward slopes of the Mariana trench: implications for its structure and evolution. *J. Geophys. Res.* **88**, 7411–7428.

Bloomer, S.H. and Fisher, R.L. (1987) Petrology and geochemistry of igneous rocks from the Tonga Trench, a non-accreting plate boundary. *J. Geol.* **95**, 469–495.

Bloomer, S.H. and Hawkins, J.W. (1983) Gabbroic and ultramafic rocks from the Mariana Trench: an island arc ophiolite. In: *Tectonic and Geologic Evolution of the SE Asian Seas and Islands, Am. Geophys. Union, Geophys. Monogr.* **27**, 234–316.

Bloomer, S.H., Natland, J.H. and Fisher, R.L. (1989) Mineral relationships in gabbroic rocks from fracture zones of Indian Ocean Ridges: evidence for extensive fractionation, parental diversity, and boundary-layer recrystallization. In: Saunders, A.D. and Norry, M.J. (eds) *Magmatism in the Ocean Basins, Geol. Soc. London, Spec. Publ.* **42**, 107–124.

Boettcher, A.L. and O'Neil, J.R. (1980) Stable isotope, chemical and petrographic studies of high-pressure amphiboles and micas: evidence for metasomatism in the mantle source regions of alkali basalts and kimberlites. *Am. J. Sci.* **280A**, 594–621.

Bogdanov, Y.A. and Ploshko, V. (1968) Igneous and metamorphic rocks from the abyssal Romanche depression. *Doklady Akademii Nauk, SSSR* **177**, 173–176.

Bohlke, J.K., Alt, J.C. and Muehlenbachs, K. (1984) Oxygen isotope–water relations in altered deep-sea basalts: low temperature mineralogical controls. *Can. J. Earth Sci.* **21**, 67–77.

Bohlke, J.K., Honnorez, J. and Honnorez-Guerstein, B.M. (1980) Alteration of basalts from site 396B, DSDP: petrographic and mineralogic studies. *Contrib. Mineral. Petrol.* **73**, 341–364.

Bohlke, J.K., Honnorez, J., Honnorez-Guerstein, B.M., Muehlenbachs, K. and Peterson, N. (1981) Heterogeneous alteration of the upper oceanic crust: correlation of rock chemistry, magnetic properties and O isotope ratios with alteration patterns in basalts from site 396B, DSDP. *J. Geophys. Res.* **86**, 7935–7950.

Bonatti, E. (1965) Palagonite, hyaloclastites and alteration of volcanic glass in the ocean. *Bull. Volcanol.* **28**, 257–269.

Bonatti, E. and Harrison, C.G.A. (1976) Hot lines in the Earth's mantle. *Nature, London* **263**, 402–404.

Bonatti, E. and Harrison, C.G.A. (1988), Eruption styles of basalt in oceanic spreading ridges and seamounts: effect of magma temperature and viscosity. *J. Geophys. Res.* **93**, 2967–2980.

Bonatti, E. and Michael, P.J. (1989) Mantle peridotites from continental rifts to ocean basins to subduction zones. *Earth Planet. Sci. Lett.* **91**, 297–311.

Bonatti, E., Honnorez, J. and Ferrara, G. (1970) Equatorial mid-Atlantic ridge: petrologic and Sr isotopic evidence for an alpine type rock assemblage. *Earth Planet. Sci. Lett.* **9**, 247–256.

Bonatti, E., Honnorez, J. and Ferrara, G. (1971) Peridotite–gabbro–basalt complex from the equatorial Mid-Atlantic Ridge. *Phil. Trans. R. Soc., London* **A268**, 385–402.

Bonatti, E., Honnorez, J., Kirst, P. and Radicati, F. (1975) Metagabbros from the Mid-Atlantic Ridge at 6°N: contact-hydrothermal-dynamic metamorphism beneath the axial valley *J. Geol.* **83**, 61–78.

Bonatti, E., Ottonello, G. and Hamlyn, P.R. (1986) Peridotites from the island of Zabargad (St. John's), Red Sea: petrology and geochemistry. *J. Geophys. Res.* **21**, 599–631.

Bostrom, K. & Peterson, M.N.A. (1966) Precipitates from hydrothermal exhalations on the East Pacific Rise. *Econ. Geol.* **61**, 1258–1265.

Bott, M.P.H. (1965) The upper mantle beneath Iceland. *Geophys. J. R. Astron. Soc.* **9**, 275–277.

Bott, M.P.H. (1988) A new look at the causes and consequences of the Iceland hot-spot. In: Morton, A.C. and Parson, LM. (eds) *Early Tertiary Volcanism and the Opening of the North East Atlantic, Geol. Soc. London, Spec. Publ.* **39**, 15–23.

Bottinga, Y., Weill, D. and Richet, P. (1982) Density calculations for silicate liquids, I: revised method for aluminosilicate compositions. *Geochim. Cosmochim. Acta* **46**, 909–919.

Boudier, F. (1978) Structure and petrology of the Lanzo peridotite massif (Piedmont Alps). *Bull. Geol. Soc. Am.* **89**, 1574–1591.

Boudier, F. and Coleman, R.G. (1981) Cross section through the peridotite in the Samail ophiolite, southeastern Oman mountains. *J. Geophys. Res.* **86**, 2573–2592.

Boudier, F. and Nicolas, A. (1972) Fusion partielle gabbroique dans la lherzolite de Lanzo. *Schweiz. Mineralog. Petrograph. Mitteil.* **52**, 39–56.

Boudier, F. and Nicolas, A. (1985) Harzburgite and lherzolite subtypes in ophiolitic and oceanic environments. *Earth Planet. Sci. Lett.* **76**, 84–92.

Bougault, H. (1974) Distribution of first series transition elements in rocks recovered during DSDP Leg 22 in the north eastern Indian Ocean. In: von der Borch, C.C., Sclater, J.G. *et al.* (eds) *Initial Reports of the Deep Sea Drilling Project* **22**, 449–457.

Bougault, H. and Cande, S.C. (1985) Background, objectives and summary of principal results Deep Sea Drilling project Sites 556–564. In: Bougault, H., Cande, S.C. *et al.* (eds) *Initial Reports of the Deep Sea Drilling Project* **82**, 5–16.

Bougault, H. and Hekinian, R. (1974) Rift valley in the Atlantic ocean near 36°50′N. Petrology and geochemistry of the basaltic rocks. *Earth Planet. Sci. Lett.* **24**, 249–261.

Bougault, H. and Treuil, M. (1980) Mid-Atlantic Ridge: zero-age geochemical variations between Azores and 22°N. *Nature, London* **286**, 209–212.

Bowen, N.L. (1927) *The Evolution of the Igneous Rocks*, Princeton University Press, Princeton, 332 pp.

Bowen, N.L. and Schairer, J.F. (1935) The system MgO–FeO–SiO_2. *Am. J. Sci.* **29**, 151–217.

Bowers, T.S. and Taylor, H.P. (1985) An integrated chemical and stable-isotope model of the origin of mid-ocean ridge hot spring systems. *J. Geophys. Res.* **90**, 12583–12606.

Bowers, T.S., Campbell, A.C., Measures, C., Spivack, A.J. and Edmond, J.M (1988) Chemical controls on the composition of vent fluids at 13-11°N and 21°N, East Pacific Rise. *J. Geophys. Res.* **93**, 4522–4537.

Bowin, C.O., Nalwalk, A.J. and Hersey, J.B. (1966) Serpentinised peridotites from the north wall of the Puerto Rico trench. *Bull. Geol. Soc. Am.* **77**, 257–270.

Boyd, F.R. (1989) Compositional distinction between oceanic and cratonic lithosphere. *Earth Planet. Sci. Lett.* **96**, 15–26.

Boyd, F.R. and England, J.L. (1960) Apparatus for phase equilibrium measurements up to 50 kilobars and temperatures up to 1750°C. *J. Geophys. Res.* **65**, 741–748.

Boyd, S.R., Mattey, D.P., Pillinger, C.T., Milledge, H.J., Mendelssohn, M. and Seal, M. (1987) Multiple growth events during diamond genesis: an integrated study of carbon and nitrogen isotopes and nitrogen aggregation state in coated stones. *Earth Planet. Sci. Lett.* **86**, 341–353.

Bratt, S.R. and Purdy, G.M. (1984) Structure and variability of ocean crust on the flanks of the East Pacific Rise between 11° and 13°N. *J. Geophys. Res.* **89**, 5111–5125.

Bratt, S.R. and Solomon, S.C. (1984) Compressional and shear wave structure of the East Pacific Rise at 11°20′N: constraints from three-component ocean bottom seismometer data. *J. Geophys. Res.* **89**, 6095–6110.

Brodholt, J.P. and Batiza, R. (1989) Global systematics of unaveraged mid-ocean ridge basalt compositions: comment on "global correlations of ocean ridge basalt chemistry with axial depth and crustal thickness" by E.M. Klein and C.H. Langmuir. *J. Geophys. Res.* **94**, 4231–4240.

Brongniart, A. (1813) Essai d'une classification mineralogiques des roches melanges. *J. Mines, Paris* **119**, 5–48.

Brooks, C.K., Jakobsson, S.P. and Campsie, J. (1974) Dredged basaltic rocks from the seaward extension of the Reykjanes and Snaefellsnes volcanic zone, Iceland. *Earth Planet. Sci. Lett.* **22**, 320–327.

Brouxel, M., Tatsumoto, M. and Clague, D.A. (1988) Sr, Nd and Pb isotopes of spinel lherzolite xenoliths, Koloa Volcanics, Kauai, Hawaii. *Trans. Am. Geophys. Union, EOS* **69**, 1517.

Brown, J.R. and Karson, J.A. (1988) Variation in axial processes on the Mid-Atlantic Ridge: the Median valley of the MARK area. *Mar. Geophys. Res.* **10**, 109–138.

Brown, L. (1984) Applications of accelerator mass spectrometry. *Ann. Rev. Earth Planet. Sci.* **12**, 39–69.

Brown, P.R.L. and Ellis, A.J. (1970) The Ohaki-Broadlands hydrothermal area, New Zealand: mineralogy and related chemistry. *Am. J. Sci.* **269**, 97–130.

Browning, P. (1984) Cryptic variation within the cumulate sequences of the Oman ophiolite: magma chamber depth and petrological implications. In: Gass, I.G., Lippard, S.J. and Shelton, A.W. (eds) *Ophiolites and Oceanic Lithosphere, Geol. Soc. London, Spec. Publ.* **13**, 71–82.

Brueckner, H.K. (1974) Mantle Rb/Sr and $^{87}Sr/^{86}Sr$ ratios from clinopyroxenes from Norwegian garnet peridotites and pyroxenites. *Earth Planet. Sci. Lett.* **24**, 26–32.

Brueckner, H.K., Zindler, A., Seyler, M. and Bonatti. E. (1988) Zabargad and the isotopic evolution of the sub-Red Sea mantle and crust. *Tectonophysics* **150**, 163–176.

Brunn, J.H. (1959) La dorsale medio-atlantic et les epanchements ophiolitiques. *Compt. Rendu Soc. Geol. France* **8**, 234–236.

Bryan, W.B. (1972) Morphology of quench crystals in submarine basalts. *J. Geophys. Res.* **77**, 5812–5819.

Bryan, W.B. (1979) Regional variations and petrogenesis of basalt glasses from the FAMOUS area, Mid-Atlantic Ridge. *J. Petrol.* **20**, 293–325.

Bryan, W.B. (1983) Systematics of model phenocryst assemblages in submarine basalts: petrologic implications. *Contrib. Mineral. Petrol.* **83**, 62–74.

Bryan, W.B. and Dick, H.J.B. (1982) Contrasted abyssal basalt liquidus trends: evidence for mantle heterogeneity. *Earth Planet. Sci. Lett.* **58**, 15–26.

Bryan, W.B. and Moore, J.G. (1977) Compositional variations of young basalts in the Mid-Atlantic Ridge rift valley near lat. 36°49′N. *Bull. Geol. Soc. Am.* **88**, 556–570.

Bryan W.B. and Thompson, G. (1977) Basalts from DSDP Leg 37 and the FAMOUS area: compositional and petrogenetic comparisons. *Can. J. Earth Sci.* **14**, 875–885.

Bryan, W.B., Thompson, G., Frey, F.A. and Dickey, J.S. (1976) Inferred settings and differentiation in basalts from the Deep Sea Drilling Project. *J. Geophys. Res.* **81**, 4285–4304.

Bryan, W.B., Thompson, G. and Ludden, J.N. (1981) Compositional variation in normal MORB from 22–25°N: Mid-Atlantic Ridge and Kane Fracture Zone. *J. Geophys. Res.* **86**, 11815–11836.

Bryan, W.B., Thompson, G. and Michael, P.J. (1979) Compositional variation in a steady-state zoned magma chamber: Mid-Atlantic Ridge at 36°50′N. *Tectonophysics* **55**, 63–85.

Buck, W.R. and Parmentier, E.M. (1986) Convection beneath young oceanic lithosphere: implications for thermal structure and gravity. *J. Geophys. Res.* **91**, 1961–1979.

Burke, K. and Wilson, J.T. (1976) Hot spots on the earth's surface. *Sci. Am.* **235**, 46–57.

Burke, K., Kidd, W.S.F. and Wilson, J.T. (1973) Plumes and concentric plume traces of the Eurasian plate. *Nature, London* **241**, 128.

Burnett, M.S., Orcutt, J.A. and McClain, J.S. (1985) Further refraction evidence for a crustal magma chamber. *Trans. Am. Geophys. Union EOS* **66**, 1091.

Byerly, G. (1980) The nature of differentiation trends in some volcanic rocks from the Galapagos Spreading Center. *J. Geophys. Res.* **85**, 3797–3810.

Byerly, G.R. and Wright, T.L. (1978) Origin of major element chemical trends in DSDP Leg 37 basalts, Mid-Atlantic Ridge. *J. Volcanol. Geotherm. Res.* **3**, 229–279.

Byerly, G.R., Melson, W.G. and Vogt, P.R. (1976) Rhyolites, andesites, ferro-basalts and ocean tholeiites from the Galapagos spreading center. *Earth Planet. Sci. Lett.* **30**, 215–221.

Byers, C.D., Christie, D.M., Muenow, D.W. and Sinton, J.M. (1984) Volatile contents and ferric-ferrous ratios of basalt, ferrobasalt, andesite and rhyodacite glass from the Galapagos 95.5°W propagating rift. *Geochim. Cosmochim. Acta* **48**, 2239–2245.

Byers, C.D., Garcia, M.O. and Muenow, D.W. (1986) Volatiles in basaltic glasses from the East Pacific Rise at 21°N: implications for MORB sources and submarine lava flow morphology. *Earth Planet. Sci. Lett.* **79**, 9–20.

Cambon, P., Joron, J.L., Bougault, H. and Treuil, M. (1980) Leg 55, Emperor Seamounts—trace elements in transitional tholeiites, alkali basalts and hawaiites: mantle heterogeneity and magmatic processes. In: Jackson, E.D., Koizumi, I. *et al.* (eds) *Initial Reports of the Deep Sea Drilling Project* **55**, 585–597.

Campbell, A.C., Bowers, T.S., Measures, C., Falkner, K.K., Khadem, M. and Edmond, J.M. (1988) A time series of vent fluid compositions from 21°N East Pacific Rise (1979, 1981, 1985) and Guyamas Basin, Gulf of California (1982, 1985). *J. Geophys. Res.* **93**, 4537–4549.

Campbell, I.H. and Griffiths, R.W. (1990) Implications of mantle plume structure for the evolution of flood basalts. *Earth Planet. Sci. Lett.* **99**, 79–93.

Cann, J.R. (1969) Spilites from the Carlsberg Ridge, Indian Ocean. *J. Petrol.* **10**, 1–19.

Cann, J.R. (1970) New model for the structure of the oceanic crust. *Nature, London* **226**, 928–930.

Cann, J.R. (1971) Petrology of basement rocks from Palmer Ridge, N.E. Atlantic. *Phil. Trans. R. Soc., London* **A268**, 605–618.

Cann, J.R. (1974) A model for oceanic crustal structure developed. *Geophys. J.R. Astron. Soc.* **39**, 169–187.

Cann, J.R. (1979) Metamorphism in the ocean crust. In: Talwani, M., Harrison, C.G. and Hayes, D.E. (eds) *Deep Drilling Results in the Atlantic Ocean: Ocean Crust, Am. Geophys. Union, Maurice Ewing Ser.* **2**, 230–238.

Cann, J.R. and Funnell, B.M. (1967) Palmer Ridge: a section through the upper part of the oceanic crust. *Nature, London* **213**, 661–664.

Caress, D.W., Menard, H.W. and Hey, R.N. (1988) Eocene reorganization of the Pacific-Farallon spreading center north of the Mendocino Fracture Zone. *J. Geophys. Res.* **93**, 2813–2838.

Carmichael, I.S.E. (1964) The petrology of Thingmuli, a Tertiary volcano in eastern Iceland. *J. Petrol.* **5**, 321–325.

Carsola, A.J. and Dietz, R.S. (1952) Submarine geology of two flat-topped northeast pacific seamounts. *Am. J. Sci.* **250**, 481–497.

Carter, J.L. (1970) Mineralogy and chemistry of the Earth's upper mantle based on the partial fusion–partial crystallization model. *Bull. Geol. Soc. Am.* **81**, 2021–2034.

Casey, J.F. (1980) Geology of the southern half of the North Arm Mountain Massif, Bay of Islands ophiolite complex, southwestern Newfoundland. Unpublished PhD thesis, University of New York at Albany, 308 pp.

Casey, J.F., Elthon, D.L., Siroky, F.X., Karson, J.A. and Sullivan, J. (1985) Geochemical and geological evidence bearing on the origin of the Bay of islands and coastal complex ophiolites of western Newfoundland. *Tectonophysics* **116**, 1–40.

Castillo, P. (1988) The Dupal anomaly as a trace of the upwelling lower mantle. *Nature, London* **336**, 667–670.

Castillo, P., Batiza, R. and Stern, R.J. (1986) Petrology and geochemistry of Nauru Basin igneous complex: large volume, off-ridge eruptions of MORB-like basalts during the Cretaceous. In: Moberly, R., Schlanger, S.O. *et al.* (eds) *Initial Reports of the Deep Sea Drilling Project* **89**, 555–576.

Castillo, P., Batiza, R., Vanko, P., Malavassi, E., Barquero, J. and Fernandez, E. (1988). Anomalously young volcanoes on old hot-spot traces. I: Geology and petrology of Cocos Island. *Bull. Geol. Soc. Am.* **100**, 1400–1414.

Castillo, P., Carlson, R.W. and Batiza, R. (1991) Origin of Nauru Basin igneous complex: Sr, Nd and Pb isotope and REE constraints. *Earth Planet Sci. Lett.* **103**, 200–213.
Chaffey, D.J., Cliff, R.A. and Wilson, B.M. (1989) Characterization of the St. Helena magma source. In: Saunders, A.D. and Norry, M.J. (eds) *Magmatism in the Ocean Basins, Geol. Soc. London, Spec. Publ.* **42**, 257–276.
Chase, C.B. (1981) Ocean island Pb: two stage histories and mantle evolution. *Earth Planet. Sci. Lett.* **52**, 277–284.
Chase, C.G. (1979) Asthenospheric counterflow: a kinematic model. *Geophys. J. R. Astron. Soc.* **56**, 1–18.
Chayes, F. (1965) Titanium and aluminium content of oceanic and circum-oceanic basalt. *Mineral. Mag.* **34**, 126–131.
Chen, C.Y. (1987) Lead isotope constraints on the origin of Hawaiian basalts. *Nature, London* **327**, 49–52.
Chen, C.Y. and Frey, F.A. (1983) Origin of Hawaiian tholeiite and alkali basalt: geochemical evidence for mixing of primitive mantle with a MORB reservoir. *Nature, London* **302**, 785–789.
Chen, C.Y. and Frey, F.A. (1985) Trace element and isotopic geochemistry of lavas from Haleakala volcano, east Maui, Hawaii: implications for the origin of Hawaiian basalts. *J. Geophys. Res.* **90**, 8743–8768.
Chernyseva, V.I. (1971) Greenstone altered rocks of rift zones in median ridges of Indian Ocean. *Int. Geol. Rev.* **13**, 903–913.
Christensen, N.I. and Salisbury, M.H. (1972) Seafloor spreading, progressive alteration of layer 2 basalts and associated changes in seismic velocities. *Earth Planet. Sci. Lett.* **15**, 367–375.
Christie, D.M. and Sinton, J.M. (1981) Evolution of abyssal lavas along propagating segments of the Galapagos spreading center. *Earth Planet. Sci. Lett.* **56**, 321–335.
Christie, D.M., and Sinton, J.M. (1986) Major element constraints on melting, differentiation and mixing of magmas from the Galapagos 95.5°W propagating rift system. *Contrib. Mineral. Petrol.* **94**, 274–288.
Christie, D.M., Charmichael, I.S.E. and Langmuir, C.H. (1986) Oxidation state of mid-ocean ridge basalt glasses. *Earth Planet. Sci. Lett.* **79**, 397–411.
Christensen, N.I. and Salisbury, M.H. (1975) Structure and constitution of the lower oceanic crust. *Rev. Geophys. Space Phys.* **13**, 57–86.
Church, W.R. (1972) Ophiolite: its definition, origin as oceanic crust, and mode of emplacement in orogenic belts, with special reference to the Appalachians. *Depart. Energ. Mines Resource. Can.* **42**, 71–85.
Clague, D.A. (1981) Linear island and seamount chains, aseismic ridges and intraplate volcanism: results from DSDP. In: Warme, J.E., Douglas, R.G. and Winterer, E.L. (eds) *The Deep Sea Drilling Project: a Decade of Progress, SEPM Spec. Publ.* **32**, 7–22.
Clague, D.A. (1987) Hawaiian alkaline volcanism. In: Fitton, J.G. and Upton, B.G.J. (eds) *Alkaline Igneous Rocks, Geol. Soc. London, Spec. Publ.* **30**, 227–252.
Clague, D.A. and Bunch, T.E. (1976) Formation of ferrobasalt at east Pacific mid-ocean spreading centers. *J. Geophys. Res.* **81**, 4247–4256.
Clague, D.A. and Dalrymple, G.B. (1987) The Hawaiian-Emperor volcanic chain, Part I: Geologic evolution. In: Decker, R.W., Wright, T.L. and Stauffer, P.H. (eds) *Volcanism in Hawaii, US Geol. Surv. Prof. Pap.* **1350**, 5–54.
Clague, D.A. and Frey, F.A. (1980) Trace element geochemistry of tholeiitic basalts from Site 433C, Suiko Seamount. In: Jackson, E.D., Koizumi, I. *et al.* (eds) *Initial Reports of the Deep Sea Drilling Project* **55**, 559–569.
Clague, D.A. and Frey, F.A. (1982) Petrology and trace element geochemistry of the Honolulu Volcanic Series, Oahu: implications for the oceanic mantle below Hawaii. *J. Petrol.* **23**, 447–504.
Clague, D.A. and Straley, P.F. (1977) Petrological nature of the oceanic Moho. *Geology* **5**, 133–136.
Clague, D.A., Frey, F.A., Thompson, G. and Rindge, S. (1981) Minor and trace element geochemistry of volcanic rocks dredged from the Galapagos spreading center: role of crystal fractionation and mantle heterogeneity. *J. Geophys. Res.* **86**, 9469–9482.
Clague, D.A., Jackson, E.D. and Wright, T.L. (1980) Petrology of Hualalai volcano, Hawaii: implications for mantle composition. *Bull. Volcanol.* **43**, 641–656.
Cochran, J.R. and Talwani, M. (1978) Gravity anomalies, regional elevation and the deep structure of the North Atlantic. *J. Geophys. Res.* **83**, 4907–4924.

Cohen, R.S. and O'Nions, R.K. (1982) Identification of recycled continental material in the mantle from Sr, Nd and Pb isotope investigation. *Earth Planet. Sci. Lett.* **61**, 73–84.

Coish, R. (1977) Ocean floor metamorphism in the Betts Cove ophiolite, Newfoundland. *Contrib. Mineral Petrol.* **60**, 255–270.

Cole, J.W. (1984) Taupo-Rotorua depression: an ensialic marginal basin of the North island, New Zealand. In: Kokelaar, B.P. and Howells, M.F. (eds) *Marginal Basin Geology, Geol. Soc. London, Spec. Publ.* **16**, 109–120.

Coleman, R.G. (1971) Plate tectonic emplacement of upper mantle peridotites along continental edges. *J. Geophys. Res.* **76**, 1212–1222.

Coleman, R.G. (1977) *Ophiolites: Ancient Oceanic Lithosphere*? Springer-Verlag, New York, 229 pp.

Coleman, R.G. and Peterman, Z.E. (1975) Oceanic plagiogranite. *J. Geophys. Res.* **80**, 1099–1108.

Collier, J. and Sinha, M.C. (1990) Seismic images of a magma chamber beneath the Lau Basin back-arc spreading centre. *Nature*, **346**, 646–648.

Coombs, D.S. (1963) Trends and affinities of basaltic magmas and pyroxenes as illustrated on the diopside–olivine–silica diagram. *Mineral. Soc. Am. Spec. Pap.* **1**, 227–250.

Coombs, D.S. and Wilkinson, J.G.F. (1969) Lineages and fractionation trends in undersaturated volcanic rocks from the East Otago province (New Zealand) and related rocks. *J. Petrol.* **10**, 440–501.

Cooper, A.K., Marlow, M.S. and Scholl, D.W. (1977) The Bering Sea—a multifarious marginal basin. In: Talwin, M. and Pitman, W.C. (eds) *Island Arcs, Deep Sea Trenches and Back-arc Basins, Am. Geophys. Union, Maurice Ewing Ser.* **1**, 437–450.

Corliss, J.B. (1970) Mid-ocean ridge basalts. I: The origin of submarine hydrothermal systems; II: Regional diversity along the Mid-Atlantic Ridge. Unpubl. PhD thesis, University of California, San Diego.

Corliss, J.B. (1971) The origin of metal-bearing submarine hydrothermal solutions. *J. Geophys. Res.* **76**, 8128–8138.

Corrigan, G. and Gibb, F.G.F. (1979) The loss of Fe and Na from a basaltic melt during experiments using the wire-loop method. *Mineral. Mag.* **43**, 121–126.

Courtney, R.C. and White, R.S. (1986) Anomalous heat flow and geoid across the Cape Verde Rise: evidence of dynamic support from a thermal plume in the mantle. *Geophys. J. R. Astron. Soc.* **87**, 815–867.

Cox, A. and Hart, R.B. (1986) *Plate Tectonics: How it Works*, Blackwell Scientific Publications, Oxford, 392 pp.

Cox, K.G. (1980) A model for flood basalt volcanism. *J. Petrol.* **21**, 629–650.

Craig, H. (1953) The geochemistry of the stable carbon isotopes. *Geochim. Cosmochim. Acta* **3**, 53–92.

Craig, H. and Lupton, J.E. (1976) Primordial neon, helium and hydrogen in oceanic basalts. *Earth Planet. Sci. Lett.* **31**, 369–385.

Crane, K. (1985) The spacing of rift axis highs: dependence on diapiric processes in the underlying asthenosphere. *Earth Planet. Sci. Lett.* **72**, 405–414.

Creager, K.C. and Jordan, T.H. (1984) Slab penetration in the lower mantle. *J. Geophys. Res.* **89**, 3031–3049.

Creager, K.C. and Jordan, T.H. (1986) Slab penetration into the lower mantle beneath the Mariana and other island arcs of the northwest Pacific. *J. Geophys. Res.* **91**, 3573–3589.

Crough, S.T. (1983) Hot spot swells. *Ann. Rev. Earth Planet. Sci.* **11**, 165–193.

Curray, J.R., Moore, D.G., Lawver, L.A., Emmel, F.J., Raitt, R.W., Henry, M. and Keickhefer, R. (1979) *Mem. Am. Assoc. Petr. Geol.* **29**, 189–198.

Curray, J.R., Moore, D.G. *et al.* (eds) (1982) *Initial Reports of the Deep Sea Drilling Project* **64**, 1313pp.

Czamanske, G.K. and Moore, J.G. (1977) Composition and phase chemistry of sulfide globules in basalts from the Mid-Atlantic Ridge rift valley near 37°N lat. *Bull. Geol. Soc. Am.* **88**, 587–599.

Dalrymple, G.B., Clague, D.A., Garcia, M.O. and Bright, S.W. (1981), Petrology and K-Ar ages of dredged samples from Laysan Island and Northampton Bank volcanoes, Hawaiian Ridge, and evolution of the Hawaiian-Emperor chain. *Bull. Geol. Soc. Am.* **92**, 884–933.

Dalrymple, G.B., Clague, D.A., Vallier, T.L. and Menard, H.W. (1987) $^{40}Ar/^{39}Ar$ age, petrology and tectonic significance of some seamounts in the Gulf of Alaska. In: Keating, B.H., Fryer, P., Batiza, R. and Boehlert, G.W. (eds) *Seamounts, Islands and Atolls, Am. Geophys. Union, Geophys. Monogr.* **43**, 297–315.

Daly, R.A. (1903) The geology of Ascutney Mountain, Vermont. *Bull. US Geol. Surv.* **209**, 1–113.

Dalziel, I.W.D. (1981) Back-arc extension in the southern Andes: a review and critical reappraisal. *Phil. Trans. R. Soc. London* **A300**, 319–335.

Dalziel, I.W.D., de Wit, M.J. and Plamer, K.F. (1974) Fossil margin basin in the southern Andes. *Nature* (*London*) **250**, 291–294.

Darwin, C. (1845) *Journal of Researches during the Voyage of H.M.S. "Beagle"* Nelson, London, 543 pp.

Davidson, J.P. (1987) Crustal contamination versus subduction zone enrichment: examples from the Lesser Antilles and implications for mantle source compositions of island arc volcanic rocks. *Geochim. Cosmochim. Acta* **51**, 2185–2198.

Davies, G.F. (1990) Mantle plumes, mantle stirring and hotspot chemistry. *Earth Planet. Sci. Lett.* **99**, 94–109.

Davies, G.R., Norry, M.J., Gerlach, D.C. and Cliff, R.A. (1989) A combined chemical and Pb–Sr–Nd isotope study of the Azores and Cape Verde hot spots: the geodynamical implications. In: Saunders, A.D. and Norry, M.J. (eds) *Magmatism in the Ocean Basins, Geol. Soc. London, Spec. Publ.* **42**, 231–255.

Davies, H.L. (1968) Papuan ultramafic belt, *Proceedings of the 23rd International Geological Congress, Prague*, Vol. 1, 209–220.

Davies, H.L. (1971) Peridotite–gabbro–basalt complex in eastern Papua: an overthrust plate of oceanic mantle and crust. *Depart. Natl. Devel. Bull. Bureau Min. Resource. Geol. Geophys.* **128**, 48 pp.

Davis, E.E. and Karsten, J.L. (1986) On the cause of the asymmetric distribution of seamounts about the Juan de Fuca ridge: ridge crest migration over heterogeneous asthenosphere. *Earth Planet. Sci. Lett.* **79**, 385–396.

Decker, R.W., Wright, T.L. and Stauffer, P.H. (1987) Volcanism in Hawaii. *US Geol. Surv. Prof. Pap. no. 1350.*

Deffeyes, K.S. (1970) The axial valley: a steady state feature of the terrain. In: Johnson, H. and Smith, B.L. (eds) *Megatectonics of Continents and Oceans*, Rutgers University Press, pp. 194–222.

Deines, P. (1980) The carbon isotopic composition of diamond: relationship to diamond shape, colour, occurrence and vapour composition. *Geochim. Cosmochim. Acta* **44**, 943–961.

Delaney, J.R., Mogk, D.W. and Mottl, M.J. (1988) Quartz-cemented breccias from the Mid-Atlantic Ridge: samples of a high salinity hydrothermal upflow zone. *J. Geophys. Res.* **92**, 9175–9192.

Delaney, J.R., Muenow, D.W. and Graham, D.G. (1978) Abundance and distribution of water, carbon and sulphur in the glassy rims of submarine pillow basalts. *Geochim. Cosmochim. Acta* **42**, 581–594.

DePaolo, D.J. (1988) *Neodymium Isotope Geochemistry: An Introduction*, Springer-Verlag, Berlin, 187 pp.

DesMarais, D.J. (1986) Carbon abundance measurements in oceanic basalts: the need for a consensus. *Earth Planet. Sci. Lett.* **79**, 21–26.

DesMarais, D.J. and Moore, J.G. (1984) Carbon and its isotopes in mid-oceanic basaltic glasses. *Earth Planet. Sci. Lett.* **69**, 48–57.

Detrick, R.S. and Crough, S.T. (1978) Island subsidence, hotspots and lithospheric thinning. *J. Geophys. Res.* **83**, 1236–1244.

Detrick, R.S., Buhl, P., Vera, E., Mutter, J., Orcutt, J., Madsen, J. and Brocher, T. (1987) Multichannel seismic imaging of a crustal magma chamber along the East Pacific Rise. *Nature, London* **326**, 35–41.

Detrick, R.S., Fox, P.J., Kastens, K., Ryan, W.B.F. and Karson, J. (1984) A Sea Beam survey of the Kane Fracture Zone and the adjacent Mid-Atlantic Ridge rift valley. *Trans. Am. Geophys. Union, EOS* **65**, 1106.

Detrick, R.S., Fox, P.J., Schulz, N., Pockalny, R., Kong, L., Mayer, L. and Ryan, W.B.F. (1990a) Geological and tectonic setting of the Mark area. In: Detrick, R., Honnorez, J., Bryan, W.B., Juteau, T. *et al.* (eds) *Proceedings of the Ocean Drilling Program, Initial Reports* (*Part A*) **106/109**, 15–22.

Detrick, R.S., Mutter, J., Buhl, P. and Kim, I.I. (1990b) No evidence from multichannel reflection data for a crustal magma chamber in the MARK area on the Mid-Atlantic Ridge. *Nature, London* **347**, 61–64.

Detrick, R.S., Sclater, J.G. and Thiede, J. (1977) The subsidence of aseismic ridges. *Earth Planet. Sci. Lett.* **34**, 185–196.

Detrick, R.S., White, R.S., Courtney, R.C. and Von Herzen, R.P. (1989) Heat flow on midplate swells. In: Wright, J.A. and Louden, K.E. (eds) *Handbook of Seafloor Heat Flow*, CRC Press, Boca Raton, FL, pp. 169–190.

DeVries Klein, G., Kobayashi, K. *et al.* (eds) (1980) *Initial Reports of the Deep Sea Drilling Project* **58**, 1022 pp.

Dewey, J. (1980) Episodicity, sequence and style at convergent plate boundaries. In: Strangeway, D.W. (ed) *The Continental Crust and its Mineral Deposits, Geol. Assoc. Can. Spec. Pap.* **20**, 553–573.

Dewey, J.F. and Bird, J.M. (1971) Origin and emplacement of the ophiolite suite: Appalachian ophiolites in Newfoundland. *J. Geophys. Res.* **76**, 3179–3206.

Dewey, J.F. and Burke, K. (1974) Hot spots and continental break-up: implications for collisional orogeny. *Geology* **2**, 57–60.

Dick, H.J.B (1977) Partial melting in the Josephine peridotite: the effect of mineral composition and its consequences for geothermometry and geobarometry. *Am. J. Sci.* **277**, 760–801.

Dick, H.J.B. (1980) Vesicularity of Shikoku Basin basalt: a possible correlation with the anomalous depth of back-arc basins. In: DeVries Klein, G., Kobayashi, K. *et al.* (eds) *Initial Reports of the Deep Sea Drilling Project* **58**, 895–904.

Dick, H.J.B. (1982) The petrology of two back-arc basins of the northern North Philippine Sea. *Am. J. Sci.* **282**, 644–700.

Dick, H.J.B. (1989) Abyssal peridotites, very slow spreading ridges and ocean ridge magmatism. In: Saunders, A.D. and Norry, M.J. (eds) *Magmatism in the Ocean Basins, Geol. Soc. London, Spec. Publ.* **42**, 71–105.

Dick, H.J.B. and Bullen, T. (1984) Chromian spinel as a petrogenetic indicator in abyssal and alpine-type peridotites and spatially related lavas. *Contrib. Mineral. Petrol.* **86**, 54–76.

Dick, H.J.B. and Fisher, R.L. (1984) Mineralogic studies of the residues of mantle melting: abyssal and alpine type peridotites. In: Kornprobst, J. (ed) *Kimberlites II: The Mantle and Crust-Mantle Relationships*, Elsevier, Amsterdam, pp. 295–308.

Dick, H.J.B., Bullen, T. and Bryan, W.B. (1984a) Mineralogic variability of the uppermost mantle along mid-ocean ridges. *Earth Planet. Sci. Lett.* **69**, 88–106.

Dick, H.J.B., Fisher, R.L. and Bryan, W.B. (1984b) Mineralogic variability of the uppermost mantle along mid-ocean ridges. *Earth Planet. Sci. Lett.* **69**, 88–106.

Dick, H.J.B., Marsh, N.G. and Bullen, T.D. (1980) DSDP Leg 58 abyssal basalts from the Shikoku Basin: their petrology and major element geochemistry. In: DeVries Klein, G., Kobayashi, K. *et al.* (eds) *Initial Reports of the Deep Sea Drilling Project* **58**, 843–872.

Dickey, J.S. (1968) Eclogitic and other inclusions in the Mineral Breccia Member of the Deborah volcanic formation at Kakanui, New Zealand. *Am. Mineral.* **53**, 1304–1319.

Dickey, J.S., Frey, F.A., Hart, S.R. and Watson, E.B. (1977) Geochemistry and petrology of dredged basalts from Bouvet Triple Junction, South Atlantic. *Geochim. Cosmochim. Acta* **41**, 1105–1118.

Dickin, A.P. (1981) Isotope geochemistry of the Tertiary igneous rocks from the Isle of Skye, NW Scotland. *J. Petrol.* **22**, 155–190.

Dixon, S. and Rutherford, M. (1979) Plagiogranites as late-stage immiscible liquids in ophiolite and mid-ocean ridge suites: an experimental study, *Earth Planet. Sci. Lett.* **45**, 45–60.

Dmitriev, L., Sobolev, A.V. and Suschevskaya, N.M. (1979) The primary melt of the oceanic tholeiite and the upper mantle composition. In: Talwani, M., Harrison, C.G.A. and Haynes, D.E (eds) *Deep Drilling Results in the Atlantic: Ocean Crust, Am. Geophys. Union, Maurice Ewing Ser.* **2**, 302–313.

Dmitriev, L., Sobolev, A.V., Uchanov, A.V., Malysheva, T.V. and Melson, W.G. (1984) Primary differences in oxygen fugacity and depth of melting in the mantle source regions for oceanic basalts. *Earth Planet. Sci. Lett.* **70**, 303–310.

Donaldson, C.H. (1976) An experimental investigation of olivine morphology. *Contrib. Mineral. Petrol.* **57**, 187–213.

Donaldson, C.H. (1979) Composition changes in basalt melt contained in a wire loop of Pt80 Rh20: effect of temperature, time and oxygen fugacity. *Mineral. Mag.* **43**, 115–119.

Donaldson, C.H. and Brown, R.W. (1977) Refractory megacrysts and magnesium-rich melt inclusions within oceanic tholeiites: indicators of magma mixing and parental magma composition. *Earth Planet. Sci. Lett.* **37**, 81–89.

Donato, G.D.I., Joron, J.L., Treuil, M. and Loubet, M. (1990) Geochemistry of zero age N-MORB from Hole 648B, ODP, Legs 106–109, MAR 22°N. In: Detrick, R., Honnorez, J., Bryan, W.B., Juteauo, T. *et al.* (eds) *Proceedings of the Ocean Drilling Program, Initial Reports* (*Part A*) **106/109**, 57–65.

Donnelly, T.W., Pritchard, R.A., Emmermann, R. and Puchelt, H. (1979a) The aging of oceanic crust: synthesis of the mineralogical and chemical results of Deep Sea Drilling project, Legs 51 through 53. In: Donnelly, T., Francheteau, J., Bryan, W.B., Robinson, P.T., Flower, M., Salisbury, M. *et al.* (eds) *Initial Reports of the Deep Sea Drilling Project*, **51**, **52**, **53**, 1563–1577.

Donnelly, T.W., Thompson, G. and Robinson, P.T. (1979b) Very low-temperature hydrothermal alteration of the oceanic crust and the problem of fluxes of potassium and magnesium. In: Talwani, M., Harrison, C.G. and Hayes, D.E. (eds) *Deep Drilling Results in the Atlantic: Ocean Crust, Am. Geophys. Union, Maurice Ewing Ser.* **2**, 369–382.

Donnelly, T.W., Thompson, G. and Salisbury, M.H (1979c) The chemistry of altered basalts at site 417, Deep Sea Drilling Project Leg 51. In: Donnelly, T., Francheteau, J., Bryan, W.B., Robinson, P.T., Flower, M., Salisbury, M. *et al.* (eds) *Initial Reports of the Deep Sea Drilling Project* **51**, **52**, **53**, 1319–1330.

Dosso, L. and Murthy, V.R. (1980) A Nd isotopic study of the Kerguelen islands: inferences on enriched oceanic mantle sources. *Earth Planet. Sci. Lett.* **48**, 268–276.

Dosso, L., Vidal, P., Cantagrel, J.M., Lameyre, J., Marot, A. and Zimine, S. (1979) "Kerguelen: continental fragment or oceanic island?": petrology and isotopic geochemistry evidence. *Earth Planet. Sci. Lett.* **43**, 46–60.

Dowty, E. (1980) Crystal growth and nucleation theory and the numerical simulation of igneous crystallization. In: Hargraves, R.B. (ed) *Physics of Magmatic Processes*, Princeton University Press, pp. 421–485.

Duke, J.M. (1976) Distribution of the period four transition elements among olivine, calcic clinopyroxene and mafic silicate phases: experimental results. *J. Petrol.* **17**, 499–521.

Duncan, R.A. (1978) Geochronology of basalts from the Ninetyeast Ridge and continental dispersion in the eastern Indian Ocean. *J. Volcanol. Geotherm. Res.* **4**, 283–305.

Duncan, R.A. (1984) Age progressive volcanism in the new England seamounts and the opening of the central Atlantic Ocean. *J. Geophys. Res.* **89**, 9980–9990.

Duncan, R.A. and Clague, D.A. (1985) Pacific plate motion recorded by linear volcanic chains. In: Narin, A.E., Stehli, F.G. and Uyeda, S. (eds) *The Ocean Basins and Margins V: 7A The Pacific Ocean*, Plenum Press, New York. pp. 89–122.

Duncan, R.A. and Rhodes, J.M. (1978) Residual glasses and melt inclusions in basalts from DSDP Legs 45 and 46: evidence of magma mixing. *Contrib. Mineral. Petrol.* **67**, 417–431.

Duncan, R.A., McCulloch, M.T., Barsczus, H.G. and Nelson, D.R. (1986) Plume versus lithosphere source for melts at Ua Pou, Marquesas Islands. *Nature, London* **322**, 534–538.

Dungan, M.A. and Ave'Lallement, H.G. (1977) Formation of small dunite bodies by metasomatic transformation of harzburgite in the Canyon Mountain ophiolite, northeast Oregon. *Bull. Oregon. Depart. Geol. Min. Ind.* **96**, 109–128.

Dupre, B. and Allegre, C.J. (1983) Pb–Sr isotope variations in Indian Ocean basalts and mixing phenomena. *Nature, London* **303**, 142–146.

Dupre, B., Lambert, B. and Allegre, C.J. (1982) Isotopic variations within a single oceanic island. *Nature, London* **299**, 620–622.

Dupuy, C., Vidal, P., Barsczus, H.G. and Chauvel, C. (1987) Origin of basalts from the Marquesas Archipelago (south central Pacific ocean): isotope and trace element constraints. *Earth Planet. Sci. Lett.* **82**, 145–152.

Dvorak, J.J., Okamura, A.T., English, T.T., Koyanagi, R.Y., Nakata, J.S., Sako, M.K., Tanigawa, W.T. and Yamashita, K.M. (1986) Mechanical response of the south flank of Kilauea volcano, Hawaii to intrusive events along the rift systems. *Tectonophysics* **124**, 193–209.

Dzurisin, D., Koyanagi, R.Y. and English, T.T. (1984) Magma supply and storage at Kilauea volcano, Hawaii, 1956–1983. *J. Volcanol. Geotherm. Res.* **21**, 188–206.

Eaby, J.S., Clague, D.A. and Delaney, J.R. (1984) Sr isotopic variations along the Juan de Fuca Ridge. *J. Geophys. Res.*, **89**, 7883–7890.

Edmond, J.M., Corliss, J.B. and Gordon, L.I (1979a) Ridge crest hydrothermal metamorphism at the Galapagos spreading center and reverse weathering. In: Talwani, M., Harrison, C.G. and Hayes, D.E. (eds) *Deep Drilling Results in the Atlantic Ocean: Ocean Crust, Am. Geophys. Union, Maurice Ewing Ser.* **2**, 383–390.

Edmond, J.M., Measures, C., Mangum, B., Grant, B., Sclater, F.R., Collier, R., Hudson, A., Gordon, L.I. and Corliss, J.B. (1979b) On the formation of metal-rich deposits at ridge crests. *Earth Planet. Sci. Lett.* **46**, 19–30.

Edmond, J.M., Measures, C., McDuff, R.E., Chan, L.H., Collier, R., Grant, B., Gordon, L.I. and Corliss, J.B. (1979c) Ridge crest hydrothermal activity and the balances of the major and minor elements in the ocean: the Galapagos data. *Earth Planet. Sci. Lett.* **46**, 1–18.

Edmond, J.M., von Damm, K., McDuff, R.E. and Measures, C.I. (1981) Chemistry of hot springs on the East Pacific Rise and their effluent dispersal (a review). *Nature* (*London*) **297**, 187–191.

Eggler, D.H., Fahlquist, D.A. and Herndon, J.M. (1973) Ultrabasic rocks from the Cayman Trough, Caribbean Sea. *Bull. Geol. Soc. Am.* **84**, 2133–2138.

Einarsson, P. (1991) Earthquakes and present tectonics in Iceland. *Tectonophysics* **189**, 261–280.

Einarsson, P. and Eiriksson, J. (1982) Earthquake fractures in the districts land and the Rangarvellir in the South Iceland Seismic Zone. *Jokull* **32**, 113–119.

Einarsson, P., Bjornsson, S., Foulger, G., Stefansson, R. and Skaftadottir, Th. (1981) Seismicity pattern in the South Iceland Seismic Zone. In: *Earthquake Predictions – an International Review, Am. Geophys. Union, Maurice Ewing Ser.* **4**, pp. 141–151.

Einarsson, P., Klein, F.W. and Bjornsson, S. (1977) The Borgarfjordur earthquakes of 1974 in west Iceland. *Bull. Seismol. Soc. Am.* **67**, 187.

Einsele, G., Gieskes, J.M., Curray, J., Moore, D.G. *et al.* (1980) Intrusion of basaltic sills into highly porous sediments, and resulting hydrothermal activity. *Nature, London* **283**, 441–445.

Ellsworth, W.L. and Koyanagi, R.Y. (1977) Three-dimensional crust and mantle structure of Kilauea Volcano, Hawaii. *J. Geophys. Res.* **82**, 5379–5394.

Elthon, D. (1979) High magnesia liquids as the parental magma for ocean floor basalts. *Nature, London* **278**, 514–518.

Elthon, D. (1981) Metamorphism in oceanic spreading centers. In: Emiliani, E. (ed), *The Sea* (Vol. 7), *The Oceanic Lithosphere*, Wiley, New York, 285–303.

Elthon, D. (1983) Isomolar and isostructural pseudo-liquidus phase diagrams for oceanic basalts. *Am. Mineral.* **68**, 506–511.

Elthon, D. (1984a) Plagioclase buoyancy in oceanic basalts: chemical effects. *Geochim. Cosmochim. Acta* **48**, 753–768.

Elthon, D. (1984b) Petrology of gabbroic rocks from the Mid-Cayman Rise spreading center. *J. Petrol.* **92**, 658–682.

Elthon, D. (1986) Comments on "Composition and depth of origins of primary mid-ocean ridge basalts" by D.C. Presnall and J.D. Koones. *Contrib. Mineral. Petrol.* **94**, 253–256.

Elthon, D. (1987) Petrology of gabbroic rocks from the Mid-Cayman Rise spreading center. *J. Geophys. Res.* **92**, 658–682.

Elthon, D. (1989) Pressure of origin of primary mid-ocean ridge basalts. In: Saunders A.D. and Norry, M.J. (eds) *Magmatism in the Ocean Basins, Geol. Soc. London, Spec. Publ.* **42**, 125–136.

Elthon, D. (1990) The petrogenesis of primary mid-ocean ridge basalts. *Rev. Aquat. Sci.* **2**, 27–53.

Elthon, D. and Casey, J. (1985) The very depleted nature of certain primary mid-ocean ridge basalts, *Geochim. Cosmochim. Acta* **49**, 289–298.

Elthon, D. and Scarfe, C.M. (1984) High-pressure phase equilibria of a high-magnesia basalt and genesis of primary oceanic basalts. *Am. Mineral.* **69**, 1–15.

Elthon, D. and Stern, C. (1978) Metamorphic petrology of the Sarmiento ophiolite complex, Chile. *Geology* **6**, 464–468.

Elthon, D., Casey, J.F. and Komor, S.C. (1982) Mineral chemistry of ultramafic cumulates from the North Arm Mountain massif of the Bay of Islands ophiolite: evidence for high pressure crystal fractionation of oceanic basalts. *J. Geophys. Res.* **87**, 8717–8734.

Emery, K.D. and Uchupi, E. (1984) *The Geology of the Atlantic Ocean*, Springer-Verlag, New York, 1050 pp.

Engel, C.G. and Engel, A.E.J. (1963) Basalts dredged from the north-eastern Pacific Ocean. *Science* **140**, 1321–1324.

Engel, C.G. and Fisher, R.L. (1975) Granitic to ultramafic rock complexes of the Indian Ocean ridge system, western Indian Ocean. *Bull. Geol. Soc. Am.* **86**, 1553–1578.

Engel, A.E.J., Engel, C.G. and Havens, R.G. (1965) Chemical characteristics of oceanic basalts and the upper mantle. *Bull. Geol. Soc. Am.* **76**, 719–734.

Epp, D. (1984) Possible peturbations to hotspot traces and implications for the origin and structure of the Line Islands. *J. Geophys. Res.* **89**, 11273–11286.

Erlank, A.J. and Kable, E.J.D. (1976) The significance of incompatible elements in Mid-Atlantic Ridge basalts from 45°N with particular reference to Zr/Nb. *Contrib. Mineral. Petrol.* **54**, 281–291.

Everts, P., Koerfer, L.E. and Schwartzbach, T. (1972) Neue K/Ar-Datierungen islaendischer Basalte: Vorlaeufige mitteiglung. *Neues Jahrb. Geol. Palaeontol. Monatsh.* **5**, 280–284.

Ewart, A., and Hawkesworth, C.J. (1987) The Pleistocene-Recent Tonga-Kermadec arc lavas: interpretation of new isotopic and rare earth data in terms of a depleted mantle source model. *J. Petrol.* **28**, 495–530.

Ewart, A., Bryan, W.B. and Gill, J.B. (1973) Mineralogy and geochemistry of the younger volcanic islands of Tonga, S.W. Pacific. *J. Petrol.* **14**, 429–465.

Ewing, J. and Houtz, R. (1979) Acoustic stratigraphy and structure of the oceanic crust. In: Talwani, M., Harrison, G.C.A. and Hayes, D.E. (eds) *Deep Drilling results in the Atlantic Ocean: Ocean Crust, Am. Geophys. Union, Maurice Ewing Ser.* **2**, 43–51.

Exley, R.A., Boyd, S.R., Mattey, D.P. and Pillinger, C.T. (1987) Nitrogen isotope geochemistry of basaltic glasses: implications for mantle degassing and structure. *Earth Planet. Sci. Lett.* **81**, 163–174.

Exley, R.A., Mattey, D.P., Clague, D.A. and Píllinger, C.T. (1986a) Carbon isotope systematics of Loihi Seamount and MORB glasses. *Earth Planet. Sci. Lett.* **78**, 189–199.

Exley, R.A., Mattey, D.P., Pillinger, C.T. and Sinton, J.M. (1986b) Carbon isotope geochemistry of basalt glasses from the Lau and North Fiji marginal basins. *Terra Cognita* **6**, 324.

Eysteinsson, H. and Hermance, J.F. (1985) Magnetotelluric measurements across the eastern neovolcanic zone in South Iceland. *J. Geophys. Res.* **90**, 10093–10103.

Falloon, T.J. and Green, D.H. (1987) Anhydrous partial melting of MORB pyrolite and other peridotite compositions at 10 kbars: implications for the origin of primitive MORB glasses. *Contrib. Mineral. Petrol.* **37**, 181–219.

Falloon, T.J. and Green, D.H. (1988) Anhydrous partial melting of peridotite from 8 to 35 kbar and the petrogenesis of MORB. In: Menzies, M.A. and Cox, K.G. (eds) *Oceanic and Continental Lithosphere: Similarities and Differences, J. Petrol. Spec. Publ.* **29**, 379–414.

Faure, G. and Powell, J.L. (1972) *Strontium Isotope Geology*, Springer-Verlag, Berlin, 188 pp.

Feigenson, M.D. (1986) Constraints on the origin of Hawaiian lavas. *J. Geophys. Res.* **91**, 9383–9393.

Fenner, C.N. (1929) The crystallization of basalts. *Am. J. Sci.* **18**, 225–253.

Fenner, C.N. (1931) The residual liquids of crystallizing magmas. *Mineral. Mag.* **22**, 539–560.

Feraud, G., Kaneoka, I. and Allegre, C.J. (1980) K/Ar ages and stress pattern in the Azores: dynamic implications. *Earth Planet. Sci. Lett.* **46**, 275–286.

Fine, G. and Stolper, E. (1986) Dissolved carbon dioxide in basaltic glasses: concentrations and speciation. *Earth Planet. Sci. Lett.* **76**, 263–278.

Fisher, D.E. (1985) Noble gases from oceanic island basalts do not require an undepleted mantle source. *Nature (London)* **316**, 716–718.

Fisher, D.E. (1986) Rare gas abundances in MORB. *Geochim. Cosmochim. Acta* **50**, 2531–2541.

Fisher, D.E. (1989) Evaluation of rare gas data in relation to oceanic magmas. In: Saunders, A.D. and Norry, M.J. (eds) *Magmatism in the Ocean Basins, Geol. Soc. London, Spec. Publ.* **42**, 301–311.

Fisher, R.L., Dick, H.J.B., Natland, J.H. and Meyer, P.S. (1986) Mafic/ultramafic suites of the slowly spreading Southwest Indian Ridge: Protea exploration of the Antarctic plate boundary, 24°–47°E. *Ofioliti* **11**, 147–178.

Fisher, R.V. and Schmincke, H.U. (1984) *Pyroclastic Rocks*, Springer-Verlag, Berlin, 472 pp.

Fisk, M.J., Duncan, R.A., Baxter, A.N., Greenough, J.D., Hargraves, R.B., Tatsumi, Y. and Shipboard Scientific Party (1989) Reunion hotspot magma chemistry over the past 65 m.y.: results from Leg 115 of the Ocean Drilling Program. *Geology* **17**, 934–937.

Fisk, M.R. (1984) Depths and temperature of mid-ocean ridge magma chambers and the composition of their source magmas. In: Gass, I.G., Lippard, S.J. and Shelton, A.W. (eds) *Ophiolites and Oceanic Lithosphere, Geol. Soc. London, Spec. Publ.* **13**, 17–23.

Fisk, M.R. (1986) Basalt magma interaction with harzburgite and the formation of high-magnesium andesites. *Geophys. Res. Lett.* **13**, 467–470.

Fisk, M.R. and Bence, A.E. (1980) Experimental crystallization of chrome spinel in FAMOUS basalt 527-1-1. *Earth Planet. Sci. Lett.* **48**, 113–123.

Fisk, M.R., Schilling, J.-G. and Sigurdsson, H. (1980) An experimental investigation of Iceland and Reykjanes Ridge tholeiites: 1 Phase relations. *Contrib. Mineral. Petrol.* **74**, 361–374.

Fiske, R.S. and Jackson, E.D. (1972) Orientation and growth of Hawaiian volcanic rifts: the effect of regional structure and gravitational stresses. *Proc. R. Soc. London* **A329**, 299–326.

Fitton, J.G. and James, D. (1986) Basic volcanism associated with intraplate linear features. *Phil. Trans R. Soc. London* **A317**, 253–266.

Flagler, P.A. and Spray, J.G. (1991) Generation of plagiogranite by amphibolite anatexis in oceanic shear zones. *Geology* **19**, 70–73.

Fleet, A.J. and McKelvey, B.C. (1978) Eocene explosive submarine volcanism, Ninetyeast Ridge, Indian Ocean. *Mar. Geol.* **26**, 73–97.

Flovenz, O.G. (1980) Seismic structure of the Icelandic crust above layer three and the relation between body wave velocity and the alteration of the basaltic crust. *J. Geophys.* **47**, 211–220.

Flower, M.F.J. (1973) Evolution of basaltic and differentiated lavas from Anjovan, Comores Archipelago. *Contrib. Mineral. Petrol.* **38**, 1057–1087.

Flower, M.F.J. (1980) Accumulation of calcic plagioclase in oceanic tholeiite: an indication of spreading rat? *Nature, London* **287**, 530–532.

Flower, M.F.J. (1981a) Binary mixing in ocean-ridge spreading segments. *Nature, London* **292**, 45–47.

Flower, M.F.J. (1981b) Thermal and kinematic controls on ocean ridge magma fractionation: contrasts between Atlantic and Pacific spreading axes. *J. Geol. Soc. London* **138**, 695–712.

Flower, M.F.J. (1982) Cryptocumulate tholeiite: an indication of magma mixing at an intermediate rate spreading axis. *Nature, London* **299**, 542–545.

Flower, M.F.J. and Robinson, P.T. (1979) Evolution of the FAMOUS ocean ridge segment: evidence from submarine and deep sea drilling investigations. In: Talwani, M., Harrison, C.G. and Hayes, D.E. (eds) *Deep Drilling Results in the Atlantic Ocean: Ocean Crust, Am. Geophys. Union, Maurice Ewing Ser.* **2**, 314–330.

Flower, M.F.J., Robinson, P.T., Schmincke, H.U. and Ohnmacht, W. (1977) Magma fractionation system beneath the Mid-Atlantic Ridge at 36–37°N. *Contrib. Mineral. Petrol.* **64**, 167–195.

Flower, M.F.J., Schmincke, H.U. and Bowmen, H. (1976) Rare earth and other trace elements in historic Azorean lavas. *J. Volcanol. Geotherm. Res.* **1**, 127–148.

Flower, M.F.J., Schmincke, H.U. and Thompson, R.N. (1975) Phlogopite stability and the $^{87}Sr/^{86}Sr$ step in the basalts along the Reykjanes Ridge. *Nature, London* **254**, 404–405.

Floyd, P.A. (1986) Petrology and geochemistry of oceanic intraplate sheet-flow basalts, Nauru Basin, Deep Sea Drilling Project Leg 89. In: Moberly, R., Schlanger, S.O. *et al.* (eds) *Initial Reports of the Deep Sea Drilling Project* **89**, 471–497.

Floyd, P.A. (1989) Geochemical features of intraplate oceanic plateau basalts. In: Saunders, A.D. and Norry, M.J. (eds) *Magmatism in the Ocean Basins, Geol. Soc. London, Spec. Publ.* **42**, 215–230.

Floyd, P.A. and Tarney, J. (1979) First-order alteration chemistry of Leg 49 basement rocks. In: Luyendyk, B.P., Cann, J.R., *et al.* (eds) *Initial Reports of the Deep Sea Drilling Project* **49**, 693–708.

Floyd, P.A. and Winchester, J.A. (1978) Identification and discrimination of altered and metamorphosed volcanic rocks using immobile elements. *Chem. Geol.* **21**, 291–306.

Fodor, R.V., Husker, J.W. and Kumar, N. (1977) Petrology of oceanic rocks from an aseismic ridge: implications for the origin of the Rio Grande Rise, South Atlantic Ocean. *Earth Planet. Sci. Lett.* **35**, 225–233.

Fornari, D.J. and Campbell, J.F. (1987) Submarine topography around the Hawaiian Islands. *US Geol. Surv. Prof. Pap.* **1350**, 109–124.

Fornari, D.J., Batiza, R. and Luckman, M.A. (1987) Seamount abundances and distribution near the East Pacific Rise 0°–24°N based on Sea Beam data. In: Keating, B.H., Fryer, P., Batiza, R. and Boehlert, G.W. (eds) *Seamounts, Islands and Atoll. Am. Geophys. Union, Geophys. Monogr.* **43**, 13–20.

Fornari, D.J., Gallo, D. G. Edwards, M.H., Madsen, J.A., Perfit, M.R. and Shor, A.N. (1989) Structure and topography of the Siqueiros transform fault system: evidence for the development of intra-transform spreading centres. *Mar. Geophys. Res.* **11**, 263–299.

Fornari, D.J., Garcia, M.O., Tyce, R.C. and Gallo, D.G. (1988c) Morphology and structure of Loihi seamount based on Seabeam sonar mapping. *J. Geophys. Res.* **93**, 15227–15238.

Fornari, D.J., Perfit, M.R., Allan, J.F. and Batiza, R. (1988a) Small-scale heterogeneities in depleted mantle sources: near-ridge seamount lava geochemistry and implications for mid-ocean ridge magmatic processes. *Nature, London* **331**, 551–513.

Fornari, D.J., Perfit, M.R., Allan, R., Batiza, R., Haymon, R., Barone, A., Ryan, W.B.F., Smith, T.,

Simkin, T. and Luckman, M.A. (1988b) Geochemical and structural studies of the Lamont seamounts: seamounts as indicators of mantle processes. *Earth Planet. Sci. Lett.* **89**, 63–83.

Fornari, D.J., Perfit, M.R., Malahoff, A. and Embly, R. (1983) Geochemical studies of abyssal lavas recovered by DSRV Alvin from Eastern Galapagos Rift, Inca Transform, and Ecuador Rift. 1: Major element variations in natural glasses and spatial distribution of lavas. *J. Geophys. Res.* **88**, 10519–10529.

Fornari, D.J., Ryan, W.B.F. and Fox, P.J. (1984) The evolution of craters and calderas on young seamounts: insights from Sea MARC-1 and Seabeam sonar surveys of a small seamount group near the axis of the East Pacific Rise at ~10°N. *J. Geophys. Res.* **89**, 11069–11083.

Foucher, J.-P., Le Pichon, X. and Sibuet, J.-C. (1982) The ocean-continent transition in the uniform lithosphere stretching model: role of partial melting in the mantle. *Phil. Trans. R. Soc. London* **A305**, 27–43.

Foulger, G.R. and Toomey, D.R. (1989) Structure and evolution of the Hengill-Graensdalur volcanic complex, Iceland: geology, geophysics and seismic tomography. *J. Geophys. Res.* **94**, 17511–17522.

Fowler, C.M.R. (1976) Crustal structure of the Mid-Atlantic Ridge crest at 37°N. *Geophys. J. R. Astron. Soc.* **47**, 459–491.

Fowler, C.M.R. and Matthews, D. (1974) Seismic refraction experiments on the Mid-Atlantic Ridge in the FAMOUS area. *Nature, London* **249**, 752–754.

Fox, P.J. and Opdyke, N.D.(1973) Geology of the ocean crust: magnetic properties of oceanic rocks. *J. Geophys. Res.* **78**, 5139–5154.

Fox, P.J., Schreiber, E. and Peterson, J.J. (1973) Geology of the oceanic crust: compressional wave velocities of oceanic rocks. *J. Geophys. Res.* **78**, 5155–5172.

Francheteau, J. and Ballard, R.D. (1983) The East Pacific Rise near 21°N and 13°N, and 20°S: inferences for along-strike variability of axial processes of the mid-ocean ridge. *Earth Planet. Sci. Lett.* **64**, 93–116.

Francis, D. (1986) The pyroxene paradox in MORB glasses—a signature of picritic parental magmas? *Nature, London* **319**, 586–589.

Francis, D.M. (1976) The origin of amphibole in lherzolite xenoliths from Nunivak Island, Alaska. *J. Petrol.* **17**, 357–378.

Francis, R.D. (1980) On the fractionation of sulfur, copper and related transition elements in silicate liquids. Unpublished PhD thesis, University of California, San Diego.

Francis, T.J.G. (1969) Upper mantle structure along the axis of the Mid-Atlantic Ridge near Iceland. *Geophys. J. R. Astron. Soc.* **17**, 502–507.

Francis, T.J.G. (1973). The seismicity of the Reykjanes Ridge. *Earth Planet. Sci. Lett.* **18**, 119–124.

Francis, T.J.G. (1975). A new interpretation of the 1968 Fernandina Caldera collapse and its implications for the Mid-Ocean Ridges. *Geophys. J. R. Astron. Soc.* **39**, 301–318.

Francis, T.J.G. (1981) Serpentinization faults and their role in the tectonics of slow spreading ridges. *J. Geophys. Res.* **86**, 11616–11622.

Frey, F.A. and Clague, D.A. (1983) Geochemistry of diverse basalt types from Loihi Seamount, Hawaii: petrogenetic implications. *Earth Planet. Sci. Lett.* **66**, 337–355.

Frey, F.A. and Green, D.H. (1974) The mineralogy, geochemistry and origin of lherzolite inclusions in Victorian basalts. *Geochim. Cosmochim. Acta* **38**, 1023–1059.

Frey, F.A. and Roden, M.F. (1987) The mantle source for the Hawaiian Islands: constraints from the lavas and ultramafic inclusions. In: Menzies, M. and Hawkesworth, C. (eds) *Mantle Metasomatism*, Academic Press, London, pp. 423–463.

Frey, F.A. and Sung, C.M. (1974) Geochemical results for basalts from Sites 253 and 254. In: Davies, T.A. Luyendyk, B.P. *et al.* (eds) *Initial Reports of the Deep Sea Drilling Project* **26**, 567–572.

Frey, F.A., Bryan, W.B. and Thompson, G. (1974) Atlantic Ocean floor, geochemistry and petrology of basalts from leg 2 and 3 of the Deep Sea Drilling Project. *J. Geophys. Res.* **79**, 5507–5527.

Frey, F.A. Dickey, J.S., Thompson, G., Bryan, W.B. and Davies, H.L. (1980) Evidence for heterogeneous primary MORB and mantle sources, NW Indian Ocean. *Contrib. Mineral. Petrol.* **90**, 18–28.

Frey, F.A., Green, D.H. and Roy, S.D. (1978) Integrated models of basalt petrogenesis: a study of quartz tholeiites to olivine melilitites from south eastern Australia utilizing geochemical and experimental petrological data *J. Petrol.* **19**, 463–513.

Frey, F.A., Haskin, M.A., Poetz, J.A. and Haskin, L.A. (1968) Rare earth abundances in some basic rocks. *J. Geophys. Res.* **73**, 6085–6098.

Fryer, P. (1981) Petrogenesis of basaltic rocks from the Mariana Trough. Unpublished PhD thesis, University of Hawaii, 157 pp.

Fryer, P. and Hussong, D. (1981) Seafloor spreading in the Mariana Trough: results of Leg 60 drill site selection surveys. In: Hussong, D.M., Uyeda, S. *et al.* (eds) *Initial Reports of the Deep Sea Drilling Project* **60**, 45–55.

Fryer, P., Sinton, J.M. and Philpotts, J.A. (1981) Basaltic glasses from the Mariana Trough. In: Hussong, D.M., Uyeda, S. *et al.* (eds) *Initial Reports of the Deep Sea Drilling Project* **60**, 601–609.

Fujii, T. (1989) Genesis of mid-ocean ridge basalts. In: Saunders, A.D. and Norry, M.J. (eds) *Magmatism in the Ocean Basins, Geol. Soc. London, Spec. Publ.* **42**, 137–146.

Fujii, T. and Bougault, H. (1983) Melting relations of a magnesian abyssal tholeiite and the origin of MORBs. *Earth Planet. Sci. Lett.* **62**, 283–295.

Fujii, T. and Kushiro, I. (1977) Melting relations and viscosity of an abyssal tholeiite. *Carnegie Inst. Washington Yearbook* **76**, 461–465.

Fujii, T. and Scarfe, C.M. (1985) Compositions of liquids coexisting with spinel lherzolite at 10 kbar and the genesis of MORBs. *Contrib: Mineral, Petrol.* **90**, 18–28.

Fujiimaki, H., Tatsumoto, M. and Aoki, K. (1984) Partition coefficients of Hf, Zr and REE between phenocrysts and groundmass. *J. Geophys. Res.* **89**, 662–672.

Furuta, T. and Tokuyama, H. (1983) Chromian spinels in Costa Rica Rift basalts, Deep Sea Drilling Project 505—a preliminary interpretation of electron microprobe analyses. In: Cann, J.R., Langseth, M., Honnorez, J., Von Herzen, R.P., White, S.M. *et al.* (eds) *Initial Reports of the Deep Sea Drilling Project* **69**, 805–810.

Galer, S.J.G. and O'Nions, R.K. (1985) Residence time of thorium, uranium and lead in the mantle with implications for mantle convection. *Nature, London* **316**, 778–782.

Galer, S.J.G. and O'Nions, R.K. (1986) Magmagenesis and the mapping of chemical and isotopic variations in the mantle. *Chem. Geol.* **56**, 45–61.

Garcia, M., Liu, N.W.K. and Muenow, D.W. (1979) Volatiles in submarine volcanic rocks from the Mariana island arc and trough. *Geochim. Cosmochim. Acta* **43**, 305–312.

Garcia, M.O. and Wolfe, E.W. (1988) Petrology of the erupted lava. In: Wolfe, E.W. (ed) *The Puu Oo eruption of Kilauea Volcano, Hawaii, episodes 1 through 20, January 3, 1983 through June 8, 1984. US Geol. Survey Prof. Paper* **1463**, 127–143.

Gass, I.G. (1968) Is the Troodos massif of Cyprus a fragment of Mesozoic ocean floor? *Nature, London* **220**, 39–42.

Gass, I.G. (1970) The evolution of volcanism in the junction area of the Red Sea, Gulf of Aden and Ethiopian Rifts. *Phil. Trans. R. Soc., London* **A267**, 369–388.

Gass, I.G. and Smewing, J.D. (1981) Ophiolites: obducted oceanic lithosphere. In: Emiliani, C. (ed) *The Sea*, (Volume 7), *The Oceanic Lithosphere*, Wiley-Interscience, New York, pp. 339–362.

Gass, I.G., Chapman, D.S., Pollack, H.N. and Thorpe, R.S. (1978) Geological and geophysical parameters of mid-plate volcanism. *Phil. Trans. R. Soc. London* **288A**, 581–597.

Gass, I.G., Lippard, S.J. and Shelton, A.W. (eds) 1984. *Ophiolites and Oceanic Lithosphere, Geol. Soc. London Spec. Publ.* **13**, 413pp.

Gast, P.W. (1965) Terrestrial ratio of potassium to rubidium and the composition of Earth's mantle. *Science* **147**, 858–860.

Gast, P.W. (1968) Trace element fractionation and the origin of tholeiitic and alkaline magma types. *Geochim. Cosmochim. Acta* **32**, 1057–1086.

Gee, J., Staudigel, H. and Natland, J.H. (1991). Geology and petrology of Jasper seamount. *J. Geophys. Res.* **96**, 4083–4106.

Gerlach, T.M. and Thomas, D.M. (1986) Carbon and sulphur isotopic composition of Kilauea parental magma. *Nature, London* **319**, 480–483.

Ghiroso, M.S. (1985) Chemical mass transfer in magmatic processes. II: Thermodynamic relations and numerical algorithms. *Contrib. Mineral. Petrol.* **90**, 121–141.

Ghiorso, M.S. and Carmichael, I.S.E. (1985) Chemical mass transfer in magmatic process, II: Applications in equilibrium crystallization, fractionation and assimilation. *Contrib. Mineral. Petrol.* **90**, 121–141.

Gill, J.G. (1976) Composition and age of Lau Basin and Ridge volcanic rocks: implications for the evolution of an interarc basin and remnant arc. *Bull. Geol. Soc. Am.* **87**, 1384–1395.

Gill, J.G. (1981) *Orogenic Andesites and Plate Tectonics*, Springer-Verlag, Berlin, 390pp.

Gillis, K.M. and Robinson, P.T. (1985) Low temperature alteration of the extrusive sequence, Troodos ophiolite, Cyprus. *Can. Mineral.* **23**, 431–441.

Gillis, K.M. and Robinson, P.T. (1988) Distribution of alteration zones in the upper oceanic crust. *Geology* **16**, 262–266.

Gillis, K.M. and Robinson, P.T. (1990) Patterns and processes of alteration in the lavas and dykes of the Troodos ophiolite, Cyprus. *J. Geophys. Res.* **95**, 12523–12548.

Gillis, K.M. and Thompson, G. (1990) Comparison of oceanic metamorphism at mid-ocean ridges and arcs: Mid-Atlantic Ridge versus Troodos ophiolite. *Ophiolite Genesis and Evolution of Oceanic Lithosphere*, UNESCO-Sulan Quaboos University Symposium, Muscat.

Girardeau, J., Mercier, J.C.C. and Wang, X. (1985) Petrology of the mafic rocks of the Xigae ophiolite, Tibet: implications for the genesis of the ocean lithosphere. *Contrib. Mineral. Petrol.* **90**, 309–321.

Graham, A.L. and Nicholls, G.D. (1969) Mass spectrographic determination of lanthanide element contents in basalts. *Geochim. Cosmochim. Acta* **33**, 555–568.

Graham, D.M. (1987) Deep seabed mapping/survey systems. *Sea Technol. November*, 29–33.

Graham, D.M., Zindler, A., Kurz, M.D., Jenkins, W.J., Batiza, R. and Staudigel, H. (1988) Helium, lead, strontium and neodymium isotope constraints on magma genesis and mantle heterogeneity beneath young Pacific seamounts. *Contrib. Mineral. Petrol.* **99**, 446–463.

Grand, S.P. (1987) Tomographic inversion for shear velocity beneath the North American plate. *J. Geophys. Res.* **92**, 14065–14090.

Green, D.H. (1972) Magmatic activity as the major process in the chemical evolution of the Earth's crust and mantle. *Tectonophysics* **13**, 47–71.

Green, D.H. and Ringwood, A.E. (1967) The stability fields of aluminous pyroxene peridotite and their relevance to upper mantle structure. *Earth Planet. Sci. Lett.* **3**, 151–160.

Green, D.H., Falloon, T.J. and Taylor, W.R. (1987) Mantle-derived magmas—roles of variable source peridotite and variable C-M-O fluid compositions. In: Mysen, B.O. (ed) *Magmatic Processes: Physicochemical Principles, Geochem. Soc. Spec. Publ.* **1**, pp. 139–154.

Green, D.H., Hibberson, W.O. and Jaques, A.L. (1979) Petrogenesis of mid-ocean ridge basalts. In: McElhinney, H.W. (ed) *The Earth: Its Origin, Structure and Evolution*, Academic Press, New York, pp. 265–299.

Green, T.H. and Pearson, N.J. (1986) Ti-rich accessory phase saturation in hydrous mafic-felsic compositions at high P, T. *Chem. Geol.*, **54**, 185–201.

Greenbaum, D. (1972a) The geology and evolution of the Troodos plutonic complex and associated chromite deposits, Cyprus. Unpublished PhD thesis, University of Leeds, 142pp.

Greenbaum, D. (1972b) Magmatic processes at ocean ridges: evidence from the Troodos Massif, Cyprus. *Nature, London* **238**, 18–21.

Griffiths, R.W. (1986) The differing effect of compositional and thermal buoyancies on the evolution of mantle diapirs. *Phys. Earth Planet. Int.* **33**, 304–317.

Griffiths, R.W. and Campbell, I.H. (1990) Stirring and structure in mantle starting plumes. *Earth Planet. Sci. Lett.* **99**, 66–78.

Grinenko, V.A., Dimitriev, L.V., Migdisov, A.A. and Sharasikin, A.Y. (1975) Sulphur contents and isotopic compositions from igneous and metamorphic rocks from Mid-Ocean Ridges. *Geochemistry* **12**, 132–137.

Grove, T.L. (1981) Use of FePt alloys to eliminate the iron loss problem in 1 atm. gas mixing experiments: theoretical and practical considerations. *Contrib. Mineral. Petrol.* **78**, 298–304.

Grove, T.L. and Bryan, W.B. (1983) Fractionation of pyroxene-phyric MORB at low pressure: an experimental study. *Contrib. Mineral. Petrol.* **84**, 293–309.

Grove, T.L. and Bryan, W.B. (1984) Low pressure fractional crystallization of MORB and the significance of composition vs. frequency diagrams. In: Dungan, M.A., Grove, T.L. and Hildreth, W. (eds) *Open Magmatic systems, Conference Proceedings*, 60–62.

Grove, T.L., Gerlach, D.C. and Sando, T.W. (1982) Origin of calc-alkaline series lavas at Medicine Lake Volcano by fractionation, assimilation and mixing. *Contrib. Mineral Petrol.* **80**, 160–182.

Guillon, J-H. (1975) Les massifs peridotitiques de Nouvelle-Caledonie. Type d'Appareil ultrabasique stratiforme de chaine recente. *Memoires ORSTOM* **76**, 120pp.

Haggerty, J.A., Schlanger, S.O. and Premoli Silva, I. (1982) Late Cretaceous and Eocene volcanism in the southern Line islands and implications for hotspot theory. *Geology* **10**, 433–437.

Hajash, A. (1975) Hydrothermal processes along mid-ocean ridges: an experimental investigation. *Contrib. Mineral. Petrol.* **53**, 205–226.

Hall, M. (1876) Note upon a portion of basalt from Mid-Atlantic. *Mineral. Mag.* **1**, 1–3.

Hemelin, B., Dupre, B. and Allegre, C.J. (1984) Lead-strontium isotopic variation along the East Pacific Rise and the Mid-Atlantic Ridge. *Earth Planet. Sci. Lett.* **67**, 340–350.

Hamlyn, P.R. and Bonatti, E. (1980) Petrology of mantle-derived ultramafics from the Owen Fracture Zone, Northwest Indian Ocean: implications for the nature of the oceanic upper mantle. *Earth Planet. Sci. Lett.* **48**, 65–79.

Hanan, B.B. and Schilling, J.G. (1989) Easter microplate evolution. *J. Geophys. Res.* **94**, 7432–7448.

Hanson, G.N. (1977) Geochemical evolution of the suboceanic mantle. *J. Geol. Soc. London* **134**, 235–253.

Hanson, G.N. (1989) An approach to trace element modelling using a simple igneous system as an example. In: Lipin, B.R. and Mckay, G.A. (eds); *Geochemistry and Mineralogy of Rare Earth Elements, Rev. Mineral.* **21**, pp. 79–98.

Harding, A.J., Orcutt, J.A., Kappus, M.E., Vera, E.E., Mutter, J.C., Buhl, P., Detrick, R.S. and Brocher, T.M. (1989) The structure of young oceanic crust at 13°N on the East Pacific Rise from expanding spread profiles. *J. Geophys. Res.* **94**, 12163–12186.

Harmon, R.S. and Hoefs, J. (1986) S-isotopic relationships in late Cenozoic destructive plate margin and continental intraplate volcanic rocks. *Terra Cognita* **6**, 182.

Harper, G.D. and Bowman, J.R. and Kuhns, R. (1988) A field chemical and stable isotope study of subseafloor metamorphism of the Josephine ophiolite, California-Oregon. *J. Geophys. Res.* **93**, 4625–4656.

Harrison, C.G.A. and Bonatti, E. (1981) The oceanic lithosphere. In: Emiliani, C. (ed) *The Sea*, (vol. 7), *The Oceanic Lithosphere*, Wiley-Interscience, New York, pp. 21–48.

Harrison, J.C. and Brisbin, W.C. (1959) Gravity anomalies off the west coast of North America, 1: Seamount Jasper *Bull. Geol. Soc. Am.* **70**, 929–934.

Harrison, T.M. and Watson, E.B. (1984) The behaviour of apatite during crystal anatexis: equilibrium and kinetic considerations. *Geochim. Cosmochim. Acta* **48**, 1467–1477.

Hart, R.A. (1970) Chemical exchange between sea-water and deep ocean basalts. *Earth Planet. Sci. Lett.* **9**, 269–279.

Hart, R.A. (1973) A model for chemical exchange in the basalt–seawater system of oceanic layer II. *Can. J. Earth Sci.* **10**, 799–816.

Hart, R.A. (1976) Progressive alteration of the oceanic crust. In: Yeats, R.S., Hart, S.R., *et al.* *Initial Reports of the Deep Sea Drilling Project* **34**, 433–437.

Hart, S.R. (1969) K, Rb, Cs contents and K/Rb, K/Cs ratios of fresh and altered submarine basalts. *Earth Planet. Sci. Lett.* **6**, 295–303.

Hart, S.R. (1971) K, Rb, Cs, Sr and Ba contents and Sr isotope ratios of ocean floor basalts. *Phil. Trans. R. Soc. London* **A268**, 573–587.

Hart, S.R. (1976) LIL-element geochemistry, Leg 34 basalts. In: Yeats, R.S., Hart, S.R. *et al.*, (eds) *Initial Reports of the Deep Sea Drilling Project* **34**, 283–288.

Hart, S.R. (1984) A large-scale isotope anomaly in the southern hemisphere mantle, *Nature, London* **309**, 753–757.

Hart, S.R. (1988) Heterogeneous mantle domains: signatures, genesis and mixing chronologies. *Earth Planet. Sci. Lett.* **90**, 273–296.

Hart. S.R. and Davis, K.E. (1978) Nickel partitioning between olivine and silicate melt. *Earth Planet. Sci. Lett.* **40**, 203–219.

Hart, S.R. and Nalwalk, A.M. (1970) K, Rb, Cs and Sr relationships in submarine basalts from the Puerto Rico Trench. *Geochim. Cosmochim. Acta* **34**, 145–156.

Hart, S.R. and Schilling, J.-G. (1973) The geochemistry of basalts from Iceland and the Reykjanes Ridge, *Carnegie Inst., Washington Yearbook* **72**, 259–262.

Hart, S.R. and Staudigel, H. (1978) Oceanic crust: age of hydrothermal alteration. *Geophys. Res. Lett.* **5**, 1009–1012.

Hart, S.R. and Staudigel, H. (1982) The control of alkalis and uranium in sea water by ocean crust alteration. *Earth Planet. Sci. Lett.* **58**, 202–212.

Hart, S.R. and Staudigel, H. (1989) Isotopic characterization and identification of recycled components. In: Hart, S.R. and Gulen, L. (eds) *Crust–Mantle Recycling at Convergent Zones*, D. Kluwer Academy, Dordrecht, Netherlands, pp.15–28.

Hart, S.R., Dymond, J., Hogan, L. and Schilling, J.G. (1983) Mantle plume noble gas component in glassy basalts from Reykjanes Ridge. *Nature, London* **305**, 403–407.

Hart, S.R., Glassley, W.E. and Karig, D.E. (1972) Basalts and seafloor spreading behind the Mariana Island arc. *Earth Planet. Sci. Lett.* **15**, 12–18.

Hart, S.R., Schilling, J.-G. and Powell, J.L. (1973) Basalts from Iceland and along the Reykjanes Ridge: Sr isotope geochemistry. *Nature* (*Phys. Sci.*) (*London*) **246**, 104–107.

Haseby, K., Fujii, N. and Uyeda, S. (1970) Thermal processes under island arcs. *Tectonophysics* **10**, 335–355.

Hawkesworth, C.J., Norry, M.J., Roddick, J.C. and Vollmer, R. (1979) $^{143}Nd/^{144}Nd$ and $^{87}Sr/^{86}Sr$ ratios from the Azores and their significance in the LIL-element enriched mantle. *Nature, London* **280**, 28–31.

Hawkesworth, C.J., O'Nions, R.K., Pankhurst, R.J. and Evensen, N.M. (1977) A geochemical study of island-arc and back-arc theoleiites from the Scotia Sea. *Earth Planet. Sci. Lett.* **36**, 253–262.

Hawkesworth, C.J., Rogers, N.W., van Calstern, P.W.C. and Menzies, M.A. (1984) Mantle enrichment processes. *Nature, London* **311**, 331–335.

Hawkesworth, C.J., van Calsteren, P., Rogers, N.W. and Menzies, M.A. (1987) Isotope variations in Recent volcanics: a trace element perspective. In: Menzies, M.A. and Hawkesworth, C.J. (eds) *Mantle metasomatism*, Academic Press, London, 365–388.

Hawkins, J.W. (1976) Petrology and geochemistry of basaltic rocks of the Lau Basin. *Earth Planet. Sci. Lett.* **28**, 283–297.

Hawkins, J.W. (1977) Petrological and geochemical characteristics of marginal basin basalts: In: Talwani, M. and Pitman, W.C. (eds) *Island Arcs, Deep Sea Trenches and Back-Arc Basins, Am. Geophys. Union, Maurice Ewing Ser.* **1**, 367–377.

Hawkins, J.W. (1980) Petrology of back-arc basins and island arcs: their possible role in the origin of ophiolites. In: Panayiotou, A. (ed) *Ophiolites—Proceedings of the International Ophiolite Symposium*, Cyprus, 1979, pp. 244–254.

Hawkins, J.W. and Melchior, J.T. (1983) Petrology of basalts from Loihi Seamount, Hawaii. *Earth Planet. Sci. Lett.* **66**, 356–368.

Hawkins, J.W. and Melchior, J.T. (1985) Petrology of Mariana Trough and Lau Basin basalts. *J. Geophys. Res.* **90**, 11431–11468.

Hawkins, J.W., Lonsdale, P.F. and Batiza, R. (1987) Petrologic evolution of the Louisville Seamount chain. In: Keating, B.H., Fryer, P., Batiza, R. and Boehlert, G.W. (eds) *Seamounts, Islands and Atolls, Am. Geophys. Union, Geophys. Monogr.* **43**, 235–254.

Haxby, W.F. and Weissel, J.K. (1986) Evidence for small scale convection from Seasat altimeter data. *J. Geophys. Res.* **91**, 3507–3520.

Hay, R.L. and Iijima, A. (1968a) Petrology of palagonite tuffs of Koko Craters, Oahu, Hawaii. *Contrib. Mineral. Petrol.* **17**, 141–154.

Hay, R.L. and Iijima, A. (1968b) Nature and origin of palagonite tuffs of the Honolulu Group on Oahu, Hawaii, *Mem. Geol. Soc. Am.* **116**, 331–376.

Hekinian, R. (1968) Rocks from the mid-ocean ridge in the Indian Ocean. *Deep-Sea Res.* **15**, 195–213.

Hekinian, R. (1971) Chemical and mineralogical differences between abyssal hill basalts and ridge tholeiites in the Eastern Pacific Ocean. *Mar. Geol.* **11**, 77–91.

Hekinian, R. (1974) Petrology of the Ninetyeast Ridge (Indian Ocean) compared to other aseismic ridges. *Contrib. Mineral. Petrol.* **43**, 125–147.

Hekinian, R. (1982) *Petrology of the Ocean Floor*, Elsevier, Amsterdam, 393pp.

Hekinian, R. and Aumento, F. (1973) Rocks from the Gibbs Fracture Zone and the Minia Seamount near 53°N in the Atlantic Ocean. *Mar. Geol.* **14**, 47–72.

Hekinian, R. and Thompson, G. (1976) Comparative geochemistry of volcanics from rift valleys, transform faults and aseismic ridges. *Contrib. Mineral. Petrol.* **57**, 145–162.

Hekinian, R., Moore, J.G. and Bryan, W.B. (1976) Volcanic rocks and processes of the Mid-Atlantic Ridge rift valley near 36°49′N. *Contrib. Mineral. Petrol.* **58**, 83–110.

Hekinian, R., Thompson, G. and Bideau, D. (1989) Axial and off-axis heterogeneity of basaltic rocks from the East Pacific Rise at 12°35′N′12°51′N and 11°26′–11°30′N. *J. Geophys. Res.* **94**, 17437–17464.

Helgason, J. (1984) Frequent shifts of the volcanic zone in Iceland. *Geology* **12**, 212–216.

Helgason, J. (1985) Shifts of the plate boundary in Iceland: some aspects of Tertiary volcanism. *J. Geophys. Res.* **90**, 10084–10092.

Helgason. J. (1989) The Fjallgardar volcanic ridge in NE Iceland: an aborted early stage plate boundary or a volcanically dormant zone. In: Saunders, A.D. and Norry, M.J. (eds) *Magmatism in the Ocean Basins, Geol. Soc. London, Spec. Publ.* **42**, pp. 201–213.

Hellman, P.L. and Green, T.H. (1979) The role of sphene as an accessory phase in the high-pressure partial melting of hydrous mafic compositions. *Earth Planet. Sci. Lett.* **42**, 191–201.

Helz, R.T. (1973) Phase relations of basalts in their melting at $P_{H_2O} = 5\,Kb$. Part II: Melt compositions. *J. Petrol.* **17**, 139–193.

Hermance, J.F. and Grillot, L.R. (1970) Correlation of magnetotelluric, seismic and temperature data from southwest Iceland. *J. Geophys. Res.* **75**, 6582–6591.

Hermond, C., Conomines, M., Fourcade, S., Allegre, C.J., Oskarsson, N. and Javoy, M. (1988) Thorium, strontium and oxygen isotopic geochemistry in recent tholeiites from Iceland: crustal influences on mantle-derived magmas. *Earth Planet. Sci. Lett.* **87**, 273–285.

Herrmann, A.G. (1970) Yttrium and lanthanides. In: Wedepohl, K.H. (ed) *Handbook of Geochemistry*, II-5, Chapter 39/57-71-C, C1–C10.

Herron, T.J., Purdy, G.M. and Pomeroy, P. (1979) Preliminary studies of long range low frequency propagation across the Oroczo Fracture Zone and East Pacific Rise using data collected during the ROSE project. *Trans. Am. Geophys. Union, EOS* **60**, 888.

Hess, H.H. (1955) Serpentinites, orogeny and epeirogeny. *Geol. Soc. Am. Spec. Pap.* **62**, 391–408.

Hess, H.H. (1962) History of ocean basins. In: Engel, A.E.J., James, H.L. and Leonard, B.F. (eds), *Petrological Studies: a Volume in Honor of A.F. Buddington*, Geological Society of America, Boulder, Colorado, pp. 599–620.

Hess, H.H. and Otalara, G. (1964) Mineralogical and chemical composition of the Mayaguez serpentinite cores. In: Burk, C.A. (ed), *Natl. Acad. Sic./Natl. Res. Counc. Publ. 1188*.

Hey, R.N. (1977a) A new class of pseudofaults and their bearing on plate tectonics: a propagating rift model. *Earth Planet. Sci. Lett.* **37**, 321–325.

Hey, R.N. (1977b) Tectonic evolution of the Cocos-Nazca spreading center. *Bull. Geol. Soc. Am.* **88**, 1404–1420.

Hey, R.N. and Wilson, D.S. (1982) Propagating rift explanation for the tectonic evolution of the northeast Pacific—the pseudo movie. *Earth Planet. Sci. Lett.* **58**, 167–188.

Hey, R.N., Kleinrock, M.C., Miller, S.P., Atwater, T.M. and Searle, R.C. (1986) Sea Beam/deep-tow investigation of an active oceanic propagating rift system, Galapagos, 95.5°W. *J. Geophys. Res.* **91**, 3369–3393.

Hey, R.N., Naar, D.F., Kleinrock, M.C., Phipps-Morgan, W.J., Morales, E. and Schilling, J.G. (1985) Microplate tectonics along a superfast seafloor spreading system near Easter Island. *Nature, London* **317**, 320–325.

Hey, R.N., Sinton, J.M. and Duennebier, F.K. (1989) Propagating rifts and spreading centers. In: Winterer, E.L., Hussong, D.M. and Decker, R.W. (eds), *The Geology of North America*, vol. N: *The Easter Pacific Ocean and Hawaii*, Geological Society of America, Boulder, Colorado 127–139.

Hill, R. and Roeder, P. (1974) The crystallization of spinel from basaltic liquid as a function of oxygen fugacity. *J. Geol.* **82**, 709–729.

Himmelberg, G.R. and Loney, R.A. (1973) Petrology of the Vulcan Peak alpine-type peridotite, southwestern Oregon. *Bull. Geol. Soc. Am.* **84**, 1585–1600.

Hodges, F.N. and Papike, J.J. (1976) DSDP Site 334: magmatic cumulates from oceanic layer 3. *J. Geophys. Res.* **81**, 4135–4151.

Hoefs, J. (1973) *Stable Isotope Geochemistry*, Springer-Verlag, Berlin, 145pp.

Hofmann, A.W. (1986) Nb in Hawaiian magma: constraints on source composition and evolution. *Chem. Geol.* **57**, 17–30.

Hofmann, A.W. and Hart, S.R. (1978) An assessment of local and regional isotopic equilibrium in the mantle. *Earth Planet. Sci. Lett.* **38**, 44–62.

Hofmann, A.W. and White, W.M. (1982) Mantle plumes from ancient oceanic crust. *Earth Planet. Sci. Lett.* **57**, 421–436.

Hofmann, A.W., Feigenson, M.D. and Raczek, I. (1984) Case studies on the origin of basalt. III: Petrogenesis of the Mauna Ulu eruption, Kilauea, 1969-1971. *Contrib. Mineral. Petrol.* **88**, 24–35.

Hofmann, A.W., Jochum, K.P., Senfert, M. and White, W.M. (1986) Nb and Pb in oceanic basalts: new constraints on mantle evolution. *Earth Planet. Sci. Lett.* **79**, 33–45.

Hole, M.J., Saunders, A.D., Marriner, G.F. and Tarney, J. (1984) Subduction of pelagic sediment: implications for the origin of Ce-anomalous basalts from the Mariana islands. *J. Geol. Soc. London* **141**, 453–472.

Holland, H.D. (1978) *The Chemistry of the Atmosphere and Oceans*, Wiley, New York, 351pp.

Hollister, C.D., Glenn, M.F. and Lonsdale, P.F. (1978) Morphology of seamounts in the western Pacific and Philippine Basin from multi-beam sonar data. *Earth Planet. Sci. Lett.* **41**, 405–418.

Honnorez, J. (1967) La palagonitization: l'alteration sous-marine du verre volcanique basique de Palagonia (Sicile). Unpublished PhD thesis, University of Bruxelles.

Honnorez, J. (1972) La palagonitization: l'alteration sous-marine du verre volcanique basique de Palagonia (Sicile). *Vulkaninstitut Immanuel Friedlander, Zurich* **9**, 1–132.

Honnorez, J. (1981) The aging of the oceanic crust at low temperature. In: Emiliani, E (ed) *The Sea* (Vol. 7), *The Oceanic Lithosphere*, pp. 525–587.

Honnorez, J., Bohlke, J.K. and Honnorez-Guerstein, B.M. (1978) Petrographical and geochemical study of the low temperature submarine alteration of basalt from Hole 396B, Leg 46. In: Dmitriev, L., Heirtzler, J., *et al.* (eds) *Initial Reports of the Deep Drilling Project* **46**, 299–329.

Honnorez, J., Emmermann, R., Hubberton, H.W., Laverne, C. and Muehlenbachs, K. (1983) Alteration processes in layer 2 basalts, DSDP hole 504B, Costa Rica Rift. In: Cann, J.R., Langseth, M.G., Honnorez, J., von Herzen, R.P., White, S.M., *et al.* (eds) *Initial Reports of the Deep Sea Drilling Project* **69**, 509–546.

Hooper, P.R. (1990) The timing of crustal extension and the eruption of continental flood basalts. *Nature, London* **345**, 246–249.

Hopson, C.A. and Franno, C.J. (1977) Igneous history of the Point Sal ophiolite, Southern California. *Bull. State Oregon Depart. Mines Ind.* **95**, 161–183.

Hopson, C.A., Coleman, R.G., Gregory, R.T., Pallister, J.S. and Bailey, E.H. (1981) Geologic section through the Samail ophiolite and associated rocks, southeastern Oman. *J. Geophys. Res.* **86**, 2527–2540.

Houseman, G.A. (1983) The deep structure of ocean ridges in a convecting mantle. *Earth Planet. Sci. Lett.* **64**, 283–294.

Huebner, J.S. (1971) Buffering techniques for hydrostatic systems at elevated pressures. In: Ulmer, G.C. (ed) *Research Techniques for High Pressure and High Temperature*, Springer-Verlag, New York, pp. 123–177.

Huebner, J.S. (1987) Use of gas mixture at low pressure to specify oxygen and other fugacities of furnace temperatures. In: Ulmer, G.C. and Barnes, H.L. (eds) *Hydrothermal Experimental Techniques*, Wiley, New York, pp. 20–60.

Humphris, S.E. and Thompson, G. (1978a) Hydrothermal alteration of oceanic basalts by seawater. *Geochim. Cosmochim. Acta* **42**, 107–125.

Humphris, S.E. and Thompson, G. (1978b) Trace element mobility during hydrothermal alteration of oceanic basalts. *Geochim. Cosmochim. Acta* **42**, 127–136.

Humphris, S.E. and Thompson, G. (1983) Geochemistry of rare earth elements in basalts from the Walvis Ridge: implications for its origin and evolution. *Earth Planet. Sci. Lett.* **66**, 223–242.

Humphris, S.E., Thompson, G., Schilling, J.G. and Kingsley, R.H. (1985) Petrological and geochemical variations along the Mid-Atlantic Ridge between 46°S and 32°S: influence of the Tristan da Cunha mantle plume. *Geochim. Cosmochim. Acta* **49**, 1445–1464.

Huppert, H.E. and Sparks, R.S.J. (1980) Restrictions on the compositions of mid-ocean ridge basalts: a fluid dynamical investigation. *Nature, London* **286**, 46–48.

Huppert, H.E. and Sparks, R.S.J. (1984) Double-diffusive convection due to crystallization in magmas. *Ann. Rev. Earth Planet. Sci.* **12**, 11–37.

Hussong, D.M. and Uyeda, S. (1981) Tectonic processes and the history of the Mariana arc: a synthesis of the results of Deep Sea Drilling Project Leg 60. In: Hussong, D.M., Uyeda, S. *et al.*, (eds) *Initial Reports of the Deep Sea Drilling Project* **60**, 909–929.

Hutchison, R. (1974) The formation of the Earth. *Nature, London* **250**, 556–558.

Hutchison, R. and Gass, I.G. (1971) Mafic and ultramafic inclusions associated with undersaturated basalt on Kod Ali island, southern Red Sea. *Contrib. Mineral. Petrol.* **31**, 94–101.

Irvine, T.N. (1977) Definition of primitive liquid compositions for basic magmas. *Carnegie Institute of Washington Yearbook*, **76**, 454–461.

Irving, A.J. and Frey, F.A. (1984) Trace element abundances in megacrysts and their host basalts: constraints on partition coefficients and megacryst genesis. *Geochim. Cosmochim. Acta* **48**, 1201–1221.

Irving, I., Robertson, W.A. and Aumento, F. (1970) The Mid-Atlantic Ridge near 45°N, VI: Remanent intensity, susceptibility and iron content of dredged samples. *Can. J. Earth Sci.* **7**, 226–238.

Irvine, T.N. (1979) Rocks whose composition is determined by crystal accumulation and sorting. In: Yoder, H.S. (ed) *The Evolution of the Igneous Rocks: 50th Anniversary Perspective*, Princeton University Press, Princeton, pp. 245–306.

Ito, E. and Anderson, A.T. (1983) Submarine metamorphism of gabbros from the Mid-Cayman Rise: petrographic and mineralogic constraints on hydrothermal processes at slow spreading ridges. *Contrib. Mineral. Petrol.* **82**, 371–388.

Ito, E. and Stern, R.J. (1981) Mariana Trough basalt: O, Nd and Sr isotope analyses. *Carnegie Inst. Washington Yearbook* **80**, 449–455.

Ito, E., White, W.M. and Gopel, C. (1987) The O, Sr, Nd, and Pb isotope geochemistry of MORB. *Chem. Geol.* **62**, 157–176.

Iyer, H.M. (1984) Geophysical evidence for the locations, shapes and sizes, and internal structures of magma chambers beneath regions of Quaternary volcanism. *Phil. Trans. R. Soc. London* **A310**, 473–510.

Jackson, E.D. and Wright, T.L. (1970) Xenoliths in the Honolulu Volcanic Series, Hawaii. *J. Petrol.* **11**, 405–430.

Jackson, E.D., Bargar, K.E., Fabbi, B.P. and Heropoulos, C. (1976) Petrology of the basaltic rocks drilled on Leg 33 of the Deep Sea Drilling Project. In: Schlanger, S.O, Jackson, E.D. *et al.* (eds) *Initial Reports of the Deep Sea Drilling Project* **33**, 571–630.

Jackson, E.D., Silver, E.A. and Dalrymple, G.B. (1972) Hawaiian-Emperor Chain and its relation to Cenozoic circum-Pacific tectonics. *Bull. Geol. Soc. Am.* **83**, 601–618.

Jackson, E.D., Swanson, D.A., Koyanagi, R.Y. and Wright, T.L. (1975) The August and October 1968 east rift eruptions of Kilauea Volcano, Hawaii, *US Geol. Surv. Prof. Pap.* **890**, 33 pp.

Jagoutz, E., Palme, H., Baddenhausen, H., Blum, K., Cendales, M., Dreibus, G., Spettel, B., Lorenz, V. and Wanke, H. (1979) The abundance of major, minor and trace elements in the earth's mantle as derived from primitive ultramafic nodules. *Proceedings of the 10th Lunar and Planetary Science Conference* 2031–2050.

Jakobsson, S.P. (1972a) On the consolidation and palagonization of the tephra of Surtsey volcanic island, Iceland. *Reykjavik Pap. Mus. Nat. Hist.* **60**, 1–8.

Jakobsson, S.P. (1972b) Chemistry and distribution pattern of recent basaltic rocks in Iceland. *Lithos* **5**, 365–386.

Jakobsson, S.P. (1978) Environment factors controlling the palagonization of the Surtsey tephra, Iceland. *Bull. Geol. Soc. Denmark* **27**, 91–105.

Jakobsson, S.P. (1979a) Outline of the petrology of Iceland. *Jokull* **29**, 57–73.

Jakobsson, S.P. (1979b) Petrology of Recent basalts of the Eastern Volcanic Zone, Iceland. *Acta Natur. Islandica* **26**, 103.

Jakobsson, S.P., Jonsson, J. and Shido, F. (1978) Petrology of the Reykjanes Peninsula, Iceland, *J. Petrol.* **19**, 669–705.

James, D.E. (1981) The combined use of oxygen and radiogenic isotopes as indicators of crustal contamination. *Ann. Rev. Earth. Planet. Sci.* **9**, 311–344.

Jancin, M., Young, K.D., Voight, B., Aronson, J.L. and Saemundsson, K. (1985) Stratigraphy and K/Ar ages across the west flank of the northeast Iceland axial rift zone, in relation to the 7 Ma volcanotectonic reorganization of Iceland. *J. Geophys. Res.* **90**, 9961–9985.

Jaques, A.L. and Chappell, B.W. (1980) Petrology and trace element geochemistry of the Papuan ultramafic belt. *Contrib. Mineral. Petrol.* **75**, 55–70.

Jaques, A.L. and Green, D.H. (1979) Determination of liquid compositions in high-pressure melting of peridotite. *Am. Mineral.* **64**, 1312–1321.

Jaques, A.L. and Green, D.H. (1980) Anhydrous melting of peridotite at 0–15 Kb pressure and the genesis of tholeiitic basalts. *Contrib. Mineral. Petrol.* **73**, 287–310.

Javoy, M., Pineau, F. and Demaiffe, D. (1984) Nitrogen and carbon isotopic compositions in the diamonds of Mbuji Mayi (Zaire). *Earth Planet. Sci. Lett.* **88**, 399–412.

Johannesson, H. (1980) Stratigraphy and evolution of rift zones, western Iceland *Natturufraedingurinn*, **50**, 13–31 (*in Icelandic*).

Johnson, G.L., Vogt, P.R. and Avery, O.E. (1971). Evolution of the Norwegian Basin. In: Delany,

F.M. (ed) *Geology of the East Continental Margin* (Vol. 2) *Europe. Rep. Inst. Geol. Sci.* **70/14**, 53.

Johnson, H.P., Karsten, J.L., Delaney, J.R., Davis, E.E., Currie, R.G. and Chase, R.L. (1983) A detailed study of the Cobb offset of the Juan de Fuca Ridge: evolution of a propagating rift. *J. Geophys. Res.* **88**, 2297–2315.

Johnson, K.T.M. and Sinton, J.M. (in press) Petrology, tectonic setting and the formation of back-arc basalts in the northern Fiji Basin. *Geol. Jahrb.*

Johnson, K.T.M., Dick, H.J.B. and Shimizu, N. (1990) Melting in the oceanic upper mantle: an ion microprobe study of diopsides in abyssal peridotites. *J. Geophys. Res.* **95**, 2661–2678.

Jordan, T.H., Menard, H.W. and Smith, D.K. (1983) Density and size distribution of seamounts in the eastern Pacific inferred from wide-beam sounding data. *J. Geophys. Res.* **88**, 10508–10518.

Jordan, T.N. (1979) Mineralogies, densities and seismic velocities of garnet lherzolites and their geophysical implications. In: Boyd, F.R. and Meyer, H.O.A. (eds) *The Mantle Sample: Inclusions in Kimberlites and other Volcanics, Proceedings of the 2nd International Kimberlite Conference, 2*: Washington (Amer. Geophys. Union), 1–14.

Joron, T.L., Bollinger, C. and Quisefit, J.P. (1980) Trace elements in Cretaceous basalts at 25°N in the Atlantic ocean: alteration, mantle composition and magmatic processes. In: Donnelly, T., Francheteau, J., Bryan, W., Robinson, P.T., Flower, M., Salisbury, M. *et al.* (eds) *Initial Reports of the Deep Sea Drilling Project* **51, 52, 53**, 1087–1098.

Jurdy, D.M. (1979) Relative plate motions and the formation of marginal basins. *J. Geophys. Res.* **84**, 6796–6802.

Jurdy, D.M. and Stefanick, M. (1983) Flow models for back-arc spreading. *Tectonophysics* **99**, 191–200.

Juster, T.C., Grove, T.L. and Perfit, M.R. (1989) Experimental constraints on the generation of FeTi basalts, andesites and rhyodacites at the Galapagos Spreading Center, 85°W and 95°W. *J. Geophys. Res.* **94**, 9251–9274.

Kappel, E.S. and Normark, W.R. (1987) Morphometric variability within the axial zone of the southern Juan de Fuca Ridge: interpretation from SeaMarc II, SeaMarc I and deep-sea photography. *J. Geophys. Res.* **11**, 11291–11302.

Karig, D.E. (1971) Origin and development of marginal basins in the Western Pacific. *J. Geophys. Res.* **76**, 2542–2561.

Karig, D.E. (1974) Evolution of arc systems in the Western Pacific. *Ann. Rev. Earth Planet. Sci.* **2**, 51–78.

Karig, D.E. and Kay, R.W. (1981) Fate of sediments on the descending plate of convergent margins. *Phil. Trans. R. Soc. London* **A301**, 233–251.

Karig, D.E., Anderson, R.N. and Bibee, L.D. (1978) Characteristics of back-arc spreading in the Mariana Trough. *J. Geophys. Res.* **83**, 1213–1226.

Karsten, J.L. (1988) Spatial and temporal variations in the petrology, morphology and tectonics of a migrating spreading center: the Endeavour segment, Juan de Fuca Ridge. PhD dissertation, University of Washington, Seattle, 329 pp.

Karsten, J.L. and Delaney, J.R. (1989) Hot spot-ridge crest convergence in the Northeast Pacific. *J. Geophys. Res.* **94**, 700–712.

Karsten, J.L., Delaney, J.R., Rhodes, J.M. and Liias, K.A. (1990) Spatial and temporal evolution of a magmatic system beneath the Endeavour segment, Juan de Fuca ridge: 2. Petrologic *constraints. J. Geophy. Res.* **95**, 19235–19256.

Kastens, K.A., Ryan, W.B.F. and Fox, P.J. (1986) The structural and volcanic expression of a fast-slipping ridge-transform-ridge plate boundary: Sea MARC I photographic surveys at the Clipperton Transform Fault. *J. Geophys. Res.* **91**, 3469–3488.

Kaula, W.M. (1973) Global gravity and tectonics. In: Cox, A. (ed) *Plate tectonics and geomagnetic reversals.* Freeman and Co., San Francisco, 500–510.

Kay, R.W. (1980) Volcanic arc magmas: implications of a melting-mixing model for element recycling in the crust-upper mantle system. *J. Geol.* **88**, 497–522.

Kay, R.W. (1984) Elemental abundances relevant to identification of magma sources. *Phil. Trans. R. Soc. London* **A310**, 535–547.

Kay, R.W. and Gast, P.W. (1973) The rare earth content and origin of alkali-rich basalts. *J. Geol.* **81**, 653–682.

Kay, R.W., Hubbard, N.J. and Gast, P.W. (1970) Chemical characteristics and origin of ocean ridge volcanic rocks. *J. Geophys. Res.* **75**, 1583–1613.

Keating, B.H., Fryer, P., Batiza, R. and Boehlert, G.W. (eds) (1987). *Seamounts, Islands and Atolls,*

Am. Geophys. Union, Geophys. Monog. **43**, Washington, DC, 405 pp.

Kempe, D.R.C. (1974) The petrology of the basalts, Leg 26. In: Davies, T.A., Luyendyk, B.P. *et al.* (eds) *Initial Reports of the Deep Sea Drilling Project* **26**, 465–503.

Kempton, P.D. (1987) Metasomatism and enrichment in the lithosphere. In: Menzies, M.A. and Hawkesworth, C.J. (eds) *Mantle Metasomatism*, Academic Press, New York, pp. 45–90.

Kempton, P.D., Autio, L.K., Rhodes, J.M., Holdaway, M.J., Dungan, M.A. and Johnson, P. (1985) Petrology of basalts from Hole 504B, Deep Sea Drilling Project Leg 83. In: Anderson, R.N., Honnorez, J., Becker, K., *et al.* (eds) *Initial Reports of the Deep Sea Drilling Project* **83**, 129–164.

Kent, G.M., Harding, A.J. and Orcutt, J.A. (1990) Evidence for a smaller magma chamber beneath the EPR at 9°30′N. *Nature, London* **344**, 650–653.

Keunen, Ph.H. (1950) *Marine Geology*, Wiley, New York, 568 pp.

Khan, M.A., Jan, M.Q., Windley, B.F., Tarney, J. and Thirlwall, M.F. (1988) The Chilas mafic-ultramafic igeneous complex: the root of the Kohistan island arc in the Himalaya of northern Pakistan. *Geol. Soc. Am. Spec. Pap.* **232**, 75–94.

Kharin, G.S., Litvin, V.M. and Rudenko, M.V. (1976) Volcanic rocks and their role in the bottom structure of the Atlantic Ocean. In: Aoki, H. and Iizuka, S. (eds) *Volcanoes and Tectonosphere*, Tokai University Press, Tokyo, pp. 61–74.

Kidd, R.G.W. (1977) A model for in process of formation of the upper oceanic crust. *Geophys. J. Astron. Soc.* **50**, 149–183.

Kidd, R.G.W. and Cann, J.R. (1974) Chilling statistics indicate an ocean floor spreading origin for the Troodos Complex, Cyprus, *Earth Planet. Sci. Lett.* **24**, 151–155.

Kirkpatrick, R.J. (1979) Processes of crystallization in pillow basalts. Hole 396B, DSDP Leg 46. In: Dmitriev, L., Heirtzler, J., *et al.* (eds) *Initial Reports of the Deep Sea Drilling Project* **46**, 271–282.

Kirkpatrick, R.J., Clague, D.A. and Friesen, W. (1980) Petrology and geochemistry of volcanic rocks, DSDP Leg 55. In: Jackson, E.D., Koizumi, I. *et al.* (eds) *Initial Reports of the Deep Sea Drilling Project* **55**, 509–557.

Kilinc, A., Carmichael, I.S.E., Rivers, M.L. and Sack, R.O. (1983). The ferric ferrous ratio of natural silicate liquids equilibrated in air. *Contrib. Mineral. Petrol.* **83**, 136–140.

Klein, A.M. and Langmuir, C.H. (1987) Global correlation of ocean ridge basalt chemistry with axial depth and crustal thickness. *J. Geophys. Res.* **92**, 8089–8115.

Klein, A.M., Langmuir, C.H., Zindler, A., Staudigel, H. and Hehelin, B. (1988) Isotope evidence of a mantle convection boundary at the Australian-Antarctic discordance. *Nature, London* **333**, 623–629.

Kleinrock, M.C., Searle, R.C. and Hey, R.N. (1989) Tectonics of the failing spreading system associated with the Galapagos 95.5°W propagator. *J. Geophys. Res.* **94**, 13839–13858.

Kokelaar, P., Howells, M.F., Bevins, R.E., Roach, R.A. and Dunkley, P.N. (1984) The Ordovician marginal basin of Wales. In: Kokelaar, B.P. and Howells, M.F. (eds) *Marginal Basin Geology, Geol. Soc. London Spec. Publ.* **16**, 245–269.

Kristmannsdottir, H. (1976) Types of clay minerals in hydrothermally altered basaltic rocks, Reykjanes, Iceland. *Jokull* **26**, 30–39.

Kroenke, L., Scott, R. *et al.* (eds). (1980) *Initial Reports of the Deep Sea Drilling Project* **69**, 820 pp.

Kuno, H. (1968) Differentiation of basaltic types. In: Hess, H.H. and Poldervaart, A. (eds) *Basalts* (Vol. 2), Interscience, New York, pp. 623–688.

Kurz, M.D., Garcia, M.O., Frey, F.A. and O'Brien, P.A. (1987) Temporal helium isotopic variations within Hawaiian volcanoes: basalts from Mauna Loa and Haleakala. *Geochim. Cosmochim. Acta.* **51**, 2905–2914.

Kurz, M.D., Jenkins, W.J. and Hart, S.R. (1982). Helium isotopic systematics of oceanic islands and mantle heterogeneity. *Nature, London* **297**, 43–47.

Kurz, M.D., Jenkins, W.J., Hart, S.R. and Clague, D. (1983) Helium isotopic variations in volcanic rocks from Loihi Seamount and the island of Hawaii. *Earth Planet. Sci. Lett.* **66**, 388–406.

Kunz, M.D., Jenkins, W.J., Schilling, J.G. and Hart, S.R. (1982) Helium isotopic variations in the mantle beneath the central North Atlantic Ocean. *Earth Planet. Sci. Lett.* **58**, 1–14.

Kushiro, I. (1968) Compositions of magmas formed by partial zone melting of the earth's upper mantle. *J. Geophys. Res.* **73**, 619–634.

Kushiro, I. (1973) Origin of some magmas in oceanic and circumoceanic regions. *Tectonophysics* **17**, 211–222.

Kushiro, I. (1980) Viscosity, density and structure of silicate melts at high pressure, and their petrological applications. In: Hargraves, R.B. (ed) *Physics of Magmatic Processes*, Princeton University Press, Princeton, pp. 93–120.

Kushiro, I. and Thompson, R.N. (1972) Origin of some abyssal tholeiites from the Mid-Atlantic Ridge. *Carnegie Inst. Washington Yearbook* **71**, 403–406.

Kuznir, N.J. (1980) Thermal evolution of the oceanic crust: its dependency on spreading rate and effect on crustal structure. *Geophys. J. R. Astron. Soc.* **61**, 167–181.

Kyser, T.K. (1986) Stable isotope variations in the mantle. In: Valley, J.W., Taylor, H.P. and O'Neil, J.R. (eds) *Stable Isotopes in High Temperature Geological Processes, Mineral. Soc. Am. Rev. Mineral.* **16**, 141–164.

Kyser, T.K. and O'Neil, J.R. (1984) Hydrogen isotope systematics of submarine basalts. *Geochim. Cosmochim Acta* **48**, 2123–2133.

Kyser, T.K., Cameron, W.E. and Nibset, E.G. (1986) Boninite petrogenesis and alteration history: constraints from stable isotope compositions of boninites from Cape Vogel, New Caledonia and Cyprus. *Contrib. Mineral. Petrol.* **92**, 1–5.

Lacey, A., Ockendon, J.R. and Turcotte, D.L. (1981) On the geometrical form of volcanoes. *Earth Planet. Sci. Lett.* **54**, 139–143.

Lachenbruch, C. (1975) Dynamics of a passive spreading center. *J. Geophys. Res.* **81**, 1883–1902.

Langmuir, C.H. (1990) Ocean ridges spring surprises. *Nature, London* **334**, 585–586.

Langmuir, C.H. and Bender, J.F. (1984) The geochemistry of oceanic basalts in the vicinity of transform faults: observations and implications. *Earth Planet. Sci. Lett.* **69**, 107–127.

Langmuir, C.H. and Hanson, G.N. (1980) An evaluation of major element heterogeneity in the mantle sources of basalts. *Phil. Trans. R. Soc. London* **A297**, 383–407.

Langmuir, C.H. and Plank, T. (1988) Quantitative reevaluation of magma chamber processes and melting regime shape. *Chem. Geol.* **70**, 153–163.

Langmuir, C.H., Bender, J.F. and Batiza, R. (1986) Petrologic and tectonic segmentation of the East Pacific Rise, 5°30′N–14°30′N. *Nature, London* **322**, 422–427.

Langmuir, C.H., Bender, J.F., Bence, A.E. and Hanson, G.N. (1977) Petrogenesis of basalts from the FAMOUS area, Mid-Atlantic, Ridge. *Earth Planet. Sci. Lett.* **36**, 133–156.

Langseth, M.G. and von Herzen, R.P. (1970) Heat flow through the floor of the world oceans. In: Maxwell, A.E. (ed) *The Sea* (Vol. 4) *New Concepts of Sea Floor Evolution*, Interscience, New York, pp. 299–352.

Langseth, M.G., Cann, J.R., Natland, J.H. and Hobart, M. (1983) Geothermal phenomena at the Costa Rico Rift. In: Cann, J.R., Langseth, M.G., Honnorez, J., von Herzen, R.P., White, S.M., *et al.* (eds) *Initial Reports of the Deep Sea Drilling Project* **69**, 5–30.

Lanphere, M.A. (1983) $^{87}Sr/^{86}Sr$ ratios for basalt from Loihi Seamount, Hawaii. *Earth Planet. Sci. Lett.* **66**, 380–387.

Lanphere, M.A., Dalrymple, G.B. and Clague, D.A. (1980) Rb–Sr systematics of basalts from the Hawaiian-Emperor volcanic chain. In: Jackson, E.D., Koizumi, I. *et al.* (eds) *Initial Reports of the Deep Sea Drilling Projects* **55**, 695–706.

LaTraille, S.L. and Hussong, D.M. (1980) Crustal structure across the Mariana Island arc. In: Hayes, D.E. (ed) *The Tectonic and Geologic Evolution of Southeast Asia Seas and Islands, Am. Geophys. Union Monogr.* **23**.

Laughton, A.S. and Whit Marsh, R.B. (1975) The Azores Gibralter plate boundary. In: Kristjanson, L. (ed) *Geodynamics of Iceland and the North Atlantic Area*, Reidal, Dordrecht, pp. 63–68.

Lawrence, J.R. (1979) Temperatures of formation of calcite veins in the basalts from DSDP holes 417A and 417D. In: Donnelly, T., Francheteau, J., Bryan, W., Robinson, P.T., Flower, M., Salisbury, M. *et al.* (eds) *Initial Reports of the Deep Sea Drilling Project*, **51, 52, 53**, 1183–1184.

Lawver, L.A. and Hawkins, J.W. (1978) Magnetic anomalies and crustal dilation in the Lau Basin. *Earth Planet. Sci. Lett.* **33**, 27–33.

Le Bas, M.J., LeMaitre, R.W., Streckeisen, A. and Zanettin, B. (1986) A chemical classification of volcanic rocks based on the total alkali-silica diagram. *J. Petrol.* **27**, 745–750.

Le Roex, A.M., Dick, H.J.B., Erlank, A.J., Reid, A.M., Frey, F.A. and Hart, S.R. (1983) Geochemistry, mineralogy and petrogenesis of lavas erupted along the Southwest Indian Ridge between the Bouvet Triple Junction and 11 degrees East. *J. Petrol.* **24**, 267–318.

Le Roex, A.M., Dick, H.J.B., Reid, A.M., Frey, F.A., Erlank, A.J. and Hart, S.R. (1985) Petrology and geochemistry of basalts from the American-Antarctic Ridge, Southern Ocean: implications for the westward influence of the Bouvet mantle plume. *Contrib. Mineral. Petrol.* **90**, 367–380.

Le Roex, A.M., Erlank, A.J. and Needham, H.D. (1981) Geochemical and mineralogical evidence for at least three distinct magma types in the FAMOUS region. *Contrib. Mineral. Petrol.* **77**, 24–37.

Leeman, W.P., Budahn, J.R., Gerlach, D.C., Smith, D.R. and Powell, B.N. (1980) Origin of Hawaiian tholeiites: trace element constraints. *Am. J. Sci.* **280A**, 794–819.

Leg 124 Shipboard Scientific Party (1989) Origin of marginal basins. *Nature, London* **338**, 380–381.

Leg 126 Scientific Drilling Party (1989) ODP Leg 126 drills the Izu-Bonin arc. *Geotimes* October, 36–38.

LeMaitre, R.W. (1962) Petrology of volcanic rocks, Gough island, South Atlantic. *Bull. Geol. Soc. Am.* **73**, 1309–1340.

Lewis, B.T.R. and Garmany, J.D. (1982) Constraints on the structure of the East Pacific Rise from seismic refraction data. *J. Geophys. Res.* **87**, 8417–8425.

Lewis, B.T.R., Robinson, P. *et al.* (1983) *Initial Reports of the Deep Sea Drilling Project* **65**, 752 pp.

Lilwall, R.C., Francis, T.T.G. and Porter, I.T. (1978) Ocean-bottom seismograph observations on the MAR near 45°N – further results. *Geophys. J. R. Astron. Soc.* **55**, 255–262.

Lin, J., Purdy, G.M., Schouten, H., Sempere, J.-C. and Zervas, C. (1990) Evidence from gravity data for focused magmatic acceretion along the Mid-Atlantic, Ridge. *Nature, London* **344**, 627–632.

Liou, J.G. and Ernst, W.G. (1979) Oceanic ridge metamorphism of the east Taiwan ophiolite. *Contrib. Mineral Petrol.* **68**, 335–348.

Lipman, P.W. and Banks, N.G. (1987) Aa flow dynamics, Mauna Loa, (1984) In: Decker, R.W., Wright, T.L. and Stauffer, P.H. (eds) *Volcanism in Hawaii, US Geol. Surv. Prof. Pap.* **1350**, 1527–1568.

Lipman, P.W., Normark, W.R., Moore, J.G., Wilson, J.B. and Gutmacker, C.E. (1988) The giant submarine Alika debris slide, Mauna Loa, Hawaii. *J. Geophys. Res.* **93**, 4279–4299.

Lippard, S.J., Shelton, A.W. and Gass, I.G. (1986) The ophiolite of Northern Oman. *Mem. Geol. London* **11**, 178 pp.

Lister, C.R.B. (1972) On the thermal balance of a mid-ocean ridge. *Geophys. J. R. Astron. Soc.* **26**, 515–535.

Lister, C.R.B. (1983) On the intermittency and crystallization mechanisms of sub-seafloor magma chambers. *Geophys. J.R. Astron. Soc.* **73**, 351–365.

Litvin, V.M. and Rudenko, M.V. (1983) Distribution of seamounts in the Atlantic. *Proc. Acad. Sci. USSR, Earth Sci.* **213**, 223–225 (English translation).

Lloyd, F.E. and Bailey, D.K. (1975) Light element metasomatism of the continental mantle: the evidence and consequences. *Phys. Chem. Earth* **9**, 389–416.

Lofgren, G. (1971) Spherulitic textures in glassy and crystalline rocks. *J. Geophys. Res.* **76**, 5635–5648.

Lofgren, G. (1974) An experimental study of plagioclase crystal morphology: isothermal crystallization. *Am. J. Sci.* **274**, 243–273.

Lofgren, G. (1980) Experimental studies on the dynamic crystallization of silicate melts. In: Hargraves, R. (ed) *Physics of Magmatic Processes*, Princeton University Press, pp. 487–551.

Loney, R.A., Himmelberg, G.R. and Coleman, R.G. (1971) Structure and petrology of the Alpine-type peridotite at Burro Mountain, California, U.S.A. *J. Petrol.* **12**, 245–309.

Long, R.E. and Mitchell, M.G. (1970) Teleseismic P-wave delay time in Iceland. *Geophys. J. R. Astron. Soc.* **20**, 41–48.

Longhi, J. (1982) Effects of fractional crystallization and cumulus processes on mineral composition trends of some lunar and terrestrial rock series. *J. Geophys.* **87**, (supplement), A54–A64.

Lonsdale, P. (1977) Regional shape and tectonics of the equatorial East Pacific Rise. *Marine Geophys. Res.* **3**, 295–315.

Lonsdale, P. (1983) Overlapping rift zones at the 5.5°S offset of the East Pacific Rise. *J. Geophys. Res.* **88**, 9393–9406.

Lonsdale, P. (1988a) Structural pattern of the Galapagos microplate and evolution of the Galapagos triple junction. *J. Geophys. Res.* **93**, 13551–13574.

Lonsdale, P. (1988b) Geography and history of the Louiseville hotspot chain in the southwest Pacific. *J. Geophys. Res.* **93**, 3078–3104.

Lonsdale, P. (1989a) The rise flank trails left by migrating offsets of the Equatorial East Pacific Rise axis. *J. Geophys. Res.* **94**, 713–743.

Lonsdale, P. (1989b) Segmentation of the Pacific-Nazca spreading center 1°–20°S. *J. Geophys. Res.* **94**, 12197–12226.

Lonsdale, P. (1989c) Geology and tectonic history of the Gulf of California. In: Winterer, E.L., Hussong, D.M. and Decker, R.W. (eds) *The Geology of North America, N: The Eastern Pacific Ocean and Hawaii*, Geological Society of America, Boulder, Colorado, pp. 499–521.

Lonsdale, P. (1991) Structural pattern of the Pacific floor offshore of Peninsular California. In: Dauphin, J.P. and Simonit, B. (eds) *The Gulf and Peninsular Provinces of California, Mem. Am. Assoc. Petrol. Geol.* **47** (in press).

Lonsdale, P. and Batiza, R. (1980) Hyaloclastite and lava flows on young seamounts examined with a submersible. *Bull. Geol. Soc. Am.* **91**, 545–554.

Lonsdale, P. and Smith, D.K. (1986) Kula plate not Kula. *Trans. Am. Geophys. Union, EOS* **67**, 1199.

Loubet, M., Sassi, R. and Donato, G.D. (1988) Mantle heterogeneities: a combined isotope and trace element approach and evidence for recycled continental crust materials in some OIB sources. *Earth Planet. Sci. Lett.* **89**, 299–315.

Louden, K.E., White, R.S., Potts, C.G. and Forsyth, D.W. (1986) Structure and seismo-tectonics of the Vema Fracture Zone. *J. Geol. Soc. London* **143**, 795–806.

Lowrie, A., Smoot, C. and Batiza, R. (1986) Are oceanic fracture zones locked and strong or weak? New evidence for volcanic activity and weakness. *Geology*, **14**, 242–245.

Luck, J-M., Birck, J.L. and Allegre, C.J. (1980) ^{187}Re–^{187}Os systematics in meteorites: early chronology of the solar system and the age of the galaxy. *Nature, London* **283**, 256–259.

Ludden, J.N. (1978) Magmatic evolution of the basaltic shield volcanoes of Réunion Island. *J. Volcanol. Geotherm. Res.* **4**, 171–198.

Ludden, J.N. and Thompson, G. (1978) Behaviour of rare earth elements during submarine weathering of tholeiitic basalt. *Nature, London* **274**, 147–148.

Ludden, J.N. and Thompson, G. (1979) An evaluation of the behavior of the rare earth elements during the weathering of sea-floor basalt. *Earth Planet. Sci. Lett.* **43**, 85–92.

Ludden, J.N., Thompson, G., Bryan, W.B. and Frey, F.A. (1980) The origin of lavas from the Ninetyeast Ridge, Eastern Indian Ocean: an evaluation of fractional crystallization models. *J. Geophys. Res.* **85**, 4405–4420.

Lupton, J.E. and Craig, H. (1975) Excess ^3He in oceanic basalts. *Earth Planet. Sci. Lett.* **26**, 133–139.

Luyendyk, B.P. (1977) Deep Sea drilling on the Ninetyeast Ridge: synthesis and a tectonic model. In: Heirtzler, J.R. *et al.* (eds) *Indian Ocean Geology and Biostratigraphy*, Am. Geophys. Union, Washington, D.C. 101–120.

Luyendyk, B.P. and Melson, W.G. (1967) Magnetic properties and petrology of rocks near the crest of the Mid-Atlantic Ridge. *Nature, London* **215**, 147–149.

Luyendyk, B.P., Bender, J.F. and Macdonald, K.C. (1977) Physiography and structure of the inner floor of the FAMOUS rift valley: observations with a deep-towed instrument package. *Bull. Geol. Soc. Am.* **88**, 648–663.

Luyendyk, B.P., Cann, J.G. and Sharman, G.S. (1979) Introduction: background and explanatory notes. In: Luyendyk, B.P., Cann, J.R. *et al.* (eds) *Initial Reports of the Deep Sea Drilling Project* **49**, 5–20.

Maaloe, S. and Aoki, K. (1977) The major element composition of the upper mantle estimated from the composition of lherzolites. *Contrib. Mineral. Petrol.* **63**, 161–173.

Maaloe, S. and Jakobsson, J. (1980) The PT phase relations of a primary oceanite from the Reykjanes peninsula, Iceland. *Lithos* **13**, 237–246.

Macdonald, G.A. (1968) Composition and origin of Hawaiian lavas. *Mem. Geol. Soc. Am.* **116**, 477–522.

Macdonald, G.A. and Katsura, T. (1964) Chemical composition of Hawaiian lavas. *J. Petrol.* **5**, 83–133.

Macdonald, K.C. (1983) Crustal processess at spreading centers. *Rev. Geophys. Space Phys.* **21**, 1441–1454.

Macdonald, K.C. (1986) The crest of the Mid-Atlantic Ridge: models of crustal generation processes and tectonics. In: Vogt, P.R. and Tucholke, B.E. (eds) *The Geology of North America, volume M: The Western North Atlantic*, Geological Society of America, Boulder, Colorado, pp. 51–68.

Macdonald, K.C. (1989) Anatomy of the magma reservoir. *Nature, London* **339**, 178–179.

Macdonald, K.C. and Fox, P.J. (1983) Overlapping spreading centres: a new kind of accretion geometry on the east Pacific Rise. *Nature, London* **302**, 55–58.

Macdonald, K.C. and Fox, P.J. (1988) The axial summit graben and cross-sectional shape of the

East Pacific Rise as indicators of axial magma chambers and recent volcanic eruptions. *Earth Planet. Sci. Lett.* **88**, 119–131.

Macdonald, K.C., Fox, P.J., Perram, L.J., Eisen, M.F., Hayman, R.M., Miller, S.P., Carbotte, S.M., Cormier, M.H. and Shor, A.N. (1988a) A new view of the mid-ocean ridge from the behaviour of ridge-axis discontinuities. *Nature, London* **335**, 217–225.

Macdonald, K.C., Hayman, R.M., Miller, S.P., Sempere, J.C. and Fox, P.J. (1988b) Deep two and Sea Beam studies of dueling propagating ridges on the east Pacific Rise near 20°40′S. *J. Geophys. Res.* **93**, 2875–2898.

Macdonald, K.C., Sempere, J.C. and Fox, P.J. (1984) East Pacific Rise from Siqueiros to Orozco fracture zones: along-strike continuity of axial neovolcanic zone and structure and evolution of overlapping spreading centers. *J. Geophys. Res.* **89**, 6049–6069.

Macdonald, K.C., Sempere, J.C. and Fox, P.J. (1986) Reply: the debate concerning overlapping spreading centres and mid-ocean ridge processes. *J. Geophys. Res.* **91**, 10501–10511.

Macdonald, R., Sparks, R.S.J., Sigurdsson, H., Mattey, D.P., McGarvie, D.W. and Smith, R.L. (1987) The 1875 eruption of Askja volcano, Iceland: combined fractional crystallization and selective contamination in the generation of the rhyolitic magma. *Mineral. Mag.* **51**, 183–203.

MacPhearson, G.J. (1983). The Snow Mountain volcanic complex: an on-land seamount in the Franciscan terrain, California. *J. Geol.* **91**, 73–92.

Mahoney, J.J. (1987) An isotopic survey of Pacific oceanic plateaux: implication for their nature and origin. In: Keating, B.H., Fryer, P., Batiza, R. and Boehlert, G.W. (eds) *Seamounts, Islands and Atolls, Am. Geophys. Union, Geophys. Monogr.* **43**, 207–220.

Mahoney, J.J., Macdougall, J.D., Lugmair, G.W. and Gopalan, K. (1983) Kerguelen hot spot source for Rajmahal Traps and Ninetyeast Ridge? *Nature, London* **303**, 385–389.

Mahoney, J.J., Natland, J.H., White, W.M., Poreda, R., Bloomer, S.H., Fisher, R.L. and Baxter, A.N. (1989) Isotopic and geochemical provinces of the western Indian Ocean spreading centers. *J. Geophys. Res.* **94**, 4033–4052.

Malahoff, A., McMurty, G.M., Wiltshire, J.C. and Yeh, H.W. (1982) Geology and geochemistry of hydrothermal deposits from active submarine volcano Loihi, Hawaii, *Nature, London* **289**, 234–239.

Mammerickx, J. and Sandwell, D. (1986) Rifting of old ocean lithosphere. *J. Geophys. Res.* **91**, 1975–1988.

Mammerickx, J. and Sharman, G.F. (1988) Tectonic evolution of the North Pacific during the Cretaceous quiet period. *J. Geophys. Res.* **93**, 3009–3024.

Mammerickx, J., Naar, D.F. and Tyce, R.L. (1988) The Mathematician paleoplate. *J. Geophys. Res.* **93**, 3025–3040.

Manuel, O.K. and Sabu, D.D. (1981) The noble gas record of the terrestrial planets. *Geochem. J.* **15**, 245–267.

Marriner, G.F., Norry, M.J. and Gibson, I.L. (1982) The petrology and geochemistry of the Agua de Pau volcano, Sao Miguel, Azores. In: *Proceedings of the 1980 International Symposium on the Activity of Oceanic Islands (IAVCEI), Univ. Azores Rev.* **3**, 159–173.

Marsh, N.G., Saunders, A.D., Tarney, J. and Dick, H.J.B. (1980) Geochemistry of basalts from the Shikoku and Daito Basins, Deep Sea Drilling Project Leg 58. In: DeVries Klein, G., Kobayashi, K. *et al.* (eds) *Initial Reports of the Deep Sea Drilling Project* **58**, 805–842.

Marshall, M. and Cox, A. (1971) Magnetism of pillow basalts and their petrology. *Bull Geol. Soc. Am.* **82**, 537–552.

Mathez, E. (1976) Sulfur solubility and magmatic sulfides in submarine basalt glass. *J. Geophys. Res.* **81**, 4269–4276.

Mathez, E. (1980) Sulfide relations in Hole 418A flows and sulfur contents of glasses. In: Donnelly, T., Francheteau, J., Bryan, W.B., Robinson, P., Flower, M., Salisbury, M. *et al.* (eds) *Initial Reports of the Deep Sea Drilling Project* **51**, **52**, **53**, 1069–1085.

Mattey, D.P. (1982) Minor and trace element geochemistry of volcanic rocks from Truk, Ponape and Kusaie, eastern Caroline Islands: evolution of a young hotspot trace across old ocean crust. *Contrib. Mineral. Petrol.* **80**, 1–13.

Mattey, D.P. (1987) Carbon isotopes in the mantle. *Terra Cognita* **7**, 31–37.

Mattey, D.P., Carr, R.H., Wright, I.P. and Pillinger, C.P. (1984) Carbon isotopes in submarine basalts. *Earth Planet. Sci. Lett.* **70**, 196–206.

Mattey, D.P., Exley, R.A. and Pillinger, C.T. (1989a) Carbon isotopic composition of coexisting fluid and dissolved species in basalt glass. *Geochim. Cosmochim. Acta* **53**, 2377–2386.

Mattey, D.P., Exley, R.A., Menzies, M.A., Pillinger, C.T., Porcelli, D., Galer, S. and O'Nions, R.K. (1989b) Relation between C, He, Sr and Nd isotopes in mantle diopsides. In: *Kimberlites and Related Rocks.* (Vol. 2), *Geol. Soc. Australia, Spec. Publ.* **14**, 913–921.

Mattey, D.P., Marsh, N.G. and Tarney, J. (1980) The geochemistry, mineralogy and petrology of basalts from the West Philippine and Parece Vela Basins and from the Palau-Kyushu and West Mariana Ridges, Deep Sea Drilling Project Leg 59. In: Kroenke, L., Scott, R. *et al.* (eds) *Initial Reports of the Deep Sea Drilling Project* **59**, 753–800.

Matthews, D.H. (1971) Altered basalts from Shallow Bank, an abyssal hill in the N.E. Atlantic, and from a nearby seamount. *Phil. Trans. R. Soc. London* **A268**, 551–571.

Mattinsson, J.M. (1979) Lead isotopes studies of basalts from IPOD Leg 49. In: Luyendyk, B.P., Cann, J.R. *et al.* (eds) *Initial Reports of the Deep Sea Drilling Project* **49**, 721–726.

Mattiolo, G.S., Bake, M.B., Rutter, M.J. and Stolper, E.S. (1989) Upper mantle oxygen fugacity and its relationship to metasomatism. *J. Geol.* **97**, 521–536.

Mayer, L.A., Ryan, W.B.F., Detrick, R.S., Fox, P.J., Kong, L. and Manchester, K. (1985) Structure and tectonics on the Mid-Atlantic Ridge south of the Kane Fracture Zone based on SeaMARC I and Sea Beam site surveys. *Trans. Am. Geophys. Union, EOS* **66**, 1092.

Mazor, E., Heymann, D. and Anders, E. (1970). Noble gases in carbonaceous chondrites. *Geochim. Cosmochim. Acta.* **34**, 781–824.

McBirney, A. and Nakamura, Y. (1974) Immiscibility in late-stage magmas of the Skaergaard intrusion. *Carnegie Inst. Washington Yearbook* **73**, 348–352.

McCarthy, J., Mutter, J.C., Morton, J.L., Sleep, N.H. and Thompson, G.A. (1988) Relic magma chamber structures preserved within the Mesozoic North Atlantic crust? *Bull. Geol. Soc. Am.* **10**, 1423–1436.

McClain, J.S., Orcutt, J.A. and Burnett, M. (1985) The East Pacific Rise in cross section: a seismic model. *J. Geophys. Res.* **90**, 8627–8640.

McClain, K.J. and Lewis, B.T.R. (1982) Geophysical evidence for the absence of a crustal magma chamber under the northern Juan de Fuca Ridge: a contrast with ROSE results. *J. Geophys. Res.* **87**, 8477–8489.

McCulloch, M.T., Gregory, R.T., Wasserburg, G.J. and Taylor, H.P. (1980) A neodymium, strontium and oxygen isotopic study of Cretaceous Samail ophiolite, and implications for the petrogenesis and seawater-hydrothermal alteration of oceanic crust. *Earth Planet. Sci. Lett.* **46**, 201–211.

McDougall, I. and Duncan, R.A. (1980) Linear volcanic chains: recording plate motions? *Tectonophysics* **63**, 276–295.

McGarvie, D.W. (1984). Torfajokull: a volcano dominated by magma mixing. *Geology* **12**, 685–688.

McGarvie, D.W., Macdonald, R., Pinkerton, H. and Smith, R.L. (1990) Petrogenetic evolution of the Torfajokull Volcanic Complex, Iceland. II: The role of magma mixing. *J. Petrol.* **31**, 461–482.

McKay, G.A. (1986) Crystal/liquid partitioning of REE in basaltic systems: extreme fractionation of REE in olivine. *Geochim. Cosmochim. Acta* **50**, 69–79.

McKenzie, D.P. (1969) Speculations on the causes and consequences of plate motions. *Geophys. J. R. Astron. Soc.* **18**, 1–32.

McKenzie, D.P. (1984) The generation and compaction of partially molten rock. *J. Petrol.* **25**, 713–765.

McKenzie, D.P. (1985a) ^{230}Th–^{238}U disequilibrium and the melting process beneath ridge axes. *Earth Planet. Sci. Lett.* **72**, 149–157.

McKenzie, D.P. (1985b) The extraction of magma from the crust and mantle. *Earth Planet. Sci. Lett.* **74**, 81–91.

McKenzie, D.P. and Bickle, M.J. (1988) The volume and composition of melt generated by extension of the lithosphere. *J. Petrol.* **29**, 625–679.

McKenzie, D.P. and O'Nions, R.K. (1983) Mantle reservoirs and ocean island basalts. *Nature, London* **301**, 229–231.

McKenzie, D.P., Parsons, B. and Talwani, M. (1990) Gravity, topography and thermal convection inside the Earth. In: V-Gram, *Magellan Bulletin about Venus and the Radar Mapping Mission*, 5–13.

McKenzie, D.P., Watts, A.B., Parson, B. and Roufosse, M. (1980) Planform of mantle convection beneath the Pacific Ocean. *Nature, London* **288**, 442–446.

Meijer, A. (1976) Pb and Sr isotopic data bearing on the origin of volcanic rocks from the Mariana island arc system. *Bull. Geol. Soc. Am.* **87**, 1358–1369.

Melchior, J. (1981) Chromite, olivine and glass compositions from Mariana back-arc basin basalts: estimates of oxidation state of the magmas. *Trans. Am. Geophys. Union, EOS* **62**, 408.

Melson, W.G. and O'Hearn, (1979) Basaltic glass erupted along the Mid-Atlantic Ridge between 0–37°N: relationships between composition and latitude. In: Talwani, M., Hay, W. and Ryan, W.B.F. (eds) *Deep Sea Drilling results in the Atlantic Ocean: Ocean Crust, Am. Geophys. Union, Maurice Ewing Ser.* **2**, 273–284.

Melson, W.G. and Thompson, G. (1971) Petrology of a transform fault zone and adjacent ridge segments. *Phil. Trans. R. Soc. London* **A268**, 423–441.

Melson, W.G. and Thompson, G. (1973) Glassy abyssal basalts, Atlantic seafloor near St. Paul's Rocks: petrology and composition of secondary clay minerals. *Bull. Geol. Soc. Am.* **84**, 703–716.

Melson, W.G. and van Andel, Tj.H. (1966) Metamorphism in the mid-Atlantic ridge, 22°N latitude. *Mar. Geol.* **4**, 165–186.

Melson, W.G., Bowen, V.T., van Andel, Tj.H. and Siever, R. (1966) Greenstones from the central valley of the mid-Atlantic Ridge. *Nature, London* **209**, 604–605.

Melson, W.G., Byerly, G.R., Nelen, J.A., O'Hearn, T., Wright, T.L. and Vallier, T. (1979) A catalog of the major element chemistry of abyssal volcanic glasses. *Min. Sci. Invest. Smithsonian Contrib. Earth Sci.* **19**, 31–60.

Melson, W.G., Hart, S.R. and Thompson, G. (1972) St. Paul's Rocks, equatorial Atlantic: petrogenesis, radiometric ages and implications on sea-floor spreading. *Mem. Geol. Soc. Am.* **132**, 241–272.

Melson, W.G., Jarosewich, E., Bowen, V.T.R. and Thompson, G. (1967) St. Peter and St. Paul's rocks: a high temperature, mantle-derived intrusion. *Science* **155**, 1532–1535.

Melson, W.G., Thompson, G. and van Andel, Tj.H. (1968) Volcanism and metamorphism in the mid-Atlantic Ridge, 22°N latitude. *J. Geophys. Res.* **73**, 5925–5931.

Melson, W.G., Vallier, T.L., Wright, T.L., Byerly, G. and Nelen, J. (1976) Chemical diversity of abyssal volcanic glass erupted along Pacific, Atlantic and Indian Ocean seafloor spreading centers. *Monogr. Am. Geophys. Union* **19**, 351–368.

Menard, H.W. (1964) *Marine Geology of the Pacific*, McGraw Hill, New York, 272 pp.

Menard, H.W. (1978) Fragmentation of the Farallon plate by pivoting subduction. *J. Geol.* **86**, 99–110.

Menard, H.W. (1984) Origin of guyots: the Beagle to Seabeam. *J. Geophys. Res.* **89**, 11117–11123.

Menzies, M. (1983) CO_2-rich mantle below eastern Australia: REE, Sr and Nd isotopic study of Cenozoic alkaline magmas and apatite-rich xenoliths, Southern Highlands Province, New South Wales, Australia. *Earth Planet. Sci. Lett.* **65**, 287–302.

Menzies, M. (1984) Chemical and isotopic heterogeneities in orogenic and ophiolitic peridotites. In: Gass, I.G., Lippard, S.J. and Shelton, A.W. (eds) *Ophiolites and Oceanic Lithosphere, Geol. Soc. London Spec. Publ.* **13**, 231–240.

Menzies, M. (1990) Archaean, Proterozoic and Phanerozoic lithosphere. In: Menzies, M. (ed) *Continental Mantle*, Oxford University Press, 255 pp.

Menzies, M. and Allen, C. (1974) Plagioclase lherzolite residual mantle relationships within two eastern Mediterranean ophiolites. *Contrib. Mineral. Petrol.* **45**, 197–213.

Menzies, M. and Hawkesworth, C.J. (1987) Upper mantle processes and composition. In: Nixon, P.H. (ed) *Mantle Xenoliths*, Wiley, Chichester, pp. 725–738.

Menzies, M. and Murthy, V.R. (1976) Sr isotopic composition of clinopyroxenes from some Mediterranean alpine lherzolites. *Geochim. Cosmochim. Acta* **40**, 1577–1581.

Menzies, M. and Murthy, V.R. (1980a) Nd and Sr isotope geochemistry of hydrous mantle nodules and their host alkali basalts: implications for local heterogeneities in metasomatically veined mantle. *Earth Planet. Sci. Lett.* **46**, 323–334.

Menzies, M. and Murthy, V.R. (1980b) Mantle metasomatism as a percursor to the genesis of alkaline magmas—isotopic evidence. *Am. J. Sci.* **280A**, 622–638.

Menzies, M.A., Rogers, N., Tindle, A. and Hawkesworth, C.J. (1987) Metasomatic and enrichment processes in lithospheric peridotites, an effect of asthenosphere–lithosphere interaction. In: Menzies, M.A. and Hawkesworth, C.J. (eds) *Mantle Metasomatism*, Academic Press, New York, pp. 313–364.

Mevel, C. (1979) Mineralogy and chemistry of secondary phases in low temperature altered basalts from DSDP legs 51, 52, and 53. In: Donnelly, T., Francheteau, J., Bryan, W.B.,

Robinson, P.T., Flower. M., Salisbury, M. *et al.* (eds) *Initial Reports of the Deep Sea Drilling Project* **51**, **52**, **53**, 1299–1312.

Mevel, C. (1981) Occurrence of pumpellyite in hydrothermally altered basalts from the Vema Fracture Zone. *Contrib. Mineral. Petrol.* **76**, 386–393.

Mevel, C. (1987) Evolution of oceanic gabbros from DSDP Leg 82: influence of the fluid phase on metamorphic crystallizations. *Earth Planet. Sci. Lett.* **83**, 67–79.

Meyer, P.S., Sigurdsson, H. and Schilling, J.-G. (1985) Petrological and geochemical variations along Iceland's neovolcanic zones. *J. Geophys. Res.* **90**, 10043–10072.

Meyer, P.S. and Shibata, T. (1989) Complex zoning in plagioclase feldspars from ODP Site 648. In: Detrick, R., Honnorez, J., Bryon, W.B., Juteau, T. *et al.* (eds) *Proc. ODP Scientific Results* **106/109**, TX 123–142.

Michael, P.J. and Bonatti, E. (1985) Peridotite composition from the North Atlantic: regional and tectonic variations and implications for partial melting. *Earth Planet. Sci. Lett.* **73**, 91–104.

Michael, P.J. and Schilling, J.-G. (1989) Chlorine in mid-ocean ridge magmas: evidence for assimilation of seawater-influenced components. *Geochim. Cosmochim. Acta* **53**, 3131–3144.

Middlemost, E.A.K. (1985) *Magmas and Magmatic Rocks: An Introduction to Igneous Petrology*, Longman, Harlow.

Miyashiro, A. (1973) *Metamorphism and Metamorphic Belts*, Wiley, New York, 429 pp.

Miyashiro, A. and Shido, F. (1980) Differentiation of gabbros in the Mid-Atlantic Ridge. *Geochem. J.* **14**, 145–154.

Miyashiro, A., Shido, F. and Ewing, M. (1969) Diversity and origin of abyssal tholeiite from the Mid-Atlantic Ridge near 24° and 30° north latitude. *Contrib. Mineral. Petrol.* **23**, 38–52.

Miyashiro, A., Shido, F. and Ewing, M. (1970) Crystallization and differentiation in abyssal tholeiites and gabbros from mid-ocean ridges. *Earth Planet. Sci. Lett.* **7**, 361–365.

Miyashiro, A., Shido, F. and Ewing, M. (1971) Metamorphism in the Mid-Atlantic Ridge near 24°N and 30°N. *Phil. Trans. R. Soc. London* **A268**, 589–603.

Miyashiro, A., Shido, F. and Kanehira, K. (1979) Metasomatic chloritization of gabbros in the Mid-Atlantic Ridge near 30°N. *Mar. Geol.* **31**, 47–52.

Molnar, P. and Atwater, T. (1978) Interarc spreading and Cordilleran tectonics as alternatives related to the age of subducted ocean lithosphere. *Earth Planet. Sci. Lett.* **41**, 330–340.

Molnar, P. and Stock, J. (1987) Relative motions of hotspots in the Pacific, Atlantic and Indian oceans since late Cretaceous time. *Nature, London* **327**, 587–591.

Monnereau, M. and Cazenave, A. (1988) Variation of the apparent compensation depth of hotspot swells with age of plate. *Earth Planet. Sci. Lett.* **91**, 179–197.

Moorbath, S. and Walker, G.P.L. (1965) Strontium isotope investigation of igneous rocks from Iceland. *Nature, London* **207**, 837–840.

Moore, D.G. (1973) Plate-edge deformation and crustal growth, Gulf of California structural province. *Bull. Geol. Soc. Am.* **84**, 1883–1906.

Moore, J.G. (1966) Rate of palagonitization of submarine basalt adjacent to Hawaii. *US Geol. Surv. Prof. Pap.* **550**, 163–171.

Moore, J.G. and Fiske, R.S. (1969) Volcanic substructure inferred from dredge samples and ocean-bottom photographs, Hawaii. *Bull. Geol. Soc. Am.* **80**, 1191–1201.

Moore, J.G. and Schilling, J.-G. (1973) Vesicles, water and sulphur in Reykjanes Ridge basalts. *Contrib. Mineral. Petrol.* **41**, 105–118.

Moore, J.G., Clague, D.A. and Normark, W.R. (1982) Diverse basalt types from Loihi seamount, Hawaii. *Geology* **10**, 88–92.

Moores, E.M. (1969) Petrology and structure of the Vourinos ophiolite complex, northern Greece. *Geol. Soc. Am. Spec. Pap.* **118**, 1–74.

Moores, E.M. and Vine, T.J. (1971) The Troodos massif, Cyprus and other ophiolites as oceanic crust: evaluation and implications. *Phil. Trans. R. Soc. London* **A268**, 443–466.

Morel, J.M. and Hekinian, R. (1980) Compositional variations of volcanics along segments of recent spreading centers. *Contrib. Mineral. Petrol.* **72**, 425–436.

Morgan, W.J. (1971) Convection plumes in the lower mantle. *Nature, London* **230**, 42–43.

Morgan, W.J. (1972a) Plate motions and deep mantle convections. *Mem. Geol. Soc. Am.* **132**, 7–22.

Morgan, W.J. (1972b) Deep mantle convection plumes and plate tectonics. *Bull. Am. Assoc. Petrol. Geol.* **56**, 203–213.

Morgan, W.J. (1978) Rodriquez, Darwin, Amsterdam... a second type of hot spot island. *J. Geophys. Res.* **83**, 5355–5360.

Morgan, W.J. (1981) Hotspot tracks and the opening of the Atlantic and Indian Oceans. In: Emiliani, C. (ed) *The Sea* (Vol. 7) *The Oceanic Lithosphere*. Wiley Interscience, New York, pp. 443–487.

Morgan, W.J. (1983) Hotspot tracks and the early rifting of the Atlantic. *Tectonophysics* **94**, 123–139.

Morris, J.D. and Hart, S.R. (1983) Isotopic and incompatible element constraints on the genesis of island arc volcanics from Cold Bay and Amak Island, Aleutians, and implications for mantle structure, *Geochim. Cosmochim. Acta* **47**, 2015–2030.

Morris, J.D. and Tera, F. (1989) ^{10}Be and ^{9}Be in mineral separates and whole rocks from volcanic arcs: implications for sediment subduction. *Geochim. Cosmochim. Acta* **53**, 3197–3206.

Morton, J.L. and Sleep, N.H. (1985) Seismic reflections from a Lau Basin magma chamber. In: Scholl, D.W. and Vallier, T.L. (eds) *Geology and Offshore Resources of Pacific Island Arcs—Tonga Region*, Circum-Pacific Council for Energy & Mineral Resources, Houston, pp. 441–453.

Mottl, M.J. (1983) Metabasalts, axial hot springs and the structure of hydrothermal systems at mid-ocean ridges. *Bull. Geol. Soc. Am.* **94**, 161–180.

Mottl, M.J. and Holland, H.D. (1978) Chemical exchange during hydrothermal alteration of basalt by seawater. I: Experimental results for major and minor components of seawater. *Geochim. Cosmochim. Acta.* **42**, 1103–1115.

Mottl, M.J. and Seyfried, W.E. (1980) Subseafloor hydrothermal systems: rock vs. seawater dominated. In: Rona, P.A. and Lowell, R.P. (eds) *Seafloor Spreading Centers: Hydrothermal Systems*, Dowden, Philadelphia, pp. 66–82.

Mottl, M.J., Holland, H.D. and Corr, R.F. (1979) Chemical exchange during hydrothermal alteration of basalt by seawater. II: Experimental results for Fe, Mn and sulfur species. *Geochim. Cosmochim. Acta.* **43**, 869–884.

Muehlenbachs, K. (1986) Alteration of the oceanic crust and ^{18}O history of seawater. In: Valley, J.W., Taylor, H.P. and O'Neil, J.R. (eds) *Stable Isotopes in High Temperature Geological Processes, Mineral. Soc. Am. Rev. Mineral.* **16**, pp. 425–444.

Muehlenbachs, K. and Clayton, R.H. (1972) Oxygen isotope geochemistry of submarine greenstones. *Can. J. Earth Sci.* **9**, 172–184.

Muehlenbachs, K. and Clayton, R.H. (1976) Oxgyen isotope composition of the oceanic crust and its bearing on seawater. *J. Geophys. Res.* **81**, 4365–4369.

Muenow, D.W., Garcia, M.O., Aggrey, K.E., Bednarz, U. and Schmincke, H.U. (1990) Volatiles in submarine glasses as a discriminant for tectonic origin: implications for the Troodos ophiolite. *Nature, London* **343**, 159–161.

Muenow, D.W., Liu, N.W.K., Garcia, M.O. and Saunders, A.D. (1980) Volatiles in submarine volcanic rocks from the spreading axis of the East Scotia Sea back-arc basin. *Earth Planet. Sci. Lett.* **47**, 272–278.

Muir, I.D. and Tilley, C.E. (1964) Basalts from the northern part of the rift zone of the Mid-Atlantic Ridge. *J. Petrol.* **5**, 272–279.

Murray, J. and Reynard, A.F. (1891) Deep sea deposits. *Report on Scientific Results of the Voyage of HMS* Challenger, Ch. 5, HM Stationery Office, London.

Mysen, B.O. and Kushiro, I. (1977) Compositional variation of coexisting phases with degree of melting of peridotite in the upper mantle. *Am. Mineral.* **62**, 843–856.

Nafziger, R.H., Ulmer, G.C. and Woerman, E. (1971) Gaseous buffering for control of oxygen fugacity at one atmosphere. In: Ulmer, G.C. (ed) *Research Techniques for High Pressure and High Temperature*, Springer-Verlag, New York, pp. 9–41.

Nakamura, N. (1974) Determination of REE, Ba, Fe, Mg, Na and K in carbonaceous and ordinary chondrites. *Geochim. Cosmochim. Acta.* **38**, 757–776.

Natland, J.H. (1979) Crystal morphologies in basalts from DSDP Site 395, 23°N, 46°W, Mid-Atlantic Ridge. In: Melson, W.G., Rabinowitz, P.D. *et al.* (eds) *Initial Reports of the Deep Sea Drilling Project* **45**, 423–445.

Natland, J.H. (1980a) Crystal morphologies in basalts dredged and drilled from the East Pacific Rise near 9°N and the Siqueiros Fracture Zone. In: Rosendahl, B.R., Hekinian, R., *et al.* (eds) *Initial Reports of the Deep Sea Drilling Project* **54**, 605–633.

Natland, J.H. (1980b) Effects of axial magma chambers beneath spreading centers on the compositions of basaltic rocks. In: Rosendahl, B.R., Hekinian, R., *et al.* (eds) *Initial Reports of the Deep Sea Drilling Project* **54**, 833–850.

Natland, J.H. (1989) Partial melting of a lithologically heterogeneous mantle: inferences from crystallization histories on magnesian abyssal tholeiites from the Siqueiros Fracture Zone. In: Saunders, A.D. and Norry, M.J. (eds) *Magmatism in the Ocean Basins, Geol. Soc. London Spec. Publ.* **42**, 41–70.

Natland, J.H. (in press) Magmatic oxides and sulfides in gabbroic rocks from ODP Hole 735B and the later development of the liquid line of descent. In: Robinson. P., Von Herzen, R.P. *et al.* (eds) *Proceedings of the Ocean Drilling Program, Initial Reports* (*Part A*) 118.

Natland, J.H. and Melson, W.G. (1980) Compositions of basaltic glasses from the East Pacific Rise and the Siqueiros Fracture Zone, near 9°N. In: Rosendahl, B.R., Hekinian, R., *et al.* (eds) *Initial Reports of the Deep Sea Drilling Project* **54**, 705–724.

Natland, J.H. and Tarney, J. (1982) Petrological evolution of the Mariana arc and back-arc basin systems: a synthesis of drilling results in the South Philippine Sea. In: Hussong, D.M., Uyeda, S., *et al.* (eds) *Initial Reports of the Deep Sea Drilling Project* **60**, 877–908.

Natland, J.H., Adamson, A.C., Laverne, C., Melson, W.G. and O'Hearn, T. (1983) A compositionally nearly steady state magma chamber at the Costa Rica Rift: evidence from basalt glass and mineral data, Deep Sea Drilling Project Sites 501, 504 and 505. In: Cann, J.R., Langseth, M.G., Honnorez, J., Von Herzen, R.P., White, S.M. *et al.* (eds) *Initial Reports of the Deep Sea Drilling Project* **69**, 811–858.

Natland, J.H., Bloomer, S.H., Fisher, R.L., Mahoney, J., O'Hearn, T. and Melson, W.G. (in press). Petrogenesis of abyssal tholeiites from the Indian Ocean.

Natland, J., Langmuir, C.H., Bender, J., Batiza, R. and Hopson, C. (1986) Petrologic systematics in the vicinity of the 9°N non-transform offset, East Pacific Rise. *Trans. Am. Geophys. Union, EOS* **67**, 1254.

Natland, J.H., Meyer, P.S., Dick, H.J.B. and Bloomer, S.H. (in press) Magmatic oxides and sulfides in gabbroic rocks from ODP Hole 735B and the later development of the liquid line of descent. In: Robinson, P.T., Von Herzen, R.P. *et al.* (eds) *Proceedings of the Ocean Drilling Program, Initial Reports* (*Part A*) **118**.

Naughton, J.J., Macdonald, G.A. and Greenberg, V.A. (1980) Some additional potassium-argon ages of Hawaiian rocks: the Maui volcanic complex of Moloki, Maui, Lanai and Kahoolawe. *J. Volcanol. Geotherm. Res.* **7**, 339–355.

Neal, C.R. (1988) The origin and composition of metasomatic fluids and amphiboles beneath Malaita, Solomon Islands. In: Menzies, M.A. and Cox, K.G. (eds) *Oceanic and Continental Lithosphere: Similarities and Differences, J. Petrol. Spec. Publ.* **29**, pp. 149–180.

Neal, C.R. and Davidson, J. (1989) An unmetasomatized source for the Malaitan alnoite (Soloman Islands): petrogenesis involving zone refining, megacryst fractionation and assimilation of the oceanic lithosphere. *Geochim. Cosmochim. Acta.* **53**, 1975–1990.

Needman, H.D. and Francheteau, J. (1974) Some characteristics of the rift valley in the Atlantic ocean near 36°N 48′. *Earth Planet. Sci. Lett.* **22**, 29–43.

Nicholls, G.D. (1965) Basalts from the deep ocean floor. *Mineral. Mag.* **34**, 373–381.

Nicholls, G.D. and Bowen, V.T. (1961) Natural glass from beneath red clay on the floor of the Atlantic. *Nature, London* **192**, 156–157.

Nicholls, G.D., Nalwalk, D. and Hayes, E.E. (1964) The nature and composition of rock samples dredged from the Mid-Atlantic Ridge between 22°N and 52°N. *Mar. Geol.* **1**, 333–340.

Nicholls, I.A., Ferguson, J., Jones, H., Marks, G.P. and Mutter, J.C. (1981) Ultramafic blocks from the ocean floor southwest of Australia. *Earth Planet. Sci. Lett.* **56**, 362–374.

Nicolas, A. (1986) A melt extraction model based on structural studies in mantle peridotites. *J. Petrol.* **27**, 999–1022.

Nicolas, A. (1989) *Structure of Ophiolites and Dynamics of Oceanic Lithosphere*, Kluwer Academic, Dordrecht, 367pp.

Nicolas, A. and Jackson, E.D. (1972) Repartition en deux provinces des peridotites des chaines alpines logeant la Mediterranee: implications geotectonique. *Schweize. Mineral. Petrograph. Mitteil.* **52**, 479–495.

Nielsen, R.L. (1988) A model for the simulation of combined major and trace element liquid lines of descent. *Geochim. Cosmochim. Acta.* **52**, 27–38.

Neilsen, R.L. and Dungan, M.A. (1983) Low pressure mineral-melt equilibria in natural anhydrous mafic systems. *Contrib. Mineral. Petrol.* **84**, 310–326.

Nisbet, E.G. and Fowler, C.M.R. (1978) The Mid-Atlantic Ridge at 37 and 45°N: some geophysical and petrological constraints. *Geophys. J. R. Astron. Soc.* **54**, 631–660.

Nisbet, E.G. and Pearce, J.A. (1973) TiO_2 and a possible guide to past oceanic spreading rates. *Nature, London* **246**, 468–469.

Nixon, P.H. and Boyd, F.R. (1979) Garnet-bearing lherzolites and discrete nodule suites from the Malaita alnoite, Solomon Islands, S.W. Pacific, and their bearing on oceanic mantle composition and geotherm. In: Boyd, F.R. and Meyer, H.O.A. (eds) *The Mantle Sample: Inclusions in Kimberlites and Other Volcanics*, American Geophysical Union, Washington, D.C. pp. 400–423.

Nixon, P.H. and Coleman, P.J. (1978) Garnet-bearing lherzolites and discrete nodule suites from the Malaita alnoite, Solomon Islands, and their bearing on the nature and origin of the Ontong Java Plateau. *Bull. Austral. Soc. Explor. Geophys.* **9**, 103–107.

Nixon, P.H. and Neal, C.R. (1987) Ontong–Java Plateau: deep-seated xenoliths from thick oceanic lithosphere. In: Nixon, P.H. (ed) *Mantle Xenoliths*, Wiley, Chichester, pp. 336–345.

Nixon, P.H., Mitchell, R.H. and Rogers, N. (1980) Petrogenesis of alnoitic rocks from Solomon Islands, Melanesia. *Mineral. Mag.* **43**, 587–596.

Norry, M.J. and Fitton, J.G. (1983) Compositional differences between oceanic and continental basic lavas and their significance. In: Hawkesworth, C.J. and Norry, M.J. (eds) *Continental Basalts and Mantle Xenoliths*, Shiva Publishing, Nantwich, pp.5–19.

Nur, A. & Ben-Avraham, Z. (1982) Oceanic plateaus, the fragmentation of continents and mountain building. *J. Geophys. Res.* **87**, 3644–3661.

Ode, H. (1957) Mechanical analysis of the dyke pattern of the Scottish Peaks area, Colorado. *Bull. Geol. Soc. Am.* **68**, 567–576.

O'Donnell, T.H. and Presnall, D.C. (1980) Chemical variations of the glass and mineral phases in basalts dredged from 25°-30°N along the Mid-Atlantic Ridge. *Am. J. Sci.* **280A**, 845–868.

O'Hara, M.J. (1965) Primary magmas and the origin of basalts. *Scot. J. Geol.* **1**, 19–40.

O'Hara, M.J. (1968a) Are ocean floor basalts primary magma? *Nature, London* **220**, 683–686.

O'Hara, M.J. (1968b) The bearing of phase equilibria studies in synthetic and natural systems on the origin and evolution of basic and ultrabasic rocks. *Earth Sci. Rev.* **4**, 69–133.

O'Hara, M.J. (1973) Non-primary magmas and dubious mantle plume beneath Iceland. *Nature, London* **243**, 507–508.

O'Hara, M.J. (1975) Is there an Icelandic mantle plume? *Nature, London* **253**, 708–710.

O'Hara, M.J. (1977) Geochemical evolution during fractional crystallization of a periodically refilled magma chamber. *Nature, London* **266**, 503–507.

O'Hara, M.J. (1985) The importance of the "shape" of the melting regime during partial melting of the mantle. *Nature, London* **314**, 58–62.

O'Hara, M.J. and Mathews, R.E. (1981) Geochemical evolution in an advancing, periodically-tapped, continuously fractionated magma chamber. *J. Geol. Soc. London* **138**, 237–277.

Okal, E.A and Batiza, R. (1987) Hotspots: the first 25 years. In: Keating, B.H., Fryer, P., Batiza, R. and Boehlert, G.W. (eds) *Seamounts, Islands and Atolls, Am. Geophys. Union, Geophys. Monogr.* **43**, pp. 1–11.

Olson, P. and Singer, H.A. (1985) Creeping plumes. *J. Fluid Mech.* **158**, 511–531.

Olson, P., Silver, P.G. and Carlson, R.W. (1990) The large-scale structure of convection in the Earth's mantle. *Nature, London* **344**, 209–215.

O'Nions, R.K. and Gronvold, K. (1973) Petrogenetic relationships of acid and basic rocks in Iceland: Sr isotopes and REE in late and postglacial volcanics. *Earth Planet. Sci. Lett.* **19**, 397–409.

O'Nions, R.K. and Pankhurst, R.J. (1973) Secular variation in the Sr isotope composition of Icelandic volcanic rocks. *Earth Planet. Sci. Lett.* **21**, 13–21.

O'Nions, R.K. and Pankhurst, R.J. (1974) Petrogenetic significance of isotope and trace element variations in volcanic rocks from the Mid-Atlantic. *J. Petrol.* **15**, 603–634.

O'Nions, R.K. and Pankhurst, R.J. (1976) Sr isotopes and REE geochemistry of DSDP Leg 37 basalts. *Earth Planet. Sci. Lett.* **32**, 255–261.

O'Nions, R.K., Evensen, N.M. and Hamilton, P.J. (1978) Differentiation and evolution of the mantle. *Phil. Trans. R. Soc. London* **A297**, 479–493.

O'Nions, R.K., Hamilton, P.J. and Evenson, N.M. (1977) Variations in $^{143}Nd/^{144}Nd$ and $^{87}Sr/^{86}Sr$ ratios in oceanic basalts. *Earth Planet. Sci. Lett.* **21**, 13–22.

O'Nions, R.K., Pankhurst, R.J., Freideiissen, I.B. and Jakobsson, S.P. (1973) Strontium isotopes and rare earth elements in basalts from the Heimaey and Surtsey volcanic eruptions. *Nature, London* **243**, 213–214.

O'Nions, R.K., Pankhurst, R.J. and Gronvold, K. (1976) Nature and development of basalt magma sources beneath Iceland and the Reykjanes Ridge. *J. Petrol.* **17**, 315–338.

Orcutt, J.A., Kennett, B.L.N. and Dorman, L.M. (1976) Structure of the East Pacific Rise from ocean bottom seismometer survey. *Geophys. J. R. Astron. Soc.* **45**, 305–320.

Orcutt, J.A., McClain, J.S. and Burnett, M. (1984) Evolution of the oceanic crust: results from recent seismic experiments. In: Gass, I.G., Lippard, S.J. and Shelton, A.W. (eds) *Ophiolites and Oceanic Lithosphere, Geol. Soc. London, Spec. Publ.* **13**, 7–16.

Osborn, E.F. (1959) Role of oxygen pressure in the crystallization and differentiation of basaltic magma. *Am. J. Sci.* **257**, 609–647.

Oskarsson, N., Steinthorsson, S. and Sigvaldason, H. (1985) Iceland geochemical anomaly: origin, volcanotectonics, chemical fractionation and isotope evolution of the crust. *J. Geophys. Res.* **90**, 10011–10025.

Ottonello, G., Piccardo, G.B. and Ernst, W.G. (1979) Petrogenesis of some Ligurian peridotites. Part II: Rare earth element geochemistry. *Geochim. Cosmochim. Acta* **43**, 1273–1284.

Otufuji, Y. and Matsuda, T. (1983) Paleomagnetic evidence for the clockwise rotation of southwest Japan. *Earth Planet. Sci. Lett.* **62**, 349–359.

Otufuji, Y. and Matsuda, T. (1984) Timing and rotational motion of southwest Japan inferred from palaeomagnetism. *Earth Planet. Sci. Lett.* **70**, 373–382.

Oversby, V.M. (1970) The isotopic composition of lead in iron meteorites. *Geochim. Cosmochim. Acta* **34**, 77–88.

Oxburgh, E.R. (1980) Heat flow and magma genesis. In: Hargraves, R.B. (ed) *Physics of Magmatic Processes*, Princeton University Press, pp. 161–199.

Oxburgh, E.R. and Turcotte, D.L. (1970) The thermal structure of island arcs. *Bull. Geol. Soc. Am.* **81**, 1665–1688.

Oxburgh, E.R. and Turcotte, D.L. (1974) Membrane tectonics and the East African rift. *Earth Planet. Sci. Lett.* **22**, 133–140.

Ozima, M. (1975) Ar isotopes and earth atmosphere evolution models. *Geochim. Cosmochim. Acta* **39**, 1127–1134.

Packham, G.H. and Falvey, D.A. (1971) An hypothesis for the formation of marginal seas in the western Pacific. *Tectonophysics* **11**, 79–110.

Palacz, Z.A. and Saunders, A.D. (1986) Coupled trace element and isotope enrichment in the Cook-Austral-Samoa Islands, southwest Pacific. *Earth Planet. Sci. Lett.* **79**, 270–280.

Pallister, J.S. and Hopson, C.A. (1981) Samail ophiolite plutonic suite: field relations, phase variation, cryptic variation and layering, and a model of a spreading ridge magma chamber. *J. Geophys. Res.* **86**, 2593–2644.

Palmason, G. (1971) Crustal structure of Iceland from explosion seismology. *Soc. Sci. Islandica Rit* **40**, 187–198.

Palmason, G. (1973) Kinematics and heat flow in a volcanic rift zone, with application to Iceland. *Geophys. J. R. Astron. Soc.* **33**, 451–481.

Palmason, G. and Saemundsson, K. (1974) Iceland in relation to the Mid-Atlantic Ridge. *Ann. Rev. Earth Planet. Sci.* **2**, 25–50.

Palme, H. and Nickel, K.G. (1985) Ca/Al ratio and composition of the earth's upper mantle. *Geochim. Cosmochim. Acta* **49**, 2123–2132.

Park, K.-H. and Staudigel, H. (1990) Radiogenic isotope ratios and initial seafloor alteration in submarine Serocki volcano basalts. In: Detrick, R., Honnorez, J., Bryan, W.B., Juteau, T. *et al.* (eds) *Proceedings of the Ocean Drilling Program, Initial Reports (Part A)* **106/109**, 117–121.

Parmentier, E.M. and Forsyth, D.W. (1985) Three-dimensional flow beneath a slow-spreading ridge axis: a dynamic contribution to the deepening of the median valley toward fracture zones. *J. Geophys. Res.* **90**, 678–684.

Parmentier, E.M., Turcotte, D.L. and Torrance, K.E. (1975) Numerical experiments on the structure of mantle plumes. *J. Geophys. Res.* **80**, 4417–4424.

Parsons, B. and Daly, S. (1983) The relationship between surface topography, gravity anomalies and temperature structure of convection. *J. Geophys. Res.* **88**, 1129–1144.

Parsons B. and Sclater, J.G. (1977) Analysis of the variation of ocean floor bathymetry and heat flow with age. *J. Geophys. Res.* **82**, 803–827.

Paster, T.P. (1968) Petrologic variations within submarine basalt pillows of the South Pacific-Antarctic ocean. Unpublished PhD thesis, Florida State University.

Patchett, P.J. (1981) Evolution of continental crust and mantle heterogeneity: evidence from Hf isotopes. *Contrib. Mineral. Petrol.* **78**, 279–297.

Pearce, J.A. (1983) Role of the sub-continental lithosphere in magma genesis at active continental margins. In: Hawkesworth, C.J. and Norry, M.J. (eds) *Continental Basalts and Mantle Xenoliths*, Shiva Publications, Nantwich, pp. 230–249.

Pearce, J.A., Alabaster, T., Shelton, A.W. and Searle, M.P. (1981) The Oman ophiolite as a Cretaceous arc-basin complex: evidence and implications. *Phil. Trans. R. Soc. London* **A300**, 299–317.

Pearce, J.A., Lippard, S.J. and Roberts, S. (1984) Characteristics and tectonic significance of supra-subduction zone ophiolites. In: Kokelaar, B.P. and Howells, M.F. (eds) *Marginal Basin Geology, Geol. Soc. London, Spec. Publ.* **16**, pp. 77–94.

Pedersen, J.J., Fox, P.J. and Schreiber, E. (1974) Newfoundland ophiolites and the geology of the oceanic layer. *Nature, London* **247**, 194–196.

Pedersen, R.B. and Malpas, J. (1984) The origin of plagiogranites from the Karmoy ophiolite, western Norway. *Contrib. Mineral. Petrol.* **88**, 36–52.

Perfit, M.R. and Fornari, D.J. (1983) Geochemical studies of abyssal lavas recovered by DSRV Alvin from eastern Galapagos rift, Inca transform and Ecuador rift. 2: phase chemistry and crystallization history. *J. Geophys. Res.* **88**, 10530–10550.

Perfit, M.R., Fornari, D.J., Malahoff, A. and Embley, R.W. (1983) Geochemical studies of abyssal lavas recovered by DSV Alvin from eastern Galapagos rift, Inca transform and Equador rift. 3: trace element abundances and petrogenesis. *J. Geophys. Res.* **88**, 10551–10572.

Perfit, M.R., Gust, D.A., Bence, A.E., Arculus, R.J. and Taylor, S.R. (1980) Chemical characteristics of island arc basalts: implications for mantle sources. *Chem. Geol.* **30**, 227–256.

Peterman, Z.E. and Hedge, C.E. (1971) Related strontium isotopic and chemical variations in oceanic basalts. *Bull. Geol. Soc. Am.* **82**, 493–499.

Philpotts, J.A., Schnetzler, C.C. and Hart, S.R. (1969) Submarine basalts: some K, Rb, Sr, Ba, rare earths, H_2O and CO_2 data bearing on their alteration, modification by plagioclase and possible source materials. *Earth Planet. Sci. Lett.* **7**, 293–299.

Pineau, F. and Javoy, M. (1983) Carbon isotopes and concentrations in mid-ocean ridge basalts. *Earth Planet. Sci. Lett.* **62**, 239–257.

Pineau, F., Javoy, M. and Bottinga, Y. (1976) $^{13}C/^{12}C$ ratios of rocks and inclusions in popping rocks of the Mid-Atlantic ridge and their bearing on the problems of deep-seated carbon. *Earth Planet. Sci. Lett.* **29**, 413–421.

Piper, J.D.A. (1973) Volcanic history and tectonics of the north Langjokull region, central Iceland. *Can. J. Earth Sci.* **10**, 164–175.

Pollack, H.N., Gass, I.G., Thorpe, R.S. and Chapman, D.S. (1981) On the vulnerability of lithospheric plates to mid-plate volcanism: reply to comments by P.R. Vogt. *J. Geophys. Res.* **86**, 961–966.

Polve, M. and Allegre, C.J. (1980) Orogenic lherzolite complexes studies by ^{87}Sr-^{86}Sr: a clue to understanding the mantle convection processes. *Earth Planet. Sci. Lett.* **51**, 71–93.

Poreda, R. (1985) Helium-3 and deuterium in back-arc basalts: Lau Basin and Mariana Trough. *Earth Planet. Sci. Lett.* **73**, 244–254.

Poreda, R. and Craig, H. (1989) Helium isotope ratios in circum-Pacific volcanic arcs. *Nature, London* **338**, 473–478.

Poster, C.K. (1973) Ultrasonic velocities in rocks from the Troodos Massif, Cyprus. *Nature, London* **243**, 2–3.

Preda, R. and di Brozolo, F.R. (1984) Neon isotope variations in Mid-Atlantic Ridge basalts. *Earth Planet. Sci. Lett.* **69**, 277–289.

Presnall, D.C. and Brenner, N.L. (1974) A method for studying iron silicate liquids under reducing conditions with negligible iron loss. *Geochim Cosmochim. Acta.* **38**, 1785–1788.

Presnall, D.C. and Hoover, J.D. (1984) Composition and depth of origin of primary mid-ocean ridge basalts. *Contrib. Mineral. Petrol.* **87**, 170–178.

Presnall, D.C. and Hoover, J.D. (1986) Composition and depth of origin of primary mid-ocean ridge basalts—reply to D. Elthon. *Contrib. Mineral. Petrol.* **94**, 257–261.

Presnall, D.C. and Hoover, J.D. (1987) High pressure phase equilibrium constraints on the origin of mid-ocean ridge basalts. In: Mysen, B.O. (ed) *Magmatic Processes: Physicochemical Principles, Geochem. Soc. Spec. Publ.* **1**, pp. 75–89.

Presnall, D.C., Dixon, J.R., O'Donnell, T.H. and Dixon, S.A. (1979) Generation of mid-ocean ridge tholeiites. *J. Petrol.* **20**, 3–35.

Price, R.C., Kennedy, A.K., Riggs-Sneeringer, M. and Frey, F.A. (1986) Geochemistry of basalts from the Indian ocean triple junction: implications for the generation and evolution of Indian Ocean ridge basalts. *Earth Planet. Sci. Lett.* **78**, 379–396.

Prichard, H.M. (1985) The Shetland ophiolite. In: Gee, D.G. and Stur, B.A. (eds) *The Caledonide Orogen—Scandinavia and Related Areas*, Wiley, New York, pp. 1173–1184.

Pringle, M., Staudigel, H. and Gee, J. (1991) Geochronology of Jasper Seamount: seven million years of volcanism. *Geology* **19**, 364–368.

Prinz, M., Keil, K., Green, J.A., Reid, A.M., Bonatti, E. and Honnirez, J. (1976) Ultramafic and mafic dredge samples from the Equatorial Mid-Atlantic Ridge and fracture zone. *J. Geophys. Res.* **81**, 4087–4103.

Puchett, H., Emmermann, R. and Srivastava, R.K. (1977) Rare earth and other trace elements in basalts from the Mid-Atlantic Ridge 36°N DSDP Leg 37. In: Aumento, F., Melson, W.G. *et al.* (eds) *Initial Reports of the Deep Sea Drilling Project* **37**, 581–590.

Purdey, G.M., Rabinowitz, P.D. and Veltrop, J.J.A. (1979) The Kane Fracture Zone in the central North Atlantic. *Earth Planet. Sci. Lett.* **45**, 429–434.

Purdy, M. (1987) New observations of shallow seismic structure of young oceanic crust. *J. Geophys. Res.* **93**, 9351–9362.

Quick, J. (1981) Petrology and petrogenesis of the Trinity periodotite, an upper mantle diapir in the eastern Klamath Mountains, northern California. *J. Geophys. Res.* **86**, 11837–11864.

Quon, S.H. and Ehlers, E.G. (1963) Rocks of the northern part of the Mid-Atlantic Ridge. *Bull. Geol. Soc. Am.* **74**, 1–7.

Raitt, R.W. (1963) The crustal rocks. In: Hill, M.N. (ed) *The Sea* (Vol. 3), Wiley Interscience, New York, pp. 85–102.

Ramberg, B., Gray, D.F. and Reynolds, R.G. (1977) Tectonic evolution of the FAMOUS area of the Mid-Atlantic Ridge, lat. 35°50′ to 37°20′. *Bull. Geol. Soc. Am.* **88**, 609–620.

Ramberg, I.B. and Van Andel, T.H. (1977) Morphology and tectonic evolution of the rift valley at latitude 36°30′N, Mid-Atlantic Ridge, *Bull. Geol. Soc. Am.* **88**, 577–586.

Rea, D.K. and Vallier, T.L. (1983) Two Cretaceous volcanic episodes in the western Pacific Ocean. *Bull. Geol. Soc. Am.* **94**, 1430–1437.

Reddy, V.V., Subbarao, K.V., Reddy, G.R., Matsuda, J. and Hekinian, R. (1978) Geochemistry of volcanics from the Ninetyeast Ridge and its vicinity in the Indian Ocean. *Mar. Geol.* **26**, 99–117.

Reid, I., Orcutt, J.A. and Prothero, W.A. (1977) Seismic evidence for a narrow zone of partial melting underlying the East Pacific Rise at 21°N. *Bull. Geol. Soc. Am.* **88**, 678–682.

Reid, J.B. and Woods, G.A. (1978) Oceanic mantle beneath the southern Rio Grande rift. *Earth Planet. Sci. Lett.* **41**, 303–316.

Reisberg, L. and Zindler, A. (1986) Extreme isotopic variations in the upper mantle: evidence from Ronda. *Earth Planet. Sci. Lett.* **81**, 29–45.

Rhodes, J.M., Dungan, M.A., Blanchard, D.P. and Long, P.E. (1979) Magma mixing at mid-ocean ridges: evidence from basalts drilled near 22°N on the Mid-Atlantic Ridge. *Tectonophysics* **55**, 35–61.

Ribe, N.M. (1985) The generation and composition of partial melts in the earth's mantle. *Earth Planet. Sci. Lett.* **73**, 361–376.

Rice, S., Langmuir, C.H., Bender, J.F., Hanson, G.N., Bence, A.E. and Taylor, S.R. (1980) Basalts from DSDP holes 417A & 417D, fractionated melts of a LREE depleted source. In: Donnelly, T., Francheteu, J., Bryan, W., Robinson, P.T., Flower, M., Salisbury, M. *et al.* (eds) *Initial Reports of the Deep Sea Drilling Project* **51**, **52**, **53**, 1099–1112.

Richard, P. and Allegre, C.J. (1980) Neodymium and strontium isotope study of ophiolite and orogenic lherozolite petrogenesis. *Earth Planet. Sci. Lett.* **47**, 65–74.

Richards, M.A., Duncan, R.A. and Courtillot, V.E. (1989) Flood basalts and hot-spot tracks: plume heads and tails. *Science* **246**, 103–107.

Richardson, C.J., Cann, J.R., Richards, H.G. and Cowan, J.G. (1987) Meta-depleted root zones of the Troodos ore-forming hydrothermal systems, Cyprus. *Earth Planet. Sci. Lett.* **84**, 243–253.

Richardson, S.H., Erlank, A.J., Duncan, A.R. and Reid, D.L. (1982) Correlated Nd, Sr and Pb isotope variation in Walvis Ridge basalts and implications for the evolution of their mantle source. *Earth Planet. Sci. Lett.* **59**, 327–342.

Richardson, S.H., Hart, S.R. and Staudigel, H. (1980) Vein mineral ages for old oceanic crust. *J. Geophys. Res.* **85**, 7195–7200.

Richter, F.M. and McKenzie, D.P. (1984) Dynamic models for melt segregation from a deformable matrix. *J. Geol.* **92**, 729–740.

Ridley, W.I., Reid, A.M. and Bass, M.N. (1973) The geology and petrology of basalts from Leg 6 of the Deep Sea Drilling Project. *Trans. Am. Geophys. Union, EOS* **54**, 132.

Ringwood, A.E. (1962) A model for the upper mantle. *J. Geophys. Res.* **67**, 857–867 & 4473–4477.

Ringwood, A.E. (1974) The petrological evolution of island arc systems. *J. Geol. Soc. London* **130**, 183–204.

Ringwood, A.E. (1979) *Origin of the Earth and the Moon*, Springer-Verlag, New York, 295pp.

Ringwood, A.E. (1982) Phase transformations and differentiation in subducted lithosphere: implications for mantle dynamics, basalt petrogenesis and crustal evolution. *J. Geol.* **90**, 611–643.

Rison, W. and Craig, H. (1983) Helium isotopes and mantle volatiles in Loihi seamount and Hawaiian island basalts and xenoliths. *Earth Planet. Sci. Lett.* **66**, 407–426.

Robinson, P., Von Herzen, R. *et al.* (1989) *Proceedings of the Ocean Drilling Program, Initial Reports (Part A)* **118**, 580pp.

Robinson, P.T., Flower, M.F.J., Schmincke, H.U. and Ohnmacht. W. (1977) Low temperature alteration of oceanic basalts, DSDP Leg 37, In: Aumento, F., Melson, W.G. *et al.* (eds) *Initial Reports of the Deep Sea Drilling Project* **37**, 775–793.

Roden, M.F., Frey, A. and Clague, D.A. (1984) Geochemistry of tholeiitic and alkalic lavas from the Koolan Range, Hawaii: implications for Hawaiian volcanism. *Earth Planet. Sci. Lett.* **69**, 141–158.

Roden, M.F., Frey, F.A. and Francis, D.M. (1985) An example of consequent mantle metasomatism in peridotite inclusions from Nunivak Island, Alaska. *J. Petrol.* **25**, 546–577.

Roden, M.F., Hart, S.R., Frey, F.A. and Melson, W.G. (1984) Sr, Nd and Pb isotopic and REE geochemistry of St. Paul's Rocks: the metamorphic and metasomatic development of an alkali basalt mantle source. *Contrib. Mineral. Petrol.* **85**, 3876–3890.

Roedder, E. (1979) Silicate liquid immiscibility in magmas. In: Yoder, H.S. (ed) *The Evolution of the Igneous Rocks*, Princeton University Press, pp. 15–57.

Roeder, P.L. (1974) Activity of iron and iron solubility in basaltic liquids. *Earth Planet. Sci. Lett.* **23**, 397–410.

Roeder, P.L. and Emslie, R.F. (1970) Olivine-liquid equilibria. *Contrib. Mineral. Petrol.* **29**, 275–289.

Rona, P.A. (1978a) Criteria for recognition of hydrothermal mineral deposits in oceanic crust. *Econ. Geol.* **73**, 135–160.

Rona, P.A. (1978b) Magnetic signatures of hydrothermal alteration and volcanogenic mineral deposits in oceanic crust. *J. Volcanol. Geotherm. Res.* **3**, 219–225.

Rona, P.A. (1988) Hydrothermal mineralization at oceanic ridges. *Can. Mineral.* **26**, 431–465.

Rona, P. and Gray. D.F. (1980) Structural behaviour of fracture zones symmetric and asymmetric about a spreading axis: Mid-Atlantic Ridge (lat. 23°N to 27°N) *Bull. Geol. Soc. Am.* **91**, 485–494.

Rona, P.A., Thompson, G., Mottl, M.J., Karson, J.A., Jenkins, W.J., Graham, D., Mallette, M., von Damm, K. and Edmond, J.M. (1984) Hydrothermal activity at the TAG hydrothermal field, Mid-Atlantic Ridge crest at 26°N. *J. Geophys. Res.* **89**, 11365–11377.

Rosencrantz, E.J. (1980) The geology of the northern half of North Arm Massif, Bay of Islands Ophiolite complex with application to upper ocean crust lithology, structure and genesis. Unpublished PhD thesis, University of New York at Albany, 250pp.

Rosendahl, B.R. (1976) Evolution of oceanic crust II: constraints, implications and inferences. *J. Geophys. Res.* **81**, 5305–5314.

Rosendahl, B.R., Raitt, R.W., Dorman, L.M., Bibee, L.D., Hussong, D.M. and Sutton, G.H. (1976) Evolution of oceanic crust. 1: A physical model of the East Pacific Rise crest derived from seismic refraction data. *J. Geophys. Res.* **81**, 5294–5304.

Rozanova, T.V. and Baturin, G.N. (1971) Hydrothermal ore shows in the floor of the Indian Ocean. *Oceanology* **11**, 874–879.

RRISP (Reykjanes Ridge Icelandic Seismic Project Working Group) (1980) Reykjanes Ridge Icelandic Seismic Experiment (RRISP 77). *J Geophys.* **47**, 228–283.

Ryan, M.P. (1987) Neutral buoyancy and the mechanical evolution of magmatic systems. In: Mysen, B.O. (ed) *Magmatic Processes: Physicochemical Principles, Geochem. Soc. Spec. Publ.* **1**, 259–287.

Ryan, M.P. (1988) The mechanics and three-dimensional internal structure of active magmatic systems: Kilauea volcano, Hawaii. *J. Geophys. Res.* **93**, 4213–4248.

Ryan, M.P., Koyanagi, R. and Fiske, R.S. (1981) Modelling the three-dimensional structure of macroscopic magma transport systems: application to Kilauea volcano, Hawaii. *J. Geophys. Res.* **86**, 7111–7130.

Sack, R.O., Walker, D. and Carmichael, I.S.E. (1987) Experimental petrology of alkalic lavas: constraints on cotectics of multiple saturation in natural liquids. *Contrib. Mineral. Petrol.* **96**, 1–23.

Saemundsson, K. (1967) Vulkanismum und tectonik des Hengill-Gebietes in Sudwest-Island. (English summary). *Acta Natur. Islandica* **2**, 101.

Saemundsson, K. (1974) Evolution of the axial rifting zone in northern Iceland and the Tjomes Fracture Zone. *Bull. Geol. Soc. Am.* **85**, 495–504.

Saemundsson, K. (1978) Fissure swarms and central volcanoes of the neovolcanic zones in Iceland. *Geol. J. Spec. Issue* **10**, 415–432.

Saemundsson, K. (1979) Outline of the geology of Iceland. *Jokull* **29**, 7–28.

Sager, W.W. and Bleil, U. (1987) Latitudinal shift of Pacific hotspot during the late Cretaceous and early Tertiary. *Nature, London* **326**, 488–490.

Sakai, H., DesMarais, D.J., Ueda, A. and Moore, J.G. (1984) Concentrations and isotope ratios of carbon, nitrogen and sulphur in ocean floor basalts. *Geochim. Cosmochim. Acta* **48**, 2433–2441.

Salisbury, M. and Christensen, N. (1978) The seismic velocity structure of a traverse through the Bay of Islands ophiolite complex, Newfoundland, an exposure of ocean crust and upper mantle. *J. Geophys. Res.* **83**, 805–817.

Salters, V.J.M. and Shimizu, N. (1988) World-wide occurrence of HHSE-depleted mantle. *Geochim. Cosmochim. Acta* **52**, 2177–2182.

Sarda, P., Staudacher, T. and Allegre, C.J. (1985) $^{40}Ar/^{38}Ar$ in MORB glasses: constraints on atmosphere and mantle evolution. *Earth Planet. Sci. Lett.* **72**, 367–375.

Sato, H. (1979) Segregation vesicles and immiscible liquid droplets in ocean floor basalt of Hole 396B, IPOD/DSDP Leg 46. In: Dmitriev, L., Heirtzler, J. *et al.* (eds) *Initial Reports of the Deep Sea Drilling Project* **46**, 283–286.

Sato, H., Aoki, K., Okamoto, K. and Fujita, B. (1978) Petrology and chemistry of basaltic rocks from hole 396B, IPOD/DSDP Leg 46. In: Lewis, B.T.R., Robinson, P. *et al.* (eds) *Initial Reports of the Deep Sea Drilling Project* **46**, 115–141.

Saunders, A.D. (1983) Geochemistry of basalts recovered from the Gulf of California during Leg 65 of the Deep Sea Drilling Project. In: Lewis, B.T.R., Robinson, P.T. *et al.* (eds) *Initial Reports of the Deep Sea Drilling Project* **65**, 591–621.

Saunders, A.D. (1984) The rare earth element characteristics of igneous rocks from the ocean basins. In: Henderson. P. (ed) *Rare Earth Element Geochemistry*, Elsevier, Amsterdam, pp. 205–236.

Saunders, A.D. and Tarney, J. (1979) The geochemistry of basalts from a back-arc spreading centre in the East Scotia Sea. *Geochim Cosmochim. Acta.* **43**, 555–572.

Saunders, A.D. and Tarney, J. (1984) Geochemical characteristics of basaltic volcanism within back-arc basins. In: Kokelaar, B.P. and Howells, M.F. (eds) *Marginal Basin Geology, Geol. Soc. London Spec. Publ.* **16**, pp. 59–76.

Saunders, A.D., Fornari, D.J., Joron, J-L., Tarney, J. and Treuil, M. (1982b) Geochemistry of basic igneous rocks, Gulf of California, Deep Sea Drilling Project Leg 64. In: Curray, J.R., Moore, D.G. *et al.* (ed) *Initial Reports of the Deep Sea Drilling Project* **64**, 595–642.

Saunders, A.D., Fornari, D.J. and Morrison, M.A. (1982a) The composition and emplacement of basaltic magmas produced during the development of continent-margin basins: the Gulf of California, Mexico. *J. Geol. Soc. London* **139**, 335–346.

Saunders, A.D., Norry, M.J. and Tarney, J. (1988) Origin of MORB and chemically-depleted mantle reservoirs: trace element constraints. In: Menzies, M.A. and Cox, K.G. (eds) *Oceanic and Continental Lithosphere: Similarities and Differences, J. Petrol. Spec. Publ.* **29**, 415–445.

Saunders, A.D., Tarney, J., Marsh, N.G. and Wood, D.A. (1980) Ophiolites as ocean crust or marginal basin crust: a geochemical approach. In: Panayiotou, A. (ed) *Ophiolites—Proceedings of the International Ophiolite Conference, 1979, Nicosia, Cyprus*, pp. 193–204.

Saunders, A.D., Tarney, J., Stern, C.R. and Dalziel, I.W.D. (1979) Geochemistry of Mesozoic marginal basin floor igneous rocks from southern Chile. *Bull. Geol. Soc. Am.* **90**, 237–258.

Saunders, A.D., Tarney, J. and Weaver, S.D. (1980) Transverse variations across the Antarctic

Peninsula: implications for the genesis of calc-alkali magmas. *Earth Planet. Sci. Lett.* **46**, 344–360.

Schafer, von, K. (1972) Transform faults in Iceland. *Geol. Runds.* **61**, 942–950.

Scheidegger, K. (1973) Temperatures and compositions of magmas ascending along mid-ocean ridges. *J. Geophys. Res.* **78**, 3340–3355.

Scheidegger, K.F. and Corliss, J.B. (1981) Petrogenesis and secondary alteration of upper layer 2 basalts of the Nazca Plate. *Mem. Geol. Soc. Am.* **154**, 131–145.

Schiffman, P. and Smith, B.M. (1988) Petrology and oxygen isotope geochemistry of a fossil seawater hydrothermal system within the Solea Graben, northern Troodos ophiolite, Cyprus. *J. Geophys. Res.* **93**, 4612–4624.

Schiffman, P., Smith, B.M., Varga, R.J. and Moores. E.M. (1987) Geometry, conditions and timing of off-axis hydrothermal metamorphism and ore deposition in the Solea Graben. *Nature, London* **325**, 423–425.

Schilling, J.-G. (1971) Sea-floor evolution: rare earth evidence. *Phil. Trans. R. Soc. London* **A268**, 663–706.

Schilling, J.-G. (1973a) After mantle plume: rare earth evidence. *Nature* (*Phys. Sci.*), *London* **242**, 2–5.

Schilling, J.-G. (1973b) Iceland mantle plume: geochemical study of Reykjanes Ridge. *Nature, London* **242**, 565–571.

Schilling, J.-G. (1975) Azores mantle blob: rare earth evidence. *Earth Planet. Sci. Lett.* **25**, 103–115.

Schilling, J.-G. (1985) Upper mantle heterogeneities and dynamics. *Nature, London* **314**, 62–67.

Schilling, J.-G. and Noe-Nygaard, A. (1974) Faeroe-Iceland plume: rare earth evidence. *Earth Planet. Sci. Lett.* **24**, 1–14.

Schilling, J.-G. and Winchester, J.W. (1969) Rare earth contribution to the origin of Hawaiian lavas. *Contrib. Mineral. Petrol.* **23**, 27–37.

Schilling, J.-G., Bergeron, M.B. and Evans, R. (1980) Halogens in the mantle beneath the North Atlantic. *Phil. Trans. R. Soc. London* **A297**, 147–178.

Schilling, J.-G., Kingsley, R.H. and Bergeron, M.B. (1977) Rare earth abundances in DSDP sites 332, 334 and 335, and inferences on the Azores mantle blob activity with time. In: Aumento, F., Melson, W.G. *et al.* (eds) *Initial Reports of the Deep Sea Drilling Project* **37**, 591–597.

Schilling, J.-G., Sigurdsson, H., Davis, A.N. and Hey, R.N. (1985b) Easter microplate evolution. *Nature, London* **317**, 325–331.

Schilling, J.-G., Sigurdsson, H. and Kingsley, R.H. (1978) Skagi and western neovolcanic zones in Iceland. 2: Geochemical variations. *J. Geophys. Res.* **83**, 3983–4002.

Schilling, J.-G., Thompson, G., Kingsley, R. and Humphris, S. (1985a) Hotspot-migrating ridge interaction in the South Atlantic. *Nature, London* **313**, 187–191.

Schilling, J.-G., Zajac, M., Evans, R., Johnston, T., White, W., Devine, J.D. and Kingsley, R. (1983) Petrologic and geochemical variations along the Mid-Atlantic Ridge from 29°N to 73°N. *Am. J. Sci.* **283**, 510–586.

Schlanger, S.O., Campbell, J.F. and Jackson, M.W. (1987) Post-Eocene subsidence of the Marshall Islands recorded by drowned atolls on Harrie and Sylvania guyots. In: Keating, M.H., Fryer, P., Batiza, R. and Boehlert, G.W. (eds) *Seamounts, Islands and Atolls, Am. Geophys. Union, Geophys. Monogr.* **43**, pp. 165–174.

Schlanger, S.O., Jenkyns, H. and Premoli-Silva, I. (1981) Volcanism and vertical tectonics in the Pacific basin related to global Cretaceous transgressions. *Earth Planet. Sci. Lett.* **52**, 435–449.

Schmincke, H.U. (1973) Magmatic evolution and tectonic regime in the Canary, Madeira and Azores island groups. *Bull. Geol. Soc. Am.* **84**, 633–648.

Schmincke, H.U. and Weibel, M. (1972) Chemical study of rocks from Madeira, Porto Santo, Sao Miquel and Terceira. *N. Jahrb. Mineral. Abhandl.* **117**, 253–259.

Schouten, H. and Klitgord, K.D. (1983) Overlapping spreading centres on the East Pacific Rise. *Nature, London* **503**, 549–550.

Schouten, H. and White, R.S. (1980) Zero-offset fracture zones. *Geology* **8**, 175–179.

Schouten, H., Denham, C. and Smith, W. (1982) On the quality of marine magnetic anomaly sources and sea-floor topography. *Geophys. J. R. Astron. Soc.* **70**, 245–260.

Schouten, H., Klitgord, K.D. and Whitehead, J.A. (1985) Segmentation of mid-ocean ridges. *Nature, London* **317**, 225–229.

Sclater, J.G. and Parsons, B. (1981) Oceans and continents: similarities and differences in the mechanism of heat loss. *J. Geophys. Res.* **86**, 11535–11552.
Sclater, J.G., von Herzen, R.P., Williams, D.L., Anderson, R.N. and Klitgord, K. (1974) The heat flow low on the flank of the Galapagos spreading center. *Geophys. J. R. Astron. Soc.* **38**, 609–626.
Scott, D.R. and Stevenson, D.J. (1986) Magma ascent by porous flow. *J. Geophys. Res.* **91**, 9283–9296.
Scott, D.R. and Stevenson, D.J. (1989) A self-consistent model of melting, magma migration and buoyancy-driven circulation beneath mid-ocean ridges. *J. Geophys. Res.* **94**, 2973–2988.
Scott, R.B. and Hajash, A. (1976) Initial submarine alteration of basaltic pillow lavas: a microprobe study. *Am. J. Sci.* **276**, 480–501.
Scott, S.D. (1985) Seafloor polymetallic sulfide deposits: modern and ancient. *Mar. Mining*, **5**, 191–212.
Searle, R.C. (1976a) Lithospheric structure of the Azores Plateau from Rayleigh-wave dispersion. *Geophys. J. R. Astron. Soc.* **44**, 537–546.
Searle, R.C. (1976b) Spreading rates in Iceland. *Nature, London* **261**, 75.
Searle, R.C. (1979) Side-scan sonar studies of the North Atlantic fracture zones. *J. Geol. Soc. London* **136**, 283–292.
Searle, R.C. (1984) GLORIA survey of the East Pacific Rise near 3.5°S: tectonics and volcanic characteristics of a fast spreading mid-ocean rise. *Tectonophysics* **101**, 319–344.
Searle, R.C. and Laughton, A.S. (1977) Sonar studies of the Mid-Atlantic Ridge and Kurchatov Fracture Zone. *J. Geophys. Res.* **82**, 5313–5328.
Searle, R.C. and Whitmarsh, R. (1978) The structure of King's Trough, north-east Atlantic, from bathymetric, seismic and gravity studies. *Geophys. J. R. Astron. Soc.* **53**, 259–287.
Sekine, T., Irifune, T., Ringwood, A.E. and Hibberson, W.O. (1986) High-pressure transformation of ecologite to garnetite in subducted oceanic crust. *Nature, London* **319**, 584–586.
Self, S. and Gunn, B.M. (1976) Petrology, volume and age relations of alkaline and saturated peralkaline volcanics from Terceira, Azores. *Contrib. Mineral Petrol.* **54**, 293–313.
Sempere, J.-C. and Macdonald, K.C. (1986a) Deep tow studies of the overlapping spreading centers at 9°03′N on the East Pacific Rise. *Tectonics* **5**, 881–900.
Sempere, J.-C. and Macdonald, K.C. (1986b) Overlapping spreading centres: implications from crack growth simulation by the displacement discontinuity method. *Tectonophysics* **5**, 151–163.
Sempere, J.-C., Purdy, G.M. and Schouten, H. (1990) Segmentation of the MAR between 24°N and 30°40′N. *Nature, London* **344**, 427–431.
Sen, G. (1982) Composition of basaltic liquids generated from a partially depleted lherzolite at 9 kbar pressure. *Nature, London* **299**, 336–338.
Sen, G. (1987) Xenoliths associated with the Hawaiian hot spot. In: Nixon, P.H. (ed) *Mantle Xenoliths*, Wiley, Chichester, pp. 359–375.
Serri, G., Hebert, R. and Hekinian, R. (1985) Chemistry of ultramafic tectonites and ultramafic to gabbroic cumulates from the major ocean basins and the northern Apennine ophiolites. *Ofioliti* **10**, 63–76.
Seyfried, W.E. (1987) Experimental and theoretical constraints on hydrothermal alteration processes at mid-ocean ridges. *Ann. Rev. Earth Sci.* **15**, 317–335.
Seyfried, W.E. and Bischoff, J.L. (1977) Hydrothermal transport of heavy metals by seawater: the role of seawater/basalt ratio. *Earth Planet. Sci. Lett.* **34**, 71–77.
Seyfried, W.E. and Bischoff, J.L. (1979) Low temperature basalt alteration by seawater: an experimental study at 70°C and 150°C. *Geochim. Cosmochim. Acta* **43**, 1937–1947.
Seyfried, W.E. and Bischoff, J.L. (1981) Experimental seawater–basalt interaction at 300°C, 500 bars, chemical exchange, secondary mineral formation and implication for transport of heavy metals. *Geochim. Cosmochim. Acta.* **45**, 135–147.
Seyfried, W.E., Mottl, M.J. and Bischoff, J.L. (1978) Seawater/basalt ratio effects on the chemistry and mineralogy of spilites from the ocean floor. *Nature, London* **275**, 211–213.
Shand, S.J. (1949) Rocks of the mid-Atlantic ridge. *J. Geol.* **57**, 89–92.
Shaw, H.R. (1973) Mantle convection and volcanic periodicity in the Pacific: evidence from Hawaii. *Bull. Geol. Soc. Am.* **84**, 1505–1526.
Shaw, H.R. and Jackson, E.D. (1973) Linear island chains in the Pacific: results of thermal plumes or gravitational anchors? *J. Geophys. Res.* **78**, 8634–9652.
Shcherbakov, S.A. and Savelyeva, G.N. (1984) Structures of ultramafic rocks of the Mariana

Trench and the Owen Fracture Zone. *Geotectonics* **18**, 159–167.

Shido, F., Miyashiro, A. and Ewing, M. (1974) Compositional variation in pillow lavas from the mid-Atlantic ridge. *Mar. Geol.* **16**, 177–190.

Shih. C.-Y. (1972) The rare earth geochemistry of oceanic igneous rocks. Unpublished PhD thesis, Columbia University, New York, 151pp.

Shimizu, N. and Arculus, R.J. (1975) Rare earth element concentrations in a suite of basanitoids and alkali olivine basalts from Grenada, Lesser Antilles. *Contrib. Mineral. Petrol.* **50**, 231–240.

Shimokawa, T. and Masuda, A. (1972) Rare-earths in Icelandic neovolcanic rocks. *Contrib. Mineral. Petrol.* **37**, 39–46.

Sigurdsson, H. (1970) The petrology and chemistry of the Setburg volcanic region and the intermediate and acid rocks of Iceland. Unpublished PhD thesis, University of Durham.

Sigurdsson, H. (1977) Generation of Icelandic rhyolites by melting of plagiogranites in the oceanic layer. *Nature, London* **269**, 25–28.

Sigurdsson, H. (1981) First-order major element variation in basalt glasses from the Mid-Atlantic Ridge: 29°N to 73°N. *J. Geophys. Res.* **86**, 9483–9502.

Sigurdsson, H. and Brown, G.M. (1970) An unusual enstatite-forsterite basalt from Kolbeinsey Island, north of Iceland. *J. Petrol.* **11**, 205–220.

Sigurdsson, H. and Schilling, J.-G. (1976) Spinels in Mid-Atlantic Ridge basalts: chemistry and occurrence. *Earth Planet. Sci. Lett.* **29**, 7–20.

Sigurdsson, H. and Sparks, R.S.J. (1978) Lateral magma flow within rifted Icelandic crust. *Nature, London* **274**, 126–130.

Sigurdsson, H. and Sparks, R.S.J. (1981) Petrology of acid and mixed magma ejecta from the 1875 eruption of Askja, Iceland. *J. Petrol.* **22**, 41–84.

Sigurdsson, H., Schilling, J.-G. and Meyer, P.S. (1978) Skagi and Langjokull volcanic zones in Iceland. 1: Petrology and structure. *J. Geophys. Res.* **83**, 3971–3982.

Sigvaldson, G.E. (1974) Basalts from the centre of the assumed Icelandic mantle plume. *J. Petrol.* **15**, 497–524.

Sigvaldson, G.E., Steinthorsson, S. and Oskarsson, N. (1976) The simultaneous production of basalts enriched and depleted in large lithophile trace ions (LIL) within the same fissure swarms in Iceland. *Bull. Soc. Geol. France* **18**, 863–867.

Silver, P.G., Carlson, R.W. and Olson, P. (1988) Deep slab, geochemical heterogeneity and the large-scale structure of mantle convection: investigation of an enduring paradox. *Ann. Rev. Earth Planet. Sci.* **16**, 477–541.

Simkin, T. (1972) Origin of some flat-topped volcanoes and guyots. *Mem. Geol. Soc. Am.* **132**, 183–193.

Sinha, M.C. and Louden, K.E. (1983) The Oceanographer Fracture Zones—I. Crustal structure from seismic refraction studies. *Geophys. J. R. Astron. Soc.* **75**, 713–736.

Sinha, M.C., Patel, P.D., Unsworth, M.J., Owen, T.R.E. and MacCormack, M.R.G. (1990) An active source electromagnetic sounding system for marine use. *Mar. Geophys. Res.* **12**, 59–68.

Sinton, J.M. (1979) Ultramafic inclusions and high pressure xenocrysts in submarine basanitoid equatorial Mid-Atlantic Ridge. *Contrib. Mineral. Petrol.* **70**, 49–57.

Sinton, J.M. and Byerly, G.R. (1980) Mineral compositions and crystallization trends in Deep Sea Drilling Project Holes 417D and 418A. In: Donnelly, T., Francheteau, J., Bryan, W., Robinson, P., Flower, M., Salisbury, M., *et al.* (eds) *Initial Reports of the Deep Sea Drilling Project* **51**, **52**, **53**, 1039–1054.

Sinton, J.M. and Fryer, P. (1987) Mariana Trough lavas from 18°N: implications for the origin of back-arc basin basalts. *J. Geophys. Res.* **92**, 12782–12802.

Sinton, J.M., Price, R.C., Johnson, K.T.M., Staudigel, H. and Zindler, A. (in press) Petrology and geochemistry of submarine lavas from the Lau and North Fiji back-arc basins. In: Kroenke, L.W. (ed) *Basin Formation, Ridge Crest Processes and Metallogenesis in the North Fiji Basin, Circum-Pacific Council for Energy and Mineral Resources, Earth Science Series*, AAPG, Tulsa.

Sinton, J.M., Wilson, D.S., Christie, D.M., Hey, R.N. and Delaney, J.R. (1983) Petrologic consequences of rift propagation on oceanic spreading centers. *Earth Planet. Sic. Lett.* **62**, 193–207.

Sleep, N.H. (1975) Formation of ocean crust: some thermal constraints. *J. Geophys. Res.* **80**, 4037–4042.

Sleep, N.H. (1978) Thermal structure and kinematics of mid-ocean ridge axes: some implications to basaltic volcanism. *Geophys. Res. Lett.* **5**, 426–428.

Sleep, N.H. (1984) Tapping of magmas from ubiquitous mantle heterogeneities: an alternative to mantle plumes? *J. Geophys. Res.* **89**, 10029–10041.

Sleep, N.M. and Toksoz, N. (1971) Evolution of marginal basins. *Nature, London* **233**, 548–550.

Smith, D.K. (1988) Shape analysis of Pacific seamounts. *Earth Planet. Sci. Lett.* **90**, 457–466.

Smith, D.K. and Jordan, T.H. (1988) Seamount statistics in the Pacific Ocean. *J. Geophys. Res.* **93**, 2899–2918.

Smith, W.H.F., Staudigel, H., Watts, A.B. and Pringle, M. (1989) The Magellan Seamounts: early Cretaceous record of the South Pacific isotopic and thermal anomaly. *J. Geophys. Res.* **94**, 10501–10523.

Smoot, N.C. (1982) Guyots in the mid-Emperor chain mapped with multibeam sonar. *Mar. Geol.* **47**, 153–163.

Somers, M.L., Carson, R.M., Revie, J.M., Edge, R.H., Barrow, B.J. and Andrews, A.G. (1978) GLORIA II—an improved long range sidescan sonar. *Oceanlogy International, Proceedings of Offshore Instrumentation and Communications*, Institute of Electrical Engineering, pp.15–24.

Sparks, R.S.J., Huppert, H.E. and Turner, J.S. (1984) The fluid dynamics of evolving magma chambers. *Phil. Trans. R. Soc. London* **A310**, 511–534.

Sparks, R.S.J., Meyer, P. and Sigurdsson, H. (1980) Density variation amongst mid-ocean ridge basalts: implications for magma mixing and the scarcity of primitive magmas. *Earth Planet. Sci. Lett.* **46**, 419–430.

Spence, D.A. and Turcotte, D.L. (1985) Magma driven propagation of cracks. *J. Geophys. Res.* **90**, 575–580.

Spiegelman, M. and McKenzie, D. (1987) Simple 3-D models for melt extraction at mid-ocean ridges and island arcs. *Earth Planet. Sci. Lett.* **83**, 137–152.

Spooner, E.T.C. and Fyfe, W.S. (1973) Sub-seafloor metamorphism, heat and mass transfer. *Contrib. Mineral. Petrol.* **42**, 287–304.

Spooner, E.T.C., Chapman, H.J. and Smewing, J.D. (1977) Strontium isotopic contamination and oxidation during ocean floor hydrothermal metamorphism of the ophiolitic rocks of the Troodos Massif, Cyprus. *Geochim. Cosmochim Acta.* **41**, 873–890.

Spray, J.G. (1982) Mafic segregations in ophiolite mantle sequences. *Nature, London* **299**, 524–528.

Spray, J.G. (1984) Possible causes and consequences of upper mantle decoupling and ophiolite emplacement. In: Gass, I.G., Lippard, S.J. and Shelton, A.W. (eds) *Ophiolites and Oceanic Lithosphere, Geol. Soc. London Spec. Publ.* **13**, pp. 255–268.

Spray, J.G. (1988) Thrust-related metamorphism beneath the Shetlands Islands oceanic fragment, northeast Scotland. *Can. J. Earth Sci.* **25**, 1760–1776.

Spray. J.G. (1989) Upper mantle segregation processes: evidence from alpine-type peridotites. In: Saunders, A.D. and Norry, M.J. (eds) *Magmatism in the Ocean Basins, Geol. Soc. London, Spec. Publ.* **42**, pp. 29–40.

Spudich, P. and Orcutt, J. (1980) A new look at the seismic velocity structure of the oceanic crust. *Rev. Geophys. Space Phys.* **18**, 627–645.

Spulber, S.D. and Rutherford, M.J. (1983) The origin of rhyolite and plagiogranite in oceanic crust: an experimental study. *J. Petrol.* **24**, 1–25.

Stakes, D.S. and O'Neil, J.R. (1982) Mineralogy and stable isotope geochemisty of hydrothermal altered oceanic rocks. *Earth Planet. Sci. Lett.* **57**, 285–304.

Stakes, D.S. and Scheidegger, K.F. (1981) Temporal variations in secondary minerals from Nazca plate basalts, diabases, and microgabbros. *Mem. Geol. Soc. Am.* **154**, 109–130.

Stakes, D.S., Shervais, J.W. and Hopson, C.A. (1984) The volcanic–tectonic cycle of the FAMOUS and AMAR valleys, Mid-Atlantic Ridge (36°47′N): evidence from basalt glass and phenocryst compositional variations for a steady state magma chamber beneath the valley midsections, AMAR 3. *J. Geophys. Res.* **89**, 6995–7028.

Staudacher, T. and Allegre, C.J. (1982) Terrestrial xenology. *Earth Planet. Sci. Lett.* **60**, 389–406.

Staudigel, H. and Schmincke, H.U. (1984) The Pliocene seamount series of la Palma/Canary Island. *J. Geophys. Res.* **89**, 11195–11215.

Staudigel, H., Bryan, W.B. and Thompson, G. (1979) Chemical variation in glass–whole rock pairs from individual cooling units in holes 417D and 418A. In: Donnelly, T., Francheteau, J., Bryan, W.B., Robinson, P.T., Flower, M., Salisbury, M. *et al.* (eds) *Initial Reports of the Deep Sea Drilling Project* **51**, **52**, **53**, 977–986.

Staudigel, H., Hart, S.R. and Richardson, S.H. (1981) Alteration of the oceanic crust: processes and timing. *Earth Planet. Sci. Lett.* **52**, 311–327.

Staudigel, H., Hart, S.R., Schmincke, H.U. and Smith, B.M. (1989) Cretaceous ocean crust at DSDP sites 417 and 418: carbon uptake from weathering versus loss by magmatic outgassing. *Geochim. Cosmochim. Acta.* **53**, 3091–3094.

Staudigel, H., Zindler, A., Hart, S.R., Leslie, T., Chen, C.Y. and Clague, D. (1984) The isotopic systematics of a juvenile intraplate volcano: Pb, Nd and Sr isotope ratios of basalts from Loihi Seamount, Hawaii. *Earth Planet. Sci. Lett.* **69**, 13–29.

Steinmann, G. (1927) Die ophiolithischen zonen in den Mediterranen Kettengebrige. *Proceedings of the 14th International Geological Congress, Madrid* **2**, pp. 638–667.

Steinmetz, L., Whitmarsh, R.B. and Morevia, V.S. (1977) Upper mantle structure beneath the Mid-Atlantic Ridge north of the Azores based on observations of compressional waves. *Geophys. J. R. Astron. Soc.* **50**, 353–380.

Steithorsson, S. (1978) Tephra layers in drill hole core from the Vatnajokull ice cap. *Jokull* **27**, 2–27.

Steithorsson, S., Oskarsson, N. and Sigvaldason, G.E. (1985) Origin of alkali basalts in Iceland: a plate tectonic model. *J. Geophys. Res.* **90**, 10027–10042.

Stern, C.R. (1979) Open and closed system fractionation within two Chilean ophiolites and the tectonic implications. *Contrib. Mineral. Petrol.* **68**, 243–258.

Stern, C.R. (1980) Geochemistry of Chilean ophiolites: evidence for the compositional evolution of the mantle source of back-arc basins. *J. Geophys. Res.* **85**, 955–966.

Stern, C.R. and Wyllie, P.J. (1975) Effect of iron absorption by noble-metal capsules on phase boundaries in rock-melting experiments at 30 kbars. *Am. Mineral.* **60**, 681–689.

Stern, R.J., Smoot, N.C. and Rubin, M. (1984) Unzipping of the Volcano Arc, Japan. *Tectonophysics* **102**, 153–174.

Stille, P., Unruh, D.M. and Tatsumoto, M. (1986) Pb, Sr, Nd and Hf isotope constraints on the origin of Hawaiian basalts and evidence for a unique mantle source. *Geochim. Cosmochim. Acta.* **50**, 2303–2319.

Stoeser, D.B. (1975) Igneous rocks from Leg 30 of the Deep Sea Drilling Project. In: Andrews, J.E., Packham, G. *et al.* (eds) *Initial Reports of the Deep Sea Drilling Project* **30**, 401–444.

Stolper, E. (1980) A phase diagram for mid-ocean ridge basalt: preliminary results and implications for petrogenesis. *Contrib. Mineral. Petrol.* **74**, 13–27.

Stolper, E. and Walker, D. (1980) Melt density and the average composition of basalt. *Contrib. Mineral. Petrol.* **74**, 7–12.

Stolz, A.J. and Davis, G.R. (1988) Chemical and Sr, Nd and Pb isotopic evidence for the role of fluids with contrasting compositions in the metasomatism of spinel lherzolite xenoliths from south-eastern Australia. In: Menzies, M. and Cox, K.G. (eds) *Continental and Oceanic Lithosphere, J. Petrol. Spec. Vol.* **29**, pp. 303–330.

Storey, B.C. and Mair, B.F. (1982) The composite floor of the Cretaceous back-arc basin of South Georgia. *J. Geol. Soc. London* **139**, 729–737.

Storey, M., Saunders, A.D., Tarney, J., Leat, P., Thirlwall, M.F., Thompson, R.N., Menzies, M.A. and Marriner, G.F. (1988) Geochemical evidence for plume-mantle interactions beneath Kerguelen and Heard Islands, Indian Ocean. *Nature, London* **336**, 371–374.

Storey, M., Wolff, T.A., Norry, M.J. and Marriner, G.F. (1989) Origin of hybrid lavas from Agua de Pau volcano, Sao Miguel, Azores. In: Saunders, A.D. and Norry, M.J. (eds) *Magmatism in the Ocean Basins, Geol. Soc. London Spec. Publ.* **42**, pp. 161–180.

Strong, D.F. (1972) Petrology of lavas from Grande Comore. *J. Petrol.* **13**, 181–217.

Styles, P. and Gerdes, K.D. (1983) St. John's Island (Red Sea): a new geophysical model and its implications for the emplacement of ultramafic rocks in fracture zones and at continental margins. *Earth Planet. Sci. Lett.* **65**, 353–368.

Sun, S.S. (1980) Lead isotopic study of young volcanic rocks from mid-ocean ridges, ocean islands and island arcs. *Phil. Trans. R. Soc. London* **A297**, 409–445.

Sun, S.S. (1985) Ocean islands—plums or plumes? *Nature, London* **316**, 103–104.

Sun, S.S. and Jahn, B. (1975) Lead and strontium isotopes in post-glacial basalts from Iceland. *Nature, London* **255**, 527–530.

Sun, S.S. and McDonough, W.F. (1989) Chemical and isotopic systematics of ocean basalts: implications for mantle composition and processes. In: Saunders, A.D. and Norry, M.J. (eds) *Magmatism in the Ocean Basins, Geol. Soc. London Spec. Publ.* **42**, pp. 313–345.

Sun, S.S., Nesbitt, R. and Sharaskin, A. (1979) Geochemical characteristics of MORB. *Earth Planet. Sci. Lett.* **44**, 119–138.

Sun, S.S., Tatsumoto, M. and Schilling, J.-G. (1975) Mantle plume mixing along the Reykjanes Ridge axis: lead isotopic evidence. *Science* **190**, 13–17.

Swanson, D.A. (1972) Magma supply rate at Kilauea volcano, 1952–1971. *Science* **175**, 169–170.

Swanson, D.A., Duffield, W.A. and Fiske, R.S. (1976a) Displacement of the south flank of Kilauea Volcano: the result of forceful intrusion of magma into rift zones. *US Geol. Surv. Prof. Pap.* **963**, 39pp.

Swanson, D.A., Jackson, J.B., Koyanagi, R.Y. and Wright, T.L. (1976b) The February 1969 east rift eruptions of Kilauea Volcano, Hawaii. *US Geol. Surv. Prof. Pap.* **891**, 30pp.

Takahashi, E. and Kushiro, I. (1983) Melting of a dry peridotite at high pressures and basalt magma genesis. *Am. Mineral.* **68**, 859–879.

Takahashi, E. and Scarfe, C.F. (1985) Melting of peridotite to 14 GPa and the genesis of komatiite. *Nature, London* **335**, 566–568.

Talwani, M. and Eldholm, O. (1977) Evolution of the Norwegian-Greenland Sea. *Bull. Geol. Soc. Am.* **88**, 969–999.

Talwani, M., Windish, C.C. and Langseth M. (1971) Reykjanes Ridge crest: a detailed geophysical study. *J. Geophys. Res.* **76**, 473–517.

Tamaki, K. and Larson, R.L. (1988) The Mesozoic tectonic history of the Magellan microplate in the western central pacific. *J. Geophys. Res.* **93**, 2857–2874.

Tanaka, T. and Aoki, K.P. (1979) Petrogenetic implications of REE and Ba data on mafic and ultramafic inclusions from Itinome-gata, Japan. *J. Geol.* **89**, 369–390.

Tapponier, P. and Francheteau, T. (1978) Necking of the lithosphere and mechanics of slowly accreting plate boundaries. *J. Geophys. Res.* **83**, 3955–3970.

Tarney, J. and Windley, B.F. (1981) Marginal basins through geological time. *Phil. Trans. R. Soc. London* **A301**, 217–232.

Tarney, J., Dalziel, I.W.D. and de Wit, M.J. (1976) Marginal basin "rocas verdes" complex from Southern Chile: a model for Archaean greenstone belt formation. In: Windley, B.F. (ed) *The Early History of the Earth*, Wiley, London, pp. 131–146.

Tarney, J., Saunders, A.D., Mattey, D.P., Wood, D.A. and Marsh, N.G. (1981) Geochemical aspects of back-arc spreading in the Scotia Sea and Western Pacific. *Phil. Trans. R. Soc. London.* **A300**, 263–285.

Tarney, J., Saunders, A.D. and Weaver, S.D. (1977) Geochemistry of volcanic rocks from the island arcs and marginal basins of the Scotia Arc region. *Am. Geophys. Union Maurice Ewing Ser.* **1**, 367–378.

Tarney, J., Saunders, A.D., Weaver, S.D., Donnellan, N.C.B. and Hendry, G.L. (1979) Minor element geochemistry of basalts from Leg 49. In: Luyendyk, B.P., Cann, J.R. *et al.* (eds) *Initial Reports of the Deep Sea Drilling Project* **49**, 660–685.

Tarney, J., Wood, D.A., Saunders, A.D., Cann, J.R. and Varet, J. (1980) Heterogeneity in the North Atlantic: evidence from Deep Sea Drilling. *Phil. Trans. R. Soc. London* **A297**, 179–202.

Tatsumi, Y., Hamilton, D.L. and Nesbitt, R.W. (1986) Chemical characteristics of fluid phase released from a subducted lithosphere and origin of arc magmas: evidence from high-pressure experiments and natural rocks. *J. Volcanol. Geotherm. Res.* **29**, 293–309.

Tatsumoto, M., Hedge, C.E. and Engel, A.E.J. (1965) Potassium, rubidium, strontium, thorium, uranium and the ratio of strontium-87 to strontium-86 in oceanic tholeiitic basalt. *Science* **180**, 886–888.

Taylor, P.T., Wood, C.A. and O'Hearn, T.J. (1980) Morphological investigations of submarine volcanism: Henderson seamount. *Geology* **8**, 390–395.

Tera, F., Brown, L., Morris, J., Sacks, I.S., Klein, J. and Middleton, R. (1986) Sediment incorporation in island-arc magmas; inferences from ^{10}Be. *Geochim. Cosmochim. Acta.* **50**, 535–550.

Thayer, T.P. (1963) Flow-layering in alpine peridotite-gabbro complexes. *Mineral. Soc. Am. Spec. Pap.* **1**, 55–61.

Thayer, T.P. (1969) Peridotite-gabbro complexes as keys to petrology of mid-ocean ridges. *Bull. Geol. Soc. Am.* **80**, 1515–1522.

Thayer, T.P. (1980) Syncrystallization and subsolidus deformation in ophiolite peridotite and gabbro. *Am. J. Sci.* **280A**, 269–283.

Thomson, C.W. and Murray, J. (1891) Report on the scientific results of the voyage of HMS Challenger (2 volumes), HM Stationery Office, London.

Thompson, G. (1973) A geochemical study of the low temperature interaction of seawater and oceanic igneous rock. *Trans. Am. Geophys. Union* **54**, 1015–1019.

Thompson, G. (1983) Hydrothermal fluxes in the ocean. In: Riley, J.P. and Chester, R. (eds) *Chemical Oceanography* (Vol. 8), Academic Press, London, pp. 271–337.

Thompson, G. (1984) Basalt-seawater interaction. In: Rona, P.A., Bostrom, K., Laubier, L. and Smith, K.L. (eds) *Hydrothermal Processes at Seafloor Spreading Centers*, Plenum, New York, pp. 225–278.

Thompson, G. and Humphris, S.E. (1974) Petrology and geochemistry of rocks from the Walvis Ridge: Deep Sea Drilling Project Leg 74, sites 525, 527 and 528. In: Moore, T.C., Rabinowitz, P.D. *et al.* (eds) *Initial Reports of the Deep Sea Drilling Project* **74**, 755–764.

Thompson, G. and Humphris, S.E. (1980) Silicate mineralogy of basalts from the East Pacific Rise and Siqueiros Fracture Zone: Deep Sea Drilling Project Leg 54. In: Rosendahl, B.R., Hekinian, R. *et al.* (eds) *Initial Reports of the Deep Sea Drilling Project* **54**, 651–669.

Thompson, G. and Melson, W.G. (1972) The petrology of oceanic crust across fracture zones in the Atlantic ocean: evidence of a new kind of seafloor spreading. *J. Geol.* **80**, 526–538.

Thompson, G., Bryan, W.B., Ballard, R.O., Hamuro, K. and Melson, W.G. (1985) Axial processes along a segment of the East Pacific Rise 10°–12°N. *Nature, London* **318**, 429–433.

Thompson, G., Bryan, W.B., Frey, F.A. and Sung, C.M. (1974) Petrology and geochemistry of basalts and related rocks from Sites 214, 215, 216, DSDP Leg 22, Indian Ocean. In: von der Borch, C.C., Sclater, J.G. *et. al.* (eds) *Initial Reports of the Deep Sea Drilling Project* **22**, 459–468.

Thompson, G., Bryan, W.B. and Humphris, S.E. (1989) Axial volcanism on the East Pacific Rise and Siqueiros Fracture Zone: Deep Sea Drilling Project Leg 54. In: Saunders, A.D. and Norry, M.J. (eds) *Magmatism in the Ocean Basins, Geol. Soc. London Spec. Publ.* **42**, pp. 181–200.

Thompson, G., Bryan, W.B. and Melson, W.G. (1980) Geological and geophysical investigation of the Mid-Cayman Rise spreading center: geochemical variation and petrogenesis of basalt glasses. *J. Geol.* **88**, 41–55.

Thompson, G., Humphris, S.E., Schroeder, B., Sulanowska, M. and Rona, P.A. (1988) Active vents and massive sulfides at 26°N (TAG) and 23°N (Snakepit) on the Mid-Atlantic Ridge. *Can. Mineral.* **26**, 697–711.

Thompson, G., Humphris, S.E. and Schilling, J.G. (1983) Petrology and geochemistry of basaltic rocks from the Rio Grande Rise, South Atlantic: Deep Sea Drilling Project, Leg 72, Hole 516F. In: Barker, P.F., Carlson, R.L., Johnson, D.A. *et al.* (eds) *Initial Reports of the Deep Sea Drilling Project* **72**, 457–466.

Thompson, R.N. (1982) Magmatism of the British Tertiary volcanic province. *Scot. J. Geol.* **18**, 49–107.

Thompson, R.N. (1987) Phase-equilibria constraints on the genesis and magmatic evolution of oceanic basalts. *Earth Sci. Rev.* **24**, 161–210.

Thompson, R.N. and Flower, M.F.J. (1971) One-atmosphere melting and crystallization relations of lavas from Anjouan, Comores Archipelago, western Indian Ocean. *Earth Planet. Sci. Lett.* **12**, 97–107.

Thompson, R.N. and Kushiro, I. (1972) The oxygen fugacity within graphite capsules in piston-cylinder apparatus at high pressures. *Carnegie Inst. Washington Yearbook* **71**, 615–616.

Thompson, R.N. and Tilley, C.E. (1969) Melting and crystallization relations of Kilauean basalts of Hawaii. The lavas of the 1959–60 Kilauea eruption. *Earth Planet. Sci. Lett.* **5**, 469–477.

Thorarinsson, S. (1974) On the topography of the volcanic zones in Iceland (Abstract). In: Kristjansson, L. (ed) *Geodynamics of the Iceland and the North Atlantic Regions*, D. Reidel, Amsterdam, pp. 203–205.

Thurber, C.H. (1984) Seismic detection of the summit magma complex of Kilauea Volcano, Hawaii. *Science* **223**, 165–167.

Tiezzi, L.J. and Scott, R.B. (1980) Crystal fractionation in a cumulate gabbro, Mid-Atlantic Ridge, 26°N. *J. Geophys. Res.* **85**, 5438–5454.

Tighe, S., Fox, P.J., Detrick, R.S., Tyce, R., Langmuir, C.H., Mutter, J. and Ryan, W.B.F. (1988) *East Pacific Rise Data Synthesis*, Joint Oceanographic Institutes Incorporated, Washington, D.C. (3 volumes with set of charts).

Tilley, C.E. (1947) The dunite-mylonites of St. Paul's Rocks (Atlantic). *Am. J. Sci.* **246**, 483–491.

Tilley, C.E. (1966) A note on the dunite (peridotite) mylonites of St. Paul's Rocks (Atlantic). *Geol. Mag.* **103**, 120–123.

Tivey, M.A., Schouten, H., Sempere, J.C. and Woolridge, A. (1989) Implications of the 3-D structure of the TAG magnetic anomaly on the Mid-Atlantic Ridge. *Trans. Am. Geophys. Union, EOS* **70**, 455.

Toksoz, N. and Bird, P. (1977) Formation and evolution of marginal basins and continental plateaus. In: Talwani, M. and Pitman, W.C. (eds) *Island Arcs, Deep Sea Trenches and Back-arc Basins, Am. Geophys. Union, Maurice Ewing Ser.* **1**, pp. 379–393.

Tormey, D.R., Grove, T.L. and Bryan, W.B. (1987) Experimental petrology of normal MORB near the Kane Fracture Zone: 22°–25°N, Mid-Atlantic Ridge. *Contrib. Mineral. Petrol.* **96**, 121–139.

Tracy, R.J. (1980) Petrology and significance of an ultramafic xenolith suite from Tahiti. *Earth Planet. Sci. Lett.* **48**, 80–96.

Tryggvasson. E. (1964) Arrival times of P-waves and upper mantle structure. *Bull. Seismol. Soc. Am.* **54**, 727–736.

Turcotte, D.L. and Oxburgh, E.R. (1967) Finite amplitude convection cells and continental drift. *J. Fluid Mech.* **28**, 29–42.

Turcotte, D.L. and Oxburgh, E.R. (1978) Intraplate volcanism. *Phil. Trans. R. Soc. London* **288A**, 561–579.

Tyce, R.C. (1987) Deep seafloor mapping systems—a review. *J. Mar. Technol. Soc.* **20**, 17–27.

Ulmer, P. and Luth, R.W. (1988) The graphite-COH fluid equilibrium in P, T, fO_2 space: an experimental determination at high pressure. *Trans. Am. Geophys. Union, EOS* **69**, 512.

Upton, B.G.J. and Wadsworth, W.J. (1966) The basalts of Reunion Island, Indian Ocean. *Bull. Volcanol.* **29**, 7–24.

Urey, H.C. (1947) The thermodynamic properties of isotopic substances. *J. Chem. Soc.* **11**, 562–581.

Uyeda, S. and Kanamori, H. (1979) Back-arc opening and the mode of subduction. *J. Geophys. Res.* **84**, 1049–1061.

Uyeda, S. and Miyashiro, A. (1974) Plate tectonics and the Japanese Islands: a synthesis. *Bull. Geol. Soc. Am.* **85**, 1159–1170.

Vallier, T.L., Dean, W.E., Rea, D.K. and Thiede, J. (1983) Geologic evolution of the Hess Rise, Central North Pacific Ocean. *Bull. Geol. Soc. Am.* **94**, 1289–1307.

Vance, D., Stone, J.O.H. and O'Nions, (1989) He, Sr and Nd isotopes in xenoliths from Hawaii and other ocean islands. *Earth Planet. Sci. Lett.* **96**, 147–160.

Verma, S.P., Schilling, J.-G. and Waggoner, D.G. (1983) Neodymium isotopic evidence for Galapagos hotspot-spreading center system evolution. *Nature, London* **306**, 654–657.

Viereck, L., Flower, M.F.J., Hertogen, J., Schmincke, H.U. and Jenner, G.A. (1989) The genesis and significance of N-MORB sub-types. *Contrib. Mineral. Petrol.* **102**, 112–126.

Vine, F.J. and Matthews, D.H. (1963) Magnetic anomalies over oceanic ridges. *Nature, London* **199**, 947–949.

Vink, G.E. (1984) A hotspot model for Iceland and the Voring Plateau. *J. Geophys. Res.* **89**, 9949–9959.

Vogt, P.R. (1971) Asthenosphere motion recorded by the ocean floor, south of Iceland. *Earth Planet. Sci. Lett.* **13**, 153–160.

Vogt, P.R. (1972) The Faeroe-Iceland-Greenland aseismic ridge and the western boundary undercurrent. *Nature, London* **239**, 79–81.

Vogt, P.R. (1974a) Volcano height and plate thickness. *Earth Planet. Sci. Lett.* **23**, 337–348.

Vogt, P.R. (1974b) Volcano spacing, fractures and thickness of the lithosphere. *Earth Planet. Sci. Lett.* **21**, 235–253.

Vogt, P.R. (1974c) The Iceland phenomenon: imprints of a hotspot on the ocean crust, and implications for flow beneath the plates. In: Kristjansson, L. (ed) *Geodynamics of Iceland and the North Atlantic Regions*, D. Reidel, Amsterdam, pp. 105–126.

Vogt, P.R. (1976) Plumes, subaxial pipe flow, and topography along the mid-ocean ridge. *Earth Planet. Sci. Lett.* **29**, 309–325.

Vogt, P.R. (1979) Amplitudes of oceanic magnetic anomalies and the chemistry of oceanic crust: synthesis and review of "magnetic telechemistry". *Can. J. Earth Sci.* **16**, 2236–2262.

Vogt, P.R. (1981) On the applicability of thermal conduction models to mid-plate volcanism: comments on a paper by Gass *et al. J. Geophys. Res.* **86**, 950–960.

Vogt, P.R. and Avery, O.E. (1974) Detailed magnetic surveys in the north east Atlantic and Labrador Sea. *J. Geophys. Res.* **79**, 363–389.

Vogt, P.R. and Johnson, G.L. (1973) Magnetic telechemistry of oceanic crust? *Nature, London* **245**, 373–375.

Vogt, P.R. and Smoot, N.C. (1984) The Geisha Guyots: multibeam bathymetry and morphometric interpretation. *J. Geophys. Res.* **89**, 11085–11107.

Volpe, A.M., Macdougall, D. and Hawkins, J.W. (1987) Mariana trough basalts (MTB): trace element and Sr–Nd isotopic evidence for mixing between MORB-like and arc-like melts. *Earth Planet. Sci. Lett.* **82**, 241–254.

Volpe, A.M., Macdougall, D. and Hawkins, J.W. (1988) Lau Basin basalts (LBB): trace element and Sr–Nd isotopic evidence for heterogeneity in back-arc basin mantle. *Earth Planet. Sci. Lett.* **90**, 174–186.

Von Damm, K.L. (1988) Systematics of and postulated controls on submarine hydrothermal solution chemistry. *J. Geophys. Res.* **93**, 4551–4561.

Wacker, J.F. (1986) Noble gases in the diamond-free ureilite, ASPHA 78019: the role of shock and nebular processes. *Geochim. Cosmochim. Acta.* **50**, 633–642.

Wager, L.R. (1960) The major element variation of the layered series of the Skaergaard intrusion and a re-estimation of the average composition of the hidden layered series and the successive residual magmas. *J. Petrol.* **1**, 364–398.

Wager, L.R. (1968) Rhythmic and cryptic layering in mafic and ultramafic plutons. In: Hess, H.H. and Poldervaart, A. (eds) *Basalts* (Vol. 2), Interscience, New York, pp. 573–622.

Wager, L.R. and Brown, G.M. (1967) *Layered Igneous Rocks*, W.H. Freeman, San Franciso, 588 pp.

Wager, L.R., Vincent, E.A. and Smales, A.A. (1957) Sulphides in the Skaergaard intrusion. *Econ. Geol.* **52**, 855–903.

Walker, D. and DeLong, S.E. (1982) Soret separation of mid-ocean ridge basalt magma. *Contrib. Mineral. Petrol.* **79**, 231–240.

Walker, D., Shibata, T. and DeLong, S.E. (1979) Abyssal tholeiites from the Oceanographer Fracture Zone, II: Phase equilibria and mixing. *Contrib. Mineral. Petrol.* **70**, 111–125.

Walker, D.A. (1989) Seismicity of the interiors of plates in the Pacific Basin. *Map from Hawaiian Institute of Geophysics*, University of Hawaii, Honolulu, Hawaii.

Walker, G.P.L. (1963) The Breiddalur central volcano, eastern Iceland, *Q. J. Geol. Soc. London* **119**, 29–63.

Walker, G.P.L. (1974) Eruptive mechanisms in Iceland. In: Kristjansson, L. (ed) *Geodynamics of Iceland and the North Atlantic regions*, D. Reidel, Amsterdam, pp. 189–202.

Walker, G.P.L. (1975) Excess spreading axis and spreading rate in Iceland. *Nature, London* **255**, 468–470.

Wanke, H. (1981) Constitution of terrestrial planets. *Phil. Trans. R. Soc. London* **A303**, 287–302.

Ward, P.L. (1971) New interpretations of the geology of Iceland. *Bull. Geol. Soc. Am.* **82**, 2991–3012.

Watkins, N.D. and Paster, T.P. (1971) The magnetic properties of igneous rocks from the ocean floor. *Phil. Trans. R. Soc. London* **A268**, 507–550.

Watson, E.B. (1976) Glass inclusions as samples of early magmatic liquid: determinative method and application to a South Atlantic basalt. *J. Volcanol. Geotherm. Res.* **1**, 73–84.

Watson, E.B. (1979) Apatite saturation in basic to intermediate magmas. *Geophys. Res. Lett.* **6**, 937–940.

Watts, A.B. (1984) Introduction to seamount special section. *J. Geophys. Res.* **89**, 11066–11068.

Watts, A.B., McKenzie, D.P., Parsons, B.E. and Roufosse, M. (1985) The relationship between gravity and bathymetry in the Pacific Ocean. *Geophys. J. R. Astron. Soc.* **83**, 263–298.

Watts, A.B., Weissel, J.K., Duncan, R.A. and Larson, R.L. (1988) Origin of the Louiseville Ridge and its relationship to the Eltanin Fracture Zone system. *J. Geophys. Res.* **93**, 3051–3077.

Weaver, B.L., Tarney, J. and Saunders, A.D. (1985) Geochemistry and mineralogy of basalts recovered from central North Atlantic. In: Bougault, H., Cande, S.C. *et al.* (eds) *Initial Reports of the Deep Sea Drilling Project* **82**, 395–419.

Weaver, B.L., Wood, D.A., Tarney, J. and Joron, J.L. (1987) Geochemistry of ocean island basalts from the South Atlantic: Ascension, Bouvet, St. Helena, Gough, and Tristan da Cunha. In: Fitton, J.G. and Upton, B.G.J. (eds) *Alkaline Igneous Rocks, Geol. Soc. London Spec. Publ.* **30**, pp. 253–267.

Weaver, S.D., Saunders, A.D., Pankhurst, R.J. and Tarney, J. (1979) A geochemical study of magmatism associated with the initial stages of back-arc spreading: Quaternary volcanics of Bransfield Strait, South Shetland islands. *Contrib. Mineral. Petrol.* **68**, 151–169.

Wegener, A. (1929) *Die Entstehung der Kontinente und Ozeane*, 4th Ed., Methuen, London, 212pp.

Weis, D., Bassias, Y., Gautier, I. and Mennessier, J-P. (1989) Dupal anomaly in existence 115 Ma ago: evidence from isotopic study of the Kerguelen Plateau (South Indian ocean). *Geochim. Cosmochim. Acta* **53**, 2125–2131.

Weissel, J.K. (1981) Magnetic lineations in marginal basins of the western Pacific. *Phil. Trans. R. Soc. London* **A300**, 223–247.

Welke, H., Moorbath, S., Cumming, G.L. and Sigurdsson, H. (1968) Lead isotope studies on igneous rocks from Iceland. *Earth Planet. Sci. Lett.* **4**, 221–231.

White, R.S. (1984) Atlantic oceanic crust: seismic structure of a slow spreading ridge. In: Gass, I.G., Lippard, S.J. and Shelton, A.W. (eds) *Ophiolites and Oceanic Lithosphere, Geol. Soc. London Spec. Publ.* **13**, pp. 101–111.

White, R.S. (1988a) The Earth's crust and lithosphere. In: Menzies, M.A. and Cox, K.G. (eds) *Oceanic and Continental Lithosphere: Similarities and Differences, J. Petrol. Spec. Publ.* **29**, pp. 1–10.

White, R.S. (1988b) A hotspot model for early Tertiary volcanism in the N. Atlantic. In: Morton, A.C. and Parson, L.M. (eds) *Early Tertiary Volcanism and the Opening of the North East Atlantic, Geol. Soc. London Spec. Publ.* **39**, pp. 3–13.

White, R.S. (1989) Asthenospheric control on magmatism in the ocean basins. In: Saunders, A.D. and Norry, M.J. (eds) *Magmatism in the Ocean Basins, Geol. Soc. London Spec. Publ.* **42**, pp. 17–27.

White, R.S. (1990) Initiation of the Iceland plume and opening of the North Atlantic. In: Tankard, A.J. and Balkwill, H.R. (eds) *Extensional Tectonics and Stratigraphy of the North Atlantic Margins, Mem. Am. Assoc. Petrol. Geol.* **46**, pp. 149–154.

White, R.S. and McKenzie, D. (1989a) Volcanism at rifts. *Sci. Am.* **260**, 62–71.

White, R.S. and McKenzie, D. (1989b) Magmatism at rift zones: the generation of volcanic continental margins and flood basalts. *J. Geophys. Res.* **94**, 7685–7729.

White, R.S., Detrick, R.S., Mutter, J.C., Buhl, P., Minshull, T.A. and Morris, E. (1990) New seismic images of oceanic crustal structure. *Geology* **18**, 462–465.

White, R.S., Detrick, R.S., Sinha, M.C. and Cormier, M.H. (1984) Anomalous seismic crustal structure of oceanic fracture zones. *Geophys. J. R. Astron. Soc.* **79**, 779–798.

White, R.S., Spence, G.D., Fowler, S.R., McKenzie, D., Westbrook, G.K. and Bowen, A.N. (1987) Magmatism at rifted margins. *Nature, London* **330**, 439–444.

White, W.M. (1985) Sources of oceanic basalts: radiogenic isotopic evidence. *Geology* **13**, 115–118.

White, W.M. and Dupre, B. (1986) Sediment subduction and magma genesis in the Lesser Antilles: isotope and trace element constraints. *J. Geophys. Res.* **91**, 5927–5940.

White, W.M. and Hofmann, A.W. (1982) Sr and Nd isotope geochemistry of oceanic basalts and mantle evolution. *Nature, London* **296**, 821–825.

White, W.M. and Schilling, J.-G. (1978) The nature and origin of geochemical variation in Mid-Atlantic Ridge basalts from the central North Atlantic. *Geochim. Cosmochim. Acta* **42**, 1501–1516.

White, W.M., Hoffman, A.W. and Puchelt, H. (1987) Isotope geochemistry of Pacific mid-ocean ridge basalt. *J. Geophys. Res.* **92**, 4881–4893.

White, W.M., Schilling, J.G. and Hart, S.R. (1976) Evidence for the Azores mantle plume from strontium isotope geochemistry of the central North Atlantic. *Nature, London* **263**, 659–663.

White, W.M., Tapia, M.D.M. and Schilling, J.G. (1979) The petrology and geochemistry of the Azores Islands. *Contrib. Mineral. Petrol.* **69**, 201–213.

Whitehead, J.A (1986) Buoyancy-driven instabilities of low-viscosity zones as models of magma-rich zones. *J. Geophys. Res.* **91**, 9303–9314.

Whitehead, J.A. and Luther, D.S. (1975) Dynamics of laboratory diapir and plume models. *J. Geophys. Res.* **80**, 705–717.

Whitehead, J.A., Dick, H.J.B. and Schouten, H. (1984) A mechanism for magmatic accretion under spreading centres. *Nature, London* **312**, 146–148.

Whitmarsh, R.B. and Calvert, A.J. (1986) Crustal structure of Atlantic fracture zones, I: The Charlie-Gibbs F.Z. *Geophys. J. R. Astron. Soc.* **85**, 107–138.

Wickman, F.E. (1956) The cycle of carbon and the stable carbon isotopes. *Geochim. Cosmochim. Acta* **9**, 136–148.

Wilkinson, J.F.G. (1982) The genesis of mid-ocean ridge basalt. *Earth Sci. Rev.* **18**, 1–57.

Williams, D.L. and von Herzen, R.P. (1974) Heat loss from the earth: new estimate. *Geology* **2**, 327–338.

Wilshire, H.G., Meyer, C.E., Nakata, J.K., Calk, L.C., Shervais, J.W., Nielson, J.E. and Schwarzmann, E.C. (1985) Mafic and ultramafic rocks of the western United States. *US Geol. Surv. Open File Rep.* No. 85.

Wilson, D.S., Clague, D.A., Sleep, N.H. and Morton, J.L. (1988) Implications of magma convection from the size and temperature of magma chambers at fast spreading ridges. *J. Geophys. Res.* **93**, 11974–11984.

Wilson, J.T. (1963a) A possible origin of the Hawaiian Islands. *Can. J. Phys.* **41**, 863–870.

Wilson, J.T. (1963b) Evidence from islands on the spreading of ocean floors. *Nature, London* **197**, 536–538.

Wilson, J.T. (1965) A new class of faults and their bearing on continental drift. *Nature, London* **207**, 343–347.

Wilson, J.T. (1973) Mantle plumes and plate tectonics. *Tectonophysics* **19**, 149–164.

Wilson, R.A.M. (1959) The geology of the Xeros-Troodos area: *Mem. Geol. Surv. Cyprus.* **1**, 1–135.

Wiseman, J.D.H. (1937) Basalts from the Carlsberg Ridge, Indian Ocean. *Scientific Reports of the John Murray Expedition*, British Museum (Natural History) (Vol. 3), pp. 1–28.

Wolery, T.J. and Sleep, N.H. (1976) Hydrothermal circulation and geochemical flux at mid-ocean ridges. *J. Geol.* **84**, 249–276.

Wolery, T.J. and Sleep, N.H. (1989) Interactions of geochemical cycles with the mantle. In: Gregor, C.B., Garrels, R.M., Mackenzie, F.T. and Maynard, J.B. (eds) *Chemical Cycles in the Evolution of the Earth*, Wiley, New York, pp. 77–103.

Wood, D.A. (1976) Spatial and temporal variations in the trace element geochemistry of the eastern Iceland flood basalt succession. *J. Geophys. Res.* **81**, 4353–4360.

Wood, D.A. (1979) A variably veined suboceanic upper mantle—genetic significance for mid-ocean ridge basalts from geochemical evidence. *Geology* **7**, 499–503.

Wood, D.A., Joron, J.-L., Marsh, N.G., Tarney, J. and Treuil, M. (1980a) Major and trace element variations in basalts from the North Philippine Sea drilled by IPOD Leg 58: a comparative study of back-arc basalts with lava series from Japan and mid-ocean ridges. In: DeVries Klein, G., Kobayashi, K., *et al.* (eds) *Initial Reports of the Deep Sea Drilling Project* **58**, 873–894.

Wood, D.A., Joron, J.-L. and Treuil, M. (1979a) A re-appraisal of the use of trace elements to classify and discriminate between magma series erupted in different tectonic settings. *Earth Planet. Sci. Lett.* **45**, 326–336.

Wood, D.A., Joron, J.L., Treuil, M., Norry, M. and Tarney, J. (1979b) Elemental and Sr isotope variations in basic lavas from Iceland and the surrounding ocean floor. The nature of mantle source inhomogeneities. *Contrib. Mineral. Petrol.* **70**, 319–339.

Wood, D.A., Marsh, N.G., Tarney, J., Joron, J.-L., Fryer, P. and Treuil, M. (1981) Geochemistry of igneous rocks recovered from a transect across the Mariana trough, arc, forearc and trench, Sites 453 to 461, DSDP Leg 60. In: Hussong, D.M., Uyeda, S., *et al.* (eds) *Initial Reports of the Deep Sea Drilling Project* **60**, 611–645.

Wood, D.A., Mattey, D.P., Joron, J.-L., Marsh, N.G., Tarney, J. and Treuil, M. (1980b) A geochemical study of 17 selected samples from basement cores recovered at Sites 447, 448, 449, 450 and 451, Deep Sea Drilling Project Leg 59. In: Kroenke, L., Scott, R. *et al.* (eds) *Initial Reports of the Deep Sea Drilling Project* **59**, 743–752.

Wood, D.A., Tarney, J., Varet, J., Saunders, A.D., Bougault, H., Joron, J.-L., Treuil, M. and Cann, J.R., (1979c) Geochemistry of basalts drilled in the North Atlantic by IPOD Leg 49: implications for mantle heterogeneity. *Earth Planet. Sci. Lett.* **42**, 77–97.

Wood, D.A., Varet, J., Bougault, H., Corre, O., Joron, J.-L., Treuil, M., Bizouard, H., Norry, M.J., Hawkesworth, C.J. and Roddick, J.C. (1979d) The petrology, geochemistry and mineralogy of North Atlantic basalts: a discussion based on IPOD Leg 49. In: Luyendyk, B.P., Cann, J.R., *et al.* (eds) *Initial Reports of the Deep Sea Drilling Project* **49**, 597–655.

Woodhead, J.D. and Fraser, D.G. (1985) Pb, Sr and ^{10}Be isotopic studies of volcanic rocks from the northern Mariana islands. Implications for magma genesis and crustal recycling in the western Pacific. *Geochim. Cosmochim. Acta* **49**, 1925–1930.

Woodhead, J.D., Harmon, R.S. and Fraser, D.G. (1987) O, S, Sr and Pb isotope variations in volcanic rocks from the Northern Mariana Islands: implications for crustal recycling in intra-oceanic arcs. *Earth Planet. Sci. Lett.* **83**, 39–52.

Woodhouse, J.H. and Dziewonski, A.M. (1984) Mapping the upper mantle: three-dimensional modelling of earth structure by inversion of seismic waveforms. *J. Geophys. Res.* **89**, 5953–5986.

Woolridge, A.L., Haggarty, J.E., Rona, P.A. and Harrison, C.G.A. (1990) Magnetic properties and opaque mineralogy of rocks from selected seafloor hydrothermal sites. *J. Geophys. Res.* **95**, 12351–12374.

Wright, T.L. (1971) Chemistry of Kilauea and Mauna Loa in space and time. *US Geol. Surv. Prof. Pap.* **735**, 40pp.

Wright, T.L. (1984) Origin of Hawaiian tholeiite: a metasomatic model. *J. Geophys. Res.* **89**, 3233–3252.

Wright, T.L. and Clague, D.A. (1989) Petrology of Hawaiian lavas. In: *Decade of N. American Geology*, Geological Society of America, Boulder, Colorado.

Wright, T.L. and Fiske, R.S. (1971) Origin of the differentiated and hybrid lavas of Kilauea volcano. *J. Petrol.* **12**, 1–65.

Wright, T.L. and Helz, R.J. (1987) Recent advances in Hawaiian petrology and geochemistry. *US Geol. Surv. Prof. Pap.* **1350**, 625–641.

Wright, T.L. and Swanson, D.A. (1987) The significance of observations at active volcanoes: a review and annotated bibliography of studies at Kilauea and Mount St. Helens. In: Mysen, B.O. (ed) *Magmatic Processes: Physicochemical Principles, Geochem. Soc. Spec. Publ.* **1**, 231–240.

Wright, T.L. and Tilling, R.I. (1980) Chemical variation in Kilauea eruptions, 1971–1974. In: Irving, A. (ed) *Jackson Vol. Am. J. Sci.* **280A**, 777–793.

Wright, T.L., Peck, D.L. and Shaw, H.R. (1976) Kilauea lava lakes: natural laboratories for study of cooling, crystallization and differentiation of basaltic magma. *Monogr. Am. Geophys. Union* **19**, pp. 375–392.

Wright, T.L., Swanson, D.A. and Duffield, W.A. (1975) Chemical composition of Kilauea east rift lava, 1968–1971. *J. Petrol.* **16**, 110–133.

Wyllie, P. (ed) (1967) *Ultramafic and Related Rocks*, Wiley, New York, 464pp.

Wyss, M. (1980) Hawaiian rifts and recent Icelandic volcanism: expressions of plume generated radial stress fields. *J. Geophys.* **47**, 19–22.

Yoder, H.S. (1973) Contemporaneous basaltic and rhyolitic magmas. *Am. Mineral.* **58**, 153–171.

Yoder, H.S. and Tilley, C.E. (1962) Origin of basalt magmas: an experimental study of natural and synthetic rock systems. *J. Petrol.* **3**, 342–532.

Yonover, R.N. (1989) Petrologic effects of rift failure at the Galapagos spreading center near 95.5°W including analyses of glass inclusions by laser mass spectrometry and ion microprobe. Unpublished PhD thesis, University of Hawaii, Honolulu, Hawaii, 193pp.

Yonover, R.N., Sinton, J.M. and Christie, D.M. (in press) Petrologic manifestation of rift failure the Gapalagos spreading center near 95.5°W. *J. Geophys. Res.*

Zartman, R.E. and Tera, F. (1973) Lead concentration and isotopic composition in five peridotite inclusions of probable mantle origin. *Earth Planet. Sci. Lett.* **20**, 54–66.

Zhang, D. and Clayton, R.N. (1988) Nitrogen abundances and isotopic compositions in MORB and subduction zone rocks. *Trans. Am. Geophys. Union, EOS* **69**, 510–511.

Zielinski, R.A. (1975) Trace element evaluation of a suite of rocks from Reunion Island, Indian Ocean. *Geochim. Cosmochim. Acta* **39**, 713–734.

Zielinski, R.A. and Frey, F.A. (1970) Gough island: evaluation of a fractional crystallization model. *Contrib. Mineral. Petrol.* **29**, 242–254.

Zindler, A. and Hart, S.R. (1986) Chemical geodynamics. *Ann. Rev. Earth Planet. Sci.* **14**, 493–571.

Zindler, A., Jagontz, E. and Goldstein, S. (1982) Nd, Sr and Pb isotopic systematics in a three-component mantle: a new perspective. *Nature, London* **58**, 519–523.

Zindler, A., Staudigel, H. and Batiza, R. (1984) Isotope and trace element geochemistry of young Pacific seamounts: implications for the scale of upper mantle heterogeneity. *Earth Planet. Sci. Lett.* **70**, 175–195.

Index

The manufacturer's authorised representative in the EU is Springer Nature Customer Service Centre GmbH, Europaplatz 3, 69115 Heidelberg, Germany. If you have any concerns regarding our products, please contact ProductSafety@springernature.com

Printed and bound by CPI Group (UK) Ltd, Croydon, CR0 4YY

15/07/2026

02167631-0001